Acclaim for Richard Holmes's

THE AGE OF WONDER

"Seldom have science and art been so gloriously married." —*Salon*

"*The Age of Wonder* is the long-awaited fermentation of the author's knowledge of the Romantic poets and his lifelong fascination with science." —*The Economist*

"Gives us . . . a new model for scientific exploration and poetic expression in the Romantic period. Informative and invigorating, generous and beguiling, it is, indeed, wonderful." —*The Guardian* (London)

"Holmes is certainly the man to undertake this intellectual salvage operation. . . . Ambitious. . . . Eloquent." —*The Wall Street Journal*

"Holmes's excitement at fusing long-familiar events and personages into something startlingly new is not unlike the exuberance of the age that animates his groundbreaking book." —*The New York Times*

"What Holmes has given us with this account of the Romantic scientists is, curiously enough, a thrilling new way to interpret the poets of the era. To bring new light to such a widely read group—and from the angle least expected, that of rigorous scientific study—is Holmes's considerable gift." —Poetry Foundation

"It was a singular time, and this is a singular book." —*Fortune Magazine*

"The Romantics gave us many of our notions of how science is done, which makes the subject of this book—even leaving aside the brilliance with which much of it is told—significant beyond its importance as intellectual history."
—*American Scholar*

"Romanticism and Science are justly reunited in Holmes's new book. . . . A revelation. . . . Thrilling."
—*The Independent* (London)

"Exhilarating. . . . Instructive and delightful. . . . Finely observed. . . . Generous and hugely enjoyable."
—*The Daily Telegraph* (London)

"Holmes depicts the Romantic age as a 'relay race of scientific stories.' In richly evocative prose he explores how great moments of insight, such as the discovery of Uranus, transformed the heart as well as the mind."
—*Discover Magazine*

"*The Age of Wonder* is popular history at its best, racy, readable, and well documented. . . . The scientists of that age were as Romantic as the poets. The scientific discoveries were as unexpected and intoxicating as the poems. . . . The boundless prodigality of nature inspired scientists and poets with the same feelings of wonder."
—*The New York Review of Books*

"The most flat-out fascinating book so far this year. . . . Beyond riveting." —*Time*

"In this big two-hearted river of a book, the twin energies of scientific curiosity and poetic invention pulsate on every page. . . . Luminous and horizon-expanding."
—*The New York Times Book Review*

"Well-researched and vividly written, *The Age of Wonder* will fascinate scientists and poets alike."
—*The Literary Review*

"Superlative. . . . The Romantics tapped the marvels of nature and sounded the infinite benefits of science. It's a song, if we can hear it, that can transform us today."
—*Salon*

"Holmes suffuses his book with joy, hope, and wonder. . . . Reading it is like a holiday in a sunny landscape, full of fascinating bypaths that lead to unexpected vistas."
—*The Sunday Times* (London)

"I've been fascinated by *The Age of Wonder*. . . . Holmes talks about how scientists and poets were very much aligned in the Age of Enlightenment. . . . What was discovered, whether in labs or in the cliffs of Tahiti, excited and inspired everyone."
—Yo-Yo Ma

Richard Holmes

THE AGE OF WONDER

Richard Holmes is the author of *Footsteps: Adventures of a Romantic Biographer*; *Sidetracks: Explorations of a Romantic Biographer*; *Dr. Johnson & Mr. Savage*; *Shelley: The Pursuit* (for which he received the Somerset Maugham Award); *Coleridge: Early Visions*; and *Coleridge: Darker Reflections* (a *New York Times Book Review* Editor's Choice and a National Book Critics Circle Award Finalist). He lives in England.

ALSO BY RICHARD HOLMES

One for Sorrow (poems)

Shelley: The Pursuit

Shelley on Love (editor)

Gautier: My Fantoms (translations)

Nerval: The Chimeras (with Peter Jay)

Mary Wollstonecraft and William Godwin: A Short Residence in Sweden and Memoirs (editor, Penguin Classics)

De Feministe en de Filosoof

Dr. Johnson & Mr. Savage

Coleridge: Early Visions

Coleridge: Darker Reflections

Coleridge: Selected Poems (editor, Penguin Classics)

Footsteps: Adventures of a Romantic Biographer

Sidetracks: Explorations of a Romantic Biographer

Insights: The Romantic Poets and their Circle

Classic Biographies (series editor)

THE AGE OF WONDER

THE AGE OF WONDER

How the Romantic Generation Discovered the Beauty and Terror of Science

RICHARD HOLMES

Vintage Books
A Division of Random House, Inc.
New York

To Jon Cook at Radio Flatlands

FIRST VINTAGE BOOKS EDITION, MARCH 2010

The Library of Congress has cataloged the Pantheon edition as follows:
Holmes, Richard, [date]
The age of wonder : how the romantic generation discovered the beauty and terror of science / Richard Holmes.
p. cm
Includes bibliographical references and index.
1. Science—Great Britain—History—18th century. 2. Discoveries in science—Great Britain—History—18th century. I. Title.
Q127.G4H65 2009
509.41'09033—dc22
2008049587

Vintage ISBN: 978-1-4000-3187-0

www.vintagebooks.com

Printed in the United States of America
10 9 8 7 6 5 4 3 2 1

Contents

Illustrations

Mary Shelley, by Richard Rothwell, 1840. © National Portrait Gallery, London
John Herschel aged about seven, 1799. With the kind permission of John Herschel-Shorland.
The Royal Astronomical Society's Gold Medal presented to Caroline Herschel in 1828. © The Mistress and Fellows of Girton College, Cambridge
Michael Faraday, by William Brockedon, 1831. © National Portrait Gallery, London
John Herschel, by Henry William Pickersgill, c.1835. © National Portrait Gallery, London
David Brewster. Lithograph after Daniel Maclise, c.1830. © National Portrait Gallery, London
Charles Babbage. Detail from a daguerreotype by Antoine Claudet, c.1847–51. © National Portrait Gallery, London
Charles Darwin. Albumen print by Maull & Polyblank, c.1855. © National Portrait Gallery, London
Mary Somerville, by Sir Francis Leggatt Chantre, 1832. © National Portrait Gallery, London
Louis de Bougainville. Commemorative postage stamp.
Charles Waterton, by Charles Wilson Peale, 1824. © National Portrait Gallery, London
Nature Unveiling Herself Before Science. Two bronzes by Louis Ernest Barrias, 1890.
Isaac Newton. Bronze statue by Eduardo Paolozzi, 1995. © Rob Ford/Alamy
Hubble Telescope image showing the Andromeda galaxy. © Adam Block/Science Photo Library

Two things fill my mind with ever new and increasing wonder and awe, the more often and persistently I reflect upon them: the starry heaven above me and the moral law within me...I see them in front of me and unite them immediately with the consciousness of my own existence.

IMMANUEL KANT, *Critique of Practical Reason* (1788)

~

He thought about himself, and the whole Earth,
Of Man the wonderful, and of the Stars,
And how the deuce they ever could have birth;
And then he thought of Earthquakes, and of Wars,
How many miles the Moon might have in girth,
Of Air-balloons, and of the many bars
To perfect Knowledge of the boundless Skies;
And then he thought of Donna Julia's eyes.

BYRON, *Don Juan* (1819), Canto 1, stanza 92

~

Those to whom the harmonious doors
Of Science have unbarred celestial stores...

WILLIAM WORDSWORTH, 'Lines Additional to an Evening Walk' (1794)

~

Nothing is so fatal to the progress of the human mind as to suppose our views of science are ultimate; that there are no mysteries in nature; that our triumphs are complete; and that there are no new worlds to conquer.

HUMPHRY DAVY, lecture (1810)

~

I shall attack Chemistry, like a Shark.

SAMUEL TAYLOR COLERIDGE, letter (1800)

~

...Then felt I like some watcher of the skies
When a new planet swims into his ken;
Or like stout Cortez when with wond'ring eyes
He stared at the Pacific...

JOHN KEATS, ms of sonnet (1816)

~

To the natural philosopher there is no natural object unimportant or trifling... a soap bubble... an apple... a pebble... He walks in the midst of wonders.

JOHN HERSCHEL, *A Preliminary Discourse on the Study of Natural Philosophy* (1830)

~

Yes, there is a march of Science, but who shall beat the drums of its retreat?

CHARLES LAMB, shortly before his death (1834)

Prologue

1

In my first chemistry class, at the age of fourteen, I successfully *precipitated* a single crystal of mineral salts. This elementary experiment was done by heating a solution of copper sulphate (I think) over a Bunsen burner, and leaving it to cool overnight. The next morning there it lay at the bottom of my carefully labelled test tube: a single beautiful crystal, the size of a flattened Fox's Glacier Mint, a miniature ziggurat with a faint blue opalescence, propped up against the inside of the glass (too big to lie flat), monumental and mysterious to my eyes. No one else's test tube held anything but a few feeble grains. I was triumphant, my scientific future assured.

But it turned out that the chemistry master did not believe me. The crystal was too big to be true. He said (not at all unkindly) that I had obviously faked it, and slipped a piece of coloured glass into the test tube instead. It was quite a good joke. I implored him, 'Oh, test it, sir; *just test it!*' But he refused, and moved on to other matters. In that moment of helpless disappointment I think I first glimpsed exactly what real science should be. To add to it, years later I learned the motto of the Royal Society: *Nullius in Verba* – 'Nothing upon Another's Word'. I have never forgotten this incident, and have often related it to scientific friends. They nod sympathetically, though they tend to add that I did not (as a matter of chemical fact) *precipitate* a crystal at all – what I did was to *seed* one, a rather different process. No doubt this is so. But the eventual consequence, after many years of cooling, has certainly been to precipitate this book.

2

The Age of Wonder is a relay race of scientific stories, and they link together to explore a larger historical narrative. This is my account of the second scientific revolution, which swept through Britain at the end of the eighteenth century, and produced a new vision which has rightly been called Romantic science.[1]

Romanticism as a cultural force is generally regarded as intensely hostile to science, its ideal of subjectivity eternally opposed to that of scientific objectivity. But I do not believe this was always the case, or that the terms are so mutually exclusive. The notion of *wonder* seems to be something that once united them, and can still do so. In effect there is Romantic science in the same sense that there is Romantic poetry, and often for the same enduring reasons.

The first scientific revolution, of the seventeenth century, is familiarly associated with the names of Newton, Hooke, Locke and Descartes, and the almost simultaneous foundations of the Royal Society in London and the Académie des Sciences in Paris. Its existence has long been accepted, and the biographies of its leading figures are well known.* But this second revolution was something different. The first person who referred to a 'second scientific revolution' was probably the poet Coleridge in his *Philosophical Lectures* of 1819.[2] It was inspired primarily by a sudden series of breakthroughs in the fields of astronomy and chemistry. It was a movement that grew out of eighteenth-century Enlightenment rationalism, but largely transformed it, by bringing a new imaginative intensity and excitement to scientific work. It was driven by a common ideal of intense, even reckless, personal commitment to discovery.

It was also a movement of transition. It flourished for a relatively brief time, perhaps two generations, but produced long-lasting consequences – raising hopes and questions – that are still with us today. Romantic science can be dated roughly, and certainly symbolically, between two celebrated voyages of exploration. These were Captain James Cook's first round-the-world expedition aboard the *Endeavour*, begun in 1768, and Charles Darwin's voyage to the Galapagos islands aboard the *Beagle*, begun in 1831. This is the time I have called the *Age of Wonder*, and with any luck we have not yet quite outgrown it.

The idea of the exploratory voyage, often lonely and perilous, is in one form or another a central and defining metaphor of Romantic science. That is how William Wordsworth brilliantly transformed the great

* The fine survey by Lisa Jardine, *Ingenious Pursuits: Building the Scientific Revolution* (1999), gives a vivid picture of the leading figures in the seventeenth-century scientific revolution across Europe, and includes a significant introductory essay on the emerging role of science in modern society. See also my bibliography, 'The Bigger Picture', page 485.

Enlightenment image of Sir Isaac Newton into a Romantic one. While a university student in the 1780s Wordsworth had often contemplated the full-size marble statue of Newton, with his severely close-cropped hair, that still dominates the stone-flagged entrance hall to the chapel of Trinity College, Cambridge. As Wordsworth originally put it, he could see, a few yards from his bedroom window, over the brick wall of St John's College,

> The Antechapel, where the Statue stood
> Of Newton, with his Prism and silent Face.

Sometime after 1805, Wordsworth animated this static figure, so monumentally fixed in his assured religious setting. Newton became a haunted and restless Romantic traveller amidst the stars:

> And from my pillow, looking forth by light
> Of moon or favouring stars, I could behold
> The Antechapel where the Statue stood
> Of Newton, with his prism and silent face,
> The marble index of a Mind for ever
> Voyaging through strange seas of Thought, alone.[3]

Around such a vision Romantic science created, or crystallised, several other crucial conceptions – or misconceptions – which are still with us. First, the dazzling idea of the solitary scientific 'genius', thirsting and reckless for knowledge, *for its own sake and perhaps at any cost.* This neo-Faustian idea, celebrated by many of the imaginative writers of the period, including Goethe and Mary Shelley, is certainly one of the great, ambiguous creations of Romantic science which we have all inherited. Closely connected with this is the idea of the 'Eureka moment', the intuitive inspired instant of invention or discovery, for which no amount of preparation or preliminary analysis can really prepare. Originally the cry of the Greek philosopher Archimedes, this became the 'fire from heaven' of Romanticism, the other true mark of scientific genius, which also allied it very closely to poetic inspiration and creativity. Romantic science would seek to identify such moments of singular, almost mystical vision in its own history. One of its first and most influential examples was to become the story of the solitary, brooding Newton in his orchard, seeing an apple fall and 'suddenly' having his vision of universal gravitation.

This story was never told by Newton at the time, but only began to emerge in the mid-eighteenth century, in a series of memoirs and reminiscences.*

The notion of an infinite, mysterious Nature, waiting to be discovered or seduced into revealing all her secrets, was widely held. Scientific instruments played an increasingly important role in this process of revelation, allowing man not merely to extend his senses passively – using the telescope, the microscope, the barometer – but to intervene actively, using the voltaic battery, the electrical generator, the scalpel or the air pump. Even the Montgolfier balloon could be seen as an instrument of discovery, or indeed of seduction.

There was, too, a subtle reaction against the idea of a purely mechanistic universe, the mathematical world of Newtonian physics, the hard material world of objects and impacts. These doubts, expressed especially in Germany, favoured a softer 'dynamic' science of invisible powers and mysterious energies, of fluidity and transformations, of growth and organic change. This is one of the reasons that the study of electricity (and chemistry in general) became the signature science of the period; though astronomy itself, once the exemplary science of the Enlightenment, would also be changed by Romantic cosmology.

The ideal of a pure, 'disinterested' science, independent of political ideology and even religious doctrine, also began slowly to emerge. The emphasis on a secular, humanist (even atheist) body of knowledge, dedicated to the 'benefit of all mankind', was particularly strong in Revolutionary France. This would soon involve Romantic science in new kinds of controversy: for instance, whether it could be an instrument of the state, in the case of inventing weapons of war. Or a handmaiden of the Church, supporting the widely held view of 'Natural theology', in which science reveals evidence of a divine Creation or intelligent design.

---

* The apple fell in his orchard at Woolsthorpe, Lincolnshire, where Newton, aged twenty-five, had retired from Cambridge during the Plague of 1665. Various versions of the story began to appear after his death in 1727. It appears in Stukeley's unpublished Memoir of Newton, originally written in 1727, but not given to the Royal Society in manuscript until 1752; in unpublished notes for a biography by his nephew John Conduit; and for the first time in print in Voltaire's *Letters on the English Nation* (1734). Part of the power of the story was that it replaced the sacred Biblical account of the Fall from Innocence in Genesis (Eve and the apple) with a secular parable of the Ascent to Knowledge. See Patricia Fara, *Newton: The Making of Genius* (2005); and for a broad visionary perspective, Jacob Bronowski's scientific classic *The Ascent of Man* (1973).

With these went the new notion of a popular science, a people's science. The scientific revolution of the late seventeenth century had promulgated an essentially private, elitist, specialist form of knowledge. Its *lingua franca* was Latin, and its common currency mathematics. Its audience was a small (if international) circle of scholars and *savants*. Romantic science, on the other hand, had a new commitment to explain, to educate, to communicate to a general public.

This became the first great age of the public scientific lecture, the laboratory demonstration and the introductory textbook, often written by women. It was the age when science began to be taught to children, and the 'experimental method' became the basis of a new, secular philosophy of life, in which the infinite wonders of Creation (whether divine or not) were increasingly valued for their own sake. It was a science that, for the first time, generated sustained public debates, such as the great Regency controversy over 'Vitalism': whether there was such a thing as a life force or principle, or whether men and women (or animals) had souls.

Finally, it was the age which challenged the elite monopoly of the Royal Society, and saw the foundation of scores of new scientific institutions, mechanics institutes and 'philosophical' societies, most notably the Royal Institution in Albemarle Street in 1799, the Geological Society in 1807, the Astronomical Society in 1820, and the British Association for the Advancement of Science in 1831.

Much of this transition from Enlightenment to Romantic science is expressed in the paintings of Joseph Wright of Derby. Closely attached to the Lunar Society, and the friend of Erasmus Darwin and Joseph Priestley, Wright became a dramatic painter of experimental and laboratory scenes which reinterpreted late-eighteenth-century Enlightenment science as a series of mysterious, romantic moments of revelation and vision. The calm, glowing light of reason is surrounded by the intense, psychological chiaroscuro associated with Georges de la Tour. This is most evident in the famous series of scientific demonstration scenes painted at the height of his career: *The Orrery* (1766, Derby City Museum and the frontispiece of this book), *The Air Pump* (1767, National Gallery, London) and *The Alchemist* (1768, Derby City Museum). But these memorable paintings also ask whether Romantic science contained terror as well as wonder: if discovery and invention brought new dread as well as new hope into the world. We have certainly inherited this dilemma.

3

The Age of Wonder aims to raise and reflect upon such questions. Yet in the end the book remains a narrative, a piece of biographical story-telling. It tries to capture something of the inner life of science, its impact on the heart as well as on the mind. In the broadest sense it aims to present scientific passion, so much of which is summed up in that child-like, but infinitely complex word, *wonder*. Plato argued that the notion of 'wonder' was central to all philosophical thought: 'In Wonder all Philosophy began: in Wonder it ends...But the first Wonder is the Offspring of Ignorance; the last is the Parent of Adoration.'[4]

Wonder, in other words, goes through various stages, evolving both with age and with knowledge, but retaining an irreducible fire and spontaneity. This seems to be the implication of Wordsworth's famous lyric of 1802, which was inspired not by Newton's prism, but by Nature's:

> My heart leaps up when I behold
> A rainbow in the sky;
> So was it when my life began;
> So is it now I am a man;
> So be it when I shall grow old,
> Or let me die!...[5]

This book is centred on two scientific lives, those of the astronomer William Herschel and the chemist Humphry Davy. Their discoveries dominate the period, yet they offer two almost diametrically opposed versions of the Romantic 'scientist', a term not coined until 1833, after they were both dead. It also gives an account of their assistants and protégés, who eventually became much more than that, and handed on the flame to the very different world of professional Victorian science. But it draws in many other lives, and it is interrupted by many episodes of scientific endeavour and high adventure so characteristic of the Romantic spirit: ballooning, exploring, soul-hunting. These were all part of the great journey.*

---

* A brief guide to the many figures who jostle into this book, some familiar but others obscure or unexpected, appears in my Cast List, page 471.

It is also held together by, as a kind of chorus figure or guide, a scientific Virgil. It is no coincidence that he began his career a young and naïve scientific traveller, an adventurer and secret journal-keeper. However, he ended it as the longest-serving, most experienced and most domineering President of the Royal Society: the botanist, diplomat and *éminence grise* Sir Joseph Banks. As a young man Banks sailed with Captain Cook round the world, setting out in 1768 on that perilous three-year voyage into the unknown. This voyage may count as one of the earliest distinctive exploits of Romantic science, not least because it involved a long stay in a beautiful but ambiguous version of Paradise – Otaheite, or the South Pacific island of Tahiti.

1

Joseph Banks in Paradise

1

On 13 April 1769, young Joseph Banks, official botanist to HM Bark *Endeavour*, first clapped eyes on the island of Tahiti, 17 degrees South, 149 degrees West. He had been told that this was the location of Paradise: a wonderful idea, although he did not quite believe it.

Banks was twenty-six years old, tall and well-built, with an appealing bramble of dark curls. By temperament he was cheerful, confident and adventurous: a true child of the Enlightenment. Yet he had thoughtful eyes and, at moments, a certain brooding intensity: a premonition of a quite different sensibility, the dreaming inwardness of Romanticism. He did not like to give way to it. So he kept good company with his shipmates, and had carefully maintained his physical fitness throughout the first eight months of the voyage. He regarded himself – 'thank god' – as in as good mental and physical trim as a man could be. When occasionally depressed, he did vigorous jumping 'rope exercises' in his cabin, once nearly breaking his leg while skipping.[1]

He was capable of working patiently for hours on end in the extremely cramped conditions on board. The quarterdeck cabin, which he shared with his friend Dr Daniel Solander, was approximately eight feet by ten. He had adopted a strict daily routine of botanical drawing, electrical experiments, animal dissections, deck-walking, bird-shooting (when available) and journal-writing. He constantly fished specimens from the sea, shot or netted wild birds, and observed meteorological phenomena, such as the beautiful 'lunar rainbows'. When his gums had begun bleeding ominously with the onset of scurvy, he had calmly treated himself with a specially pre-prepared syrup ('Dr Hume's mixture') of concentrated lemon juice, taking precisely six ounces a day.[2] Within a week he was cured.

Just occasionally young Banks's scientific enthusiasm turned to explosive impatience. When rudely prevented from carrying out any botanical field trips by the Spanish Consul at Rio de Janeiro, and confined for three weeks to the sweltering ship in the harbour at Rio, he wrote colourfully to a friend at the Royal Society: 'You have heard of Tantalus in hell, you have heard of the French man laying swaddled in linen between two of his Mistresses both naked using every possible means to excite desire. But you have never heard of a tantalized wretch who has born his situation with less patience than I have done mine. I have cursed, swore, raved, stamped.'[3] Banks did however unofficially slip over the side at night to collect wild seeds and plants, a hoard which included the exotic purple bougainvillea.

Once among the Polynesian isles, Banks spent hours at the topgallant masthead, his large form crouched awkwardly in the crow's nest, looking for landfall beneath the heavy tropical cloudbase. At night the crew would hear distant surf roaring through the dark. Now at last he gazed out at the fabled blue lagoon, the black volcanic sand, and the intriguing palm trees (Linnaeus's Arecaceae). Above the beach the precipitous hills, dense with dark-green foliage and gleaming with white streams, rose sharply to 7,000 feet. On the naval chart Banks noted that the place was marked, prosaically enough, 'Port Royal Bay, King George the Third's Island'. 'As soon as the anchors were well down the boats were hoisted out and we all went ashore where we were met by some hundreds of the inhabitants whose faces at least gave evident signs that we were not unwelcome guests, tho they at first hardly dare approach us. After a little time they became very familiar. The first who aproachd us came creeping almost on his hands and knees and gave us a green bough the token of peace.'

Taking the hint, all the British shore party pulled down green boughs from the surrounding palm trees and carried them along the beach, waving them like ceremonial parasols. Eventually they were shown an idyllic spot close by a stream, where it was indicated that they could set up camp. The green boughs were thrown down in a great pile on the sand, 'and thus peace was concluded'. Here the British settlement known as Fort Venus was to be established: 'We then walkd into the woods followd by the whole train to whom we gave beads and small presents. In this manner we walked for 4 or 5 miles under groves of Cocoa nut and Bread fruit trees loaded with a profusion of fruit and giving the most grateful shade I have ever experienced. Under these were the habitations of the people most of them without walls. In short the scene we saw was the truest picture of an Arcadia of which we were going to be Kings that the imagination can form.'

As the men walked back, feeling dangerously like royalty, the Tahitian girls draped them with flowers, offered 'all kind of civilities' and gestured invitingly towards the coconut mats spread in the shade. Banks felt, reluctantly, that since islanders' houses were 'entirely without walls' it was not quite the moment to 'put their politeness to every test'. He would not have failed to have done so 'had circumstances been more favourable'.[4]

2

Tahiti lies roughly east–west just below the 17th parallel, one of the largest of what are now the Society Islands, roughly halfway between Peru and Australia. It is shaped not unlike a figure of eight, some 120 miles ('40 leagues') in circumference. Most of its foreshores are easily accessible, a series of broad, curving bays with black volcanic sands or pinkish-white coral beaches, fringed by coconut palms and breadfruit trees. But a few hundred yards inland, the ground rises sharply into an entirely different topography. The steep, densely wooded volcanic hills lead upwards to a remote and hostile landscape of deep gullies, sheer cliffs and perilous ledges.

Contrary to legend, the *Endeavour*, commanded by Lieutenant James Cook, was not the first European ship to make landfall in Tahiti. Spanish expeditions, under Quiroz or Torres, had probably touched there in the late sixteenth century, and claimed it for Spain.[5] A previous English expedition, under Captain Wallis of the *Dolphin*, had definitely landed there in 1767, when it was described as 'romantic', and claimed for England. A French expedition under Louis-Antoine de Bougainville had anchored there the following year, and claimed it for France.

The French had racily christened Tahiti 'La Nouvelle Cythère', the New Island of Love. Banks's opposite number, the French botanist Philibert Commerson (who named the bougainvillea after his captain), had published a sensational letter in the *Mercure de France* describing Tahiti as a sexual 'Utopia'. It proved that Jean-Jacques Rousseau was right about the existence of the Noble Savage. But then, the French had only spent nine days on the island.*

Cook was more sceptical, and had every member of his crew (including the officers) examined for venereal infections four weeks before arriving, by their surgeon Jonathan Monkhouse. He issued a series of Landing Instructions, which stated that the first rule of conduct ashore was civilised behaviour: 'To Endeavour by every fair means to cultivate a Friendship with the Natives and to treat them with all Imaginable Humanity.'[6] It was no coincidence that he enshrined the ship's own name in this instruction.

Joseph Banks had his own views on Paradise. He gave a whimsical account of his first night ashore in his *Endeavour Journal.* He dined deliciously on dressed fish and breadfruit, next to a Tahitian queen, who 'did me the honour with very little invitation to squat down on the mats close by me'. However, the queen was 'ugly enough in conscience'. Banks then noticed a very pretty girl, 'with a fire in her eyes' and white hibiscus in her hair, lingering in the 'common crowd' at the door. He encouraged her to come and sit on his other side, studiously ignored the queen for the rest of the evening, and 'loaded' the Polynesian beauty with bead necklaces and every compliment he could manage. 'How this would have ended is hard to say,' he observed later. In fact the amorous party broke up abruptly when it was discovered that his friend Solander had had a snuffbox picked from his pocket, and a fellow officer had lost 'a pair of opera glasses'. It is not explained why he had brought opera glasses ashore in the first place.

* De Bougainville's account of his ship anchoring at Tahiti for the first time in April 1768 became one of the most celebrated passages in all French romantic travel-writing. 'I have to admit that it was nigh impossible to keep 400 young Frenchmen at work, sailors who had not seen a woman for six months, in view of what followed. In spite of all our precautions, a young Tahitian girl slipped aboard and placed herself on the quarterdeck immediately above one of the big hatchways, which was fully open to allow air in to the sailors sweating at the capstan below. The young girl casually let slip the only piece of cloth which covered her, and appeared to the eyes of all the crew exactly as naked Venus appeared to the Phrygian shepherd. Truly, she had the celestial form of the goddess of Love. More and more sailors and soldiers crowded to the foot of the hatchway, and no capstan was ever wound with such alacrity as on this occasion. Only naval discipline succeeded in keeping these bewitched young fellows from rioting; and indeed we officers had some little difficulty in restraining ourselves.' Bougainville, *Voyage autour du Monde* (1771, Chapter 8, 'Mouillage à Tahiti').

This thieving proved to be completely customary in Tahiti, and led to many painful misunderstandings on both sides. The first occurred the following day, when a Tahitian quite openly made off with a marine's musket, and was immediately shot dead by a punctilious guard. Banks quickly grasped that some quite different notion of property must be involved, and noted grimly: 'We retird to the ship not well pleasd with the days expedition, guilty no doubt in some measure of the death of a man who the most severe laws of equity would not have condemnd to so severe a punishment. No canoes about the ship this morning, indeed we could not expect any as it is probable that the news of our behaviour yesterday was now known every where, a circumstance which will doubtless not increase the confidence of our friends the Indians.' Nonetheless, to Banks's relief and evident surprise, good relations were restored within twenty-four hours.

The *Endeavour* expedition remained for three months on Tahiti. Its main object was to observe a Transit of Venus across the face of the sun. (Cook stated that this was the reason their settlement was named Fort Venus, though his junior officers gave a different explanation.) This was due on the morning of 3 June 1769, and there would be no other transit for the next hundred years (not until 1874). It was a unique chance to establish the solar parallax, and hence the distance of the sun from the earth. This calculation depended on observing the exact timing at which the silhouette of Venus first entered, and then exited from, the sun's disc.

Banks was not part of the astronomical team, but when the expedition's quadrant was stolen one night shortly before the transit was due, he reacted with characteristic energy and courage. He knew that without this large and exquisitely calibrated brass instrument, used to measure precise astronomical angles, the entire observation would be rendered valueless. Not waiting for Cook or his marine guards, Banks roused the expedition's official astronomer, William Green, and set off immediately on foot in pursuit of the thief. In the dizzy heat, Banks followed the trail far up into the hills, accompanied only by a reluctant Green, one unarmed midshipman and a Tahitian interpreter. They penetrated seven miles inland through the Tahitian jungle, further than any European had been before: 'The weather was excessive hot, the Thermometer before we left the tents up at 91 made our journey very tiresome. Sometimes we walk'd sometimes we ran when we imagind (which we sometimes did) that the chase was just before us till we arrivd at the top of a hill about

4 miles from the tents. From this place [the interpreter] Tubourai shew'd us a point about 3 miles off and made us understand that we were not to expect the instrument till we got there. We now considerd our situation. No arms among us but a pair of pocket pistols which I always carried; going at least 7 miles from our fort where the Indians might not be quite so submissive as at home; going also to take from them a prize for which they had ventured their lives.'[7]

Banks decided to send back the midshipman with a brief message to Cook that armed reinforcements would be welcome. Meanwhile he and Green would press on alone, 'telling him at the same time that it was impossible we could return till dark night'.

Before dusk, Banks ran the thief to ground in an unknown and potentially hostile village. A crowd quickly gathered round them, 'rudely' jostling them. Following a Tahitian custom he had already learned, Banks quickly drew out a ring on the grass, and surrounded by 'some hundreds' of faces, sat quietly down in the centre. Here, instead of threatening or blustering, he began to explain and negotiate. For some time nothing transpired. Then, piece by piece, starting with its heavy wooden deal case, the quadrant was solemnly returned. 'Mr Green began to overlook the Instrument to see if any part or parts were wanting . . . The stand was not there but that we were informd had been left behind by the thief and we should have it on our return . . . Nothing else was wanting but what could easily be repaired, so we pack'd all up in grass as well as we could and proceeded homewards.'

By the time armed marines came up, sweating and jittery, about two miles down the track, Banks had completed the transaction and made several new friends. Everyone returned peacefully to Fort Venus on the shore. For this exploit, all conducted with the greatest calm and good humour, Banks earned the profound gratitude of Cook, who noted that 'Mr Banks is always very alert upon all occasions wherein the Natives are concerned.'[8] Banks concluded mildly in his journal: 'All were, you may imagine, not a little pleased at the event of our excursion.'[9]

Banks and Cook were a seemingly ill-matched pair. They were divided by background, education, class and manners. Yet they formed a curiously effective team. Cook's cool and formal manners towards the Tahitians were balanced by Banks's natural openness and enthusiasm, which easily won friends. With their help he would gather a mass of plant and animal specimens, and make what was in effect an early anthropological study of Tahitian customs. His journal entries cover everything from clothes (or lack of them) and cookery to dancing, tattooing, sexual practices, fishing methods, wood-carving, and religious beliefs. His accounts of a dog being roasted, or a young woman having her buttocks tattooed, are frank and unforgettable. He attended Tahitian ceremonial events, slept in their huts, ate their food, recorded their customs and learned their language. He was pioneering a new kind of science. As he wrote in his journal: 'I found them to be a people so free from deceit that I trusted myself among them almost as freely as I could do in my own countrey, sleeping continually in their houses in the woods with not so much as a single companion.'[10]

3

Educated in the traditional classics at Harrow, Eton and Christ Church, Oxford, young Joseph Banks had discovered science and the natural world at the age of fourteen. Towards the end of his life he told a sort of 'conversion' story about this to his friend the surgeon Sir Everard Home. It was later enshrined by the French naturalist Georges Cuvier in his obituary speech or *Éloge* to the Institut de France. Emerging late one summer afternoon from a schoolboy swim in the Thames at Eton, the teenage Banks found himself alone on the river, all his schoolfriends gone. Walking back through the green lanes, solitary and preoccupied, he suddenly saw the mass of wildflowers along the hedgerows vividly illuminated in the slanting, golden evening light. Their beauty and strangeness came to him like a revelation. 'After some reflection, he said to himself, it is surely more natural that I should be taught to know all the productions of Nature, in preference to Greek and Latin; but the latter is my father's command and it is my duty to obey him ... He began immediately to teach himself Botany.'

Despite the stilted form of this recollection (it is in Home's words and dates from fifty years after the event), it seems that to the young Banks botany implied a kind of Romantic rebellion against his father, as well as against the standard school curriculum of classics. Even more important, it brought him into contact with a race of people who would normally have been quite invisible to a privileged Eton schoolboy such as he. These were the wise women of the country lanes and hedgerows, the gypsy herbalists who collected 'simples' or medicinal plants 'to supply the Druggist and Apothecaries shops' of Windsor and Slough. They were a strange but knowledgeable tribe, whom he soon learned to treat with respect. More than that, he paid them sixpence for every 'material piece of information' they supplied.

Banks also told Everard Home that it was his mother – not his father – who handed over her lovingly worn copy of Gerard's *Herbal*, kept 'in her dressing room', with wonderful engravings that entranced him. It is thus that he is shown in a family portrait (possibly by Zoffany): an attractively long-haired and long-legged teenager, alert and faintly insolent, confidently posed in a studded leather chair with a portfolio of botanical engravings spread before him. Just under his left elbow, extraordinarily prophetic, is a large geographer's globe in its mahogany cradle, with a rhumb-line of sunlight curving down towards the equator.

From then on Banks saw his destiny as a naturalist, and began avidly collecting rare plants, wildflowers, herbs, shells, stones, animals, insects, fish and fossils. His conversion story reveals other elements of his life and character: self-confidence, wealth, surprising sensitivity, unconventional directness, and an attraction to women. At university he made himself a disciple of the great Swedish naturalist Carl Linnaeus, the leading Enlightenment botanist of Europe. Linnaeus had redefined the taxonomy of plants by identifying them according to their reproductive organs, re-cataloguing them in Latin according to genus, species and family, and collecting an unmatched array of specimens in his gardens at Uppsala.

Finding that there was no Linnaean lecturer in botany at Oxford, Banks reacted in a characteristic way. He rode to Cambridge, begged an interview with the Professor of Botany there, John Martyn, and simply asked to be recommended the best young botanist available. He came back triumphantly with a gifted young Jewish botanist, Israel Lyons, who had agreed to teach the subject to Banks and a group of like-minded undergraduates at Oxford. Banks paid Lyons a good salary out of his own pocket. Later he recommended him to an Admiralty expedition, and he remained his friend and patron for life. Lyons was Banks's first scientific protégé. From the start Banks displayed the commanding air, as well as the charm, of a wealthy man. This trait was given free rein when his father died in 1761. At the age of eighteen he was now sole heir to large estates in Lincolnshire and Yorkshire (they included over 200 farms) which would bring him £6,000 per annum (eventually rising to over £30,000), an enormous income for the period.

The family money made Banks a complete gentleman of leisure, a potentially fatal development, and he moved with his beloved mother and his only sister, Sophia, to a large house in Chelsea, near the Physic Garden. The conventional thing would have been for him to embark, like most of his friends, on the Grand Tour of Europe. Instead, the twenty-two-year-old Banks bought himself a berth on HMS *Niger*, and embarked on a strenuous seven-month botanical tour to the bleak shores of Labrador and Newfoundland. The Professor of Botany at Edinburgh wrote to him with some astonishment that it was 'rumoured that you was going to the country of the Eskimaux Indians to gratify your taste for Natural Knowledge'.

Banks demonstrated his energy and commitment on this expedition, earning the approval of all the naval officers, including his friend Captain Constantine John Phipps, and a certain Lieutenant James Cook, who was in charge of chart-making. He wrote witty, faintly scurrilous letters to his sister Sophia, and also kept the first of his great journals, most notable for their racy style, appalling spelling and non-existent punctuation. On his return in November 1766, with a vast quantity of plant specimens (and some caoutchouc from Portugal), Banks was elected a Fellow of the Royal Society, still aged only twenty-three. He began what was to become his famous herbarium, scientific library and collection of prints and drawings. His rapidly expanding circle of scientific friends included the rakish Lord Sandwich, future head of the Admiralty, and the quiet, portly and dedicated Daniel Solander, a young Swedish botanist, trained under Linnaeus at Uppsala, who managed the Natural History section of the British Museum.

Two years later, Banks heard of the round-the-world expedition in HM Bark *Endeavour*. The ship was in fact a specially converted coastal 'cat' from Whitby, broad-beamed, shallow-draughted and immensely strong, capable of being beached for repairs, and of carrying large quantities of stores and livestock below decks (and on them). But she was little more than a hundred feet from stem to stern, and had extremely restricted quarters. She was to be commanded by Lieutenant James Cook, forty years old, lean and reserved, the tough and experienced mariner from the little port of Staithes in Yorkshire who had made his name charting the Newfoundland coast.

The expedition was organised by the Admiralty, but also partly financed by the Royal Society, which supplied £4,000 towards astronomical observations. It had four main objectives: first, the observing of the Transit of Venus on Tahiti; second, charting and exploring the Polynesian islands west of Cape Horn; third, exploring the landmasses known to lie between the 30th and 40th parallels – New Zealand (possibly the tip of a continent) and Van Diemen's Land (Tasmania), possibly part of Australia; and fourth, collecting botanical and zoological specimens from anywhere in the southern hemisphere. It also had a medical aim, to reduce the fatal outbreaks of shipboard scurvy by the use of sauerkraut and citrus fruits.

The Royal Society had already appointed as the expedition's official astronomer William Green, assistant to the Astronomer Royal, Nevil Maskelyne. Banks immediately proposed himself as its official botanist. He would finance his own eight-man natural history 'suite', including two artists, a scientific secretary, Herman Spöring, two black servants from the Lincolnshire estate, his friend Dr Solander and – characteristically – a pair of greyhounds. For these, and a mass of equipment, Banks laid out as much as £10,000, nearly two years' income. For him it was to be a voyage in search of pure knowledge, and he laid in specialist equipment which created a considerable stir. A colleague reported admiringly, and with perhaps a touch of envy, to Linnaeus in Uppsala: 'No people ever went to sea better fitted out for the purpose of Natural History; nor more elegantly. They have got a fine library of Natural History; they have all sorts of machines for catching and preserving insects; all kinds of nets, trawls, drags and hooks for coral fishing; they have even a curious contrivance of a telescope by which, put into water, you can see the bottom at a great depth.' He concluded reassuringly to Linnaeus: 'All this is owing to you and your writings.'[11]

But there was, of course, an element of imperial competition. Cook had sealed Admiralty instructions to look out, after leaving Tahiti, for a possible 'great Southern continent' lying between latitude 30 and 40 degrees South. This was much further south than those parts of Australia's eastern seaboard which were already known through the Dutch navigators. It was believed that New Zealand might form the northern tip to this continent, and that it might contain huge natural resources. If this continent existed, it had to be claimed and mapped (with a view to possible colonisation) before the French did so. The Admiralty seems to have been unaware of Antarctica.

The imperial instructions were not really so secret. Both Banks and Solander knew about them before departure, and even Linnaeus was informed.[12] Moreover, neither Banks nor Cook really believed in the mysterious southern continent. Banks made a long, sceptical journal entry as they crossed the Pacific in March 1769, concluding: 'It is however some pleasure to be able to disprove that which does not exist but in the opinions of Theoretical writers, of which sort most are who have wrote any thing about these seas without having themselves been in them. They have generaly supposd that every foot of sea which they beleivd no ship had passd over to be land, tho they had little or nothing to support that opinion but vague reports ... ' Nevertheless, he was fully aware of how little was known about the Pacific islands in general, and of the perils of circumnavigation, especially between Tahiti and Indonesia. It had nearly destroyed Bougainville's entire crew the year before.

Among the many friends Banks was leaving behind was Solander's colleague the botanist and horticulturalist James Lee, who took an intense professional interest in the Pacific voyage. Lee owned the remarkable Vineyard Nurseries at the village of Hammersmith on the Thames. He was the author of a best-selling plant manual, *An Introduction to Botany extracted from the works of Dr Linnaeus* (1760), which ran into several editions, and he advised Banks on plant-collecting. Lee also trained up young naturalists at the nurseries. Among his assistants was an eighteen-year-old Scottish Quaker, Sydney Parkinson, a quiet, observant young man, whom Banks decided to employ as his second botanical artist aboard the *Endeavour*. It was a good choice, but with tragic consequences.

Another young person in Lee's charge was twenty-year-old Harriet Blosset, to whom he was legal guardian. Lee was teaching her to study plants, and she would eagerly have signed up for the expedition herself. But of course no women were officially allowed on board His Majesty's vessels, although the French botanist Philibert Commerson had smuggled his mistress aboard Bougainville's ship, disguised as a cabin boy. It was rumoured at the nurseries that Harriet was 'desperately in love with Mr Banks', and there was a good deal of gossip about them immediately before the expedition's departure.[13] A fellow botanist, Robert Thornton, extravagantly catalogued Harriet as a young lady who 'possessed extraordinary beauty, and every accomplishment, with a fortune of ten thousand pounds. Mr Banks had often seen her, when visiting the rare plants of Lee's, and thought her the fairest among the flowers.'[14]

In fact Harriet was one of three sisters who lived with their widowed mother in Holborn. Banks does seem to have been genuinely fond of her, and subsequent events suggest there was some kind of understanding between them. Her guardian James Lee looked upon it as an unofficial engagement, which would be announced if Banks should return alive from the Pacific. There was also some joke about Harriet knitting a set of 'worked' waistcoats for Banks while he was away, patterned with wild-flowers – perhaps one for each season he was absent.[15]

Yet Banks was certainly cautious about marriage at this stage in his career, remarking drily to a friend that though he loved experiments, matrimony was 'an experiment... with uncertain consequences', and rarely brought lifelong happiness. The eve of his great voyage was certainly not the moment to try it.[16] In a rare introspective entry Banks would reflect in his journal that he would probably never see Europe again, and that there were only two people in the world who would truly miss him. 'Today for the first time we dined in Africa, and took our leave of Europe for heaven alone knows how long, perhaps for Ever; that thought demands a sigh as a tribute due to the memory of freinds left behind and they have it; but two cannot be spared, t'would give more pain to the sigher, than pleasure to those sighd for. Tis Enough that they are rememberd, they would not wish to be too much thought of by one so long to be seperated from them and left alone to the Mercy of winds and waves.'[17]

If these two were his mother and his sister Sophia, then he did not wish to sigh unduly for Harriet Blosset. A certain bluffness was in order. When asked why he did not settle for the security of the eighteenth-century Grand Tour, the object of which as Dr Johnson said was to visit the classical civilisations along the shores of the Mediterranean, he replied briskly: 'Every blockhead does that; my Grand Tour shall be one round the whole Globe.'[18]

Banks spent his last night before going aboard at the opera. Then he dined in company with Harriet Blosset at her mother's house, accompanied by a Swiss geologist, Horace de Saussure, who assumed from their behaviour that they were 'betrothed'. Saussure described Harriet as very pretty and attentive, but 'a prudent coquette', and Banks as quite reconciled to their imminent parting, and drinking rather too much champagne.[19]

When the naturalist Gilbert White, snug in his Hampshire village, heard of Banks's departure on the high seas, he wrote thoughtfully to their mutual friend Thomas Pennant: 'When I reflect on the youth and affluence of this enterprizing young gentleman I am filled with wonder to see how conspicuously the contempt of dangers, and the love of excelling in his favourite studies, stands forth in his character ... If he survives, with what delight we shall peruse his Journals, his Fauna, his Flora! If he falls by the way, I shall revere his fortitude, and contempt of pleasures and indulgences: but shall always regret him.'[20]

4

Through the brilliance of Cook's navigation, and the skill of his crew-management, the *Endeavour* arrived at Tahiti with over six weeks in which to prepare for its main task, the transit observations. Previous expeditions had often been decimated by this stage, but Cook had lost only four men, and none to disease. The crew's diet included a serving of cabbage sauerkraut 'fresh every morning [as] at Covent Garden market', and Banks had shot seabirds wherever possible for fresh meat, including several large albatross with nine-foot wingspans.

The first death was the result of an accident with an anchor chain in Madeira. The next two occurred on land, and involved Banks. A field expedition he was leading had been overtaken by a snowstorm on Tierra del Fuego. It was a grim and confused story, which revealed something of Banks's qualities in a crisis. The party of twelve men (including Green, Solander and several sailors) had first run into trouble when one of Banks's young artists, Alexander Buchan, suffered an epileptic fit. Then a sudden blizzard cut off their retreat to the ship, several hours away down the mountains, and the party became separated in a birch wood as night fell.

Overcome by the biting cold, Banks's two black servants drank a stolen bottle of rum, and lay down in the snow and refused to go on. Meanwhile Solander, always rather stout and unfit, simply collapsed. Disintegration and disaster threatened the entire expedition. As darkness came on and the temperature plummeted, Banks tried to hold them together. First he regrouped the scattered men further down the mountainside with Green, made a fire and organised a brushwood 'wigwam', where Buchan was revived. Then Banks went back through the sub-zero night, with as many hands as he could muster, to drag the half-conscious Solander down through the birch wood to safety. It was an act which cemented their friendship. Banks also sent hands to save his black servants, but they were 'immoderately drunk', and could not – or would not – be carried back to the camp.

It was now past midnight, and everyone was stunned with cold, but Banks went out again in a last attempt to save them. 'Richmond was upon his legs but not able to walk, the other lay on the ground insensible as stone.' Banks tried to light a fire, but it was doused by falling snow. It was 'absolutely impossible' to bring the two men down. Finally he laid them out on a bed of branches, covered them with brushwood, and left them, hoping they would survive the night, insulated by alcohol. Going back at dawn, he found them both dead.[21]

When the rest of the party finally returned to the *Endeavour*, Cook noted that they all retired to their hammocks except Banks. After making his report and classifying his specimens, he insisted on going out in one of the ship's small boats alone, and spent the rest of the day in the bay, a solitary figure hunched over the stern, fishing with a seine net. Cook had not blamed him for his companions' deaths; but for the first time perhaps, he felt the weight of his responsibilities.

The third death was a suicide in the Pacific. This revealed another side to Banks. He made a long, thoughtful entry over the incident, in which a young able seaman, 'remarkable quiet and industrious', had apparently jumped overboard after being accused of stealing a sealskin tobacco pouch from the captain's cabin. Banks was struck by the melancholy event, remarking thoughtfully that 'it must appear incredible to every body who is not well acquainted with the powerfull effects that shame can work upon young minds'. Cook did not pursue the incident, but it seems clear from Banks's entry that he suspected homosexual bullying by an older member of the crew.[22]

The initial days on Tahiti were obviously exciting, but curiously tense. There was the unfortunate shooting in the first week, and the scare over the quadrant in the third. Young Alexander Buchan was taken ill again, and died from what appeared to be a repeat of the epileptic fit in Tierra del Fuego. Banks wrote in his journal: 'Dr Solander Mr Sporing Mr Parkinson and some of the officers of the ship attended his funeral. I sincerely regret him as an ingenious and good young man, but his Loss to me is irretrievable, my airy dreams of entertaining my freinds in England with the scenes that I am to see here are vanishd.' Banks's comments seem curiously harsh, and suggest his instinctive sense of entitlement. 'No account of the figures and dresses of men can be satisfactory unless illustrated with figures: had providence spared him a month longer what an advantage would it have been to my undertaking. But I must submit.'[23]

This note would be repeated elsewhere in his journal. Yet the expedition's other artist, the eighteen-year-old Sydney Parkinson, had no doubts about his employer's humanity. He had witnessed how Banks had nursed Buchan in the Tierra del Fuego débâcle, and wrote a long entry in his own journal reflecting on Banks's response to the unnecessary shooting of the Tahitian over the stolen musket. 'When Mr Banks heard of the affair, he was highly displeased, saying, "If we quarrel with these Indians, we should not agree with Angels." And he did all he could to accommodate the difference, going across the river, and through the mediation of an old man, prevailed upon many of the natives to come over to us, bearing plaintain trees, which is a signal of peace among them; and clapping their hands to their breasts, cried "Tyau!", which signifies friendship. They sat down by us; sent for coa nuts; and we drank milk with them.'[24]

With the security of the entire expedition in his hands, Cook was naturally cautious. He decided that a permanent armed encampment, Fort Venus, should be built on the beach to protect the expedition ashore and assert its authority. Banks says the Tahitians approved of this, and helped with the construction. Drawings by Parkinson, though the fort's situation among the palm trees is intended to look idyllic, show a square earthen stockade surmounted by a wooden palisade with naval swivel cannons mounted along the top. The fort was fifty yards wide by thirty yards deep, commanding a stretch of river on the inland side. In front along the shore was a trading area, where boats and canoes were drawn up, but all stores and arms were kept inside under guard, except for barrels of water by the stream. There were wooden gates which were closed at dusk, with armed sentries.

Within the perimeter, Cook established an official reception area, with a flagstaff flying a large Union Jack. There was a big rectangular marquee for gatherings and feasts, surrounded by an encampment of smaller supply tents and sleeping quarters, together with a bakery, a forge and an observatory. Banks had brought his own bell tent, only fifteen feet in diameter, but obviously the most well-equipped and comfortable. It soon became a popular destination with visiting Tahitians, and there was great rivalry for invitations to dine and sleep there. He noted in his journal: 'Our little fortification is now compleat, it consists of high breastworks at each end, the front palisades and the rear guarded by the river on the bank of which are placd full Water cask[s]. At every angle is mounted a swivel and two carraige guns pointed the two ways by which the Indians might attack us out of the woods. Our sentrys are also as well releivd as they could be in the most regular fortification.'[25]

This security was regarded as important for good relations, and the fort may have been as much designed to keep the sailors in, as the Tahitians out. Cook enforced a basic naval discipline, which included having one able seaman flogged on the quarterdeck for threatening a Tahitian woman with an axe.[26] Naturally there was a night curfew, but it was not very strictly observed, especially by the officers.

The constant theft of goods, especially of anything made of metal, regularly disrupted relations between the two communities. It was theft, too, that most clearly demonstrated the cruel gulf between the two civilisations. To the Europeans theft was a violation of legal ownership, an assault on private property and wealth. To the Tahitians it was a skilful affirmation of communal resources, an attempt to balance their self-evident poverty against overwhelming European superfluity. There was no source of metal anywhere on the island. The Tahitians' hunting knives were made out of wood, their fish hooks out of mother-of-pearl, their cooking pots out of clay. The Europeans clanked and glittered with metal.

As Cook himself observed, the *Endeavour* was an enormous treasure trove of metal goods: from iron nails, hammers and carpenters' tools to the most puzzling of watches, telescopes and scientific instruments. To the Tahitians it was wholly justifiable to redistribute such items. Banks, who had to keep a watchful eye on his scientific equipment, noticeably his dissection knives and his two solar microscopes, noted: 'I do not know by what accident I have so long omitted to mention how much these people are given to theiving. I will make up for my neglect however today by saying that great and small Chiefs and common men all are firmly of opinion that if they can once get possession of any thing it immediately becomes their own.'[27] ♣

♣ A very large ethical and philosophical issue about the nature of justice, property and ownership in society evidently lurked beneath these fleeting reflections of Banks and Cook. Over the next thirty years it would be addressed in various ways by Jean-Jacques Rousseau, Adam Smith, William Godwin and Thomas Paine. Beyond that lay the whole question of imperialism and colonialism, that great, tangled Victorian inheritance, looming like a dark stormcloud on the distant horizon. For the time being the bluff innocence

Ruminating on these larger ethical questions did not allow Banks to ignore simple practical problems, like the ubiquitous flies: 'The flies have been so troublesome ever since we have been ashore that we can scarce get any business done for them; they eat the painters colours off the paper as fast as they can be laid on, and if a fish is to be drawn there is more trouble in keeping them off it than in the drawing itself.'[28] The men tried many expedients: fly swats, flytraps made of molasses, and even mosquito nets draped over Parkinson while he worked.

Much time was spent in bargaining for sexual favours. The basic currency was any kind of usable metal object: there was no need for gold or silver or trinkets. Among the able seamen the initial going rate was one ship's nail for one ordinary fuck, but hyper-inflation soon set in. The Tahitians well understood a market economy. There was a run on anything metal that could be smuggled off the ship – cutlery, cleats, handles, cooking utensils, spare tools, but especially nails. It was said that the *Endeavour*'s carpenter soon operated an illegal monopoly on metal goods, and nails were leaving the ship by the sackful.

Later in June there was a crisis when one of the *Endeavour*'s crew stole a hundredweight bag of nails, and refused to reveal its whereabouts even after a flogging: 'One of the theives was detected but only 7 nails were found upon him out of 100 Wht and he bore his punishment without impeaching any of his accomplices. This loss is of a very serious nature as these nails if circulated by the people among the Indians will much lessen the value of Iron, our staple commodity.'[29]

Cook disapproved of sexual bartering, and made attempts to regulate the trade in love-making – 'quite unsupported', he later drily observed, by any of his officers. He remained philosophical, observing, not without humour, that there was a cautionary tale told about Captain Wallis's ship the *Dolphin*: when leaving Polynesian waters two years previously, so many nails had been surreptitiously prised out of her timbers that she almost split apart in the next Pacific storm she encountered. It was only later that the full, disastrous medical consequences of this spontaneous sexual trade became apparent.

---

of this first expedition is well caught by Banks's naval biographer, Patrick O'Brian: 'In any case the thefts were not all on one side: [Captain] Wallis had taken possession of the entire island [of Tahiti] and its dependencies, which brings to mind the remark about the relative guilt of the man who steals a goose from off a common and the other who steals the common from under the goose.' Patrick O'Brian, *Joseph Banks: A Life* (1987), p.95.

Yet Cook was already aware of the terrible risk and burden of spreading venereal disease, and wrote a long entry in his journal for 6 June 1769 reflecting on them. Certainly he had taken every precaution that his own crew were free from sexual infection when they arrived. They had been examined by Mr Monkhouse, the *Endeavour*'s surgeon, and they had in effect been in shipboard quarantine for eight months. But the Tahitian 'Women were so very liberal with their favours' that venereal disease had soon spread itself 'to the greatest part of the Ship's Company'. The Tahitians themselves called it 'the British disease', and Cook thought they were probably correct, though he wondered if it was already endemic, brought either by the French or by the Spanish. 'However this is little satisfaction to them who must suffer by it in a very great degree and may in time spread itself over all the Islands of the South Seas, to the eternal reproach of those who first brought it among them.'[30] ♣

Some crew members had moral scruples from the start. Young Sydney Parkinson noted disapprovingly in his journal: 'Most of our ship's company procured temporary wives amongst the Natives, with whom they occasionally cohabited; an indulgence which even many reputed virtuous Europeans allow themselves, in uncivilised parts of the world, with impunity. As if a change of place altered the moral turpitude of fornication: and what is a sin in Europe, is only a simple innocent

---

♣ It was soon accepted that the Europeans in general were responsible. A satirical poem dedicated to Banks in 1777 had a bitterly sarcastic footnote referring to the transmission of 'the Neapolitan fever' to Tahiti, 'where from the promiscuous intercourse of the Natives, it will probably very soon annihilate them all, and in the most dreadful manner, for the honour of Christian humanity': 'An Historic Epistle from Omai to the Queen of Tahiti' (1777). In addition there is nature's revenge on marauding European crews, as described in Coleridge's ballad *The Ancient Mariner*. It is often forgotten that this poem describes the death of an entire ship's complement of 200 men (bar the Mariner) after an encounter with a terrifying and diseased woman, 'Life-in-Death':

> Her lips were red, her looks were free,
> Her locks were yellow as gold:
> Her skin was as white as leprosy,
> The Nightmare Life-in-Death was she,
> Who thicks man's blood with cold.
> (*The Ancient Mariner*, lines 190–4)

The full catastrophe of venereal disease, which devastated the Pacific populations over the next two generations, has been described by Alan Moorehead in *The Fatal Impact* (1966).

gratification in America; which is to suppose that the obligation of chastity is local, and restricted only to particular parts of the globe.'[31]

Banks appeared to have no such scruples. He made a point of leaving the camp most nights and, as he put it, 'sleeping alone in the woods'. He told himself, perhaps with the easiness of birth and privilege, that his intentions were as much botanical as amorous, and that no moral code was seriously infringed. After all, it was all *research*. Yet it is difficult to see him as a simple predator. He was clearly attractive to Tahitian women – robust, generous, good humoured – and it is striking how quickly he gained a footing (if that is the term) in Tahitian society generally.

He reached an important and lasting understanding with the Tahitian queen, Oborea. This included the pretty girl 'with fire in her eyes', who conveniently turned out to be one of the queen's personal servants, Otheothea. But it was much more than a sexual agreement. Almost uniquely, Banks was welcomed into many hidden aspects of Tahitian life, including dining, dressing and religious rituals. It also brought him his most vital contact, with one of the Tahitian 'priests' or wise men, Tupia, who taught him the language and many of the island customs.

Characteristically, Banks was virtually the only member of the *Endeavour* who bothered to learn more than a very few words of Tahitian. His journal contains a basic vocabulary. The words fall into four main sections, which perhaps suggest his particular areas of interest: first, plants and animals ('breadfruit, dolphin, coconut, parroquet, shark'); then intimate parts of the human body ('breasts, nails, shoulders, buttocks, nipples'); then sky phenomena ('sun, moon, stars, comet, cloud'); and finally qualities ('good, bad, bitter, sweet, hungry'). There are also some verbs, including those for stealing, understanding, eating, and being angry or tired. But the list cannot be very complete, since there are no words for love, laughter, music or beauty – and it would be difficult to talk Tahitian without any of these.

Banks's skill with language gave him a new role as the chief trading officer or 'marketing man' for the *Endeavour*. He established himself in a canoe drawn up on the shore outside Fort Venus, and every morning would negotiate for food and supplies. He was acutely aware of the shifting trading rates, noting on 11 May: 'Cocoa nuts were brought down so plentifully this morn that by ½ past 6 I had bought 350. This made it necessary to drop the price of them least so many being brought at once we should exhaust the country and want hereafter. Not withstanding I had before night bought more than a thousand at the rates of 6 for an amber coulourd bead, 10 for a white one, and 20 for a fortypenny nail.'

Trading also brought him into regular contact with Tahitians of every class, and helped him establish a broad base of good friendships, while Cook and the other officers remained more aloof. His journal shows him constantly enlarging his Tahitian social circle, referring to people by their names, many of them in terms of trust and affection. When this trust was broken or shaken, Banks was often mortified. He frequently blamed himself, rather than the Tahitians, for misunderstandings or false accusations of theft.

He learned the local name for the island, which he transliterated into English: 'We have now got the Indian name of the Island, *Otahite*, so therefore for the future I shall call it.' His spelling was simply based on the pronunciation 'O Tahiti'. He also found that the Tahitians had in turn transliterated their visitors' English names, but after their own fashion. 'As for our own names the Indians find so much difficulty in pronouncing them that we are forcd to indulge them in calling us what they please.' The results were rather odd, and Banks suspected that they were partially amusing nicknames. Captain Cook was 'Toote'; Dr Solander was 'Torano'; the chief mate Mr Molineux was 'Boba' (Banks guessed from his Christian name, 'Robert'); and Banks himself was 'Tapáne', which appeared to mean a drum. Whereas the English had difficulty in recognising more than a handful of Tahitians by name, Banks observed that the Tahitians were much quicker, and soon had names for 'almost every man in the ship'.[32]

Banks's new role expanded to that of civilian diplomat and social secretary. Not being an official part of Cook's naval command gave him a certain flexibility between ship and shore. He helped to arrange many of the informal dinners at Fort Venus, as well as the official visits to the ship. He was also able to partake in Tahitian ceremonies not strictly approved of by Cook. As a result, from May 1769 onwards, Banks's journal entries steadily change their character. They are still full of exquisite botanical and zoological details, but they become more and more anthropological. People begin to replace plants. The daily journal entries begin to cover an astonishing range of phenomena: tattooing, nose-flute-playing, naked wrestling, roasting dogs, surfing.

The young Linnaean collector, with his detached interest in cataloguing, dissection and taxonomy, was being transformed by his Tahitian experience. The Enlightenment botanist, the aristocratic collector and classifier, was steadily being drawn in to share another ethnic culture and its customs. His *Endeavour Journal* would become fuller for Tahiti than for any other part of the Pacific. Eventually it would expand into a long report, couched in anthropological terms, 'On the Manner and Customs of the South Sea Islands'. It would be the most detailed monograph he ever wrote.[33] Banks was becoming an ethnologist, a human investigator, more and more sympathetically involved with another community. The Tahitians are no longer 'savages', but his 'friends'. He was trying to understand Paradise, even if he did not quite believe in it.

5

The occasion of the Transit of Venus, on 3 June 1769, provided a good opportunity for Banks's new approach. In late May, Cook had set up three astronomical observation points to insure against the possible interference of localised cloud cover. Banks accompanied the furthest group of observers to the outlying island of Moorea. While recording the transit was one of the main objectives of the entire expedition, it was one which the Tahitians could not be expected to understand. Yet Banks's journal entry for 3 June 1769 shows the consideration with which he treated the islanders during this crucial piece of scientific research.

Banks had set up the instruments at a camp above the shoreline by 8 a.m., and had also provided 'a large quantity of provisions' for trade and diplomatic gifts. Leaving the telescopes, he waited down by the beach. Two large canoes appeared, carrying the king of the island, Tarróa, and his sister Nuna. Banks was standing in the shade of a tree, and immediately went down to them: 'I went out and met them and brought them very formally into a circle I had made, into which I had before sufferd none of the natives to come. Standing is not the fashion among these people. I must provide them a seat, which I did by unwrapping a turban of Indian cloth which I wore instead of a hat, and spreading it upon the ground. Upon which we all sat down and the king's present was brought Consisting of a hog, a dog and a quantity of Bread fruit Cocoa nuts &c. I immediately sent a canoe to the Observatory to fetch my present, an adze a shirt and some beads with which his majesty seemd well satisfied.'

This was a customary exchange of gifts. But Banks was determined to explain to the king what his men were doing. 'After the first Internal contact [of Venus with the sun's disc] was over I went to my Companions at the Observatory carrying with me Tarroa, Nuna and some of their cheif atendants. To them we shewd the planet upon the sun and made them understand that we came on purpose to see it. After this they went back and myself with them.'

Yet the nonchalant end of this journal entry shows that Banks was also perfectly prepared to take advantage of his privileged situation: 'At sunset I came off having purchasd another hog from the King. Soon after my arrival at the tent 3 handsome Girls came off in a canoe to see us. They had been at the tent in the morning with Tarroa. They chatted with us very freely and with very little persuasion agreed to send away their carriage and sleep in [the] tent. A proof of confidence which I have not before met with upon so short an acquaintance.'[34]

The next day Banks added mischievously: 'We prepared ourselves to depart, in spite of the intreaties of our fair companions who persuaded us much to stay.' But who was seducing whom? Who was exploiting whom? Many of Banks's most striking observations on Tahiti record behaviour which seems difficult to evaluate or interpret. Once in late April, one of his closest friends among the Tahitian women, Terapo, appeared at the gate of Fort Venus in great distress. Banks carefully recorded what followed: 'Terapo was observd to be among the women on the outside of the gate, I went out to her and brought her in, tears stood in her eyes which the moment she enterd the tent began to flow plentifully. I began to enquire the cause; she instead of answering me took from under her garment a sharks tooth and struck it into her head with great force 6 or 7 times. a profusion of Blood followd these strokes and alarmd me not a little. For two or 3 minutes she bled freely more than a pint in quantity, during that time she talkd loud in a most melancholy tone. I was not a little movd at so singular a spectacle and holding her in my arms did not cease to enquire what might be the cause of so strange an action.'

Terapo consistently refused to explain, though Banks's gesture of taking her in his arms suggests the possibility of some kind of emotional upset between them. There were several other Tahitians in the tent at the time – yet 'all talked and laughed as if nothing melancholy was going forward'. This only deepened the mystery. Terapo's recovery was no less abrupt and inexplicable: 'What surpriz'd me most of all was that as soon as the bleeding ceas'd she lookd up smiling and immediately began to collect peices of cloth which during her bleeding she had thrown down to catch the blood. These she carried away out of the tents and threw into the sea, carefully dispersing them abroad as if desirous that no one should be reminded of her action by the sight of them. She then went into the river and after washing her whole body returnd to the tents as lively and chearfull as any one in them.'[35]

Banks later discovered that this dramatic way of expressing grief was universal among the Tahitian women, and he saw many who had permanent 'grief scars' on their heads. He learned something about such things from queen Oborea's little family circle. This group – consisting of the queen, her twenty-year-old lover Obadee, her servant Otheothea (Banks's lover) and several close male friends – seems to have adopted Banks, and looked after his welfare. They frequently all came to sleep in his tent, when feasting and love-making seems to have taken place easily and indiscriminately. Sometimes this could lead to comic-opera complications, as Banks would smilingly hint in his journal.

> 21 May. Sunday, Divine service performd, at which was present Oborea, Otheothea, Obadee, &c. all behav'd very decently. After dinner Obadee, who had been for some time absent, returnd to the fort. Oborea desired he might not be let in, his countenance was however so melancholy that we could not but admit him. He looked most piteously at Oborea, she most disdainfully at him. She seems to us to act in the character of a Ninon d'Enclos who, satiated with her lover, resolves to change him at all events. The more so as I am offered, if I please, to supply his place! But I am at present otherwise engag'd; indeed was I free as air, her Majesties person is not the most desireable.

Other mishaps ensued towards the end of the month. Banks, Cook and Solander had decided on an expedition to explore the western end of the bay, and to bargain for some wild pigs rumoured to be held by the local chieftain, Dootah. Banks was followed solicitously up the coast by queen Oborea and her entourage in their large and comfortable outrigger canoes. When the expedition was benighted in chief Dootah's village (no accommodation being offered), Banks agreed to separate from the others and sleep in the queen's well-appointed canoe, which had a cabin constructed between the floats.

As he explained in his journal, he and the queen had naturally removed all their clothes. 'We went to bed early as is the custom here: I strippd myself for the greater convenience of sleeping as the night was hot. Oborea insisted that my cloths should be put into her custody, otherwise she said they would certainly be stolen. I readily submitted and laid down to sleep with all imaginable tranquility.'

The next morning Banks awoke to find almost all his kit missing – his handsome nankeen jacket with its fine brass buttons, his breeches, his waistcoat, his much-prized pistols and even his powder-horn. All had been – most unfortunately, murmured the queen – stolen in the night. After unavailing searches and appeals, Banks was faced by the prospect of a shame-faced retreat to Fort Venus with neither the promised pigs, nor his precious pistols, or even most of his clothes. Queen Oborea seems to have enacted a form of revenge. She supplied Banks with Tahitian shawls and blankets to replace his European clothes, and bade him farewell. For once, Banks was distinctly unamused: 'I made a motley apearance, my dress being half English and half Indian. Dootahah soon after made his apearance; I pressed him to recover my Jacket but neither he nor Oborea would take the least step towards it so that I am almost inclind to believe that they acted principals in the theft.'[36]

Any resentful feelings were swept aside the following afternoon. Rounding the tip of the bay, they looked out to sea and saw something wholly unexpected and 'truly surprising'. This was the astonishing and never-to-be-forgotten sight, far out on the unprotected edge of the lagoon, of a group of dark Tahitian heads bobbing amidst the enormous dark-blue Pacific waves. At first Banks thought they had been flung out of their canoes and were drowning. Then he realised that *the Tahitians were surfing.*

No European had ever witnessed – or at least recorded – this strange, extreme and quintessentially South Seas sport before. It left Banks amazed by the courage and dexterity of the Tahitian surfers, and the beauty and nonchalant grace with which they mastered the huge and terrifying Pacific rollers: 'It was in a place where the shore was not guarded by a reef as is usualy the case, consequently a high surf fell upon the shore. A more dreadfull one I have not often seen: no European boat could have landed in it and I think no Europaean who had by any means got into [it] could possibly have saved his life, as the shore was coverd with pebbles and large stones. In the midst of these breakers 10 or 12 Indians were swimming.'

Here the power of wild nature was not tamed, but was harnessed by human beings; and they evidently revelled in it. The Tahitians had developed what were clearly surfboards, constructed out of the smooth, curved ends of old canoes. They were scornful of all danger, and exultant in their physical skills. 'Whenever a surf broke near them [they] dived under it with infinite ease, rising up on the other side; but their cheif amusement was carried on by the stern of an old canoe. With this before them they swam out as far as the outermost breach, then one or two would get into it and opposing the blunt end to the breaking wave were hurried in with incredible swiftness. Sometimes they were carried almost ashore, but generally the wave broke over them before they were half way. In which case the[y] dived and quickly rose on the other side with the canoe in their hands, which was towed out again and the same method repeated.'

Most extraordinary of all, this perilous activity evidently had absolutely no practical purpose or possible use. It was nothing to do with fishing, or transport, or navigation. The Tahitians did it for the sheer, inexhaustible delight of the thing. It was a complete Paradise sport: 'We stood admiring this very wonderfull scene for full half an hour, in which time no one of the actors attempted to come ashore but all seemd most highly entertained with their strange diversion.'[37]

Some Tahitian ceremonies were carefully organised, and suitable for all the *Endeavour*'s crew, such as the afternoon of naked wrestling organised by queen Oborea. Others were less official. One morning a number of young women arrived by canoe, and were offered to Banks in a curiously provoking ceremony:

> 12 May. While I sat trading in the boat at the door of the fort a double Canoe came with several women and one man under the awning. The Indians round me made signs that I should go out and meet them ... Tupia who stood by me acted as my deputy in receiving them ... Another man then came forward having in his arms a large bundle of cloth. This he opend out and spread it piece by piece on the ground between the women and me. It consisted of nine pieces. Three were first laid. The foremost of the women, who seemd to be the principal, then stepped upon them and quickly unveiling all her charms gave me a most convenient opportunity of admiring them by turning herself gradualy round.

Further pieces of cloth were then laid out in front of Banks, and the woman stepped closer and repeated her slow, smiling, naked gyrations. No awkwardness seems to have been felt on either side. 'She then once more displayd her naked beauties and immediately marchd up to me, a man following her and doubling up the cloth as he came forwards which she immediately made me understand was intended as a present for me. I took her by the hand and led her to the tents acompanied by another woman her friend. To both of them I made presents but could not prevail upon them to stay more than an hour.'

This is clearly a seduction scene, and the unnamed Tahitian man is bartering the woman. Yet there is no gloating in Banks's entry; nor is it clear whether he took advantage of this frank proposal. Cook also witnessed this scene, and remarked that the young woman acted 'with as much Innocency as one could possibly conceive'.[38]

By mid-June Banks was increasingly prepared to abandon European inhibitions, including his clothes. He noted frequently, 'I lay in the woods last night as I very often did,' by which one can understand he was probably with Otheothea. On 10 June his journal records how he stripped off, had his body covered with charcoal and white wood ash, and danced ceremonially with a witch doctor (*Heiva*). He was joined by two naked women and a boy, and together they danced through the length of the village, past the gate of Fort Venus, and along the shore.

It must have been an extraordinary sight, the expedition's chief botanist whirling past the marine guards in the sunlight. But this Tahitian ceremony was not at all what it might have appeared to uninstructed European eyes. It was not an erotic rite, but a dance of ritual mourning. Banks and the young women were taking the part of ancestral ghosts (*Ninevehs*). 'Tubourai was the Heiva, the three others and myself were the Nineveh. He put on his dress, most Fantastical tho not unbecoming ... I was next prepard by stripping off my European cloths and putting me on a small strip of cloth round my waist, the only garment I was allowd to have, but I had no pretensions to be ashamd of my nakedness for neither of the women were a bit more coverd than myself. They then began to smut me and themselves with charcoal and water, the Indian boy was compleatly black, the women and myself as low as our shoulders. We then set out. Tubourai began by praying twice, once near the Corps again near his own house ... To the fort then we went to the surprize of our freinds and affright of the Indians who were there, for they every where fly before the Heiva like sheep before a wolf.' The dancing continued along the shore, and went on for the rest of the afternoon, 'After which we repaird home, the Heiva undressd and we went into the river and scrubbd one another till it was dark before the blacking would come off.' [39]

After eight weeks it became clear that many other officers were not integrating so well into the Tahitian way of life. One of them committed an elementary error by foolishly violating a religious taboo: 'Mr Monkhouse our surgeon met to day with an insult from an Indian, the first that has been met with by any of us. He was pulling a flower from a tree which grew on a burying ground and consequently was I suppose sacred, when an Indian came behind him and struck him; he seiz'd hold of him and attempted to beat him, but was prevented by two more who coming up seizd hold of his hair and rescued their companion after which they all ran away.'[40]

Even Captain Cook managed to create an unnecessary crisis when it was discovered that a metal fire-rake had been stolen from the fort. Determined to set an example, he impounded a score of native canoes. When the rake was swiftly returned, Cook then demanded that all other implements stolen from the camp in the last month should also be restored before he would return the canoes. It was quickly clear to Banks that Cook had here overplayed his hand with the Tahitians. The situation grew more complicated when it was learned that the canoes actually belonged to another group of islanders, who were bringing much-needed food to their relatives. They had no previous connection with the British, and were obviously not responsible for any of the thefts.

The aggrieved Tahitians appealed directly to Banks, rather than to Cook, over this blatant injustice. 'Great application was made to me in my return that some of these might be released.' For the first time Banks appeared openly critical of Cook in his journal: 'I confess had I taken a step so violent I would have seizd either the persons of the people who had stolen from us, most of whoom we either knew or shrewdly suspected, or their goods at least instead of those of people who are intirely unconcernd in the affair and have not probably interest enough with their superiors (to whoom all valuable things are carried) to procure the restoration demanded.'[41]

For several days all trading ceased, and the fish in the sequestered canoes began to rot, filling the fort with an ominous smell. Then one of the duty officers compounded their difficulties by committing another needless offence. Taking a party of sailors out to collect ballast stone for the *Endeavour*, he promptly began 'pulling down' a Tahitian burying ground. Once again the Tahitian appeal was made directly to Banks: 'To this the Indians objected much and [a] messenger came to the tents saying that they would not suffer it. I went with the 2nd Lieutenant to the place.' Banks, in his diplomatic role, eventually managed to soothe both parties, had the burying ground restored, and found a nearby riverbed where the sailors 'gatherd stones very Easily without a possibility of offending anybody'.[42]

The issue of the impounded canoes remained, however, and suggested hostile attitudes on both sides: 'The fish in the Canoes stink most immoderately so as in some winds to render our situation in the tents rather disagreable ... The market has been totaly stoppd ever since the boats were seizd, nothing being offerd to sale but a few apples; our freinds however are liberal in presents so that we make a shift to live without expending our bread.'[43]

Queen Oborea and Banks's flame Otheothea reappeared at the fort, though initially Banks thought it wiser for them to sleep outside in their canoes, and they were 'rather out of humour'.[44] The crisis was only gradually defused, as Cook allowed the canoes to be taken back three or four at a time, in return for small peace offerings. One unexpected development was that Oborea's ex-husband, known as Oamo, put in an appearance to plead for the release of the boats. To everyone's surprise, Oamo behaved very politely towards his ex-wife, and he made the most favourable impression on Banks. He showed himself to be a 'very sensible man by the shrewd questions he asks about England its manners and customs &c.'[45] But the general issue of theft and restitution was never really resolved, and relations with the Tahitians were less relaxed in the last month of the expedition's stay. Chief Dootah completely withdrew from the Europeans, claiming he had been frightened by Banks shooting for wild duck.

Food remained a source of mutual interest, and one remarkable culinary event featured a dog, which the priest Tupia killed, dressed and roasted, while Banks carefully took down the recipe. Most of the sailors were repelled, but Banks declared the results to be delicious. 'A most excellent dish he made for us who were not much prejudicd against any species of food. I cannot however promise that an European dog would eat as well, as these scarce in their lives touch animal food, Cocoa nut kernel, Bread fruit, yams &c, being what their masters can best afford to give them and what indeed from custom I suppose they preferr to any kind of food.'

Banks was also more at odds than previously with his naval companions, and there was some kind of quarrel with the insensitive surgeon Monkhouse. Banks tactfully omitted this from his journal, but young Sydney Parkinson recorded a confrontation between the two, and thought it arose over Monkhouse propositioning Otheothea. Several of Oborea's Tahitian girls had arrived at Banks's tent 'very earnest in getting themselves husbands'. They behaved 'very agreeable until bedtime, and determined to lie in Mr Banks's tent, which they accordingly did, till the Surgeon having some words with one of them ... he insisted she should not sleep there, and thrust her out'. Otheothea was then heard crying for some time in the tent. Parkinson noted dramatically: 'Mr Monkhouse and Mr Banks came to an *eclaircisement* some time after; had very high words and I expected they would have decided it by a duel, which, however, they prudently avoided.' Oborea and her retinue then left in their canoes, and would not return to the camp. 'But Mr Banks went and staid with them all night.'[46]

It was probably no coincidence that Cook now decided that he would take his botanist off on a separate expedition. This was planned as a circumnavigation of the entire island in the *Endeavour*'s small sailing boat. Its official naval objective was to chart all possible harbours, and discover any signs of previous European landings – notably French or (as it was supposed) Spanish. For Banks, however, it was a glorious scientific field expedition, and a tantalising extension of his new anthropological investigations.

Starting at Matavi Bay in the north of the island, the circumnavigation took six days. They set out with a small crew and a handful of marines at 3 a.m. on 26 June, heading eastwards. There was considerable uncertainty about their reception once they passed beyond the territory of Matavi Bay, where Oborea and Dootah had influence. One of their guides said that 'people not subject to Dootah' would kill them. Accordingly they adopted a cautious mode of advance. Banks and Cook travelled mostly on foot along the shoreline, while the pinnace, its marines armed with loaded muskets, was rowed just offshore, keeping pace and overseeing their progress. A number of native canoes followed them.

'Banks as usual explored, botanized, conversed,' noted Cook with a smile.[47] Indeed he was soon plunging inland and out of sight, claiming to be in search of specimens, waving a large butterfly net as his preferred weapon of defence. Banks thought nothing of foraging by himself ashore, once disappearing at dusk to hunt for provisions. He shot a duck and two curlews, then pressed on deeper inland. 'I went into the woods, it was quite dark so that neither people nor victuals could I find except one house where I was furnishd with fire, a breadfruit and a half, and a few *ahees* [nuts].' That night he slept under the awning of a native canoe.

Some discoveries were reassuring. In one village they found an English goose and a turkey cock which had been left behind by the *Dolphin*'s crew two years previously. 'Both of them immensely fat and as tame as possible, following the Indians every where who seemd immensely fond of them.' Other sights were less so. In a longhouse in this neighbourhood Banks spotted a rather ominous wall decoration. Proudly mounted on a semi-circular board at the end of the hut were a collection of human bones. Banks carefully inspected them – they were all under-jaw bones – no less than fifteen in all: 'They appeard quite fresh, not one at all damagd even by the loss of a Tooth.' These were evidently war trophies, and even perhaps signs of cannibalism. Banks enquired boldly, but could get no reply. 'I askd many questions about them but the people would not attend at all to me and either did not or would not understand either words or signs upon that subject.'[48] Later he learned they had been 'carried away as trophies and are usd by the Indians here in exactly the same manner as the North Americans do scalps'.[49]

Some receptions were welcoming, but deceptive. 'Many Canoes came off to meet us and in them some very handsome women who by their behaviour seemd to be sent out to entice us to come ashore, which we most readily did.' They were received in a very friendly manner by Wiverou, who was chief of the district. A splendid feast was prepared, accommodation offered, and Banks confidently paid court to the women, 'hoping to get a snug lodging by that means, as I had often done'. This is a revealing admission, and as it turned out it was wholly unjustified. As the evening drew on, and the women found Banks more importuning, 'they dropped off one by one'. He ruefully remarked that at last he found himself in the position of being 'jilted 5 or 6 times, and obliged to seek out for a lodging myself'. He slept alone in a hut, naked as was now his custom, except for a piece of Tahitian cloth thrown over his waist. For once he implies that he felt himself to be the outcast, and this rejection evidently gave him pause for thought.

Indeed, for all the apparent hospitality, their situation always remained surprisingly uncertain away from Fort Venus and the guns of the *Endeavour*. It could easily become alarming. Banks noted a tense moment on the third morning: 'About 5 O'Clock our sentry awakd us with the alarming intelligence of the boat being missing. He had he said seen her about ½ an hour before at her grapling which was about 50 yards from the shore, but that on hearing the noise of Oars he lookd out again and could see nothing of her. We started up and made all possible haste to the waterside. The morn was fine and starlight but no boat in sight. Our situation was now sufficiently disagreable: the Indians had probably attackd her first and finding the people asleep easily carried her, in which case they would not fail to attack us very soon, who were 4 in number armd with one musquet and cartouch box and two pocket pistols without a spare ball or charge of powder for them.'

For fifteen minutes the little party stood alone on the Tahitian beach, suddenly very conscious that they were white Europeans, isolated and ill-armed, on the remote beach of an island that did not belong to them. They watched the sun come up, and waited to be massacred. Then, to their immense relief, the pinnace reappeared around the point of the bay. She had simply slipped her mooring and drifted out to sea while her crew slept. They told themselves that the murderous party of attacking Tahitians had been a figment of their European fears.[50]

Other experiences were unsettling in a different way. On their last day they discovered an enormous stone '*marai*' or funeral monument, shaped like a pyramid, some forty-four feet high and nearly 300 feet wide, with steps of superbly polished white coral down both sides. This, the 'masterpiece' of Tahitian architecture on the island, was unsettling to Banks because its construction seemed technically inexplicable. 'It is almost beyond belief that Indians could raise so large a structure without the assistance of Iron tools to shape their stones or mortar to join them.'

Not far away was another mystery: a huge wicker man constructed of basketwork, evidently for some obscure sacrificial rite. 'The whole was neatly coverd with feathers, white to represent skin and black to represent hair and tattow. On the head were three protuberances which we should have calld horns but the Indians calld them tata ete, little men. The image was calld by them Maúwe; they said it was the only one of the kind in Otahite and readily attempted to explain its use. But their language was totaly unintelligible and seemed to referr to some customs to which we are perfect strangers.'

By the time of their return to Fort Venus on 1 July, Cook had completed a beautiful and lucid chart of the island, the figure of eight with its 'marshy isthmus' at the join, which would serve European mariners for generations to come, a model of clarity and accuracy. Banks had hugely increased his supply of botanical specimens, and his knowledge of the fruit and animal resources of the island. But the human mystery of Tahiti had deepened. Its history, customs, religious practices, sexual rites all challenged European understanding, and demanded a new science of explanation.

One of the most puzzling and disturbing of all the ceremonies that Banks witnessed was the tattooing of a young girl's buttocks. Tattooing was universal in Tahiti, and its function among young male warriors was self-evident. Complex patterns were worked across the legs, the upper torso, on the fingers and ankles, and around the loins: proof of a young man's courage, and also of his place in the social hierarchy. The skin was pierced with a block of sharpened wooden pins, and impregnated with a purple-black vegetable dye mixed with coconut oil. The operation was long and exquisitely painful, usually performed in stages over several months, and was itself a form of male initiation rite.

Banks could understand all this very well. What he could not understand was why women were forced to undergo it also, moreover at a cruelly young age. Was it a form of sexual initiation? Or purely decorative? Or a form of tribal identity marking? Tahitian women decorated themselves with flowers, and wore beautiful mother-of-pearl earrings, of which Banks made an entire collection. But they used very little other decorations or jewellery.

> 5 July 1769. This morn I saw the operation of Tattowing the buttocks performd upon a girl of about 12 years old. It proved as I have always suspected a most painfull one. It was done with a large instrument about 2 inches long containing about 30 teeth, every stroke of this hundreds of which were made in a minute drew blood. The patient bore this for about ¼ of an hour with most stoical resolution; by that time however the pain began to operate too stron[g]ly to be peacably endured. She began to complain and soon burst out into loud lamentations and would fain have persuaded the operator to cease. She was however held down by two women who sometimes scolded, sometimes beat, and at others coaxd her.

Banks became more and more restless as this operation proceeded. 'I was setting in the adjacent house with Tomio for an hour, all which time it lasted and was not finishd when I went away, tho very near. This was one side only of her buttocks for the other had been done some time before. The arches upon the loins upon which they value themselves much were not yet done, the doing of which they told caused more pain than what I had seen.'

Finally he could stand it no more, and went back alone to Fort Venus. He was clearly both disturbed and fascinated by the whole procedure, though he gives little away about his deeper feelings – whether he was repulsed or shocked, or even sexually excited. He later wrote: 'For this Custom they give no reason, but that they were taught it by their fore-fathers ... So essential is it esteemed to Beauty, and so disgraceful is the want of it esteemed, that every one submits to it.'[51]

On 3 July Banks made one last expedition into the interior, this time accompanied only by the surgeon Monkhouse. His choice of companion seems to have been deliberate. They pursued a river line up into the mountains, pressing on as far as they could go, painfully clambering up the riverbed, sweating and stumbling, searching for plants and minerals. On the way Banks concluded rightly that Tahiti must be volcanic in origin, 'a volcano which now no longer burns'; which also explained the fact that the Tahitian god was known as 'the Father of Earthquakes'.

Twelve miles inland, further than any previous expedition had ever penetrated, they were brought to an abrupt halt by an enormous and beautiful waterfall, surrounded by 'truly dreadful' cliffs more than a hundred feet high. Beneath it lay 'a pool so deep that the Indians said we could not go beyond it'. Here, in this enchanted but faintly menacing place, the secret heart of the Tahitian island, it seems the two men bathed and talked together, until European rivalries were happily forgotten.[52]

6

After a stay of three months, the British expedition prepared to leave in the second week of July 1769. Banks spent a whole day sowing South American fruit seeds for the Tahitians to harvest after they were gone: lemons, limes, watermelons, oranges. While he loaded his final specimens of Tahitian plants and animals aboard, he considered the possibility of taking a human representative of Paradise back to England. The matter had been raised with Tupia, the wise priest, who proposed that he himself should make the perilous journey together with his young son: 'This morn Tupia came on board, he had renewd his resolves of going with us to England, a circumstance which gives me much satisfaction. He is certainly a most proper man, well born, cheif Tahowa or priest of this Island, consequently skilld in the mysteries of their religion. But what makes him more than any thing else desireable is his experience in the navigation of these people and knowledge of the Islands in these seas. He has told us the names of above 70, the most of which he has himself been at.'[53]

Although Tupia was evidently enthusiastic to make the journey, Captain Cook would not underwrite the decision. He did not feel that the Tahitian could be signed on as an official member of the expedition, and he thought that once he was in England the Admiralty and the Crown would 'in all human probability' refuse to support him financially. Banks had no such hesitations, and resolved to be responsible for both Tupia's welfare and his upkeep, saying he was taking on Tupia as his friend and his guest. Cook agreed, and would find Tupia's help as the expedition's South Seas navigator and Polynesian translator invaluable.

Banks added a comment that seems extraordinarily revealing. He suddenly thinks of outdoing his fashionable country-house friends back in Lincolnshire with their exotic pets. 'I do not know why I may not keep [Tupia] as a curiosity, as well as some of my neighbours do lions and tygers, at a larger expence than he will probably ever put me to.' The idea that his friend and adviser could have been considered, even for a moment, as a 'curiosity', or a wild animal specimen, comes as a shock. It shows that Banks, for all his sympathy and humanity, could easily revert to his role as Linnaean collector and wealthy European landowner on a jaunt among the natives. However one explains it, the remark hangs uneasily in the air, never quite dissipated, never quite forgotten: the snake in the garden.

Nonetheless, Banks closed this entry on a more typically generous note: 'The amusement I shall have in [Tupia's] future conversation, and the benefit he will be of to this ship, as well as what he may be if another should be sent into these seas, will I think fully repay me.'[54]

There was a last-minute drama when, as Fort Venus was being dismantled, two of the marines slipped away into the woods, having said they had beautiful Tahitian wives, were content to resign His Majesty's service, and intended to stay. Cook sent out a tracking party, but also took native hostages, which caused a good deal of ill-feeling. Once again it was Banks who defused a potentially ugly situation, by agreeing to spend the last night onshore with his Tahitian friends, until the marines should return. 'At day break a large number of people gatherd about the fort many of them with weapons; we were intirely without defences so I made the best I could of it by going out among them. They wer[e] very civil and shewd much fear as they have done of me upon all occasions, probably because I never shewd the least of them, but have upon all our quarrels gone immediately into the thickest of them. They told me that our people would soon return.'

The marines did return, to everyone's huge relief, at eight o'clock that morning, and Banks watched carefully through his telescope as they were hauled aboard the *Endeavour* while the hostages were released in exchange. Once he saw they were all 'safe and sound' he discharged his own Tahitian 'prisoners' from his tent, 'making each such a present as we though[t] would please them with which some were well content'.[55] Though he does not mention it, this may also have been his last chance to spend a night with Otheothea.

The *Endeavour* finally hoisted anchor early on the morning of 13 July 1769. 'After a stay of 3 months we left our beloved Islanders with much regret,' reported Banks, with careful understatement.[56] The whole of Matavi Bay was full of Tahitian canoes. Oborea and Otheothea came aboard briefly to say tearful farewells. Banks and Tupia then climbed the rigging and stood together in the crow's nest, waving. Sydney Parkinson wrote: 'On our leaving the shore the people in the canoes set up their woeful cry – *Awai! Awai!* – and the young women wept very much. Some of the canoes came up to the side of the ship, while she was under sail, and brought us many cocoas.'[57]

7

Banks had gained a complicated impression of Paradise. As the *Endeavour* sailed westwards towards New Zealand throughout August 1769, with brief stops at other Polynesian islands (seventeen in all), he sat down in his sweltering cabin to put his reflections in some kind of order. The result was his long anthropological essay 'On the Manners and Customs of the South Sea Islands', perhaps the most original paper he ever wrote.

Tahiti was indeed a kind of Paradise: astonishingly beautiful, its people open and generous, and its way of life languid and voluptuous. But there were many darker elements: strong, even oppressive social hierarchies; endemic thieving; a strange religion haunted by ghosts and superstitions; infanticide; and warlike propensities just below the surface. Nonetheless, Banks's essay is full of his glowing memories, which would later stand him in good stead on the bleakest moments of the journey home: 'No country can boast such delightfull walks as this, the whole plains where the people live are coverd with groves of Breadfruit and cocoa nut trees without underwood; these are intersected in all directions by the paths which go from one house to the other, so the whole countrey is a shade than which nothing can be more gratefull in a climate where the sun has so powerfull an influence.'[58]

The essay is packed with technical information: Tahitian methods of cooking, boat-building, house-construction, tool-making, fishing, dancing, drum-making, navigation, weather-predicting, ceremonial dramas, tattooing (again). Banks also writes tenderly of shared meals, enchanting dresses and languid afternoons. His remarks on the innocence of Tahitian ornaments are characteristic: 'Ornaments they have very few, they are very fond of earings but wear them only in one ear. When we came they had them of their own, made of Shell, stone, berries, red pease, and some small pearls which they wore 3 tied together; but our beads very quickly supplyd their place; they also are very fond of flowers, especialy of the Cape Jasmine of which they have great plenty planted near their houses; these they stick into the holes of their ears, and into their hair, if they have enough of them which is but seldom. The men wear feathers often the tails of tropick birds stuck upright in their hair.'

There is a long passage on the beautiful cleanliness of the Tahitian body, both male and female. All Tahitians wash themselves at least three times a day in the rivers, making their skin smooth and glowing. Their teeth are dazzling white, and they remove all body hair. Banks even grew accustomed to the strange, unforgettable smell of their hair oil: 'This is made of Cocoa nut oil in which some sweet woods or flowers are infusd; the oil is most commonly very rancid and consequently the wearers of it smell most disagreably, at first we found it so but very little use reconcild me at least very compleatly to it. These people are free from all smells of mortality and surely rancid as their oil is it must be preferrd to the odoriferous perfume of toes and armpits so frequent in Europe.'

The Tahitians' simplicity and innocence (the question of theft aside) came out in innumerable ways, as for example in their attitude to alcohol: 'Drink they have none but water and cocoa nut Juice, nor do they seem to have any method of Intoxication among them. Some there were who drank pretty freely of our liquors and in a few instances became very drunk but seemd far from pleasd with their intoxication, the individuals afterwards shunning a repitition of it instead of greedily desiring it as most Indians are said to do.'[59]

The idea of sexual innocence proved more complicated for a European to accept: 'All privacy is banishd even from those actions which the decency of Europaeans keep most secret: this no doubt is the reason why both sexes express the most indecent ideas in conversation without the least emotion; in this their language is very copious and they delight in such conversation beyond any other. Chastity indeed is but little valued especialy among the midling people; if a wife is found guilty of a breach of it her only punishment is a beating from her husband. Notwithstanding this some of the Eares or cheifs are I beleive perfectly virtuous.'

What later came to be regarded as the most scandalous of all Tahitian customs, the young women's seductive courtship dance, or '*timorodee*', Banks describes with calm detachment and a certain amused appreciation: 'Besides this they dance, especially the young girls whenever they can collect 8 or 10 together, singing most indecent words using most indecent actions and setting their mouths askew in a most extrordinary manner, in the practise of which they are brought up from their earlyest childhood. In doing this they keep time to a surprizing nicety, I might almost say as true as any dancers I have seen in Europe, tho their time is certainly much more simple. This excercise is however left off as soon as they arrive at Years of maturity. For as soon as ever they have formd a connection with a man they are expected to leave of Dancing Timorodee – as it is called.'[60]

The only Tahitian practice that Banks found totally alien and repulsive was that of infanticide, which was used with regularity and without compunction as a form of birth control by couples who were not yet ready to support children. Banks could scarcely believe this, until he questioned several couples who freely admitted to destroying two or three children, showing not the slightest apparent guilt or regret. This was a different kind of innocence, one far harder to accept. Banks pursued the question, and found that the custom originated in the formation of communal groups in which sexual favours were freely exchanged between different partners: 'They are calld Arreoy and have meetings among themselves where the men amuse themselves with wrestling &c. and the women with dancing the indecent dances before mentiond, in the course of which they give full liberty to their desires.'

He also found that the *Arreoy*, and the custom of infanticide, owed their existence 'chiefly to the men'. 'A Woman howsoever fond she may be of the name of Arreoy, and the liberty attending it before she conceives, generaly desires much to forfeit that title for the preservation of her child.' But in this decision he thought that the women had not the smallest influence. 'If she cannot find a man who will own it, she must of course destroy it; and if she can, with him alone it lies whether or not it shall be preserv'd.' In that case both the man and the woman forfeited their place in the *Arreoy*, and the sexual freedoms associated with it. Moreover, the woman became known by the term '*Whannownow*', or bearer of children. This was, as Banks indignantly exclaimed, 'a title as disgracefull among these people, as it ought to be honourable in every good and well governd society'.[61]

8

The epic voyage continued for another two years. They circumnavigated the two islands of New Zealand, mapped the eastern coastline of Australia (including Botany Bay), and narrowly survived a disastrous shipwreck on the Great Barrier Reef. *

* The brief, tentative landings that took place on the coast of 'New Holland' (Australia) during May 1770, though they yielded Banks and Solander many prizes in flora and fauna, did not at the time strike Cook with anything like the significance that they would later acquire with the arrival of the First Fleet at Sydney Cove in January 1788. Cook's long entry for 6 May 1770 gives details of the 'capacious, safe and commodious' anchorage at 'Stingray Harbour' (firmly renamed 'Botany Bay' by Banks), the varied woods and the 'very beautiful birds such as Cockatoos, Lorryquetes, Parrots etc', but notes that the Aboriginal inhabitants were both reclusive and hostile, 'and we were never able to form any connection with them'. By 29 May the *Endeavour* was already tangled in the maze of perilous shoals leading to the Great Barrier Reef.

Twelve months after they left Tahiti, as they headed northwards for the Torres Strait and Indonesia, Banks looked back on all the indigenous people he had seen, in one of his rare philosophical passages. In it he comes as close to the idea of 'noble savages' as he ever would: 'Thus live these – I had almost said happy – people, content with little nay almost nothing. Far enough removd from the anxieties attending upon riches, or even the possession of what we Europeans call common necessaries: anxieties intended maybe by Providence to counterbalance the pleasure arising from the Posession of wishd for attainments, consequently increasing with increasing wealth, and in some measure keeping up the balance of happiness between the rich and the poor.'

He must have talked at length with both Cook and Solander on this subject, and Cook makes his own long entry reflecting on the artificiality of European 'civilisation'. But while Cook clung to the necessity of European forms and discipline, Banks was rather inclined to dwell on the superfluity of European needs. These were perhaps the reflections of a man who had always been used to wealth and comforts. 'From them appear how small are the real wants of human nature, which we Europeans have increasd to an excess which would certainly appear incredible to these people could they be told it. Nor shall we cease to increase them as long as Luxuries can be invented and riches found for the purchase of them; and how soon these Luxuries degenerate into necessaries may be sufficiently evincd by the universal use of strong liquors, Tobacco, spices, Tea &c. &c.'[62]

On 3 September 1770 Banks was making another reflective entry, this time on the state of the ship's company after more than two years away from England. General health was outstandingly good, discipline remained effective, and the terrors of the Great Barrier Reef had shown how magnificently the crew could still pull together in a crisis. Yet there was a growing sense of exhaustion and sickness for hearth and home. 'The greatest part of them were now pretty far gone with the longing for home which the Physicians have gone so far as to esteem a disease under the name of Nostalgia; indeed I can find hardly any body in the ship clear of its effects but the Captn Dr Solander and myself, indeed we three have pretty constant employment for our minds which I beleive to be the best if not the only remedy for it.'[63]

It was now, when three-quarters of their journey was safely done, and they had reached their first semi-Europeanised port, that real catastrophe struck. They put into Batavia on Java (now Jakarta, the capital of Indonesia), where the whole crew were progressively overcome by a lethal combination of malarial fever and dysentery. Between November 1770 and March 1771, when they reached the Cape of Good Hope, the *Endeavour* lost thirty-seven of its men, nearly half the original crew. At one point Cook was only able to muster fourteen seamen on deck. Banks's personal team was reduced from eight to four. The expedition's astronomer Green died; the scientific secretary Spöring died; Tupia and his little son Tayeto died; Monkhouse the surgeon died; Thompson the ship's cook died; Satterley the ship's carpenter died; Molineux the ship's master died; Hicks the first lieutenant died; and Banks's faithful artist, young Sydney Parkinson, died. Solander would have died too, but for Banks's unstinting nursing care.[64]

Banks himself suffered for weeks from amoebic dysentery, sometimes 'so weak as scarcely to be able to crawl downstairs', and experienced 'the pains of the Damned almost'. These deaths had a devastating effect on his memories of the expedition. Finally, within sight of England, his surviving greyhound bitch, Lady, universally loved among the crew, was heard to howl out in the night. The next morning she was found flung across a chair in the cabin, still guarding Banks's writing table, but dead.

By the time they reached London on 13 July 1771, Banks felt little exuberance. He was shattered and disorientated. The bucolic memories of Tahiti were more than two years old, and instead he was haunted by the recent horrible deaths of so many friends and shipmates. Solander was still very weak, and not out of danger. Banks's family were not in town to greet and congratulate him, but 'dispersed almost to the extremities of the Kingdom' for the summer. He wrote to his friend Thomas Pennant FRS immediately on arrival: 'A few short lines must suffice ... Mr Buchan, Mr Parkinson and Mr Sporing are all dead, as is our Astronomer, seven officers, and about a third part of the ship's crew of diseases contracted in the East Indies – not in the South Seas, where health seems to have her chief residence. Our Collections will I hope satisfy you ... I must see [my family] before I begin to arrange or meddle with anything ... *Grass* I must have in the mean time. Salt provisions and Sea air have been to me like too much hardmeat to a horse. In a few days shall be able to write more understandably. Now I am Mad, Mad, Mad. My poor brain whirls round with innumerable sensations.'[65]

His safe return was greeted tenderly by his sister Sophia at Revesby in Lincolnshire. From the bottom of her heart she thanked the 'Merciful god who has daily preserved my Dear Brother from the perils, and very great ones, of the Sea!' Her sudden outburst of piety suggests how vividly she realised the dangers that her beloved brother had consistently played down, but barely survived. On his behalf she fondly (and unavailingly) promised that he would mend his ways and his Christian faith. She could pledge that he was well-intentioned, and was one of those who 'according to their Faith, use their best Endeavours, far as in their power they can, to do the Will of the Supreme Being'.[66] Sophia may well have had reason to worry about Banks's state of mind. He spent a fortnight recovering on the family estate in Lincolnshire, but spoke little about his experiences, even to Sophia. He walked, ate, shot and slept; then ate and slept again.

On his return to London he made no attempt to get in touch with Harriet Blosset, though James Lee and Harriet's mother clearly assumed that an engagement would be announced. It was obvious now that, whatever else, his experiences had left Banks utterly unfit for a quiet, regular, married life. Some evidence for this comes indirectly from a gossiping friend of Thomas Pennant's. Even if not entirely accurate, it seems to reflect something of Banks's disturbed state of mind. 'Upon his arrival in England [Banks] took no sort of notice of Miss Blosset for the first week or nearly so ... On this Miss Blosset set out for London and wrote him a letter desiring an interview of explanation. To this Mr Banks answered by a letter of 2 or 3 sheets, professing love etc but that he found he was of too volatile a temper to marry.' They did have at least one painful meeting, when Harriet is reported to have wept and 'swooned'.[67]

Further gossip was being reported by the novelist Fanny Burney and Lady Mary Coke in August. The story of the waistcoats provided much amusement. 'Mr Morris was excessively drole according to custom; and said he hoped Mr Banks, who since his return has desired Miss Blosset will excuse his marrying her, will pay her for the materials of all the worked waistcoats she made for him during the time he was sailing round the world.'[68]

There was some talk of broken promises and scandal. One wit suggested that Banks should be 'immediately placed in the Stocks ... for this injury'.[69] A friend of James Lee's, Dr Robert Thornton, later claimed that Banks had given Harriet an engagement ring before he set out, and had made 'many solemn vows' which he now callously reneged on.

In Thornton's view it was the alluring women of Tahiti, with their free sexual practices, who had corrupted Banks's feelings and destroyed his morals. 'Some people are ill-natured enough to say that, vitiated in his taste by seeing the elegant women of Otaheite, who must indeed have *something very peculiar in their natures* to captivate such a man, upon his return, Mr Banks came indeed to see the young lady and the plants; but she found her lover now preferred a flower, *or even a butterfly*, to her superior charms.' For Harriet the three-year wait ended in 'a most mortifying disappointment'.[70]

But perhaps it was more a relief. The kindly Solander, who knew and liked Harriet and her mother, and had of course witnessed Banks's anthropological behaviour in Tahiti, gently intervened and advised both parties not to proceed.[71] Banks privately offered Harriet's guardian James Lee a 'substantial' sum of money, which was accepted as a form of dowry for her future. The amount was rumoured to be £5,000 (half the sum he had previously laid out on the expedition), which suggests that Banks was not in the least callous, but felt more than ordinary guilt; though he could well afford to be generous. Harriet Blosset soon after made a happy marriage with a virtuous and botanical clergyman, Dr Dessalis, and was 'blessed by a numerous and lovely family'.[72]

Rumours about Banks's behaviour with Tahitian girls continued to spread in London for a number of months. Whether it was really this that determined him to break off with Miss Blosset (or she with him) is not clear. Satirical poems, fictional 'letters' and amusing cartoons certainly began to circulate, in which Banks's subtropical butterfly net and microscope were put to suggestive use. In one cartoon he was shown chasing a beautiful butterfly labelled 'Miss Bl . . . '.

Whatever the truth of these stories, it is clear that Banks was a changed man on his return to England, and it took him several years to settle back into conventional modes of behaviour. But sudden fame may have been even more unsettling than his unresolved affair with Harriet Blosset. On his return to London, Banks found to his immense surprise that the expedition was being greeted as a national triumph. Alongside Captain Cook, he and Solander were being treated as celebrities.

On 10 August they were summoned to meet the King at Windsor. For Banks the formal interview turned into a long ramble round Windsor Great Park, the first of many. Royal interest in the botanical possibilities of Kew Gardens promised great things. Moreover a real friendship quickly formed between George III, aged thirty-three, and Banks, aged twenty-eight. Both men owned large landed estates, were fascinated by agriculture and science, and were embarked on public careers, young and full of hope.

Banks and Solander next spent a debriefing weekend with the First Lord of the Admiralty, Lord Sandwich, at his country retreat. Then they were formally congratulated and repeatedly dined by the Royal Society. In November they were awarded honorary doctorates by the University of Oxford. Linnaeus wrote in Banks's praise: 'I cannot sufficiently admire Mr Banks who has exposed himself to so many dangers and has bestowed more money in the services of Natural History than any other man. Surely none but an Englishman would have the spirit to do what he has done.'[73]

The newspapers and monthlies – the *Westminster Journal*, the *Gentleman's Magazine*, *Bingley's Journal* – printed articles on their adventures, and dinner invitations started to pour in. Though Captain Cook was praised, Banks and Solander had rapidly become the scientific lions. They had brought back over a thousand new plant specimens, over five hundred animal skins and skeletons, and innumerable native artefacts. They had brought back new worlds: Australia, New Zealand, but above all the South Pacific.

London society was agog. Lady Mary Coke wrote in her diary: 'The most talked of at present are Messers Banks and Solander. I saw them at Court, and afterwards at Lady Hertford's but did not hear them give any account of their voyage round the world which I am told is very *amusing*.'[74] Dr Johnson gravely discussed 'culling simples' with Banks, and offered to write a Latin motto for the ship's goat. He thought a 'happier pen' than his might even write an epic poem on the expedition. Shortly afterwards Banks was elected to Johnson's exclusive Club.[75] Boswell, biographer's pen in hand, had a 'great curiosity' to see the 'famous Mr Banks'. He described him as 'a genteel young man, very black, and of an agreeable countenance, easy and communicative, without any affectation or appearance of assuming'.[76]

Sir Joshua Reynolds painted a dashing portrait of Banks in his study, his dark hair suitably wild and unpowdered, his fur-lined jacket flung open, his waistcoat unbuttoned, a loose pile of papers from his journal under one hand, and a large globe at his elbow. The rousing inscription was from Horace: *Cras Ingens Iterabimus Aequor* – Tomorrow We'll Sail the Vasty Deep Once More.

Everyone was awaiting an official written account of the great voyage. From the time of Hakluyt such travelogues had been immensely popular, and this one was impatiently anticipated. But one of the terms of the *Endeavour* expedition was that all journals and diaries would be surrendered at the end of the voyage, and submitted to an official historian. The journals of Cook and Banks, the papers and botanical notes of Solander, the precious drawings of Buchan and Parkinson, were accordingly all handed over to a professional author, who was to prepare a three-volume account for the sum of £600.

The man chosen was fifty-six-year-old Dr John Hawkesworth, a literary scholar and professional journalist. He was evidently considered a safe pair of hands, having written a number of short biographies and successfully collaborated with Dr Johnson on two periodicals, the *Rambler* and the *Adventurer*. The misleading title of the latter, which had nothing to do with exploration, may have reinforced his apparent credentials. The subject was a gift, and the material was magnificent, if sometimes a little risqué. All that was required were accuracy, objectivity and the ability to assemble a vivid narrative. After nearly two years' labour, Hawkesworth achieved none of these.

Hawkesworth's *Account of Voyages Undertaken . . . for Making Discoveries in the Southern Hemisphere and Performed by . . . Captain Cook . . .* was published in three volumes in 1773. It was prolix, abstract, and much given to philosophical digression. Its author was easily shocked, and quick to moralise. He had no scientific or naval experience to draw on, and his views on foreign customs and native morality were prejudiced and illiberal. While digressing on the 'Noble Savage', Hawkesworth easily struck a lurid and provocative note. He wrote with delicious outrage of Tahitian dances and sexual practices. The girls danced the *timorodee* with 'motions and gestures beyond imagination wanton . . . a scale of dissolute sensuality wholly unknown to every other nation . . . and which no imagination could possibly conceive'.[77]

A second account of the expedition, *Journal of a Voyage on His Majesty's Ship, the Endeavour . . .*, also published in 1773, was based on Sydney Parkinson's journal as edited by his brother Stanfield. There had been a quarrel with Hawkesworth over the copyright of these papers, and Banks had also struggled to retrieve Parkinson's botanical illustrations from Stanfield Parkinson. Banks felt, not unreasonably, that he had paid for them as Parkinson's employer on the *Endeavour* (he had also discreetly sent £500 to Parkinson's bereaved parents). Parkinson's death in Batavia embittered and prolonged all these negotiations.

When it finally appeared, the Tahiti section of Parkinson's journal proved to be brief but strikingly vivid, and left an extremely favourable impression of Banks. Parkinson was particularly observant of small details of Tahitian life: how the natives climbed coconut trees using a rope tied between their ankles; how they kindled fire by rubbing bark; how they wove baskets and dyed clothes; how they played their flutes with the nose; how the girls wore gardenias behind their ears and danced while snapping mother-of-pearl castanets; and how in the *timorodee* the most provoking gesture they made was pouting and twisting up their lips in what Parkinson called 'the wry mouth'. It was also characteristic of young Parkinson that he had tried to learn to swim like the Tahitians, that on Banks's advice he collected Tahitian vocabulary, and that after some hesitation he had had his arms tattooed with a 'lively bluish purple' design, of which he had been inordinately proud.

Two years after his return to England, when Polynesian affairs were still the rage, Banks himself put pen to paper in a short, preliminary appreciation of the Paradise island. It took the form of a light-hearted letter entitled 'Thoughts on the Manners of the Otaheite'. It was a surprising piece, skittish and suggestive in tone, mannered in its classical references, and verging on the kind of mild pornographic *frisson* thought to be favoured by the French philosophers of Paradise: 'In the Island of Otaheite where Love is the chief Occupation, the favourite, nay almost the sole Luxury of the Inhabitants, both the bodies and souls of the women are moulded in the utmost perfection for that soft science. Idleness the father of Love reigns here in almost unmolested ease… Except in the article of Complexion, in which our European ladies certainly excell all inhabitants of the Torrid Zone, I have nowhere seen such Elegant women as those of Otaheite. Such the Grecians were from whose model the Venus of Medicis was copied, undistorted by bandages. Nature has full liberty: the growing form [develops] in whatever direction she pleases. And amply does she repay this indulgence in producing such forms as exist here [in Europe] only in marble or canvas: nay! Such as might even defy the imitation of the chissel of Phidias, or the pencil of Apelles. Nor are these forms a little aided by their Dress: not squeezed as our Women are, by a cincture scarce less tenacious than Iron.'[78] ♣

This was perhaps a glimpse of Banks the Tahitian libertine, though it was only circulated privately. The way was now open for Banks to publish his own journal, over 200,000 words in manuscript, together with some of the hundreds of beautiful illustrations and line drawings he had commissioned. A folio volume of 800 plates was planned, together with extensive journal extracts. Solander agreed to help him with the cataloguing and editorial work, and various assistants were hired, including the young Edward Jenner. It was intended as the greatest scientific publication of Banks's lifetime, his masterpiece.

9

The Pacific voyage, despite its final horrors of sickness and death, had not dampened Banks's scientific wanderlust. 'To explore is my Wish,' he wrote the following spring, 'but the Place to which I may be sent almost indifferent to me, whether the Sources of the Nile, or the South Pole are to be visited, I am equally ready to embark on the undertaking.'[79]

In summer 1772 Cook was commissioned by the Admiralty to undertake a second, much larger Pacific expedition, this time with two ships. Banks longed to go on this new adventure, and made extensive preparations and invested thousands of pounds in new botanical equipment. But perhaps celebrity had gone to his head. His plans were increasingly ambitious, and he had summoned an extraordinary range of scientific and artistic talents to accompany him, a sixteen-man team including the

---

♣ All this is quite unlike the wonderful simplicity and directness of Banks's original *Endeavour Journal*, and reminds one how delicate the balance – both moral and stylistic – already was between observation and exploitation in these early pioneering days. Banks never wrote about Tahiti again in this mode, though none of his friends (except possibly Solander) would have disapproved of this gentlemanly *jeu d'esprit*. It has to be added that this is nothing compared to the epistolary lubricities of Banks's friend Sir William Hamilton. Other influential essays on the South Sea Paradise were published by Bougainville, Denis Diderot and Jean-Jacques Rousseau at this time. Diderot's *Supplement to the Voyage of Bougainville* (written in 1772 but not published until 1777) proclaimed Tahiti as a model for the reform of sexual relations in Europe: relaxing the conventions of marriage, promulgating free love between the young, and emphasising the importance of mutual physical pleasure between partners.

chemist and radical Joseph Priestley, the painter Johann Zoffany, and the brilliant young London physician Dr James Lind (later to be Shelley's extracurricular science teacher at Eton).

Cook was quite prepared to acquiesce to all Banks's scientific requirements, and had the great cabin in his new ship, the *Resolution*, redesigned to meet them, with higher ceilings, folding work tables and fitted cabinets. His own tiny captain's cabin was moved to the rear of the *Resolution*'s quarterdeck. But the Admiralty regarded Banks's demands as unacceptably imperious, and, without warning, withdrew its authorisation. Humiliatingly, all Banks's equipment was offloaded and dumped on the quayside at Sheerness. On 20 June he received a sharp but stately letter of rebuke from his powerful friend Lord Sandwich: 'Your public spirit in undertaking so dangerous a voyage, your inattention to any expense, ... and your extensive knowledge as a naturalist, make it lamented that you are no longer one of the crew of the Resolution. But it may not be improper to set you right in one particular which you possibly may have misunderstood, and that is that you suppose the ships to have been fitted out for your use, which I own I by no means apprehend to be the case.'[80]

The Admiralty had, in effect, rejected the notion of underwriting another purely scientific voyage, for what Sandwich called 'improvements in natural knowledge'. From now on Cook's voyages were to take on more practical and empire-building objectives (though they would include testing the rival chronometers of John Harrison and John Arnold).

Lord Sandwich made it clear that Banks would have to pursue his science on his own: 'Upon the whole I hope that for the advantage of the curious part of Mankind, your zeal for distant voyages will not yet cease, I heartily wish you success in all your undertakings, but I would advise you in order to ensure success to fit out a ship yourself; that, and only that, can give you the absolute command of the whole Expedition.'[81]

So Banks commissioned his own brig, the *Sir Lawrence*, went to the Hebrides and inspected Fingal's Cave, and then sailed on to Iceland, where he made many friends, admired the geysers and volcanoes, collected lava specimens, but made few original discoveries. Back in London he continued to work with Solander on his *Endeavour Journal*, and set out his extraordinary collection of specimens at a temporary apartment in New Burlington Street. These began to attract learned visitors. The Keeper of the Ashmolean Museum in Oxford, William

Sheffield, wrote a long, astonished description of Banks's scientific treasures to Gilbert White in Hampshire.

Contrary to expectation, these were far more than just botanical specimens. They formed in effect a complete museum of Pacific culture, combining natural history with ethnology and human artefacts in a quite new way. They were housed in three enormous, overflowing rooms, each with its own theme. The first, the 'Armoury', belonged symbolically to the human male, dedicated to weapons, utensils and sailing equipment from all over the South Seas. The second was more female in theme, a huge domestic collection of clothes, headdresses, cloaks, woven cloths, ornaments and jewellery, together with 1,300 new species of plant 'never seen or heard of before in Europe'. The third room was dedicated simply to Nature in all her diversity. It contained 'an almost numberless collection of animals; quadrupeds, birds, fish, amphibia, reptiles, insects and vermes, preserved in spirits, most of them new and nondescript [unclassified] ... Add to these the choicest collection of drawings in Natural History that perhaps enriched any cabinet, public or private: 987 plants drawn and coloured by Parkinson; and 1300 or 1400 more drawn with each of them a flower, a leaf, and a portion of stalk, coloured by the same hand; besides a number of other drawings of animals, birds, fish etc ... '

The Oxford Keeper was struck by the beauty and diversity of the whole amazing collection, a glimpse into an entirely new and wonderful world. Banks had found a new role as its guardian and its promoter. 'Indeed most of these tropical islands, if we can credit our friend's description of them, are *terrestrial paradises*.'[82]

Banks's early hero Carl Linnaeus had turned collecting and displaying into something approaching a European art form. At Uppsala he planted a clock garden or 'botanical sundial', marking each hour by clumps of plants that opened only at one particular time of day (according to the strength of the sun). The time could thus be 'read' by the rotating patches of open petals, and even by the release of flower perfumes (such as tobacco plants in the early evening). However, Linnaeus's genius for taxonomy and display disguised the fact that his natural history was essentially static.♣

♣ Carl Linnaeus (1707–78) emphatically rejected evolution. His 'systematics' revealed no connecting law of growth or change, as would the transformational notion explored by several later botanists until Gregor Mendel (1822–84), patiently studying generation after generation of garden peas, gave rigour to the science of genetics. Coleridge pointed to this

Banks was now welcomed into the scientific societies in London: the Royal Society, the Society of Antiquaries, the Society of Dilettanti. He was summoned more frequently to advise the King at Kew, where from 1773 he was gratified to find himself acting as unofficial director. After the débâcle over Harriet Blosset, he began living with a young woman called Sarah Wells, and set her up in an apartment in Chapel Street, on the other side of St James's Park. Here he would meet Solander and his other friends, give noisy dinner parties and have plenty of talk of science and adventure. This *ménage* seemed an extension of his Tahitian liberties, and certainly there was no prospect of conventional betrothal or marriage. Solander refers simply to 'Mrs Wells's' charm, good nature, and delicious supplies of 'Game & Fish'.[83]

Indeed, the *Town and Country Magazine* for September 1773 claimed that 'Mr B the Circumnavigator' had an illegitimate child, but perhaps this was more confused botanical satire, as the mother was named as 'Miss B—N living in *Orchard* Street'. Nonetheless, one of Banks's close friends, the zoologist Johann Fabricius, wrote to him in November sending compliments to Sarah Wells and adding: 'What had she brought you? . . . A boy or a girl?'[84] If there was a child, Banks did not allow it to affect his free social arrangements. Sarah became known and much liked by many visiting men of science, the Swedish naturalist Johann Alströmer referring to her intelligent conversation and fondly recalling a memorable '*Soupé* at his *Maitresse*, Mistress Wells's', with Banks and Solander on riotous good form.[85]

Tahiti pursued Banks in other ways. In summer 1774 one of Cook's fleet commanders, Captain Furneaux of HMS *Adventure*, returned with a first visitor to England from the South Seas. He was entered in the ship's muster books as 'Tetuby Homey', from Huahine in the Society Islands, '22 years, Able Seaman'. This news immediately reminded Banks of all his hopes for Tupia and his son, which had been so tragically destroyed

---

difference between an organising taxonomy and a dynamic scientific principle or law in essays in *The Friend* (1819). The psychology of collecting, ordering and naming specimens could also be seen as a form of mental colonising and empire-building. 'Taxonomy, after all, is a form of imperialism. During the nineteenth century, when British naval surveys were flooding London with specimens to be classified, inserting them in their proper niches in the Linnaean hierarchy, had undeniable political overtones. Take a bird or a lizard or a flower from Patagonia or the South Seas, perhaps one that had had a local name for centuries, rechristen it with a Latin binomial, and presto! It had become a tiny British colony.' Anne Fadiman, 'Collecting Nature', in *At Large and at Small* (2007), p.19.

in Batavia in 1770. Banks and Solander hurried down to Portsmouth to greet 'Homey' in July.

There, confined to the captain's cabin, they found a tall and strikingly handsome Tahitian man, who was soon to become known in England as 'Mai' or 'Omai'. He announced that he hoped to make his fortune, and fully intended to return to Tahiti as a rich and experienced traveller, having survived the expected savagery of the English.[86] Omai turned out to be quick-witted, charming and astute. His exotic good looks, with large, soulful eyes, were much admired in English society, especially among the more racy of the aristocratic ladies.

Banks treated Omai partly as an honoured guest, and partly as an exotic specimen. The ambiguity of the attitudes displayed in his Tahitian journal was now put to the test. Banks fitted Omai up with European clothes, a brown velvet jacket, white waistcoat and grey silk breeches. He took him to dine with the Royal Society, with the Society of Philosophers (ten times), and carefully introduced him at a number of society *soirées*. Omai's bow, executed with the aplomb of a dancer, became celebrated. He quickly won all hearts, and was eventually presented by Banks to King George III at Kew. The introduction became legendary, when Omai executed a superb version of his bow, and then sprang forward to grasp the royal hand, and grinning broadly, cried out, 'How do King Tosh!'[87]

From then on he was lionised almost continually for a year. Thanks to Banks he met a host of celebrities, among them Lord Sandwich, Dr Johnson, Fanny Burney and the poet Anna Seward, who wrote a poem about him. He learned to ride, shoot, conduct flirtations, and play excellent chess – Dr Johnson never stopped teasing his friend, the learned antiquarian Giuseppe Baretti, that Omai had once checkmated him. Omai also made excellent jokes about current English fashions. Fanny Burney records his delighted and unrestrained 'Ha! Ha! Ha!' on seeing the Duchess of Devonshire's high-piled hairstyle.

Conscious of European diseases, Banks had Omai undergo Jenner's new technique of inoculation with cowpox vaccine, against the lethal smallpox. He also caused something of a scandal by absolutely refusing to teach Omai to read, or to have him instructed in any form of Christian religion. Their most happy time together came in the summer of 1775, when Banks took Omai with several friends on a field expedition to Whitby and Scarborough. They travelled up in leisurely fashion, comfortably installed in Banks's large, lumbering coach, stopping off to eat at remote country inns and botanise in the summer fields.

An imposing portrait of Omai, standing formally alongside Banks and Solander, was painted by William Parry, and displayed at the Royal Academy in 1777.[88] It again demonstrates the ambiguity of the relationship between patron and protégé. Banks points dramatically towards Omai, who stands gazing out at the viewer, wrapped in the dazzling white robes of Tahitian ceremonial dress, almost Roman in his stately demeanour. His naked feet and tattooed hands are clearly shown. Though there is an extraordinarily calm, almost aristocratic, beauty about his presence, it is not clear if he is Banks's companion or his trophy. Other portraits were painted by Sir Joshua Reynolds, who also executed an exquisite pencil drawing of Omai's head, emphasising his magnificent mane of dark hair, his large, tender eyes and finely formed mouth. Another more prosaic solo portrait was especially commissioned for John Hunter's anthropological collection, later housed at the Royal College of Surgeons.[89]

In 1777 Cook departed on his third Pacific voyage, taking Omai with him. He left behind his record of his second expedition, *A Voyage Towards the South Pole and Round the World*. The text was accompanied by extensive illustrations, including pictures of Omai, numerous botanical studies of rare plants, and sketches of the naked Tahitian dances witnessed by Banks and Parkinson. The drawings of Omai were later used by the anatomist William Lawrence in his *Lectures on the Natural History of Man* (1819).

Cook's sober book caught the public's imagination. The poet William Cowper, tucked away in his Buckinghamshire vicarage at Olney, and permanently trembling on the brink of disabling depression, found extraordinary relief and delight in imagining the great voyage southwards. To explain his sensations, Cowper invented the idea of the 'armchair traveller': 'My imagination is so captivated upon these occasions, that I seem to partake with the navigators, in all the dangers they encountered. I lose my anchor; my main-sail is rent into shreds; I kill a shark, and by signs converse with a Patagonian, and all this without moving from my fireside.'[90]

In his long, reflective poem *The Task*, Cowper accompanied Cook and Banks in his imagination. He transformed Banks, rather suitably, into an adventurous bee, busily foraging for pollen.

He travels and expatiates, as the bee
From flower to flower, so he from land to land;
The manners, customs, policy of all
Pay contribution to the store he gleans;
He sucks intelligence in every clime,
And spreads the honey of his deep research
At his return, a rich repast for me.

He travels and I too. I tread his deck,
Ascend his topmast, through his peering eyes
Discover countries, with a kindred heart
Suffer his woes and share in his escapes,
While fancy, like the finger of a clock,
Runs the great Circuit, and is still at home.[91]

Omai landed back in Tahiti in August 1777, and set up as a merchant of Western goods. He also became a sort of guide and impresario for visiting Westerners, ironically finding himself doing Banks's job in reverse, explaining European culture to the sceptical Tahitians. He sold red feathers, cooking pots and pistols, but never fully reintegrated into Tahitian society. Cowper included Omai's story in *The Task*, reflecting on the excitement of exploration, but also on the clash between European and Pacific cultures. He suggested that Omai might have become a victim of Romantic scientific enquiry, left permanently alienated from both worlds:

I cannot think thee yet so dull of heart
And spiritless, as never to regret
Sweets tasted here, and left as soon as known.
Methinks I see thee straying on the beach
And asking of the surge that bathes thy foot
If ever it has wash'd our distant shore.
I see thee weep, and thine are honest tears,
A patriot's for his country.[92]

Banks's own liberated behaviour in the years immediately following his return to London suggests that he too had been permanently affected by his Tahitian experience. A visitor to Revesby in 1776 referred to him as 'a wild eccentric character' who obviously still dreamed of his 'voyage to Otaheite', and neglected his estates.[93] He was reported to have taken a young woman – presumably Sarah Wells – on a scandalous fishing party with Lord Sandwich and his mistress Martha Ray, during which the women sang and danced while the men 'played the kettle drums' (perhaps in an attempt to recreate the Tahitian *timorodee*).

Public opinion might laugh at him as an old-fashioned libertine, as in a satirical poem that went into circulation entitled 'Mimosa or, The Sensitive Plant, Dedicated to Mr Banks'. Yet Banks genuinely believed that British society was often cruelly restrictive towards women, although he told the author Mrs Ann Radcliffe that he thought women themselves were often responsible: 'The greater part of the Evils to which your sex are liable under our present Customs of Society originate in the decisions of women ... The Penalty by which women uniformly permit the smallest deviation of a Female character from the Rigid paths of Virtue is more severe than Death & more afflicting than the tortures of the Dungeon.'[94]

But gradually the reputation of the South Seas Paradise became more complicated: innocence gave way to experience. In February 1779 Captain Cook was killed by natives on a beach in Kealakekua Bay, Hawaii, during his third Pacific voyage. Several of his own officers thought Cook himself was at least partly to blame, for his increasingly aggressive use of heavily armed beach landing-parties, and his method of seizing native hostages upon arrival. His second-in-command, Captain Charles Clerke, wrote in his report: 'Upon the whole I firmly believe matters would not have been carried to the extremities they were, had not Capt. Cook attempted to chastise a man in the midst of this multitude.' But he also noted the horror with which the British crew learned that Cook's body had been dismembered and distributed, piece by piece, among the Hawaiian chieftains across the whole island.[95] The days of the green palm-tree boughs were long over.

The news took over a year to reach Britain. The artist Philippe-Jacques de Loutherbourg executed a huge and fantastic painting of Cook's apotheosis, the bony old Yorkshireman leaning back in the arms of a grateful Britannia, who lifts him into the clouds of glory. There was no indication of the dark colonial inheritance that Cook had left behind on earth. The *Journal of Captain Cook's Last Voyage to the Pacific Ocean* was edited and published by John Rickman in 1781. Its additional material included a controversial account of Cook's violent death, and Omai's strange, alienated return to Tahiti. Both in their own ways were premonitions of the colonial tragedy that was eventually to follow.

Tahiti was rapidly turning into a legend, and a somewhat tarnished one at that. When a hugely expensive pantomime entitled *Omai, or a Trip Round the World* was successfully staged at Drury Lane in 1785, the island had started its long decline into a source of popular entertainment. The extravagant sets and titillating costumes, all designed by Loutherbourg, foreshadowed a world of grass-skirt cliché that would eventually lead to Hollywood. Shrewdly capitalising on this new-found fashion, Madame Charlotte Hayes staged in London a notorious nude 'Tahitian Review', in which 'a dozen beautiful Nymphs ... performed the celebrated rites of Venus, as practised at Tahiti'. It was said that wealthy clients could then 'anthropologically' sample the native girls (who were all of course London cockneys).

10

Meanwhile Banks established a kind of permanent scientific salon at a new house at 32 Soho Square, where his adoring sister Sophia was brought in to act as his housekeeper. The unofficial *ménage* with Sarah Wells across the park in Chapel Street continued, but perhaps under increasing sisterly protests. Her brother, Sophia felt, should begin to settle, conform to convention and become 'enlightened with the Bright Sunshine of the Gospel'.[96] Indeed, Banks never embarked on any other expedition after his voyage to Iceland in 1772. Instead he continued to develop his enormous archive of scientific papers, drawings and specimens, with the help of Solander, now his official archivist and librarian. Yet still Banks published nothing. The daring young botanist and explorer was slowly turning into a landlocked collector and administrator.

In November 1778 Banks was elected President of the Royal Society at the remarkably early age of thirty-five. Then, quite suddenly it would seem, he decided to marry, and began to pay court to a twenty-one-year-old heiress, Dorothea Hugessen, the cheery daughter of a wealthy landowner from Kent, worth (as Jane Austen would say) £14,000 a year. They married the following March at St Andrew's, Holborn, and Banks settled down to a position at the heart of the British scientific establishment for the next forty-one years. Dorothea became a much-loved companion, and proved herself a wonderful hostess at Soho Square. Surprisingly she would have no children, but she formed a close alliance with her sister-in-law Sophia. Together the two women succeeded in managing the more chaotic side of Banks's social life with great success.

This required a final parting with Sarah Wells, which was tactfully and generously managed. Solander again proved himself an avuncular go-between. He later remarked that 'Banks and Mrs Wells parted on very good terms. – She had sense enough to find he acted right, and of course she behaved very well. All her old friends visit her as formerly.' There was no mention of a child, or of any regrets that Banks may have felt. Instead, Solander added that Banks had now spruced himself up for the weekly Royal Society meetings at Crane Court, off Fleet Street, appearing in 'Full Dressed Velvet or Silk coat etc'. He would 'properly fill the President's chair'.[97]

For his presidency, Banks took as his heraldic crest the figure of a lizard. He explained his choice as follows: 'I have taken the Lizard, an Animal said to be endowed by Nature with an instinctive love of Mankind, as my Device, & have caused it to be engraved on my Seal, as a perpetual Remembrance that a man is never so well employed, as when he is labouring for the advantage of the Public; without the Expectation, the Hope or even a Wish to derive advantage of any kind from the result of his Exertions.'[98]

Yet he settled into the presidential chair uneasily. It was typical of him that, on his election, he wrote as follows to Sir William Hamilton in Naples. 'That I envy you your situation within two miles of an erupting Volcano, you will easily guess. I read your Letters with that kind of Fidgetty Anxiety which continuously upbraids for not being in a similar situation. I envy you, I pity myself, I blame myself & then begin to tumble over my Dried Plants in hopes to put such wishes out of my head. Which now I am tied by the leg to an Arm Chair, I must with diligence suppress.'[99]

In November 1780 he oversaw the historic move of the Royal Society's offices from its obscure lodgings in Crane Court to its grand new premises in the recently completed Somerset House on The Strand, in a suitably commanding position overlooking the Thames. It was now recognised as a palace of science, to and from which travellers would come and go to the ends of the earth.[100]

In 1781 Banks was knighted for his energetic scientific work as Director of the royal gardens at Kew. Over the next decade he transformed the rambling and disorganised estate along the Thames into a scientific repository and botanical haven that far outstretched anything achieved by Linnaeus. He established more than 50,000 trees and shrubs at Kew, and introduced a vast number of new and exotic species that are now regarded as native: among them magnolias, fuchsias, monkey-puzzle trees, and the evergreen sequoia.[101] He had notable successes with rare and difficult species such as the Venus fly-trap. The poet Coleridge among others refers to him as a reliable source of new exotic and experimental drugs such as Indian hemp, 'Bang' and cannabis.[102]

Yet the world of the South Pacific was drifting steadily away from Banks. His great companion and fellow voyager, the amiable, easy-going and ever-faithful Solander, was struck down by a heart attack, and died in the guest room at Soho Square in June 1782. Banks was inconsolable, and felt this loss more than any other he had experienced: it seemed to him the loss also of his own youth. He wrote tight-lipped to a mutual friend, John Lloyd: 'To write about the loss of poor Solander would be to renew both our feelings for little purpose; suffice it to say then that few men, however Exalted their pursuits, were ever more feelingly missed either in the paths of Science or of Friendship.'[103]

A little later he wrote more confidentially to Johann Alströmer, who had once shared their carefree dinners with Sarah Wells, and who was now elected President of the Swedish Academy of Sciences. 'His loss is irreplaceable. Even if I were to meet such a learned and noble man as he was, my old heart could no longer receive the impression which twenty years ago it took as effortlessly as wax, one which will not dissolve until my heart does ... I can never think of it without feeling such acute pain as makes a man shudder.'[104]

There were now fewer and fewer survivors from the original voyage to Paradise; Banks felt like the 'last of the Otaheites'. Perhaps also it was Solander's death which fatally delayed any further work on Banks's great *Endeavour* travel book. In 1785 he still wrote hopefully, seeing it as a kind of memorial to his friend: 'Solander's name will appear next to mine on the title page because everything has been brought together through our common industry. There is hardly a single clause written in it, while he lived, in which he did not have a part ... it can be completed in two months if only the engraver can be brought to put the finishing touches to it.'[105] But nothing appeared.

Banks suffered his first serious and disabling attack of gout in the summer of 1787, when he was still only forty-four. He received a sympathetic letter from the King, but neither realised how grave the affliction would become. By his fifties he was almost literally tied to his presidential chair, as he had feared and prophesied. Incapacitated by agonising swellings in his legs, the once tireless and athletic young explorer had to be pushed about his London house in a wheelchair.

His body may have been chairbound, but his spirit was increasingly airborne. In fact Banks's personal enthusiasm as the universal scientific patron largely shaped and directed the adventurous character of Romantic science, which now flowered and flourished like one of his most exotic specimens. He revealed himself as a talent-spotter of genius, encouraging expeditions to Australia, Africa, China and South America; supporting projects as diverse as telescope-building, ballooning, merino sheep-farming and weather forecasting; helping to found museums of botany, anthropology, comparative anatomy; and above all maintaining through a huge network of correspondence and personal meetings the idea of science as a truly shared and international endeavour, even in a time of war, and even in relentless (if well-mannered) competition with the French.♣

He now looked back proudly at his own voyage as something historic and exemplary, to be emulated by the next generation: 'I may flatter myself that being the first man of scientific education who undertook a voyage of discovery, and that voyage of discovery being the first that turned out satisfactorily in this enlightened age, I was in some measure the first who gave that turn to such voyages.'[106]

The great French naturalist Georges Cuvier agreed, later describing the *Endeavour* voyage as forming 'an epoch in the history of science. Natural history contracted an alliance with astronomy and exploration, and began to extend its researches over an ever-widening sphere ... Everything seemed to realise the romantic wonders of the Odyssey ... Banks displayed his astonishing energy: fatigue did not

♣ It has been conservatively calculated that Banks's correspondence ran to over 50,000 items, though these are still widely scattered in archives in Britain, America, Australia and New Zealand. See the Joseph Banks Archive Project on the Internet. There have been various recent selected editions of his correspondence. These include *The Selected Letters of Joseph Banks* (2000) and the superb new edition *The Scientific Correspondence of Sir Joseph Banks, 1765–1820*, in six volumes (2007), both edited by Neil Chambers.

depress him, nor danger deter him ... and not simply by *seeing*, but by actively *observing*, he showed his true scientific character ... Banks was always in the advance.'[107]

Banks wrote provokingly to a young man hesitating to embark on a perilous scientific expedition to the feverish shores of Java: 'I have no doubt [your family] wish to force you to adopt Sardinapalus's advice to his citizens to "Eat, drink & propagate" ... Let me hear from you how you feel inclined to prefer Ease and indulgence to Hardship and activity. I was about 23 when I began my Perigrinations; you are somewhat older, but you may be assured that if I had listened to a multitude of voices that were raised up to dissuade me from my Enterprise, I should have now been a Quiet country gentleman.'

Banks's house in the south-west corner of Soho Square soon became known as the operations centre of scientific research in Britain. It was widely recognised as such throughout Europe – especially in France, Germany and Scandinavia. His correspondence reached round the world, from Paris to New York to Moscow to Sydney. He had the ear of George III (until the King went mad). His library and herbarium were open to all; his daily ten o'clock planning breakfasts at Soho Square were famous; his house parties at his new country estate at Spring Grove in Surrey, purchased especially for the purpose, were often like international conferences.

He received visitors from all over the world, and was the patron of numerous private projects. He advised on the settlement of Australia, was made a Privy Councillor in 1797, and served on the Board of Longitude. After some early disagreements, he became the close friend of the Astronomer Royal, Nevil Maskelyne. Later he was elected President of the Africa Association (which eventually became the Royal Geographical Society), and one of the founding Vice-Presidents of the Royal Institution. He began to exercise a dominant influence over the public development of British science and exploration, encouraging royal patronage, finding funds for research projects and expeditions, and skilfully boosting their national prestige. In effect Banks became Britain's first Minister for Science.

Yet Joseph Banks never finally published his long-dreamed-of *Endeavour Voyage*, or any full account of his time in Paradise. Despite the death of his great friend Solander there is no real explanation for this failure, though perhaps it was a deliberate refusal. His journal exists in several manuscript drafts – one copied (and somewhat bowdlerised) by his sister Sophia; and there is a huge series of astonishing engravings (now archived in the Natural History Museum, London). Versions of the journal have been published by scholars, notably by J.C. Beaglehole, in facsimile by the Banks Society, and one recently put online by the University of New South Wales, Australia. But Banks's *Endeavour Voyage* may count as one of the great unfinished masterpieces of Romanticism, as mysterious in its own way as Coleridge's 'Kubla Khan', with which it bears some curious similarities, as an account of a sacred place which has been partly lost, and to which there is no return.♣

Instead, Banks seemed fated to relive his story through the extraordinary lives of his protégés. This is the genial, enabling role he plays in the fantastic series of explorations, expeditions and mind journeys that follow. His great *Endeavour* voyage had launched an Age of Wonder.

♣ Literally a Paradise Lost, in the sense that venereal disease, alcohol and Christianity had combined by the early nineteenth century to destroy the traditional social structures of Tahiti and to transform its 'pagan' innocence forever. The London Missionary Society, founded in 1810, instructed its Tahitian missionaries to 'cultivate the tenderest Compassion for the wretched condition of the Heathen, while you see them led captive to Satan at his Will. Do not resent their abominations as affronts to yourselves, but mourn over them as offensive to God.' Charles Darwin visited Tahiti on his way back from the Galapagos islands in November 1835, and later called it 'Otaheite, that fallen Paradise!' Alan Moorehead, *The Fatal Impact* (1966).

2

Herschel on the Moon

1

Shortly after his election as President of the Royal Society in 1778, Joseph Banks began to hear rumours of an unusually gifted amateur astronomer working away on his own in the West Country. These rumours first reached him through the Society's official Secretary, Sir William Watson, whose brilliant and unconventional son, a young physician based in Somerset, was a moving spirit behind the newly founded Bath Philosophical Society. Watson (junior) began sending accounts to his father of a strange maverick who owned enormously powerful telescopes (supposedly built by himself), and was making extravagant claims about the moon.

The initial reports that reached Banks were strange and somewhat sketchy. The man was called Wilhelm Hershell or William Herschel, possibly a German Jew from Dresden or Hanover.[1] One winter night in 1779 young Watson had found this Herschel in Bath, standing alone in a cobbled backstreet, viewing the moon through a large telescope. Though tall and well-dressed, and wearing his hair powdered, he was clearly an eccentric. He spoke with a strong German accent, and had no manservant accompanying him. Watson asked if he might take a look through the instrument, which he noticed was a reflector telescope, not the usual refractor used by amateurs. It was large – seven feet long – and mounted on an ingenious folding wooden frame. The whole thing was evidently home-made. To his surprise he found its resolution was better than any other telescope he had ever used. He had never seen the moon so clearly.[2]

They fell into slightly halting conversation. Watson was immediately taken by Herschel's humorous and modest manner, and soon realised that it disguised an acute unconventional intelligence. Herschel's knowledge of astronomy, though obviously self-taught, was impressive. Though he had no university education, and said he had very little mathematics, he had mastered John Flamsteed's great atlas of the constellations, read the textbooks of Robert Smith and James Ferguson, and knew a great deal about French astronomy. Above all he knew about the construction of telescopes, and the making of *specula* – mirrors. Though in his early forties, he talked of the stars with a quick, boyish enthusiasm, that betrayed intense and almost unnerving passion. Watson was so struck that he asked if he might call round to see him the very next morning.

The house at 27 Rivers Street, in the discreet upper section of Bath, was modest, and clearly Herschel was not a gentleman of leisure. The lower rooms were cluttered with astronomical equipment, but the front parlour was dominated by a harpsichord and piled high with musical scores.[3] It emerged that Herschel was a professional musician, held the post of organist at the Bath Octagon Chapel, and made ends meet by giving music lessons. He also composed, and was fascinated by the theory of musical harmony. His domestic situation was odd. He was poor, and unmarried, but Watson noticed that he spoke tenderly of a sister, who was not only his housekeeper but also his 'astronomical assistant'.[4]

Watson invited his new acquaintance to join the Bath Philosophical Society. Herschel responded with alacrity. Though hesitant to speak in public, he started submitting papers through Watson. Many of them were strange ventures into speculative cosmology and the philosophy of science. They included such subjects as 'What becomes of Light?', 'On the Electrical Fluid' and 'On the Existence of Space'.[5]

Proud of his new discovery, Watson sent what he considered the best of these early papers to his father, Sir William, at the Royal Society in London. Herschel was modestly worried that his English would not be up to the mark, and Watson tactfully corrected each paper. It was not their plain literary style, however, which caused controversy, but their content. The very first of them, 'Observations on the Mountains of the Moon', was so unconventional that it caused an unaccustomed stir when it was published in the Society's august journal *Philosophical Transactions* in spring 1780. In it Herschel claimed that with his home-made telescopes he had observed 'forests' on the lunar surface, and that the moon was 'in all probability' inhabited.

Nevil Maskelyne, the Astronomer Royal and leading cosmological light of the Royal Society, was more than a little outraged by these apparently absurd claims. He had himself established that the moon had no life-sustaining atmosphere, based on the sharpness with which it occults starlight at its edge.[6] But he was intrigued by the minute detail of Herschel's observational sketches of the moon's surface, and the apparently fantastic power of his reflector telescope. He wrote a challenging letter to Watson in Bath, questioning the seriousness of Herschel's work, and his views on the moon. It was now that Joseph Banks, always on the lookout for new and unusual scientific talent, began to pay attention.

Watson forwarded what he tactfully called 'Dr Maskelyne's extremely obliging letter' to Herschel. Clearly anxious that Herschel might take offence at the implied criticism, he urged a diplomatic response in a covering note of 5 June 1780: 'I think you would do right (pardon my giving you advice) either to add the desired improvements, or to write over again the Paper, and send it to Dr Maskelyne, who, as he is Astronomer Royal, will be pleased, I believe, with the compliment paid him, and he will present it anew to the Society.'[7]

To his relief, Herschel wrote back to the Astronomer Royal with apparent modesty on 12 June: 'I beg leave to observe Sir, that my saying that there is an absolute certainty of the Moon's being inhabited, may perhaps be ascribed to a certain Enthusiasm which an observer, but young in the Science of Astronomy, *can hardly divest himself of when he sees such Wonders before him.* And if you promise not to call me a Lunatic I will transcribe a passage from some observations begun 18 months ago, which will show my real sentiments on the subject.'[8]

The views that Herschel now expressed must have taken Maskelyne aback. Far from retracting his opinions, he emphasised his belief that 'from analogy' with the earth, and its likely conditions of heat, light and soil, the moon was 'beyond doubt' inhabited by life 'of some kind or other'. Even more provocatively, he thought that the terrestrial view of matters gave undue importance to the earth. 'When we call the Earth by way of distinction a planet and the Moon a satellite, we should consider whether we do not, in a certain sense, mistake the matter. Perhaps – and not unlikely – *the Moon is the planet and the Earth the satellite!* Are we not a larger moon to the Moon, than she is to us? ... What a glorious view of the heavens from the Moon! How beautifully diversified with [her] hills and valleys! ... Do not all the elements seem at war here, when we compare the earth with the Moon?'

That Herschel was writing somewhat mischievously to the Astronomer Royal becomes clear towards the end of his letter. Poetry gently creeps up on astronomy: 'The Earth acts the part of a Carriage, a heavenly waggon to carry about the more delicate Moon, to whom it is destined to give a glorious light in the absence of the Sun. Whereas we, as it were, travel on foot and have but a small lamp to give us light in our dark nights, and that too often extinguished by clouds.' The teasing wit in Herschel's final sally was unmistakeable: 'For my part, were I to choose between the Earth and the Moon, I should not hesitate a moment to fix upon the Moon for my habitation!'[9]

Maskelyne could not overlook this, and promptly visited Herschel in Bath, accompanied by Banks's new Secretary and confidant at the Royal Society, Dr Charles Blagden. The visit seems to have been somewhat stormy. They cross-questioned Herschel in a challenging manner, but reported back to Banks that they were strangely impressed, especially by Herschel's beautiful home-made telescopes, of which there were several. There was also the unusual matter of the sister, a small, shy, tongue-tied young woman who seemed as mad about astronomy as her brother. Her name was Caroline. There was however no reason to believe that the Herschels would achieve anything particularly original in astronomy. They were provincials, émigrés, and poor self-taught enthusiasts.

Unknown to Maskelyne, the tongue-tied Caroline Herschel had made her own brief note of this visitation from the great men of the metropolis. The 'long conversation' which Dr Maskelyne held with her brother William got nowhere, and to her 'sounded more like quarrelling'. Immediately Dr Maskelyne left the house her brother burst out laughing and exclaimed: 'That is a devil of a fellow!'[10]

Less than a year later, in March 1781, Banks was amazed to hear that William Herschel was about to revolutionise the entire world of Western astronomy. He had achieved – or possibly achieved – something that had not been done since the days of Pythagoras and the Ancient Greek astronomers. Herschel had discovered what was perhaps a new planet. If so, he had changed not only the solar system, but revolutionised the way men of science thought about its stability and creation.

2

William Herschel was born in Hanover on 15 November 1738, and his younger sister Caroline twelve years later, on 16 March 1750. Their passion for observational astronomy came absolutely to rule both their lives, although in very different ways. At its height, in the 1780s, brother and sister spent night after night, month after month, summer and especially winter, alone but together in the open air, under a changing canopy of stars and planets. Their minutely recorded telescope observations, published in over a hundred papers by the Royal Society, would change not only the public conception of the solar system, but of the whole Milky Way galaxy and the structure and meaning of the universe itself.

Herschel and his sister were deeply attached from childhood, and most of what is known about William's life is drawn from Caroline's affectionate but troubled journal or day book, which she later turned into a memoir. She once wrote: 'If I should leave off making memorandums of such events as affect, or are interesting to me, I should feel like – what I am, namely, a person that has nothing more to do in this world.'[11] ♣

William was well into his thirties when astronomy began to take over his existence. The Herschel family concern over several generations had been music, not stargazing. In early eighteenth-century Germany – then a series of city states – the profession of music – playing, singing, composing, and teaching – was as socially important as those of the law, the army or the Church. Each city court and most military regiments had their own orchestras, and those of Hanover had some of the most renowned in Europe. Their fame spread especially after the Elector of Hanover became George I of England in 1715, and composers such as Handel achieved Europe-wide status.

William and Caroline's father, Isaac, was a military bandsman with the Hanover Foot Guards. His own father had been a landscape gardener near Magdeburg in Saxony who had an amateur interest in the oboe, but who died when Isaac was only eleven. Isaac, virtually an orphan, and without proper education, also began life as a gardener on various

---

♣ Caroline eventually wrote out two versions of this memoir, the first in summer 1821, when she was seventy, and the second in 1840. She also destroyed two sections of the original record which she did not want read by other family members. A composite version was edited by her great-niece, Mrs John Herschel, and published by Murray in 1879. The manuscript still exists in the private collection of John Herschel-Shorland. The individual Memoirs have been meticulously published by Michael Hoskin, as *Caroline Herschel's Autobiographies* (2003). William wrote a 'Memorandum of My Life' when he was nearly sixty, but this was a sort of professional CV for fellow scientists, comparatively short and characteristically reserved (Herschel, *Scientific Papers* 1, p.xiii). For full details see the Bibliography.

aristocratic Prussian estates. But at twenty-one, in his own words, he 'lost all interest in horticulture', found he had a natural ear for music, and 'worked day and night to become an oboe player'. Despite advice from his elder brother to stick to gardening, he could 'no longer resist the desire to make music and to travel', and drifted first to Potsdam, then to Brunswick ('too Prussian for me'), and finally to Hanover, where the atmosphere was more liberal.[12] The Elector of Hanover was now George II of England, and more easy-going English manners were acceptable. In August 1731 Isaac joined the Hanover military band, a happy career choice which allowed him considerable freedom, until he was caught up in the Anglo-German campaigns against the French which swept mid-eighteenth-century Europe.[13]

At twenty-five Isaac fell in love with a local girl, Anna Moritzen, who came from a village just outside Hanover. She was a beautiful creature, but completely illiterate. They might not have married except that Anna got pregnant, and Isaac proved himself a man of honour. Anna later said prettily that Isaac dropped into her life 'as if from a cloud'.[14] They steadily produced one child every two years for the next twenty years – but though ten children were born only six survived, a cause of much grief. Anna adored her first-born, Jacob, above all else, and indulged him extravagantly; she also loved her first daughter, Sophie, the beauty of the family. With the remaining children she was more severe, especially with her youngest and least promising daughter, Caroline.

Anna always seemed to be struggling to control the large, unruly family during Isaac's frequent absences with his regiment. She tried to inculcate traditional German virtues: discipline, craftsmanship, thrift and family loyalty. She had no patience with 'book-learning', especially as far as her daughters were concerned. However, she accepted Isaac's ambition 'to make all his sons complete musicians', which she saw as a path to fame and money. One of William's earliest memories was of being given a tiny violin which his father had made for him, and being taught to play almost before he could hold it to his shoulder.[15]

Jacob was quickly established as 'the genius' of the family: a superb solo musician from childhood, handsome but vain and volatile, having the true 'artistic temperament'. William was quieter and steadier, more determined in his lessons, thoughtful and genial, a great reader. Caroline mostly remembered her mother's severity, and how her eldest brother and sister, Jacob and Sophie, were by contrast petted and indulged.

Possibly Isaac also found his wife Anna a little 'too Prussian'. There was something dreamy, almost unworldly, about Isaac Herschel. Alongside his music-making, it is evident that he had a certain metaphysical approach to the world. He had little formal education, but for that very reason his interests were wide and passionately pursued: they included instrument-making, reading philosophy and practising amateur astronomy. It was a combination very characteristic of the culture of Enlightenment Germany, at a time when its greatest philosopher, the young Immanuel Kant, was the son of a craftsman and harness maker. Isaac was a natural teacher, patient and good-humoured; while Anna was quick-tempered, opinionated and scornful of what she regarded as bookishness.

Caroline remembered her father taking her out into the street to see the winter stars on a clear, frosty night, 'to make me acquainted with the most beautiful constellations, after we had been gazing at a comet which was then visible'.[16] Perhaps this stayed in her mind, because finding comets would later become her particular passion. She also recalled being shown an eclipse of the sun, safely viewing it reflected in a bucket of water.[17] She added admiringly that her father loved helping William with his studies, and was particularly delighted with his 'various contrivances' – by which she meant his scientific models. (Her phrasing, like her accent, remained pleasingly Germanic to the end of her life.) Among these she particularly remembered a shining, neatly turned four-inch brass globe, 'upon which the equator and ecliptic were engraved by my brother', an object of childish wonder and admiration to her. This was an early sign of William's extraordinary manual skills, which she came to worship.[18] Her own secret desire was to become a concert singer, but she dared tell no one about this except William.

3

Perhaps the moments of paternal care were remembered because they were rare. Isaac was often away on campaign, and Anna ran the noisy, chaotic household as the growing family moved from apartment to apartment in Hanover, depending on their financial circumstances. There was great sibling rivalry. The bond between William and Caroline was strengthened by their vain and bullying brother Jacob, who as his mother's darling had become spoilt and domineering. There was also the unhappy elder sister Sophie, whose beauty led to an early marriage with a 'cruel and extravagant' husband which turned out to be a disaster.[19] Caroline – who was small and impish – claimed she was frequently whipped for disobedience both by her mother and by Jacob, starved for food, and treated as a scullion. Similarly, she said William had been endlessly teased by Jacob for his outstanding work at the garrison school in Latin, Greek, French and mathematics. He also mocked his model-making abilities. Jacob took nothing seriously but 'the science of music', in which he already considered himself (rightly) a virtuoso.[20]

At fourteen William joined the Hanover regimental band, alongside Jacob and his father. He soon learned to turn his hand to an astonishing array of instruments – the oboe, the violin, the harpsichord, the guitar and, a little later, the organ. He was also starting to compose, and had an early fascination with musical notation and the theory of harmony. Both he and Jacob appeared as young solo performers at the court of the Elector of Hanover, and their names were not forgotten.♣

Caroline also remembered long philosophical arguments at home in the evenings, when the brothers returned after concerts. She would lie awake in her bedroom, trying not to fall asleep and secretly delighting in William's quiet, calm voice steadily contradicting Jacob's furious outbursts. According to her the names of 'Leibniz and Newton' were shouted from the parlour 'with such warmth that my Mother's interference became necessary'.[21] When their father was at home these conversations on philosophical subjects became even more rowdy, and would frequently last till dawn. The combination of Leibniz and Newton suggests that William and Jacob were arguing about the rival virtues of calculus (a mathematical system invented by Leibniz) and fluxions (a similar system invented – but jealously guarded – by Newton). Both systems produced the new mathematics of curves and gradients, essential to the astronomical calculation of planetary orbits and the elongated ellipses of comets. It was an unusual household.[22]

♣ Three of William Herschel's works are currently available on CD. They are his Oboe Concertos in C major and E flat major, and his Chamber Symphony in F major (Newport Classics, Rhode Island, USA, 1995). They are notable for their light musical touch and fine, sprightly melodic lines, sometimes with a certain melancholy in the slower passages. The rapid, complex orchestration around the solo oboe in the concertos is handled with great confidence, and suggests Herschel's ability to manage patterns and counterpoint. This was a conceptual skill which he seemed to transfer (visually) to the patterning of stars and constellations. He moved from earthly music to the music of the spheres.

In November 1755 the five-year-old Caroline witnessed a strange portent of disruption in the after-shock of the Lisbon earthquake, which amazingly travelled as far as Germany. As she remembered it, the whole barracks shook. 'I saw both my parents standing aghast and speechless... my brothers came running in... all [the family] being panic-struck by the earth quake.'[23] This earthquake, which killed over 30,000 people in Lisbon and shook cities throughout Europe, seemed to many to call into question the idea of God (or Nature) as a benevolent Providence, and to be a sign that a new kind of scientific knowledge was required. Among many speculative works, it inspired Voltaire's *Candide.* Caroline always retained a superstitious horror of earthquakes, and said she felt one years later when she stood by her father's deathbed.[24]

In the spring of 1756, when William was seventeen and Caroline was six, the Hanover Foot Guards were posted to England, to serve under their ally the Hanoverian King George II. It was the outbreak of a long, desultory and financially draining conflict with the French that would become the Seven Years War, and would radically affect the fortunes of the Herschel family. Jacob tried to obtain a home posting with the court orchestra, but failed, and all the men of the family were conscripted. Caroline remembered the grim, silent bustle in the house. 'My dear father was thin and pale, and my brother William almost equally so, for he was of a delicate constitution and just then growing very fast. Of my brother Jacob I only remember his [making] difficulties at everything that was done for him.' The rest of the family, the three younger children including the baby Dietrich, were abruptly left on their own as the men departed.

Caroline's sense of this human drama is well caught in her *Memoir.* 'The troops hallooed and roared in the streets, the drums beat louder... and in a moment they were all gone. I found myself now with my Mother alone in a room all in confusion, in one corner of which my little brother Dietrich lay in his cradle; my tears flowed like my Mother's but neither of us could speak.' Then Caroline made a touching gesture towards the mother she feared. She ran and found one of her father's large cambric handkerchiefs, unfolded it, and carefully placed one corner in her weeping mother's hand, while holding onto the opposite corner herself. They were united, at least, in grief. 'This little action actually grew a momentary smile into her face.'[25]

The Hanover Regiment were stationed at Maidstone, in Kent. Jacob spent his pay on fashionable English clothes, William on English books, and Isaac on an allowance for Anna and the children. William fell in love with the country, began to learn the language, and made a small circle of English friends. For the first time there are hints that he was secretly beginning to dream of an entirely different, freer kind of life in the land of Newton, which had been adopted by his fellow German Handel. When the Hanover Guards were posted back to Germany the following year to fight the invading French armies, Jacob packed a beautifully tailored English suit, and William a copy of John Locke's *Essay Concerning Human Understanding*.[26]

Caroline remembered their return, one freezing winter evening in December 1756.[27] Her mother Anna was preparing a welcome-home dinner, and the six-year-old Caroline was sent to collect her father and brothers from the parade ground. But she missed them in the dark and confusion, and the frightened little girl had to make her own way home. 'I continued my search till I was spent with cold and fatigue, and on coming home I found them all at table, nobody greeting me but my brother.' As she remembered it, no one had even noticed her absence except William. She never forgot his reaction. 'My dear brother William threw down his knife and fork, and ran to welcome and crouched down to me, which made me forget all my grievances.'[28]

Jacob now obtained a timely discharge, but William and his father took part in the disastrous battle of Hastenbeck, which was fought against the French invaders twenty-five miles outside the city of Hanover, on 26 July 1757. The surrounding countryside was overrun by a French army of 60,000 troops under Marshal d'Estrées. The allied general, the Duke of Cumberland, beat a strategic retreat westwards towards Flanders. Hanover was occupied, and the Herschels' building had sixteen French infantrymen billeted on it.[29]

After a hasty family conference it was decided to smuggle William – still only eighteen – out of Germany altogether. Caroline recalled a fleeting, romantic glimpse of her brother's surreptitious departure as she stood anxiously by the street door, told not to call out or give him away: 'he glided like a shadow along, wrapped in a great coat, followed by my mother with a parcel containing his *accoutrements*.'[30] William slipped past the last sentinel at Herrenhausen, and made his way to Hamburg, where he took ship again for England. At the last moment he was joined by Jacob, and together the two brothers arrived, penniless refugees, in London. They supported themselves by copying musical scores, giving oboe lessons, and playing as freelance musicians in local orchestras. They gave a successful concert in Tunbridge Wells. In the evenings William read voraciously: English novels, books on mathematics and musical harmony, Robert Smith's *Harmonics* (1749) and James Ferguson's recently published and immensely popular *Astronomy Explained* (1756).[31]

By the autumn of 1759 Jacob was finding the life too hard, and slipped back to Hanover with his and William's combined savings, eventually finding employment as a court musician.[32] Now, for the first time in his life, William Herschel, aged twenty-one, was alone – but free, talented, and in the country of his choice. And with a secret gift, his genius for astronomy, hidden even from himself – but awaiting the opportunity to unfold. For the next five years he virtually disappeared from the family history.

Caroline was devastated when William went abroad. In retrospect, she realised that it was he alone who had cared for her, and in his long absence he became a sort of legendary figure. At home her misery deepened. Hanover remained occupied, and food supplies were short. She continued going to the garrison school, but was not allowed to learn arithmetic or languages, and was increasingly treated as a maidservant by the family. She remembered sewing immensely long woollen stockings, scrubbing laundry, and writing her mother's letters to her father on campaign. In fact her unusual literary ability was a rare source of pride, as she later recalled. 'My pen was taken frequently *in requisition* for writing not only my Mother's letters to my Father, but to many a poor soldier's Wife in our Neighbourhood to her Husband in the Camp; for it ought to be remembered that in the beginning of the last century very few women left country schools with having been taught to write.'[33]

Her father had been made a prisoner of war, and for some months her brother Jacob became effective head of the family. He 'woefully disarranged' the household, demanded larger rooms, and bullied his little sister. 'Poor I got many a whipping for being too awkward at supplying the place of footman or waiter.'[34] When her father finally returned from the wars in summer 1760, aged fifty-three, he was a broken man, his health permanently damaged by many months of imprisonment, asthmatic and with a heart condition.[35] He gave some private music lessons, smoked his pipe, and was largely ruled by his wife and his eldest son. He did however manage to regularise William's situation as a soldier absent without leave. On 29 March 1762 General A.F. von Sporcken signed a formal document of discharge.[36] But there was no sign of their son coming home.

Caroline's own health was bad. At the age of five she had caught smallpox, and now at eleven she caught typhus. While she was recovering her mother left her to crawl up and down stairs 'on my hands and feet like an infant' for several months.[37] The worst result of this illness and neglect was that Caroline's growth was permanently stunted. In a family of tall, lean children, she never grew much beyond five feet.[38] Moreover, her face had been permanently scarred by the smallpox. The lively, enchanting pixie that William had once known had become a silent, resentful gnome. But she also became increasingly determined and self-sufficient. She said that from the time of her recovery, 'I do not remember ever having spent a whole day in bed.'[39]

Isaac increasingly left the care of his surviving children largely in Jacob's hands: Alexander aged seventeen, Caroline aged twelve, and the youngest, Dietrich, a sweet but sickly child, aged seven. Her father would indulge Caroline ('and please himself') with a short lesson on the violin, but he told her mournfully that as she was now 'neither handsome nor rich' she could never expect to marry, and should resign herself to helping her aged parents.[40]

Her brother Jacob refused to allow her to train as a milliner, although she was encouraged to learn just enough to be able to deal with the household clothes and linen. Her father had once hoped to give her 'something like a polished education', but her mother insisted that, given the family situation, it should be practical and 'rough'; she would not even allow her to learn French, in case she developed ambitions to be a governess.[41] Similarly, little Dietrich was denied a dancing master. Anna also observed that it was 'her certain belief' that had William read less, he would never have stayed away in England.[42] When Jacob insisted that an extra servant girl be hired, she was given Caroline's room and bed to share. For Caroline, 'her destiny now seemed unalterable'. She was to be the family housekeeper, a spinster and permanent maidservant.[43] She later decided to destroy all the journals referring to her private feelings during these years of misery. She did not want to write in the fashionable Romantic mode of the personal confession. 'After reading over many pages,' she wrote to Dietrich, 'I thought it better to destroy them, and merely write down what I remember to have passed in our family at home, and abroad.'

In fact much remains of her inner life: as much perhaps as in the journals of Dorothy Wordsworth. This dramatic rejection of the record of her childhood unhappiness was really a prelude to continuing revelations of frustrations in adulthood. 'By what is to follow,' she explained, '[Dietrich] may also see how vainly his poor Sister has been struggling through her whole life ... wasting her time in the performance of such drudgeries and laborious works as her good Father never intended to see her grow up for.' This was the ultimate cause, she came to think, of the 'mortifications and disappointments which have attended me throughout a long life'. But all this was in retrospect, nearly sixty years later.[44]

In the summer of 1764, apparently without warning, to Caroline's astonished delight her brother William – 'let me say my *dearest* brother' – reappeared in Hanover.[45]

4

What had happened to him in the interval? From his intermittent letters to Jacob, and things he subsequently told Caroline, it is possible to reconstruct the outline of his adventures, though with many gaps. Against all expectation, he had not remained in London, or gone back to his friends in Kent, but had boldly struck into the remote north of England. Surprisingly he used his military contacts to obtain the post of civilian music master to the Durham militia, which was stationed under the Earl of Darlington at Richmond in Yorkshire.[46] This was as much a social engagement as a military one, and Herschel was soon completely independent, working as a freelance musician and music teacher in Leeds, Newcastle, Doncaster and Pontefract, and as an organist in Halifax. Little is known about these posts, except that he was constantly on the move, frequently lonely, and sometimes weeping with homesickness.

Every so often Jacob received letters from William with varied English postmarks, and written – with remarkable versatility – in German, French or English as the mood and subject took him. These letters were also covered in mechanical diagrams, and frequently shifted without a break from words to musical notation. They give a sense of Herschel's mind switching with extraordinary agility between different modes of expression and zones of thought – literary, mechanical, musical, philosophical.[47]

From Yorkshire on 11 March 1761 he wrote in a fit of melancholy for which he chose slightly faltering English: 'I must tell you a certain anxiety attends a vagrant life. I do daily meet with vexations and trouble and live only by hope. Many a restless night have I had; many a sigh and – I will not be ashamed to say it – many a tear.' But a fortnight later he was writing from Sunderland in sprightly French about two pretty girls he had just met – one of them '*la plus belle du monde, la Beauté elle-meme personnifée*' – whose accomplishments included excessive blushing, flirting and playing the guitar. Sadly they only met once, though Herschel later confessed that they corresponded for over a year – another indication of his loneliness, perhaps.[48]

He chose German for his philosophical reflections. All of these were thoughtful, but many of them gloomy: the stoic doctrines of Epictetus, the optimism of Leibniz ('not the least credible nor feasible'), the origins of evil, the nature of sin, the ethical (rather than the intellectual) necessity for Christian religion in European society. 'In all ages there have been philosophers who have had thoughts *above their religion*, and have been true Deists' – but it was 'impossible' in the present state of education for 'a whole nation to be true Deists'. William himself described God memorably, in German, as 'the unknowable, *must-exist Being*'.[49] With this formula he was able to set aside, for the time being at least, the problem of a personal Creator.

He had thought often about the 'immortality of the soul', but said (to Jacob at least) that he preferred not to draw any conclusions. His uncharacteristically pious explanation seems to disguise a scientific reservation that there was no 'intelligible' data on the matter: 'My feeble understanding is not capable of pushing so far into the secrets of the Almighty; *and as all those propositions have something unintelligible about them*, I think it better to remain content with my ignorance till it pleases the Creator of all things to call me to Himself and to draw away the thick curtain which now hangs before our eyes.' In fact 'pushing far into secrets' was always Herschel's natural instinct and delight.[50] Perhaps music provided a way of pondering these questions, and at this time he composed an oratorio based on Milton's *Paradise Lost*, though the manuscript score has not survived.[51]

Another way would be astronomy. For the glowing exception to these dark truths about human life was always the life of Nature, already an endless source of clarity and consolation for Herschel. 'If one observes the whole Natural World as one, one finds everything in the most Beautiful Order; it is my favourite maxim: *Tout est dans l'ordre!*'[52]

Riding between musical engagements, from one remote provincial northern town to another, often crossing over the moors alone at night, he found himself studying the panoply of stars overhead as he had done as a boy. He became well acquainted with the moon, and would later write that at this time he had intended 'to fix upon the moon for my habitation'.[53] He also later told several tales about these lonely rides, one being how on one occasion he was reading so intently that when his horse stumbled and threw him, he somersaulted over its head and landed upright still holding his book in his hand, a perfect demonstration of the Newtonian law of 'circular motion'.[54]

Herschel now began to explore further the work of James Ferguson (1710–76), a man after his own heart who had started life as an illiterate Highland farm-labourer, and had become one of the most distinguished practical astronomers and demonstrators. His *Astronomy Explained* (1756) ran to numerous popular editions, and he later vividly described in his *Autobiography* (1773) how he fell in love with astronomy. He would take a blanket out into the fields after work, and lie on his back measuring star distances and patterns with beads on a string held up over his head. He then transferred these, by the light of a stub of candle on a stone, to his first paper star-maps, spread out beside him on the grass. He said he imagined the ecliptic (the sun's curving path through the heavens) like a high road running through the stars. Gradually he taught himself astronomy and built his own telescopes. He later invented various devices for projecting constellations during his lectures, and his 'Eclipseon' for showing the various movements of the solar system.

Living in lonely bachelor lodgings, Herschel spent more and more time reading about stellar theory. He followed Robert Smith's *Harmonics* (1749) with his *Compleat System of Opticks* (1738), which contained illustrated sections on astronomical observations.[55] He began to be preoccupied with various cosmological problems: what was the relation between music, mathematics and star patterns? Was there life on the moon? What was the structure and composition of the sun? How far away were the nearest stars? What was the true size and shape of the Milky Way? Many of these problems would emerge in his earliest scientific papers, and would continue to fascinate him for the rest of his life.

He was approaching thirty, and to all appearances he was alone and adrift in a foreign land. But he was not disorganised or depressed. Much of his father's military discipline, and his own professionalism, now stood him in good stead. He worked immensely hard, with an energy and determination that never left him. His musical appointments were increasingly important, regular and better paid. At Halifax he was conducting an orchestra, playing the organ, giving singing lessons, and composing his own music. He was also learning Italian.

After a period of physical weakness in his late teens (which Caroline had anxiously remarked on), William had grown into a tall, commanding figure, with a high, intellectual forehead, and very striking dark eyes. Outwardly at least, he was cheerful and sociable. It is evident that he made friends wherever he went. At one concert he was joined by the Duke of York, brother of the new King George III, who accompanied him (rather badly) on the violoncello. On another occasion he was invited to conduct one of his own symphonies at St Cecilia's Hall in Edinburgh. At the reception afterwards, he chanced to meet the philosopher David Hume, and was promptly invited out to dinner.[56] There was something about Herschel's mixture of intensity and innocence that simply charmed people. And talented German exiles were, of course, popular.

Herschel was brought back to Hanover by a combination of circumstances. His work at Halifax had led to the first really serious opportunity of his career: the possibility of being appointed the organist of the new Octagon Chapel at Bath, when its building was completed. Bath was fast becoming the most fashionable city in England – nearly ready for Beau Brummell – and all kinds of other musical work would obviously be available there. Herschel immediately thought of his brothers, Jacob and Alexander. He had heard too that his father Isaac was ill, and not likely to live much longer. There may also have been worries about the younger children under Jacob's care – Caroline and little Dietrich.[57] At all events, the prodigal son suddenly reappeared in Hanover in the summer of 1764. He arrived saying he had just observed an eclipse of the sun as he rode over the Luneburger Heath.

Caroline was then fourteen, and her appearance following her illness must have shocked him. But there was little that he could do for her immediately, and after an absence of nearly seven years his visit to Hanover lasted a mere fortnight. It was a sober reunion. Isaac, obviously failing, could not persuade him to remain, and instead William spoke of future plans for his brothers as musicians in England. Nothing was said of Caroline at this point. William must have known it was the last time he would see his father alive.

Caroline remembered William's departure after his flying visit with grief and frustration. It was the day of her first communion, and William had particularly admired her appearance in a new black silk dress. But she was sent to church by Jacob, and not allowed to see William off. She never forgot that moment. 'The church was crowded and the door open. The Hamburg Postwagen passed at eleven, bearing away my dear brother... It was within a dozen yards from the open door; the postillion giving a *smettering* blast on his horn. Its effect on my shattered nerves, I will not attempt to describe, nor what I felt for days and weeks after.' She walked home alone, 'in feverish wretchedness', wearing her new dress and painfully aware that she was carrying the bouquet of artificial flowers that her elder sister Sophia had worn on her ill-fated wedding day.[58]

Their father died of a stroke in 1767, but William did not return for the funeral. He would not come back to Hanover for another eight years.[59]

5

William was offered the organ post in August 1766, and officially moved to Bath in December of that year. Before the chapel was opened he found a lucrative position in the famous Pump Room Band, run by the impresario James Linley. The Pump Room and Theatre was then the very height of fashionable entertainment. Linley's daughter, the singer 'Angel' Linley, later became a star at Drury Lane, and married the dramatist Richard Brinsley Sheridan.

Early on, Herschel had a quarrel with Linley over orchestral arrangements in the Pump Room, which got into the newspapers and caused a brief but diverting scandal in Bath society. The disagreements were minor – the appointment of singers, the provision of music stands – but there was some suggestion that Linley was exploiting Herschel as a German outsider. What was remarkable was the sudden revelation of Herschel's fiery temper and determination when roused. Far from conceding to Linley, he took out a series of advertisements against his concerts in the *Bath Chronicle*. He referred openly to Linley's 'low Cunning and dark Envy', and set up a competing programme with a rival diva, the Italian singer Signora Farinelli. This proved a great success.

After one season of musical warfare Linley made peace with Herschel, and their combined concerts resumed at the Pump Room, to general satisfaction. After Linley left for London, Herschel became sole director. Moreover Linley became a great admirer of Herschel, and sent his son Ozias to him to learn the violin. It was perhaps no coincidence that when Ozias went on to Oxford, he studied mathematics and astronomy.[60]

William rented a modest house five minutes' walk from the Pump Room, in the fashionable centre of Bath, at Beauford Square. He continued composing for the oboe, taught guitar, harpsichord and violin, conducted oratorios and gave singing lessons. In June 1767 he was joined by Jacob for a visit, and took up his appointment as organist and choirmaster at the Octagon Chapel, which was opened on 4 October.[61]

It was during this hectic period that his other secret passion exerted itself. In February 1766 the twenty-seven-year-old William Herschel started his first Astronomical Observation Journal. He recorded an eclipse of the moon, and the hazy appearance of Venus.[62] Hard as he worked as a musician, he was now steadily training himself as an astronomer. He devoured books on astronomical calculation, Flamsteed's star tables and Thomas Wright's cosmological speculations. He attended James Ferguson's astronomy lectures at the Pump Room in 1767, and at last met this early astronomical hero of his.[63] He spent hours star-gazing in the Beauford Square garden at night. Even when teaching his music pupils in the evening, it was said that he sometimes broke off and took them outside to look at the moon. He began to build up a small arsenal of second-hand refractor telescopes, and carefully examined their construction. He was considering what his father Isaac used to call 'one of his contrivances'.

The refractor is the classic type of straight-through telescope originally developed by Galileo, and refined by Kepler and the great seventeenth-century Dutch astronomer Christiaan Huygens. It has magnifying lenses at each end of the tube, one fixed and the other adjustable (the eyepiece), advancing or retreating to focus the image. In extendible or retractable form, it was often used by soldiers or sailors on active service, until the arrival of the binoculars. It was just such a refractor telescope that Nelson would – or would not – put to his blind eye at the battle of Copenhagen in 1798. The snapping closed of the retractable mechanism became a gesture of decision and command.

Herschel found that most refractor telescopes were satisfactory for simple low-magnification viewing of the moon or the planets. But astronomical versions were absurdly cumbersome (some up to twenty-five foot long), and almost useless for high-magnification observation of the stars. The curve or bulge in the magnifying lens acted like a prism, and broke up the white stellar light into distorting rainbow-coloured fringes at the edges. (This became known as 'chromatic aberration'. A short-sighted person can see these rainbow aberrations of starlight with the naked eye, because his pupil is also distorted at the edges.) Newton, observing this in his famous prism experiments at Cambridge, had invented an entirely different type of telescope, the *reflector*. But his, which he donated to the Royal Society, was only six inches long, with a magnifying power of forty.[64]

Confined to refractors, most eighteenth-century British astronomers had paid little attention to stellar astronomy, except where it served for navigation purposes. (The John Dollond achromatic telescope, which corrected some prismatic distortion, was only invented in 1758, and did not come into general use – as improved by his son Peter Dollond – until the turn of the century.[65]) The newly appointed Astronomer Royal, Nevil Maskelyne, based at the Greenwich Observatory, was largely concerned at this time with observing lunar eclipses, planetary transits and passing comets. His special interests lay in establishing tables for use at sea as a mariner's almanac, and in the calculation of longitude. He noted that since his seventeenth-century predecessor at Greenwich, John Flamsteed, had thoroughly mapped the heavens, he himself kept only thirty-one stars under regular observation.[66]

Since his long nights of riding over the moors, Herschel's interests had roamed far beyond the safe family of the solar system, with its restricted circuit of sun, moon and six known planets. He had the courage, the wonder and the imagination of a refugee. His whole instinct was to explore, to push out, to go beyond the boundaries. Gradually he began to think about the possibilities of Newtonian reflector telescopes. Newtonians were based on a different principle from the traditional refractors. They produced increased 'light-gathering', rather than simple magnification. As their name implies, the primary component of a reflector telescope is a large mirror, or *speculum*, highly polished and subtly curved inwards (concave) so as to gather and concentrate starlight at a much greater intensity than the lens of the naked eye. This concentrated light is then viewed through a simple adjustable eyepiece inserted into the side of the tube, the whole set-up producing wonderfully bright images and little chromatic aberration.

Instead of conventional magnification, Herschel began to think in terms of something he called 'space-penetrating power'. This was a concept he had partly developed from Robert Smith's *Opticks*.[67] Conventional eighteenth-century astronomers still studied the night sky as if it were a flat surface, or rather the interior surface of a decorated dome, inlaid with constellations. Flamsteed's beautiful *Celestial Atlas*, first published as a large decorative folio in 1729, presented the sky like this. Its second edition of 1776 still remained the standard European book of reference for stellar identification.

Each constellation was given a double-page spread, showing the mythological figures that gave them their names drawn in flat engraved outlines, as well as the known stars belonging to the group. The brighter stars were identified by their home constellation and a Greek letter of the alphabet. So Alpha Orionis, also known by its Arabic name Betelgeuse, was the bright star on the shoulder of Orion the Hunter; and Zeta Tauri (which would later catch Herschel's attention) was a third-magnitude star in Taurus the Bull.♣

But Herschel began to conceive of *deep space*. He began to imagine a telescope which might plunge deep down into the sky and explore it like a great unplumbed ocean of stars. This was something a reflector telescope might be able to do supremely well, if its concave mirror were sufficiently large. But because even small astronomical mirrors were expensive, and large ones had not yet been developed (even by London lens-makers like Dollond), Herschel realised that he would have to make them himself. Moreover, to achieve the exquisitely fine reflective surface he required, they would have to be cast in metal, not glass.

Meanwhile the other Herschel brothers began to shuttle between Bath and Hanover. Jacob came over for a brief visit in summer 1767, following

---

♣ A typical brass eighteenth-century orrery showed the sequence of six known planets: Mercury, Venus, Earth, Mars, Jupiter (with moons) and Saturn (with rings) orbiting around a central sun (sometimes operated by clockwork and illuminated by candles). Flamsteed showed all constellations – such as Herschel's early favourites Orion, Andromeda and Taurus – against mythological engravings of their signs: the Hunter, the Goddess, the Bull. His *Atlas Coelestis* catalogued 3,000 stars; the modern Hubble telescope has identified some nineteen million. But the presentation of the night sky as a curved dome of mythological constellations is still quite usual, as for instance in the magnificently restored curved ceiling of Grand Central Station, New York.

Isaac's death, but after giving virtuoso performances in the Pump Room he preferred to return to his high life in Hanover. Young Dietrich, now aged fifteen, came the following summer, and was given a fine holiday. Finally Alexander came and settled in 1770.[68] William moved to a larger house at 7 New King Street, and Alexander was given the attic rooms, while William took over the first floor and had the reception rooms redecorated and furnished with a new harpsichord for his singing and music lessons.

All the time he was evidently worrying about Caroline, and finally in the spring of 1772, after long discussion with Alexander, he wrote to Hanover to ask if Caroline (then aged twenty-one, and having reached her majority) would like to join them at Bath. Knowing the opposition his proposal would face from their mother and Jacob, William put his suggestion in the most plain and practical terms, as Caroline recalled. She should make a trial as to whether 'by his instructions I might not become a useful singer for his winter concerts and oratorios'. She could also become her brothers' housekeeper. If after two years this 'did not answer our expectations', William would send her back. Significantly, he mentioned not a word of astronomy.[69]

Caroline longed to accept. But her mother fiercely objected, and so of course did Jacob. 'I had set my heart upon this change in my situation, [but] Jacob began to turn the whole scheme to ridicule ... [although] he never heard the sound of my voice except in speaking.'[70] Caroline found her own way of stubbornly preparing for her escape. She practised singing the solo parts of oratorios 'with a gag between my teeth', so she could not be heard at home; and she secretly knitted enough cotton stockings for Dietrich to last him 'two years at least'.

Finally Herschel himself went over to Hanover, and won his mother over by pointedly promising to settle an annuity on her to pay for a maid to replace Caroline. He never succeeded in getting his elder brother's agreement, however. Jacob was away attending the Queen of Denmark at a court festival, and blustering letters arrived 'expressing nothing but regret and impatience' at the whole plan. William simply ignored them, and Caroline left 'without receiving the consent of my eldest brother'. They departed on 16 August 1772, and from this moment William became the real head of the family.

Caroline still spoke practically no English. Her elfin face, badly marked by the childhood smallpox scars, made her painfully shy. At less than five feet she was of such diminutive stature that at times she seemed like a pixie out of some German folk tale. She had an almost childlike enthusiasm, energy and sense of mischief. The one known portrait of her at this age, a charming miniature silhouette, confirms this impression. Her profile is fine, pert, almost boyish, but with full, slightly pouting lips, and a neat, very determined little chin. Her hair bubbles round her head in a mass of curls, and falls down her back, where it is secured with a ribbon. She has a sprite-like quality about her.

Caroline adored the journey to England, keeping a wide-eyed diary of the trip, like an excited teenager. In Holland her hat was gloriously blown off into a canal. At night William made her sit outside on the top of the carriage so he could show her the constellations. On the crossing to Norfolk, one of their ship's masts was carried away in a storm. Anchoring off the beach at Great Yarmouth (future home of Dickens's Lil' Emily), they were transferred with their bags to an open boat, rowed through the swell, and unceremoniously 'thrown like balls' onto the shore by two strapping English sailors.

Outside Norwich, the horses ran away with their carriage and they went 'flying into a dry ditch'. In London they walked round the streets, seeing St Paul's and the Bank, admiring the lights and examining the shops. But William would only pause outside those selling optical instruments – 'I do not think we stopped at any other.' By the time they arrived by the overnight coach in Bath, Caroline reckoned she had only slept in a bed twice in eleven days. That was what it was going to be like living with her brother. 'I was almost annihilated,' she wrote triumphantly. William covered the whole journey in one sentence in his journal. 'Set off on my return to England in company with my sister.'[71]

6

William now hustled Caroline into her new life. Summoning her to a seven o'clock breakfast, he began immediately to give her lessons in English and arithmetic, and showed her 'booking and keeping household accounts of cash received and laid out'. He said he would give her three singing lessons a day, while she practised the harpsichord, dealt with the household linen and prepared the menus. She was given rooms in the attic with Alexander, but was commanded to act as hostess in the salon.

William treated her affectionately but sternly, insisting that she go out to shop on her own in the market at Bath, even though she still only spoke a few words of English, which she had, as she put it, 'on our journey learned like a Parrot'.[72] She found herself 'alone among fishwomen, butchers, basket-men etc', and also had to contend with the 'hot-headed old Welsh woman' who cooked. She felt she was encountering a 'natural antipathy' which the lower class of the English had against foreigners.[73] But she could also be fierce herself: William's neighbour, the motherly Mrs Bullman, she quickly dismissed as 'very little better than an *Idiot*', a term much favoured by Caroline.[74]

At first she struggled against *Heimweh* (homesickness) but she showed herself unexpectedly dauntless, and gradually settled into the taxing new routine. Breakfast was shortly after 6 a.m. ('much too early for me, who would rather have remained up all night'), followed by household accounts, shopping, laundry, three-hourly singing lessons, instruction in English and arithmetic, music copying, formal practice on the harpsichord kept in the front room, and reading out loud from English novels.[75] 'By way of relaxation', she and William talked of nothing but astronomy. She never forgot 'the bright constellations with which I had made acquaintance during the fine night we spent on the Postwagen travelling through Holland'. But she also remembered that William had promised to train her up as a professional concert singer, who would one day be independent.

It took time for the full emotional *rapport* to renew itself between the tall, handsome thirty-four-year-old bachelor brother, driven and ambitious, and the shy, tiny, awkward twenty-two-year-old sister, who had never before travelled outside her native Hanover, but who was bursting with unfulfilled dreams and longings. To begin with their relationship seemed formal, almost like that between father and daughter. In many ways William was quite withdrawn – enthusiastic and talkative in the mornings, but remote in the evenings after any guests had departed. 'I seldom saw my Brother in the evening... He used to retire to bed with a basin of milk or glass of water, and Smith's *Harmonics* and *Optics*, Ferguson's *Astronomy* etc, and so went to sleep buried under his favourite authors; and his first thoughts on rising were how to obtain instruments for viewing those objects himself of which he had been reading.' At breakfast, Caroline was usually subjected to 'ample stuff for an astronomical Lecture'.[76]

William loved Caroline tenderly, but he also bullied her, in what he saw as a kindly pedagogical way. He could be an unsparing disciplinarian. She in turn adored him, but also feared him and grew impatient with him. He was always fierce in his domestic demands, and constantly required her to better herself: her English, her mathematics, her music and her astronomy. But gradually she learned to tease him and criticise him, while he came more and more to depend on her. In his daily notes and instructions he began to address her by the affectionate diminutive 'Lina', with its moonlike echo. Sometimes he even wrote it teasingly in French – '*Lina adieu*' – or transliterated into Greek letters, 'as you understand Greek'.[77] Caroline always referred to him simply as 'my dearest Brother', or else 'my beloved Brother'. For Caroline, William was initially the great liberator who had taken her out of the German house of bondage. But later their roles would subtly change. As William would observe to Nevil Maskelyne, it was not always self-evident which was the planet and which was the moon.

With the household running more smoothly, Herschel could now begin regular astronomical observations in their garden at night. Once Caroline had arrived, he found more time to explore the construction of telescopes. First he hired a two-and-a-half-foot-long Gregorian reflector telescope, which was too small; then in autumn 1772 he tried to construct an eighteen-foot refractor on the Huygens model. But its tube, which Caroline was instructed to make out of papier-mâché, was so long that it kept bending, like an elephant's trunk. They substituted one made out of tin, but it was still not satisfactory. Then he wrote to London for materials to construct a five-foot reflector, but was told that no one made glass mirrors large enough (at least five inches in diameter) to fit it. It was then that Herschel took the crucial decision to try to cast, grind and polish his own metal mirrors or *specula*. To start with, he acquired some metal grinding and polishing tools from John Michel, a Quaker astronomer who had retired to Bath nursing some strange, unacceptable ideas – such as the existence of 'black holes' in space from which light itself could not escape.

The accelerating pace of Herschel's experiments is caught in a memorandum of his purchases made over five months in 1773.

> *May 10th.* Bought a book of Astronomy, and one of Astronomical tables.
> *May 24th.* Bought an object glass of 10 foot focal length.

June 1st. Bought many eyeglasses, and tin tubes; made a pair of steps.
June 7th. Glasses paid for, and the use of a small reflector paid for.
June 14th. The hire of a 2 foot reflecting telescope for 3 months paid for.
Sept. 15th. Hired a 2 foot reflector.
Sept. 22nd. Bought tools for making a reflector. Had a metal [mirror] cast.
Oct. 2nd. Bought a 20 foot object glass, and 9 eyeglasses. Emerson's Optics. Attended private [music] scholars as usual.[78]

In June 1773 Herschel decided to attempt to make his own large reflector telescope, using metal mirrors as big as six inches in diameter.[79] It was a complicated and above all laborious task, requiring the casting, grinding and polishing of 'speculum metal', made of an alloy of white tin and brass. Three-inch mirrors were quite common, but a six-inch-diameter mirror with a precise concave surface required a technical feat that had never been achieved before. It called for a series of ingenious 'contrivances', which took Herschel back to his boyhood days, and all his old enthusiasm and ingenuity bubbled back.

The casting first required the construction of a small iron furnace and special moulds. These, Herschel found after many experiments, could best be made from a dried non-porous natural loam, formed from pounded horse-dung.♣ Once cast, the speculum metal had to be hand-ground with a solution of 'coarse emory and water' to achieve the required concave curve, and finally polished, 'with putty or oxide of tin or pitch', for hours on end to achieve an absolutely smooth reflective surface. It was an exhausting, and occasionally dangerous, physical process, needing endless trial and error. The furnace was liable to explode, and Herschel found that the polishing had to be done without pausing – sometimes for many hours on end.[80] If the polishing paused for even a few seconds in the final stages, the metal would harden and mist over, and the mirror would be useless.

♣ The use of horse-dung moulds for casting metal mirrors continued well into the twentieth century, with the 101-inch mirror of the Mount Wilson telescope in California, cast in Paris in 1920 and eventually used by Edwin Hubble to confirm Herschel's theories about the nature and distance of galaxies in 1922. See Gale Christianson, *Edwin Hubble: Mariner of the Nebulae* (1995). Precision was never easy to obtain: the mirror of the modern orbiting Hubble Space Telescope was found to be two micrometres too flat at the edges, an error corrected in the 1993 service mission, costing a total of $1.5 billion.

All the work had to be carried out at New King Street, turning the elegantly furnished house (intended of course for music-making and teaching) into a pungent, chaotic workshop. Initially Caroline was appalled at this transformation: 'To my sorrow I saw almost every room turned into a workshop. A Cabinet-maker making a Tube and stands of all descriptions in a handsome furnished drawing room! Alex putting up a huge turning machine . . . in a bedroom for turning patterns, grinding glasses & turning eye-pieces etc. At the same time Music durst not lay entirely dormant during the summer, and my Brother had frequent rehearsals at home.'[81]

Caroline was gradually becoming William's closest assistant. She was up at all hours, turning her hand to every practical need, housekeeping, shopping in the market, dealing with visiting music scholars, taking Pump Room choirs for singing practice, 'lending a hand' in the workshop, even reading aloud from inspiring fiction (in her bad accent) while William sweated over the mirror-polishing.[82] Their choice of books seems intended to relieve the monotony of the work: *Don Quixote*, the *Arabian Nights*, Sterne's *Tristram Shandy* – all tales of fantastic adventures or eccentric heroes. Caroline does not seem to have been permitted the most fantastic and eccentric of them all, William's favourite, *Paradise Lost.*♣

Sometimes she even provisioned William while he worked, literally putting drinks and bits of food into his mouth. On at least one momentous occasion, this extraordinary provisioning process lasted for sixteen hours without a break. It was as if Caroline was a mother bird feeding a demented nestling. Something of William's obsessional dedication, and Caroline's ambivalent feelings about it, come out in the way she described this in her journal: 'My time was so much taken up with copying Music

---

♣ In describing the rebel angel's enormous glowing shield, Milton contrives in *Paradise Lost* a beautiful reference to Galileo's refractor telescope and the view he achieved of the moon.

> . . . The broad circumference
> Hung on his shoulders like the Moon, whose orb
> Through optic glass the Tuscan artist views
> At evening from the top of Fiesole.
> Or in Val d'Arno, to descry new lands,
> Rivers, or mountains in her spotty globe.
> (*Paradise Lost*, Book I, lines 288–. See also
> Book III, lines 589–, and Book 5, 262–)

and practising, besides attendance on my Brother when polishing, that by way of keeping him *alife* I was even obliged to feed him by putting the Vitals by bits into his mouth – this was once the case when at the finishing of a 7 foot mirror he had not left his hands from it for 16 hours together ... And generally I was obliged to read to him when at some work which required no thinking, and sometimes lending a hand, I became in time as useful a member of the workshop as a boy might be to his master in the first year of his apprenticeship.'[83]

Much later, Victorian illustrators would make this into a comfortable domestic scene, a harmonious couple in an elegant drawing room, with convenient refreshments on a nearby table. In fact these epic polishing sessions took place downstairs, at the workbench of the unheated, stone-flagged basement in New King Street. Here William and Caroline were surrounded by tools and chemicals, and the distinct, pungent smell of the horse-dung moulds. It was dirty, monotonous and exhausting work, for which they wore rough clothes, and ignored ordinary household routines and niceties.[84]

Caroline's account is light-hearted and self-denigrating, in her usual manner, and yet faintly resentful. Her sense of herself as William's 'boy' apprentice suggests a measure of physical subordination and discipline. It also hints at an undignified negation of her sex. Here William was her 'master', not her kindly brother or patient teacher. Moreover, she saw herself as his 'first year' boy, at a time when apprenticeships normally lasted seven years. Though willingly undertaken, the work must have been frustrating and even perhaps humiliating for her. (What, for example, did she do if William needed to urinate during his epic polishing sessions?) Once again her account of the brother–sister relationship is problematic.

Milton here includes Galileo's confirmation of an imperfect, 'spotty' globe, as described in his famous treatise *The Starry Messenger* (1610). His observations of rugged lunar mountains and irregular craters proved that not all celestial objects were perfect, and so the theologians were wrong about the nature of God's creation (as well as about the movement of the earth around the sun). More subtly, Milton puts forward the notion of the moon as the earth's cosmic shield, battered by many warlike blows from meteors. A modern poet might assign that task to Jupiter. As a young man Milton claimed to have met Galileo in 1638, during his tour of Italy, and discussed the new cosmology. 'There it was that I found and visited the famous Galileo, grown old, a prisoner to the Inquisition, for thinking in astronomy otherwise than the Franciscan and Dominican licensers thought' – John Milton, *Areopagitica* (1644).

Meanwhile Herschel revealed extraordinary mechanical ability, combining the manual dexterity of a musician with almost ruthless determination and stamina. On one occasion he insisted on sharpening his instruments on the landlord's grindstone in the yard after midnight, and came back fainting, with one of his fingernails ripped off. On another, the casting exploded in the basement workshop, and a stream of white-hot metal shot across the stone floor, cracking it from end to end and nearly laming them both.

By 1774 Herschel had successfully assembled his first five-foot reflector telescope, with a home-made metal speculum mirror of six-inch diameter (about the size of a side-plate). His Observation Journal records proudly: 'December. At night I made astronomical Observations with telescope of my own construction.'[85] As if to distinguish it from the standard tubular refractors, he had a beautiful octagonal case of gleaming mahogany panels made for it by their cabinet-maker. With its bright brass eyepiece and small sighting scope, it looked like a fine piece of Georgian furniture, not unworthy of Chippendale himself.

It was immediately apparent that Herschel had created an instrument of unparalleled light-gathering power and clarity. He saw, for example, what very few astronomers even suspected: that the Pole Star – which had been the key to navigation, and the poet's traditional emblem of steadiness and singularity, for centuries – was not in fact one star at all, but *two stars*. This observation was not officially confirmed until Herschel received a letter from Joseph Banks, as President of the Royal Society, nearly ten years later, in March 1782.[86]

The first objects Herschel studied from his garden were his old travelling companion, the moon, and then two of the most prominent of the mysterious nebulae, or 'star-clouds', about which almost nothing was yet known. The first nebula was the one in the skirts of Andromeda, just visible with the naked eye as a faint primrose gaseous whorl beyond Cassiopeia; the other was in Orion, the mysterious blue star cluster, two stars down on the Hunter's sword blade. These colour-tints were immensely enhanced by Herschel's reflector, and he was soon producing wonderfully evocative colour descriptions of stars and planets. The nineteenth-century observer T.H. Webb would complain that Herschel was rather too 'partial to red tints', though whether this was a purely subjective problem, a physiological one, or down to his speculum metal being a better reflector at the long-wavelength end of the spectrum, is still open to debate. The modern Hubble images are even more cavalier about colouring deep-space objects.[87]

From the start, Herschel's observations have a note of authority, and he is ready to challenge current astronomical thinking. His Observation Journal for 4 March 1774 reads: 'Saw the lucid spot on Orion's Sword, thro' a 5½ foot reflector; its shape was not as Dr Smith has delineated in his *Optics*; tho' something resembling it ... From this we may infer that there are undoubtedly changes among the fixt stars, and perhaps from a careful observation of this spot something might be concluded concerning the Nature of it.'[88] Even at this early stage Herschel has the notion of a *changing* universe, and that nebulae might hold some clue to this mystery. Each winter between 1774 and 1780 he made detailed drawings of Andromeda and the Orion nebula to see if any alterations could be identified.[89] *

The nebulae represented a new field of sidereal or stellar astronomy. Only thirty nebulae were known in the 1740s, at the time of Herschel's birth. By the time Herschel began to study them in the mid-1770s, Charles Messier in Paris had catalogued just under a hundred. Within a decade, by the mid-1780s, Herschel would have increased this tenfold, to over a thousand nebulae.[90] No one really knew their composition, origins or distance. In general they were thought to be a few loose clouds of gas, hanging static in the Milky Way, some loose flotsam of God's creation, and of little cosmological significance. Herschel suspected that they were star clusters at immense distances, whose composition might hold a clue to an entirely new kind of universe.

---

* The Great Andromeda galaxy, now classified as M.31, is 2.8 million light years away, part of the laconically named 'Local Group' of galaxies, which includes our Milky Way. The Orion nebula, M.42, hangs just below the three stars of Orion's belt, and is a gaseous star-cluster within our own galaxy, a mere 1.6 thousand light years away, sometimes known as the Sword of Orion. The M. numbers were assigned by Herschel's contemporary, the Parisian astronomer Charles Messier, in an annual publication known as *La Connaissance des Temps*. His catalogue for 1780 held sixty-eight deep-sky objects. No astronomer yet had the least idea of the enormous distances involved, so huge that they cannot be given in terms of conventional 'length' measurement at all, but either in terms of the distance covered by a moving pulse of light in one year ('light years'), or else as a purely mathematical expression based on parallax and now given inelegantly as ' parsecs'. One parsec is 3.26 light years, but this does not seem to help much. One interesting psychological side-effect of this is that the universe became less and less easy to imagine *visually*. Stephen Hawking has remarked, in *A Brief History of Time* (1988), that he always found it a positive hindrance to attempt to visualise cosmological values.

Sometimes, to observe the northern sky, he took his telescope out into the street at the front of the house, and dictated notes to Caroline. That autumn they attended together a return series of Ferguson's astronomy lectures, given at the Pump Room by popular demand. Herschel's journal records that he was still giving eight one-hour music lessons a day, and Caroline was continuing several hours' singing practice.[91] But the music scholars were sometimes surprised by Herschel 'dropping his violin' in the middle of the last evening lesson, and jumping up to peer at some particular group of stars from the window. One startled student recalled: 'His lodgings [at New King Street] resembled an astronomer's much more than a musician's, being heaped up with globes, maps, telescopes, reflectors etc, under which his piano was hid, and the violoncello, like a discarded favourite, skulked away in a corner.' Herschel himself said that some of his pupils 'made me give astronomical instead of music lessons'.[92]

Back in Hanover, Anna and Jacob were still expressing doubts about Caroline living in England. Again, Herschel did not mention astronomy, but revealed that he had established a small millinery business on the ground floor at 5 Rivers Street, to supplement the household income, which Caroline was running successfully as well as pursuing her singing.[93] He then slipped over to reassure them in the summer of 1777, and for the first time wrote a series of confidential letters to Caroline *in English*. 'Mama is extremely well and *as I have represented things* gives her consent to you staying in England as long as you and I please. I wish very much to see my own home again [Bath], and conclude at present, remaining your affectionate Brother, Wm. Herschel . . . I hope to be in Bath about 14–16 of Sept.'[94]

Caroline was continuing her singing training, and beginning to perform regularly in Herschel's concerts at the Pump Room. But she 'could not help feeling some uneasiness', as she put it, about her future prospects, as more and more of William's time was 'filled up with Optical and Mechanical works'.[95] Once they went together to fulfil a singing engagement in Oxford, but Caroline remembered it largely for the perilous journey home, 'for the jaunt was made in a single horse Chaise, and my Brother was not famous for being a good driver'.[96]

Then William gave her ten guineas – a very considerable sum – to spend on whatever evening dress she liked, for her musical performances. She was overjoyed when the proprietor of the Bath Theatre, Mr Palmer, solemnly pronounced her to be 'an Ornament to the Stage', a compliment she never forgot.[97] On 15 April 1778 she was advertised, for the first time, as the principal solo singer in a programme of selections from Handel's *Messiah* at the Bath New Rooms. As this was Herschel's own end-of-season 'Benefit Concert', it was clearly he who promoted her. Her performance was such a success that she was offered her first solo professional engagement by a company at the Birmingham Festival for the following spring. Here at last was her chance of an independent career, at the age of twenty-eight. But after consultation with William, she turned it down, announcing that she was 'resolved only to sing in public where her brother was conducting'. Consciously or not, she had made a decision about her future with William.[98]

It may be no coincidence that the following year, 1779, Herschel began a much more serious and regular pattern of observations. He recorded: 'January. I gave up so much of my time to astronomical preparations that I reduced the number of my [music] scholars so as not to attend more than 3 or 4 a day.'[99] He had decided on his first major astronomical project: to establish a new catalogue of so-called 'double stars'.

John Flamsteed had observed over a hundred double stars, but had established no special record of them, and there were obviously many more to be found. The value of double stars was that they might provide a method of gauging the earth's distance from the rest of the Milky Way, by the measuring of *parallax*.♣

♣ As with road directions, a diagram is a much better way to explain parallax than a written sentence. But it is interesting to try. Parallax is basically a trigonometrical calculation applied to the heavens. Stellar parallax is a calculation which is obtained by measuring the angle of a star from the earth, and then measuring it again after six months. The earth's movement during that interval provides a long base line in space for triangulation. So the difference in the two angles of the same star (the parallax) after six months can be used in theory to calculate its distance. In fact single stars are so far away that they did not provide sufficient parallax to be measured with the techniques available at the time. Herschel thought that double stars might provide a more obvious parallax, by triangulating their movements against each other, as observed over six months from earth. In fact no sufficient parallax was observed until the nineteenth century, when Thomas Henderson measured the distance to the nearest star, Alpha Centauri, as 4.5 light years in 1832, and the German astronomer Friedrich Bessel measured the distance to 61 Cygni as 10.3 light years in 1838. As both published their results in 1838, there was a priority dispute. The first galactic distances were established by Edwin Hubble, using the celebrated 'red-shift' method in the 1920s. It is intriguing that towards the end of his career Hubble thought that 'red-shift' might be less reliable than he had originally supposed, and galactic distances are still an area of dispute, although the 'age' of the entire universe is currently agreed at 13.7 billion years. Andromeda, incidentally, is 'blue-shifted', and therefore approaching our Milky Way, with which it will eventually collide – or cohabit.

Although distances within the immediate solar system – to the moon and notably to the sun (using the Transit of Venus observations) – had been approximately measured, there was no general idea how far away the stars were, or what the size of the Milky Way might be. Kant, for example, assumed that Sirius (the Dog Star), because of its brightness, was probably the centre of the entire Milky Way galaxy, and possibly of the whole universe.[100] In fact it is one of our nearest stars, just over 8.7 light years away.

Most current ideas about the cosmos were small-scale. It was widely believed that the earth was a few thousand years old at most (Biblical calculations gave 6,000 years), and that the universe might stretch out a few million miles 'above' the earth. The 'fixed stars' revolved in an unchanging pattern, and their brightness or magnitude was probably a function of their size, rather than their distance. So a faint star was probably comparatively small, rather than comparatively far away – a perfectly reasonable assumption. (One of Herschel's most simple and radical ideas was to assume exactly the opposite.) The physical closeness of the stars and planets also explained their astrological 'influences'. The universe was small, closely connected, largely unchanging (except for comets), and almost intimate.

Nevertheless, the eighteenth century had been rich in speculative theories about the possibility of a 'Big Universe'. These included Thomas Wright's *Original Theory or New Hypothesis of the Universe* (1750) and Kant's *Universal Natural History of the Heavens* (1755), which first proposed – though without observational evidence – that there might be 'island universes' outside the Milky Way, that some distant stellar systems might be altering, and that the whole cosmos might be in some sense 'infinite', though it was not clear what exactly 'infinite' might mean, as hitherto it was a quality possessed only by God and mathematics. Herschel himself had added to these theoretical accounts with an early paper, eventually published by the Bath Philosophical Society, 'On the Utility of Speculative Enquiries'.

All these speculative essays assumed the high probability that extraterrestrial life existed, either within the immediate solar system, or further out among the stars. James Ferguson, for example, stated in the opening of his *Astronomy Explained* (1756) that the entire universe was evidently populated, if not positively crowded, with living forms: 'Thousands upon thousands of Suns ... attended by ten thousand times ten thousand Worlds ... peopled with myriads of intelligent beings, formed for endless progression in perfection and felicity.'[101] It was further assumed that such life forms, though not necessarily human in appearance, would have developed civilisations and sciences superior to our own. The question of whether they were 'fallen' in a religious sense, and required Redemption according to Christian doctrine, remained a moot point among astronomers, few of whom would have considered themselves as 'atheists' in any modern sense. 'An undevout astronomer is mad,' as the poet Edward Young reflected in *Night Thoughts* (1742–45).♣

However, the growing sense of the sheer scale of the universe, and the possibility that it had evolved over unimaginable time, and was in a process of continuous creation, did slowly give pause for thought. For a poet like Erasmus Darwin, in *The Botanic Garden* (1791), it put the Creator at an increasing shadowy distance from his Creation.[102]

This interest in extraterrestrial life was one of the reasons that Herschel remained so fascinated by the surface of the moon, with its mysterious mountains and craters, and dramatically shifting patterns and colours of shadow. When it was invitingly at the crescent (the best time to study surface detail), but too low to be observed from his tiny back yard, he would take his seven-foot telescope out into the cobbled street in front of the house. So it was, in December 1779, while Herschel was 'engaged in a series of observations on the lunar mountains', that a passing carriage stopped, a young gentleman sprang out, and he had his first historic meeting with Dr William Watson, junior. This was Herschel's first really important scientific contact in England, one not made until he was forty-one. Watson was only thirty-three.

♣ Young, in *Night Thoughts*, also imagined an infinitely distant planet with extraterrestrial inhabitants, as if it were some remote Pacific island, not unlike Tahiti perhaps:

> Canst thou not figure it, an Isle, almost
> Too small for notice in the Vast of being;
> Severed by mighty Seas of unbuilt Space
> From other Realms; from ample Continents
> Of higher Life, where nobler Natives dwell.
> (Edward Young, *Night Thoughts on Life, Death and Immortality*, 1742–45, 'Night IX, and Last')

7

Herschel later recalled the moment with appropriate gravity: 'The moon being in front of my house, late in the evening I brought my seven-foot reflector into the street ... Whilst I was looking into the telescope, a gentleman coming by the place where I was stationed, stopped to look at the instrument. When I took my eye off the telescope, he very politely asked if he might be permitted to look in ... and expressed great satisfaction at the view. Next morning, the gentleman, who proved to be Dr Watson, junior (now Sir William), called at my house to thank me for my civility in showing him the moon.'[103]

Caroline remembered it rather less formally. Herschel and Watson were so immediately taken with each other that very night that they burst into the house and began 'a conversation which lasted until near morning; and from that time on [Dr] Watson never missed to be waiting on our house against the hours he knew my Brother to be disengaged'.[104]

Watson warmly befriended Herschel, and encouraged his work even to the extent of helping with pounding horse-dung moulds and casting speculum mirrors. He quickly became what Caroline called '*almost* an intimate of the family'.[105] He had Herschel elected to the Bath Philosophical Society as 'optical instrument maker and mathematician' (no mention of musician), and over the next two years encouraged him to submit no fewer than thirty-one papers at its meetings. These included 'On the Utility of Speculative Enquiries', 'On the Existence of Space', and further unconventional observations on the moon. They are evidence of the extraordinary intellectual ferment that had seized upon Herschel.

His notion of the cosmos was already far from conventional, and several of these papers were what would now be called 'thought experiments'. In his 'Space' paper, delivered on 12 May 1780, he astonished his audience with his radical thoughts on time and distance: 'Huygens said that it was possible some of the fixt Stars might be so far off from us that their light tho' it travelled ever since the Creation at the inconceivable rate of 12 million of miles per minute, was not yet arrived to us. The thought is noble and worthy of a Philosopher. But [should] we call this immense distance *a mere imagination*? Can it be an *abstract Idea*? Is there no such thing as space?'[106]

In the case of his moon speculations, he raises the question whether a scientific idea has to be 'correct' to be significant. One of Herschel's most ingenious ideas was that moon craters were artificially constructed circular cities (or 'Circuses'), built especially to harness solar power for the lunar inhabitants: 'There is a reason to be assigned for circular Buildings on the Moon, which is that as the Atmosphere there is much rarer than ours and of consequence not so capable of refracting and (by means of clouds shining therein) reflecting the light of the Sun, it is natural enough to suppose that a Circus will remedy this deficiency. For in that shape of Building one half will have the direct, and the other half the reflected, light of the Sun. Perhaps, then on the Moon every town is one very large Circus?'[107] ♣

So, besides the two main projects, to record all new double stars and all new nebulae, Herschel was also embarked on a third and partly secret programme in 1779: to discover life on the moon. For some time he did not risk sending this section of the lunar paper to Maskelyne at the Royal Society, but both Watson and Caroline were aware of it. This was one of the reasons he needed to construct better telescopes.

The moon project had begun with a long entry made in his Observation Journal for 28 May 1776. He saw 'what I immediately took to be *woods* or large quantities of growing substances in the Moon'. With a certain angle of solar light, some of the lunar shadows looked like 'black soil' spread down a mountainside. Other puzzling stippled shadows, especially in the Mare Humorum, Herschel believed were enormous 'forests', made up of huge, spreading leafy canopies, or at least 'large growing substances'. Because of low lunar gravity, this gigantic 'vegetable

---

♣ This question bears on the whole nature of science history and biography. Michael Hoskin has suggested in his essay 'On Writing the History of Modern Astronomy' (1980) that most histories of science continue to be 'uninterrupted chronicles', which run along 'handing out medals to those who "got it right"'. They ignore the history of error, so central to the scientific process, and fail to illuminate science as a 'creative human activity' which involves the whole personality and has a broad social context – *Journal for the History of Astronomy* 11 (1980). To this one might add that Romanticism introduced three important themes into science biography. First, the 'Newton syndrome', the notion of 'scientific genius', in which science is largely advanced by a small number of preternaturally gifted (and usually isolated) individuals. Second, the existence of the 'Eureka moment', in which great discoveries are made without warning (or much preparation) in a sudden, blazing instant of revelation and synthesis. Third, the 'Frankenstein nightmare', in which all scientific progress is really a disguised form of destruction. See Thomas Söderqvist (ed.), *The Poetics of Scientific Biography* (2007).

Creation' was evidently 'of a much larger size on the Moon than it is here'.[108]

Similarly, he tended to believe that there were so many of the smaller moon craters that they must be artificial constructions: 'By reflecting a little on this subject I am almost convinced that those numberless small Circuses we see on the Moon are the works of Lunarians and may be called their Towns.' Nonetheless, true science required not speculation but accurate observation and telescopic proof. 'But this is no easy undertaking to make out, and will require the observation of many a careful Astronomer and the most capital Instruments that can be had. However this is what I will begin.'[109]

The light-gathering power of Herschel's seven-foot reflector allowed him to see many objects that no previous astronomers had accurately observed, or at least recorded. With Caroline taking notes at his dictation, they began to compose a new catalogue of double stars, and to develop a system of recording the exact time and position of any unusual stellar phenomena not previously catalogued by Flamsteed. By this means Herschel began to build up an extraordinary, instinctive familiarity with the patterning of the night sky, which gradually enabled him to 'sight-read' it as a musician reads a score. He would later himself use such musical analogies to explain the technique and art of observation.

In early 1781 it was decided to close down the millinery business at 5 Rivers Street. William and Caroline moved back to a substantial three-storey terraced house at 19 New King Street, where the telescope equipment was immediately set up in the fine little back garden: 'beyond its walls all [was] open as far as the river Avon'. Here, as Caroline noted modestly, 'many interesting discoveries were made'. At first she however had to remain at Rivers Street to oversee the selling off of the linen stock, and she missed the first few nights of observation in March. She subsequently recorded, with unusual care, that she did not return to New King Street until 21 March – as it turned out a historic absence.[110]

During these nights around the spring equinox Herschel was observing alone, and as well as continuing with their catalogue of double stars, he gave himself up to making drawings of Mars and Saturn. Possibly he was ranging more freely than usual, or possibly he was testing his ability to 'sight read' the sky. At all events, on Tuesday, 13 March 1781, slightly before midnight, Herschel spotted a new and unidentified disc-like object moving through the constellation of Gemini. This discovery would change his entire career, and become one of the legends of Romantic science.

It also raises an intriguing question: how soon did Herschel know – or suspect – what he had discovered? It seems from his Observation Journal at the time, that what he thought he had found was a new comet. The following laconic account appears in his 'First Observation Book' for 12 and 13 March 1781

March 12
5.45 in the morning.
Mars seems to be all over bright but the air is so frosty & undulating that it is possible there may be spots without my being able to distinguish them.
5.53 I am pretty sure there is no spot on Mars.
The shadow of Saturn lays at the left upon the ring.

Tuesday March 13
Pollux is followed by 3 small stars at about 2' and 3' [minutes of arc] distance.
Mars as usual.
In the quartile near Zeta Tauri the lowest of two is a curious either nebulous star or perhaps a Comet.
A small star follows the Comet at 2/3rds of the field's distance.[111]

There are no further remarks for these nights, and certainly no expression of excitement or anticipation. On the following night, Wednesday, 14 March, it was either cloudy, or Herschel did not bother to observe, for there is no entry. He may have been prevented by an official engagement to play the harpsichord at the Bath Theatre, or to rehearse oratorios with Caroline.[112] On 15 March there are short observations on Mars and Saturn, accompanied by some drawings of them made between 5 and 6 a.m., but nothing further about the 'curious nebulous star or comet'. On Friday, 16 March there is again no entry. But Herschel may have been reflecting on his sightings, and talking to Caroline over the weekend, for finally, on the night of Saturday, 17 March there is the first clear sign that he was definitely in pursuit of the mysterious new object.

Saturday March 17
11pm. I looked for the Comet or Nebulous Star and found that it is a Comet, for it has changed its place. I took a superficial measure 1 rev, 6 parts and found also that the small star ran along the other [cross] wire ... Position exactly measured 91'96 ...

Once Caroline had returned to New King Street on the twenty-first, there are regular entries in late March following the 'comet', and attempts to measure its diameter with William's newly designed micrometer. For example, on 28 March the Observation Book reads: '7.25 pm. The diameter of the Comet is certainly increased, therefore it is approaching.'[113] The increase in apparent size was a further indication of 'proper motion' and a solar orbit; and further proof that it could not possibly be a fixed star. But if it was a comet, there should be a slightly blurred, fiery outline and a distinct tail or 'coma'. Here Herschel's beautifully clear reflector images, even more than his high-magnification eyepieces, came into their own. In early April, some three weeks after his first sighting, Herschel made what seemed to be a definitive observation.

Friday April 6
I viewed the Comet with 460 [magnifications] pretty well defined, no appearance of any beard or tail. With 278 [magnifications] perfectly sharp and well defined.[114]

Though Herschel was scrupulously careful not to say so in his Observation Book, the sharp, round definition and the lack of any tail could only mean one thing: a new 'wanderer', or planet. What in fact he had observed was the seventh planet in the solar system, beyond Jupiter and Saturn, and the first new planet to be discovered for over a thousand years (since Ptolemy). He would name it patriotically after the Hanoverian king, 'Georgium Sidus' ('George's Star'), but it eventually became known to European astronomers as Uranus. 'Urania' was the goddess of astronomy, and the new planet was seen to mark a rebirth in her science.*

* The naming of the new planet was much disputed, and was not generally agreed until the mid-nineteenth century. Johann Bode, the editor of the authoritative *Berlin Astronomical Yearbook*, which quickly popularised the name 'Uranus', urged that a single name from classical mythology, with no national overtones, was required. With impeccable Prussian logic he pointed out that in Greek mythology Saturn (Kronos) was the

Yet there was no Eureka moment: quite the opposite. For the next few weeks there was a great deal of uncertainty about what sort of astronomical body Herschel had found. Nowhere does the word 'planet' appear in his Observation Journal for that spring of 1781, and there was no popular reporting of the news in the magazines. The following year, when the sensation was widely known, it would be very different, as Caroline remarked: 'Since the discovery of the Georgium Sidus, I believe few men of learning or consequence left Bath before they had seen and conversed with its discoverer.' But for the time being there were just endless measurements with the micrometer, 'and a fire to be kept in, and a dish of Coffee during the long nights of watching'. She added wryly: 'I undertook with pleasure what others might have thought a hardship.'[115]

On 22 March Herschel tentatively communicated his preliminary observations of 'a Comet' to William Watson, who passed them on to Nevil Maskelyne and Joseph Banks at the Royal Society.[116] Maskelyne immediately contacted other European astronomers, notably Charles Messier in Paris, asking for their opinion.[117] A week later Herschel followed this up with a direct report to the Royal Society, which was logged in the Society's 'Copy Journal Book' for 2 April. Now he expressed barely muted excitement: 'Saw the Diameter of the Comet extremely well defined and distinct; with several different powers thro' my 20 foot Newtonian reflector. It was a glorious sight, as the Comet was placed among a great number of small fixt stars that seemed to attend it.'[118]

Remembering Herschel's 'lunacies' of the previous year, Maskelyne was initially sceptical. He found great difficulty in even locating the new object with his own telescopes at Greenwich, a difficulty increased by Herschel's inability to provide the conventional mathematical coordinates. At this stage Herschel located all his stars on hand-drawn star maps – what he called 'an eye-draught' – an amateur technique that again visually recalls his familiarity with musical scores.[119] It was not until 4 April that Maskelyne wrote cautiously to Watson (still not to Herschel directly) that he had finally found the new 'star', and observed that it had just discernible 'motion'. However, he prudently, and not unreasonably, hedged his bets: 'This [the motion]

---

father of Jupiter (Zeus), and Uranus (the Greek sky god) was the father of Saturn. It is so recorded in his great *Uranographia* (1801), which became the most influential celestial atlas of the early nineteenth century, replacing Flamsteed's and cataloguing some 15,000 naked-eye stars.

convinces me it is a comet or a planet, but very different from any comet I ever read any description of or saw. This seems a Comet of a new species, very like a fixt star; but perhaps there may be more of them.' This safely covered all the options. He added a pointed postscript: 'PS I think [Herschel] should give an account of his telescope, and micrometers.'[120]

The Astronomer Royal was in a dilemma. He had no reason to accept Herschel as a reliable astronomer, and to declare a new planet prematurely might bring himself and the Royal Society into disrepute, and even ridicule. On the other hand, to reject what might be the greatest British astronomical find of the century, especially if the predatory French astronomers accepted it first (and even named it), would be even more damaging. He was also aware that Banks regarded this as a crucial moment in his presidency, and in the fostering of good relations between the Royal Society and the Crown. King George III was particularly fascinated by stars, and particularly keen to outdo the French.

Maskelyne finally chose to act as a man of science: he went back to his own telescopes, and from 6 to 22 April made his own observations. He was, after all, acting precisely according to the motto of the Royal Society itself: *Nullius in Verba* – 'Nothing upon Another's Word'. On 23 April he at last wrote directly to 'Mr William Herschel, Musician, near the Circus, Bath'. He began prudently, but ended firmly.

> *Greenwich Royal Observatory, April 23, 1781*
> Sir, I am to acknowledge my obligation to you for the communication of your discovery of the present Comet, or planet. I don't know which to call it. It is as likely to be a regular planet moving in an orbit nearly circular round the sun, as a Comet moving in a very eccentric ellipsis. I have not yet seen any coma or tail to it...

This tipped the argument towards a planet, but was not a decisive opinion. Maskelyne then went into technical details about their respective telescopes – especially the need for 'very firm stands' – and the difficulties of using micrometers to measure apparent changing diameters (and hence establish a possible planetary orbit): 'If the light of the small planet is not still, & free from scintillations, it is impossible to prove it to have any other than a spurious diameter that may arise from the faults to which the best telescopes are subject.' Nonetheless, he praised Herschel for making 'very good observations'.

Finally, in his last paragraph, he committed himself. 'On the 6th April I viewed the Comet with my 6 foot reflecting telescope and the greatest power 270, and saw it a very sensible size but not well defined. *This however showed it to be a planet and not a fixt star*, or of the same kind of fixt stars as to possessing native light with an insensible diameter. I am Sir, etc etc, N. Maskelyne.'[121]

Herschel had gained an invaluable ally. He immediately sent up a brief, masterly paper which was read at the Royal Society on 26 April. It was entitled simply 'An Account of a Comet', and was published in the *Philosophical Transactions* in June. He stated that 'between ten and eleven in the evening' of 13 March 1781 he had at once recognised a new object of 'uncommon magnitude' in Gemini, and immediately 'suspected it to be a comet'. But from the account he then gave of its magnitude, clarity of outline and 'proper motion' it was clear that Herschel was now claiming that the 'comet' was really a new planet. Though, no doubt advised by Watson, he did not actually say so. To support this, he also claimed that the object remained perfectly round, without the least appearance of comet's tail, when magnified 270, 460 and 932 times – the latter magnifications being far beyond what even Maskelyne's Greenwich telescopes could achieve. All this naturally excited even more controversy than his moon paper, and some murmurs of dissent.[122]

Maskelyne nevertheless stoutly confirmed his opinion to Banks that their dark horse, the 'musician of Bath', had made a revolutionary discovery, and had 'much merit'. Yet he could not suppress a touch of rueful irony. 'Mr Herschel is undoubtedly the most lucky of Astronomers in looking accidentally at the fixt stars with a 7 foot reflecting telescope magnifying 227 times to discover a comet of only 3' [seconds of arc] diameter, which if he had magnified only 100 times he could not have known from a fixt star ... Perhaps accident may do more for us than design could; and this makes one wish that the number of astronomers was multiplied in order to increase our chance of new discoveries.'[123] This suggestion that the discovery had been 'accidental', and that he had been 'lucky', was to grow increasingly disturbing to Herschel.[124]

Maskelyne had made public his support of Herschel just in time. On 29 April Messier wrote directly to 'Monsieur Hertsthel at Bath' from Paris, congratulating him – 'this discovery does you much honour' – and giving his opinion that this was very likely to be the seventh planet in the solar system. Messier had himself, he said modestly, discovered no fewer than eighteen comets in his lifetime, and this resembled none of them: it was 'a little planet with a diameter of 4 to 5 seconds, a whitish light like that of Jupiter, and having the appearance when seen with glasses of a star of the 6th magnitude'. He signed 'with consideration and respect' as 'Astronomer to the Navy of France, of the Academy of Sciences, France'.

As Maskelyne and Banks were only too aware, Messier's congratulations would soon carry the weight of the entire French Académie des Sciences.[125] Throughout the spring and summer months of 1781, more and more astronomers – in France, Britain, Germany, Italy and Sweden – observed the tiny moving speck, and took the view that it was indeed a planet circling in a massive ellipse beyond Saturn. These included Jacques Cassini, Henry Cavendish and Pierre Méchain. In October Anders Lexell, the celebrated Russian mathematician, wrote from his observatory far away in St Petersburg, sending a fully computed orbit and adding his congratulations. Using a series of parallax readings, he calculated that the planet was large and unbelievably remote, over sixteen times further from the sun than the earth, and twice as far out as Saturn. The size of the solar system had been doubled. Jérôme Lalande, who also computed the orbit, later said that this was the moment when the Académie des Sciences finally accepted the new planet – seven months after it had been sighted. Lalande himself suggested it should be christened 'Herschel'.

It is suggestive that it was mathematical calculation, rather than astronomical observation, which finally convinced the scientific community that a seventh planet really did exist. One of the things Lexell's calculation showed was that Herschel's vivid impression that the planet was increasing in apparent diameter throughout March and April (and therefore approaching the earth) must have been the product of his growing concentration and excitement, since it was actually getting smaller and moving away. Lexell continued to work patiently for several years on his calculations, and later came up with the revised figure of 18.93 times the distance from the sun, impressively close to the modern computer-generated figure of 19.218. (In fact, as the planet's orbit is elliptical not circular, the distance varies: at its closest it is 18.376 and at its furthest it is 20.083.)

In May, Watson proudly took Herschel up to London to meet his father Sir William, and to renew his now extremely cordial relations with Nevil Maskelyne. Together with the wealthy Deptford astronomer Alexander Aubert, they all dined with Sir Joseph Banks at the Mitre Club, the tavern much favoured by Dr Johnson. This was Herschel's first meeting with the inner circle of British astronomers, and it was a great success. There was an air of suppressed triumph and excitement. Banks, in high spirits, seized his hand, congratulated him on 'the *great* discovery', and announced that he was to be elected to the Royal Society and awarded the Copley Gold Medal forthwith – within the next fortnight![126] He claimed it as a decisive British victory over French astronomy, and the eminence of Messier, Pierre Laplace and Lalande, who had hitherto dominated European astronomy.

In fact Banks's enthusiasm had rather got the better of him. The Copley Medal and the fellowship election had to go through the Society's plodding bureaucratic procedures, which took another six months. Maskelyne used the interval to write warmly to Herschel in August: 'I hope you will do the astronomical world the favour to give a name to your new planet, which is entirely your own, and which we are so much obliged to you for the discovery of.'[127]

It was subsequently shown that 'Georgium Sidus' had actually been observed and recorded at least seventeen times between 1690 and 1781, and was even catalogued by Flamsteed. But it had always been dismissed as a minor 'fixed' star. It was only Herschel's observational genius – and the quality of his seven-foot reflector – which identified it as a large, steadily moving body in regular orbit round the sun: a true planet. And it was Maskelyne who, by promptly supporting Herschel and bringing his observations to the attention of other leading European astronomers, confirmed the discovery and had it accepted by the scientific community at large. It later became clear that Uranus was a weird blue ice giant (not 'little' as Messier thought), twice the distance of Saturn, and taking 84.3 years to complete a solar orbit. It is the only planet in the solar system which is tilted 'on its side', so its axis of rotation, or spin, is horizontal to its solar orbit.*

* It was also the first planet that was not easily visible and distinctive to the naked human eye (by colour, shape or position), and indeed it is quite frustrating to attempt to find with modern binoculars. Its presence was therefore curiously remote and mysterious, emphasising the largeness and strangeness of the new solar system (now doubled in size), but also breaking up the old, affectionate feeling for a much-loved planetary family. It is arguable that Uranus has still not fully entered into the popular mythology of the solar system, a difficulty not helped by the awkward pronunciation of its name in English, which worked better when given to the metal uranium in 1789. Herschel's son John tried to remedy this by giving Uranus's – try saying that! – two moons the feathery Shakespearean names of Titania and Oberon, from *A Midsummer Night's Dream.*

In November Banks wrote a friendly and characteristically droll letter to Herschel, asking him for details of how he made the discovery that famous night, and all the difficulties 'etc etc' it caused him. He wanted to refer to these when presenting him to the assembled members of the Royal Society in London the following month: 'Sir, The Council of the Royal Society have ordered their Annual Prize Medal to be presented to you in reward for your discovery of the new star. I must request that (as it is usual for me on that occasion to say something in commendation of the discovery) you will furnish me with such anecdotes of the difficulties you experienced etc etc ... as you may think proper to assist me in giving due praise to your industry and ability.'

Banks, in high good humour, also enjoyed putting Herschel on his mettle. '*Some* of our astronomers here incline to the opinion that it is a Planet, and not a Comet. If you are of that opinion, it should forthwith be provided with a name, or our nimble neighbours, the French, will certainly save us the trouble of Baptizing it.'[128]

Herschel, again advised by Watson, asked Banks if he could name the planet after the King, 'Georgium Sidus', a sound and self-effacing diplomatic stroke from a fellow Hanoverian.[129] But he was less easy about the continuing murmurs in some quarters of the Royal Society that his discovery had been in some sense 'accidental'. This struck at his very notion of scientific method. He wrote insistently, even angrily, to Banks just before the ceremony on 19 November: 'The new star *could not* have been found out even with the best telescopes had I not undertaken to examine every star in the heavens including such as are telescopic, to the amount of at least 8 or 10 thousand. I found it at the end of my second review after a number of observations ... The discovery cannot be said to be owing to chance only it being almost impossible that such a star should escape my notice ... The first moment I directed my telescope to the new star, I saw with a power of 227 that it differed sufficiently from other celestial bodies; and when I put on the higher powers of 460 and 932 was quite convinced it was not a fixt star.'[130]

This claim was to become a point of honour with Herschel, often repeated. In September 1782 he wrote to Lalande in Paris, stating emphatically that the discovery 'was not owing to chance'. Since he was embarked on a regular review of the sky, 'it must sooner or later fall into my way, and as it was that day the turn of the stars in that neighbourhood to be examined, I could not very well overlook it'.[131] The following year he wrote to the German astronomer Georg Christoph Lichtenberg at Göttingen, repeating that it was 'not by accident', and adding: 'when I came to Astronomy as a branch of [mathematics] I resolved to take nothing upon trust but see with my own eyes all what other men had seen before'.[132] Lichtenberg replied enthusiastically (in German): '*Mein Gott!* If I had only known, when I was for a few days in Bath in October 1775, that such a man lived there! As I am no friend of tea rooms, nor of cards or balls, I was much *ennuyéd* and spent my time at the top of the [cathedral] tower with a field glass ... '

When he came to write an autobiographic sketch for his friend Dr Charles Hutton FRS in 1809, Herschel was more insistent than ever: 'It has generally been supposed that it was a lucky accident which brought this star to my view; this is an evident mistake. In the regular manner I examined every star of the heavens, not only of that magnitude but many far inferior, it was that night *its turn* to be discovered. I had gradually perused the great Volume of the Author of Nature and was now come to the seventh Planet. Had business prevented me that evening, I must have found it the next, and the goodness of my telescope was such that I perceived its planetary disk as soon as I looked at it; and by application of my micrometer, I determined its motion in a few hours.'[133]

This claim is not entirely borne out by his original Observation Journal. His first sweep or 'review' of double stars, begun in 1779, had not revealed the Georgium Sidus, so discovery on the second was not inevitable. Nor was recognition instant when it came. The journal reveals no precise Eureka 'first moment' on 13 March, only the hardening suspicion drawn out over five days to Saturday, 17 March that the strange body had 'proper motion', but was neither a 'nebulous star' nor a 'comet', and so was very probably a new planet. But it was Nevil Maskelyne who was the first to say so explicitly in writing, in April.

Nevertheless, Herschel's discovery was an astonishing feat. It became his professional signature, and a historic moment for cosmology. It is hardly surprising that over the years he continued romantically to refine the story, and compressed his discovery into a single wondrous night, the inspired work of a glorious 'few hours'. Caroline never commented on this, although it seems clear that she was present during the critical nights of measuring between 21 March and 6 April 1781. The effect of this account was to present an engagingly romantic image of science at work: the solitary man of genius pursuing the mysterious moment of revelation.

Joseph Banks's presentation speech, when awarding the prestigious Copley Gold Medal for the best work in any scientific field during the year 1781, in front of the assembled Fellows of the Royal Society, was unreservedly complimentary to Herschel. The discovery of the new planet was the first great success of Banks's new presidency. In his most expansive and jovial mood, he accordingly projected a visionary future for Herschel's astronomy: 'Your attention to the improvement of telescopes has already amply repaid the labour which you bestowed upon them; but the treasures of heaven are well known to be inexhaustible. Who can say but your new star, which exceeds Saturn in its distance from the sun, may exceed him as much in magnificence of attendance? Who can say what new rings, new satellites, or what other nameless and numberless phenomena remain behind, waiting to reward future industry?'[134]

The award set the seal on Herschel's reputation, and reignited the general fascination with astronomy. The discovery of the seventh planet began a revolution in the popular conception of cosmology. It was widely reported in the gazettes, journals and year books published in London, Paris and Berlin at the end of 1782. Yet although all orreries were instantly out of date, it took some time for Uranus to enter into the popular imagery and iconography of the solar system.

One of the best of the new wave of popular astronomy books was John Bonnycastle's *Introduction to Astronomy in Letters to his Pupil*, which first appeared in 1786 (and went on to new expanded editions in 1788, 1811 and 1822). Bonnycastle gave the discovery of Uranus its own chapter: 'Of all the discoveries in this science, none will be thought more singular than that which has lately been made by Dr Herschell... This is a Primary Planet belonging to the solar system, which till 13th of March 1781, when it was first seen by Dr Herschell, had escaped the observation of every other astronomer, both ancient and modern...' Yet he still treated it as a puzzling novelty, its significance yet to be developed. 'This discovery, which at first appears more curious than useful, may yet be of great service to astronomy... and may produce many new discoveries in the celestial regions, by which our knowledge of the heavenly bodies, and of the immutable laws that govern the universe, will become much more extended: which is the great object of the science...'[135]

Bonnycastle's book was a thoroughly Romantic production, which included a good deal of 'illustrative' cosmological poetry from Milton, Dryden and Young. It also sported an engraved frontispiece by Henry

Fuseli. This showed the goddess of astronomy, Urania, in a diaphanous observation-dress, pointing seductively to her new star while instructing a youthful male pupil. The publisher was Joseph Johnson of St Paul's Churchyard, also the publisher of William Blake, William Godwin and Mary Wollstonecraft; and later of Wordsworth and Coleridge.

Bonnycastle was a great friend of the philosopher Godwin, and besides including poetry to illustrate his astronomical explanations, he considered the imaginative impact of the new astronomy. The 'Babylonian' writers of Egypt had increased the Biblical estimate of the earth's age from 6,000 to 400,000 years, but Bonnycastle pointed out that 'the best modern astronomers' had increased this to 'not less than 2 million years'. He thought that viewing the stars through a telescope both liberated the imagination and produced a certain kind of wonder, mixed with disabling awe or terror: 'Astronomy has enlarged the sphere of our conceptions, and opened to us a universe without bounds, where the human Imagination is lost. Surrounded by infinite space, and swallowed up in an immensity of being, man seems but as a drop of water in the ocean, mixed and confounded with the general mass. But from this situation, perplexing as it is, he endeavours to extricate himself; and by looking abroad into Nature, employs the powers she has bestowed upon him in investigating her works.'[136]

Uranus slowly became a symbol of the new, pioneering discoveries of Romantic science. An unfathomably larger universe was steadily opening up, and this gradually transformed popular notions of the size and mystery of the world 'beyond the heavens'. Indeed, the very terms 'world', 'heaven' and 'universe' began to change their meanings. It was the psychological breakthrough that Kant had predicted in his *Universal Natural History and Theory of the Heavens* back in 1755: 'We may cherish the hope that new planets will perhaps yet be discovered beyond Saturn.'[137]

Erasmus Darwin would eventually celebrate Herschel's new astronomy in his poem *The Botanic Garden* (1791), notably in the spectacular opening section of Canto 1. The discovery of Uranus inspired Darwin to evoke many other possible 'solar systems', each with its own sun and planetary family, spontaneously exploding into being after an initial 'big bang'. Here Darwin was using Newton's celestial mechanics (based on Kepler's three laws of planetary motion), but dramatising the new notion of an endless sequential creation as implied by Herschel. The creative cosmic force is 'Love' (as in the classical cosmology of Lucretius), while the Biblical God now seems content simply to initiate what is, in effect, a vast cosmological experiment, and then sit back as a passive observer.

> When Love Divine, with brooding wings unfurl'd,
> Call'd from the rude abyss the living World,
> 'Let there be Light!', proclaimed the Almighty Lord,
> Astonish'd Chaos heard the potent word;
> Through all his realms the kindling ether runs
> And the mass starts into a million Suns.
>
> Earths round each Sun with quick explosions burst,
> And second Planets issue from the first;
> Bend as they journey with projectile force,
> In bright ellipses their reluctant course;
> Orbs wheel in orbs, round centres centres roll,
> And form, self-balanced, one revolving whole.
> – Onward they move, amid their bright abode,
> Space without bound, the bosom of their God!

To this shimmering and kinetic passage, which seems to anticipate in language the music of Haydn's *Creation* (1796–98), Darwin added a long, admiring Note on 'Mr Herschel's sublime and curious account of the construction of the heavens'.[138]♣

Astronomers from all over Europe (especially France, Germany and Sweden) began to write to Herschel in Bath, asking for details about his metal specula, his high-magnification eyepieces and his observational techniques. In England there continued to be much scepticism about both his abilities and his telescopes. His replies tended to be formal, but

---

♣ *The Botanic Garden* was the best-selling long poem in English throughout the 1790s, after which its popularity suddenly declined, probably because its science was thought to be too materialist and 'French'. It was the first poem which presented itself in terms of a moving, 'photographic' image of the world: 'Gentle Reader . . . Here a Camera Obscura is presented to thy view, in which are lights and shade dancing on a white canvas, and magnified into apparent life! – if thou art perfectly at leisure for such trivial amusement, walk in and view the WONDERS of my ENCHANTED GARDEN.' Darwin's 'antique' prose style in this Prologue was an ironic foil to the dense, plain, highly informative manner of his scientific footnotes. Together these notes added up to a remarkable survey of the current state of the physical sciences in 1790.

occasionally he relaxed a little with astronomers whom he trusted, and whose skills he admired. He light-heartedly described the pains he took to set up, tune and even 'humour' his telescopes. He gave them a life of their own, and implied that he treated them like so many concert *prima donnas* (perhaps remembering La Farinelli, who had saved him at the Pump Room). To Alexander Aubert in London he wrote one of his most whimsical accounts on 9 January 1782, when enclosing his new catalogue of double stars. 'These instruments have played me so many tricks that I have at last found them out in many of their humours, and have made them *confess to me what they would have concealed*, if I had not with such perseverance and patience, *courted them*. I have tortured them with powers, flattered them with attendance to find out the critical moments when they would act, tried them with Specula of a short and long focus, a large aperture and a narrow one. It would be hard if they had not proved kind to me at last!'[139]

It is striking how frequently he now compared the art of astronomical observation to learning and playing a musical instrument. To Aubert he wrote of the need to adjust each telescope individually and 'to screw an instrument up to its utmost pitch. (As you are an Harmonist you will pardon the musical phrase.)'

Yet for some months Herschel had to continue to defend his telescopes against sceptics in the Royal Society. To the accusation that his discovery was by chance, they now added the implication that the huge powers of magnification he claimed were illusory. Particular scepticism was directed at his lens of 6,000 power, since it was calculated that a star so highly magnified would move through the viewing field of his telescope in 'less than a second', owing to the earth's rotation. Therefore it would be quite impossible to observe. Herschel replied crisply that it took all of *three* seconds, and he could follow such a star very well.[140] But to William Watson he complained that his critics evidently intended to send him 'to Bedlam', and wrote defensively on 7 January 1782: 'I do not suppose there are many persons who could even *find* a star with my [magnifying] power of 6,450; much less keep it if they had found it. Seeing is in some respects an art, which must be learnt. To make a person see with such a power is nearly the same as if I were asked to make him play one of Handel's fugues upon the organ. Many a night have I been practising to see, and it would be strange if one did not acquire a certain dexterity by such constant practice.'[141]

Watson quietly kept Banks informed of the controversy, while Banks gently temporised, suggesting that perhaps the magnifications were slightly miscalculated, but supporting Herschel against his detractors. He sent smiling presidential greetings: 'My best Compliments to Mr Herschell, with best wishes for the Sake of Science that his nights may be as Sleepless as he can wish them himself.'[142]

Alexander Aubert now firmly took Herschel's side. Thanking him for the catalogue of double stars, he remarked appreciatively on all the trouble Herschel had taken: 'but trouble is nothing to you, and the least thing we can do in return is to ... convince the world that though your discoveries are wonderful, they are not *imaginary*... Your great power of 6450 continues to astonish, your micrometer also ... Go on, my dear Sir, with courage, mind not a few barking, jealous puppies; a little time will clear up the matter and if it lays in my power you shall not be sent to Bedlam alone, for I am much inclined to be one of the party.'[143]

Herschel's next destination, as it turned out, was not Bedlam but Windsor. King George III, advised by the Astronomer Royal and the President of the Royal Society, had chosen to ignore these controversies. He summoned Herschel to court to congratulate him, but asked Banks and Maskelyne to make an independent trial of the now celebrated seven-foot telescope at the Greenwich Observatory. On 8 May Herschel left for London, his precious telescope and folding stand perilously packed into a mahogany travel-box ('to be screwed together on the spot where wanted'), accompanied by a hastily assembled trunk of equipment including his large Flamsteed atlas (marked up by Caroline), his new catalogue of double stars (similarly written up by Caroline), 'micrometers, tables, etc', and rather makeshift court dress.[144]

At Greenwich, Maskelyne was stunned by the superior quality and light-gathering power of Herschel's 'home made' mirrors. He immediately recognised that they were far more powerful than any of the official observatory telescopes, and probably than any other telescope in Europe. Maskelyne, reputed to be a jealous and illiberal man because of his supposed ill-treatment of the watchmaker John Harrison, behaved with great forthrightness and generosity to Herschel.

On 3 June 1782 Herschel wrote exuberantly to Caroline, casting aside his usual circumspect tone: 'Dear Lina ... The last two nights I have been star-gazing at Greenwich with Dr Maskelyne & Mr Aubert. We have compared our telescopes together and mine was found very superior to any at the Royal Observatory. Double stars they could not see with their instruments I had the pleasure to show them very plainly, and my [folding stand] mechanism so much approved of that Dr Maskelyne has already ordered a model to be taken from mine; and a stand to be made by it for his reflector. He is however now so much out of love with his instrument [a six-foot Newtonian] that he begins to doubt whether it *deserves* a new stand.'[145]

Banks (who had learned much about royal decorum since Tahiti) now knew that it was the perfect moment to introduce Herschel formally to the King at Windsor in May 1782. The meeting between the two Hanoverians (commoner and king, but both firmly speaking English) was a great success. Members of the King's Hanoverian entourage had already heard of the Herschel brothers as talented musicians, and His Majesty was intrigued by the change in *métier*.[146] King George, not yet mad, was renowned for his aphoristic remarks to his more talented subjects. To Edward Gibbon, for example, still deep in his six-volume history *The Decline and Fall of the Roman Empire*, he had observed archly: 'Scribble, scribble, scribble, eh, Mr Gibbon?' It was said that the King now murmured to Banks: 'Herschel should not sacrifice his valuable time to *crotchets and quavers*.'[147]

Herschel wrote swiftly to Caroline, with a note of growing excitement that had never previously appeared in his letters. 'Among Opticians and Astronomers nothing is now talked of but *what they* call my great discoveries. Alas! This shows how far they are behind, when such trifles as I have seen and done are called *great*. Let me but get at it again! I will make such telescopes & see such things – that is, I will endeavour to do so.'[148] In a later note, again using her intimate diminutive name, he added: 'You see Lina I tell you all these things, you know vanity is not my foible therefore I need not fear your Censure.'[149] He would not have feared his sister's censure a decade before.

Banks was determined to find his new astronomical protégé a salary, and if possible a suitable place. This required some diplomacy, as university professorships were for mathematicians, the post of Astronomer Royal was evidently taken, and the new post of Royal Astronomer at Kew Gardens had recently been promised to another – 'a devil of a pity'. With Banks's diplomatic nudging, the King agreed that Herschel should give up teaching music in Bath, and move to a house near Windsor, to concentrate entirely on astronomy. To achieve this, His Majesty would be pleased to create a new official post, appointing Herschel as the King's Personal Astronomer at Windsor on a salary of £200 per annum. (This was not particularly generous, but then the Astronomer Royal received only £300.) At the age of forty-three, Herschel's second career had burst into life.

After the very briefest consultation, Herschel, Caroline and their brother Alexander moved on 31 July 1782 to a large, sprawling house in the village of Datchet, positioned deep in the countryside between Slough and Windsor, just north of the river Thames. The house had large grass plots suitable for erecting telescopes, and several stables and outbuildings for the furnaces and the grinding and polishing equipment. An old laundry could be converted into an observation building. But the house itself had not been inhabited for several years, and was cold and damp. Caroline set about the huge task of cleaning and repairing.[150]

Almost immediately Herschel was commanded to bring his famous seven-foot telescope to Windsor, where it was reassembled on the terrace for everyone to view the planets. Herschel was a particular success with the three teenaged royal princesses, Charlotte, Augusta and Elizabeth. On one cloudy evening (it being an English summer) when viewing was impossible, he had the inspired idea of constructing pasteboard models of Jupiter and its four moons, and Saturn and its rings, and hanging them – illuminated by candles – from a distant garden wall on the Windsor estate. These were meticulously prepared beforehand. By ingeniously focusing down the seven-foot, he was able to show these models to the three young girls through the telescope, an early form of outdoor planetarium.[151]

Many other children of the new generation also grew up understanding the cosmos in a new way. Discovering the stars became a particular and special moment of self-discovery. The poet Coleridge remembered being taken out at night into the fields by his beloved father, the vicar and schoolmaster of Ottery St Mary in Devon, in the winter of 1781 to be shown the night sky. Coleridge was only eight, but he never forgot it. Perhaps the Reverend John Coleridge, a great follower of the monthly magazines (to which he sometimes contributed learned articles on Latin grammar), had recently read of Georgium Sidus. At all events, Coleridge treasured the memory of his father's eager demonstration of the stars and planets overhead, and the possibility of other worlds: 'I remember, that at eight years old I walked with him one evening from a farmer's house, a mile from Ottery – & he told me the names of the stars – and how Jupiter was a thousand times larger than our world – and that the other twinkling stars were Suns that had worlds rolling round them – & when I came home, he showed me how they rolled round. I heard him with profound delight & admiration; but without the least mixture of Wonder or incredulity. For from my early reading of Faery Tales, & Genii etc etc – my mind had been *habituated to the Vast*.'[152]

Such a huge, starlit prospect, inhabited by giant planets and remote classical gods, might have puzzled or alarmed a normal eight-year-old. But the striking thing is that Coleridge, who wrote many letters about his childhood and always remembered it acutely, said he felt no surprise or disbelief at all – 'not the least mixture of Wonder or incredulity' – about this revelation of the enormous scale of the universe. He felt himself already tuned to the size and mystery of the new cosmos. His Romantic sensibility – even at the age of eight – already inhabited the infinite and the inexplicable. Cosmological imagery, and especially the symbolic movement of the stars and the moon, entered deeply into his early poetry, and in a sense it came to rule the world of the Ancient Mariner and his ship.

> The moving Moon went up the sky,
> And nowhere did abide;
> Softly she was going up,
> And a star or two beside.
>
> Her beams bemocked the sultry main,
> Like April hoar-frost spread,
> But where the ship's huge shadow lay,
> The charmed water burnt alway
> A still and awful red.[153]

The prose gloss that Coleridge added to this passage almost twenty years later (1817) takes on a new resonance when compared with what we now know of Herschel's long nights of lunar observation:

> In his loneliness and fixedness he yearneth towards the journeying Moon, and the stars that still sojourn, yet still move onward; and every where the blue sky belongs to them, and is their appointed rest, and

> their native country and their own natural homes, which they enter unannounced, as Lords that are certainly expected and yet there is silent joy at their arrival.*

The young John Keats remembered an organised game at his school in Enfield, in which all the boys whirled round the playground in a huge choreographed dance, trying to imitate the entire solar system, including all the known moons (to which Herschel had by then added considerably). Unlike Newton's perfect brassy clockwork mechanism, this schoolboy universe – complete with straying comets – was a gloriously chaotic 'human orrery'.

Keats did not recall the exact details, but one may imagine seven senior boy-planets running round the central sun, while themselves being circled by smaller sprinting moons (perhaps girls), and the whole frequently disrupted by rebel comets and meteors flying across their orbits. Keats was later awarded Bonnycastle's *Introduction to Astronomy* as a senior school prize in 1811. Reading of Herschel, he enshrined the discovery of Uranus five years later in his great sonnet of 1816, 'On First Looking into Chapman's Homer'.[154]

---

* Moon and star imagery recurs in Coleridge's poetry throughout his life. One of his earliest known poems, 'To the Autumnal Moon', was a sonnet written at the age of sixteen from the lead rooftop of his London school. Many of his great West Country poems, such as 'Frost at Midnight' (1798), may be said to be suffused with moonlight. Greta Hall, where Coleridge lived at Keswick, was an old observatory, and from its leads he frequently recorded the state of moon, stars and the night sky, as well as his own little boy Hartley's comments on them. His famous poem 'Dejection' (1802) begins with the image of the 'winter-bright' new moon, with 'the old Moon in her lap', presaging a storm. When later living alone at Malta, he used a naval telescope to observe the moon and stars, and wrote many notebook entries about his inexplicable instinct to worship the moon (1805). Even such a late poem as 'Limbo', probably written at Highgate, dramatises himself as an old man gazing up at the moon in a garden. He is blind – 'a statue hath such eyes' – yet mysteriously he can still sense the moonlight pouring down on him like a benediction:

> He gazes still – his eyeless face all eye –
> As 'twere an organ full of silent sight,
> His whole face seemeth to rejoice in light!
>
> (Coleridge, *Selected Poems*, Penguin Classics, p.214)

These seem to me three of the most mysterious, moonstruck lines that Coleridge ever wrote. Perhaps he was imagining himself transformed into a sort of human telescope.

8

Once they had moved to Datchet, Herschel and his brother Alexander started an exclusive business in the manufacture of high-quality reflector telescopes. The first five of them, all seven-foot reflectors, were ordered by King George as royal gifts, and although never fully paid for by the Crown office (they were priced at a hundred guineas each), they had the invaluable effect of making Herschel the royal telescope-maker, 'By Appointment'. All telescopes, whatever their size, were individually constructed to order, took three or four weeks to make, and had an individual price, usually quoted in guineas. Herschel would supply them either in kit form or fully assembled in beautiful mahogany cases, with spare mirrors and a selection of eyepieces. Although every one was handcrafted, his immense energy achieved something like mass-production. Over the next decade he made 200 mirrors for the popular seven-foot telescope, 150 for the ten-foot, and eighty for the big twenty-five-foot, although not all of these were sold.[155]

Prices rose steadily. The renowned seven-foot telescope was usually sold in kit form for thirty guineas, but Herschel gradually raised even the kit price to a hundred guineas, the figure he quoted to the German astronomer Johann Bode in Berlin.[156] Eventually a twenty-foot in kit form sold for 600 guineas. The luxury ten-foot reflector, complete with polished mahogany case, patent adjustable stand, a selection of eyepieces and a spare mirror, cost a princely 1,500 guineas.[157] Indeed the more expensive models were sold mostly to German princes, and models also went to Lucien Bonaparte (Napoleon's brother) and the Emperor of Austria.[158] Probably the most expensive commercial telescope that Herschel ever made was commissioned by the King of Spain for £3,500, and delivered to the Madrid Observatory in 1806.[159] Scores of Herschel's telescopes were eventually sent all over England and Europe, and he personally delivered one on behalf of King George to the state observatory at Göttingen in 1786.[160]

Gradually more and more visitors began to descend on the observatory at Datchet. Caroline started to keep a neat, double-columned visitors' book, rather as if she were recording star observations, which in a sense she was. In spring 1784 the dying Dr Johnson sent the young Susannah Thrale (Mrs Thrale's third daughter) on a visit, advising her to cultivate an acquaintance with Herschel: 'He can show you in the night sky what no man before has ever seen, by some wonderful improvements he has made in the telescope. What he has to show is indeed a long way off, and perhaps concerns us little, but all truth is valuable and all knowledge pleasing in its first effects, and may subsequently be useful.'[161]

Caroline wrote vivid accounts of their routine of all-night star observations, or 'sweeps'.[162] Herschel's technique of 'sweeping' did not – as the term seems to imply – involve moving the telescope laterally, which was always a tricky operation with the bigger reflectors. Instead it was kept on the meridian, and moved slowly up and down, while the constellations turned through the field of observation as the stars moved steadily across the night sky. As this motion is caused by the earth itself rotating on its polar axis, so the telescope is effectively 'sweeping' the heavens like some immensely long broom, or the finger of a searchlight. By this method Herschel could progressively cover the entire night sky in a series of small strips, each covering about two degrees of arc.[163] The technique was far more accurate than any other stellar observation that had ever been undertaken before in the history of astronomy. But it was also immensely slow and painstaking. A complete sweep could take several years to complete.

During this time Herschel became so familiar with every part of the sky that he could identify stellar patterns, and any new objects, with amazing speed and precision. Perhaps his musical training helped him here, as much as his painfully self-taught mathematics. As he suggested himself, he could read the night sky like a skilled musician sight-reading a musical score. Or more subtly, the brain that was trained to recognise the highly complex counterpoints and harmonies of Bach or Handel could instinctively recognise analogous stellar patternings.

Herschel became fascinated by both the physics and the psychology of the observation process itself, and later wrote some of his most fascinating papers about it. From 1782 he began to record the many physical tricks his eyes could play, and also began to study the illusions of night observation. On 13 November, while trying to identify a new double star in Orion, he dictated a careful note to Caroline:

> Following 10 Orionii. I saw very distinctly double at least a dozen times pass through the whole field of view with both eyes, but was obliged to darken everything. I suspected my right eye to be tired, & know it to see objects darker. Therefore tried the left first, & saw it immediately pass

> thro' the field double several times. Saw the same afterwards with the other eye ... No star twinkled except Syrius, & those as low. The evening exceptionally fine for telescopes.[164]

The more he was challenged by professional astronomers, the more Herschel became conscious of his 'art of seeing', and how it needed explaining afresh. 'The eye is one of the most extraordinary Organs,' he repeatedly told his correspondents. Classical physiology was wrong. Visual images did not simply fall upon the optic nerve, in the same sense that they fell upon a speculum mirror. The eye constantly interpreted what it saw, especially when using the higher powers of magnification. The astronomer had to *learn* to see, and with practice (as with a musical instrument) he could grow more skilful: 'I remember a time when I could not see with a power beyond 200, with the same instrument which now gives me 460 so distinct that in fine weather I can wish for nothing more so. When you want to practise seeing (for believe me Sir, – to use a musical phrase – you must not expect to *see at sight or a livre ouvert*) apply a power something higher than what you can see well with, and go on increasing it after you have used it for some time.'[165]

Caroline later assembled an index of all Herschel's remarks on practical observation. Under 'Trials of Different Eyes and Seeings' she listed such topics as the distortion effect of 'looking long at an object etc', the need to progress from lower to higher powers of magnification, the fact that 'different eyes judge differently of [the same] colours', that 'eyes tire' without the observer noticing, and that 'we see things always smaller at first, when difficult to be seen'.[166]

Under another heading, 'Airs and Situations', she listed the particular locations and atmospheric conditions which affected a telescope. These were not always self-evident. The atmosphere itself had 'prismatic powers', and distortions could be produced by 'field breezes', viewing 'over the roof of a house', or standing 'within 6 or 8 feet of a door'. Surprisingly, because of thermal ripples rising from the ground, 'evenings tho' apparently fine, are not always good for viewing'. By contrast, 'moist air was favourable', and damp or rain, even certain kinds of fog, were 'no hindrance to seeing'. It was possible to observe in conditions of severe frost, or even falling snow, provided the mirrors were kept clear of ice.[167]

Caroline gave the term 'sweeping' a certain domestic familiarity, so that in her letters she sometimes implies she is a sort of celestial housekeeper, brushing and dusting the stars to keep them in a good state for her brother, a sort of heavenly *Hausfrau*. But perhaps she also had deeper feelings about the cosmos she was now discovering. It was no longer a mere hobby to please him. Once they had moved to Datchet, in the summer of 1782 Herschel began to train her more carefully in observation techniques, so she could become a genuine 'assistant-astronomer'. By way of encouragement he built her a special lightweight sweeper, consisting of 'a tube with two glasses' (i.e. a traditional refractor), and instructed her 'to sweep for comets'.

Initially she found working on her own in the dark rather daunting. 'I see from my Journal that I began August 22nd 1782, to write down and describe all remarkable appearance I saw in my sweeps, which were horizontal. But it was not till the last two months of the same year that I felt the least encouragement to spend the star-light nights on a grass-plot covered with dew or hoar frost, without a human being near enough to be within call.'[168] Besides, at this early stage Caroline knew 'too little of the real heavens to be able to point out every object so as to find it again without losing too much time by consulting the Atlas'. As all novice astronomers find, stars move disconcertingly rapidly through a telescopic field of vision, even that of a low-powered telescope, and can easily slip away in the few moments spent consulting a star chart and then readjusting one's eyes to night vision. (Night vision can take as long as thirty minutes to establish its full sensitivity.)

Clearly things were better for Caroline when Herschel was on hand in the garden, and not away at Windsor doing royal demonstrations. 'All these troubles were removed when I knew my brother to be at no great distance making observations with his various instruments on double stars, planets etc, and I could have his assistance immediately when I found a nebula, or cluster of stars.' In this first year Caroline found no comets, and only succeeded in identifying fourteen of the hundred or so known nebulae. She was too often interrupted by Herschel's imperious shout, when he wanted her to write down some new observation made with the large twenty-foot.[169]

Such teamwork was essential to the sweeping procedure that the Herschels developed. As William made his observations, he would call out precise descriptions of what he saw (with special attention to double stars, nebulae or comets). He would give magnitudes, colour and approximate distances and angles (using a micrometer) from other known stars within the field of view. Standing below him in the grass, and later sitting at a folding table, Caroline would meticulously note all this data down, using pen and ink and a carefully shrouded candle lantern, and consulting their 'zone clock' (a clock using a time scale related to the position of the stars, rather than the sun). Alexander Aubert would later give them a magnificent Shelton clock, with compensated brass pendulum, as a contribution to their work.[170]

With Herschel, this was not tranquil or contemplative work, as might be supposed. Caroline would 'run to the clocks, write down a memorandum, fetch and carry instruments, or measure the ground with poles etc etc of which something of the kind every moment would occur'.[171] Sometimes she would call back questions, asking for further clarifications. Most importantly she would note the exact time of each observation, using the special zone clock, which would give a precise position as each object rotated through the meridian. By this method, at no point would William have to compromise his night vision by looking at a lit page and taking his own notes.

Herschel described their sweeping methods in a paper published in April 1786, 'One Thousand New Nebulae'. Crucial to his technique was that he did not have to take his eye away from the lens, but could 'shout out' his observations while his assistant wrote them down and 'loudly repeated' them back to him. This had 'the singular advantage', as he put it, 'that the descriptions were actually writing and repeating to me while I had the object before my eye, and could at pleasure correct them'. The distinct tone of military command was emphasised by the fact that nowhere in this paper did Herschel mention that his assistant was Caroline.[172]

Standing under a night sky observing the stars can be one of the most romantic and sublime of all experiences.* But the Herschels' sweeps were

---

* It can also be oddly terrifying. A hundred years later, Thomas Hardy took up amateur astronomy for a new novel, and in his description of Swithin and Lady Constantine sharing a telescope in *Two on a Tower* (1882) he captured something of the metaphysical shock of the first experience of stellar observation. 'At night . . . there is nothing to moderate the blow which the infinitely great, the stellar universe, strikes down upon the infinitely little, the mind of the beholder; and this was the case now. Having got closer to immensity than their fellow creatures, they saw at once its beauty and its frightfulness. They more and more felt the contrast between their own tiny magnitudes and those among which they had recklessly plunged, till they were oppressed with the presence of a vastness they could not cope with even as an idea, and which hung about them like a

fantastically prolonged and demanding. In clear weather, they would often go on for six or seven hours without a break. They began at eleven at night, and often did not go to bed before dawn, in a mixed state of exhaustion and euphoria. Both slept till midday, and the house had to be kept quiet most of the morning, although Caroline often seems to have been up early, drinking coffee and writing up the night's observations in long, minute columns of figures: a sort of double book-keeping which she often referred to as 'minding the heavens'.

Observations and note-making required dogged precision and absolute concentration. It could be chill even in summer, and in winter the frost covered the grass around them, and the wind moaned through the trees. (Nevil Maskelyne had a special woollen one-piece observation suit made for him at Greenwich, with padded panels that made him look like a premonition of the Michelin Man.) Herschel took to rubbing his face and hands with raw onions to keep out the cold. When Banks came down to join them he sometimes brought oversize shoes so he could wear half a dozen pairs of stockings inside them. Caroline layered herself in woollen petticoats. Frequently it was so cold that films of ice formed on the telescope mirrors, the ink clotted in the well, and frozen beads blunted the tip of Caroline's quill.[173]

It could also be dangerous. Caroline wrote: 'I could give a pretty long list of accidents of which my Brother as well as myself narrowly escaped of proving fatal for observing with such large machineries, where all around is in darkness [and] is not unattended with danger; especially when personal safety is the last thing with which the mind is occupied at such times.'[174] The winter of 1783 was especially harsh. On one night in November that year, when William was mounted high up on the cross-bar of his twenty-foot reflector, the wind almost blew him off, and when

---

nightmare.' My own first experience with a big telescope, the 'Old Northumberland' at Cambridge Observatory, an eleven-inch refractor built in 1839, left me stunned. We observed a globular star cluster in Hercules, a blue-gold double star, Beta Cygni, and a gas cloud nebula (whose name I forgot to record, since it appeared to me so beautiful and malignant, according to my shaky notes like 'an enormous blue jellyfish rising out of a bottomless black ocean'). I think I suffered from a kind of cosmological vertigo, the strange sensation that I might fall down the telescope tube into the night and be drowned. Eventually this passed. The great Edwin Hubble used to describe an almost trance-like, Buddhist state of mind after a full night's stellar observation at Mount Wilson in California in the 1930s. See Gale Christianson, *Edwin Hubble* (1995).

he hastily clambered down the rickety structure ('the ladders had not even the braces at the bottom'), the entire wooden frame collapsed around him; workmen had to be called to release him from the wreckage of spars.[175]

On 31 December 1783, New Year's Eve, over a foot of snow had fallen, and the sky was overcast. William however postponed celebrations, and insisted on the last sweep of the year. Caroline gives the impression that he was particularly impatient, and perhaps shouting at her more than usual. 'About 10 o'clock a few stars became visible, and in the greatest hurry all was got ready for observing. My Brother at the front of the Telescope [was] directing me to make some alterations in the lateral motion.' As she hurried round the base of the telescope, 'having to run in the dark on ground covered foot deep in melting snow', she slipped and tripped over a hidden wooden stake. These stakes were used to peg down the telescope frame with guy ropes, and had large iron hooks facing vertically upwards, 'such as butchers use for hanging their joints on'.

Caroline painfully recounted what followed. 'I fell on one of these hooks which entered my right leg about six inches above the knee. My brother's call – *make haste!* – I could only answer by a pitiful cry – *I am hooked!*' She was impaled, like a fish on a barb, and could not move. Herschel was still high up on the observation platform, in complete darkness, and did not immediately realise what had happened. It seems he continued to call down through the dark, '*Make haste!*', while Caroline continued to gasp back in agony, '*I am hooked!*'[176]

Finally he grasped the situation, and called for help from the assistant who had been adjusting the telescope frame. 'He and the workman were instantly with me, but they could not lift me without leaving near 2 oz. of my flesh behind. The workman's wife was called but was afraid to do anything.' Caroline was carried back to the house, but astonishingly no doctor was called. She bandaged the wound herself, retired to bed, and proudly recorded that she was back on telescope duties within a fortnight. It seems that the extreme cold had an antiseptic effect on the large, open wound, and prevented fatal gangrene.

No doubt it was characteristic of Caroline to treat this wound lightly, and not make any fuss. Yet there is an uneasy sense throughout her account that William did not treat her with sufficient tenderness or care: 'I was obliged to be my own surgeon by applying acquabaseda and tying a kerchief about it for some days.' The local Windsor physician, Dr James

Lind, only heard about the accident a week later, 'and brought me ointment and lint and told me how to use it'. The deep wound did not heal easily, but there is still no mention of William's concern at any point. Eventually Dr Lind was called back to Datchet in early February 1784. 'At the end of six weeks I began to have some fears about my poor Limb and had Dr Lind's opinion, who on seeing the wound found it going on well; but said, if a soldier had met with such a hurt he would have been entitled to 6 weeks nursing in a hospital.'[177] It is curious that Dr Lind compared Caroline to someone in military service, and it is hard to overlook a certain note of reproach in his words.♣

Caroline surely intended some irony when she added in the *Memoir*: 'I had however the comfort to know that my Brother was no loser through this accident for the remainder of the night was cloudy and several nights afterwards afforded only a few short intervals favourable for sweeping, and until 16 January before there was any necessity for exposing myself for a whole night to the severity of the season.'

The wound had largely healed by the summer, but it would later return to give her chronic pain in old age. Her pitiful cry – '*I am hooked!*' – is curiously symbolic of her relations with her brilliant, domineering brother at this period, at a time when he was obsessed by his astronomical ideas to the exclusion of all else. Including, it might seem, his sister's well-being; although we have only her word for this.[178]

It is hardly surprising that Herschel was a little distracted. In 1784 and 1785 he drew together his most radical ideas about the cosmos, and published two revolutionary papers in the Royal Society's *Philosophical*

♣ Dr James Lind (1736–1812) was no ordinary physician. A Fellow of the Royal Society, he had been invited to accompany Captain Cook on his second voyage round the world, but instead visited Iceland with Banks, and later voyaged to China. Deeply read in classical sciences – an expert on Pliny and Lucretius – he became a physician to the royal household, and taught modern sciences part-time at Eton. He was renowned for his eccentricity and kindness. One of his last pupils was Percy Bysshe Shelley, who was delighted by his radical talk of Franklin, Lavoisier, Herschel, Davy and Godwin. Shelley made Dr Lind the scientific teacher-sage of two of his longer poems of 1817, 'Prince Athanese' and 'The Revolt of Islam'. He later told Mary Shelley: 'This man is exactly what an old man ought to be. Free, calm-spirited, full of benevolence, and even youthful ardour: his eye seems to burn with supernatural spirit beneath his brow, shaded by his venerable white locks . . . I owe that man far, ah! far more than I owe to my father.' See Richard Holmes, *Shelley: The Pursuit* (1974). Young Dr Lind was clearly the sort of man who would have admired Herschel, but he also greatly sympathised with Caroline.

Transactions. These completely transformed the commonly held idea of our solar system being surrounded by a stable dome of 'fixt stars', with a broad 'galaxy' or '*via lactae*' (meaning a 'path or stream of milk') of smaller, largely unknown stars spilt across it, roughly from east to west. This was a celestial architecture or 'construction', inspired fundamentally by the idea of a sacred temple, which had existed from the time of the Babylonians and the Greeks, and had not seriously been challenged by Flamsteed or even by Newton.[179]

'An Investigation of the Construction of the Heavens', published in June 1784, quietly set out to change this immemorial picture. It was based on all Herschel's ceaseless telescope observations, relentlessly pursued with Caroline over two years, with his new twenty-foot reflector telescope. He had identified 466 new nebulae (four times the number recently confirmed by Messier), and for the first time suggested that many, if not all, of these must be huge independent star clusters or galaxies *outside* our own Milky Way.[180]

This led him on to propose a separate, three-dimensional shape to the apparent flat 'milk stream' of the Milky Way. His proposal was based on his new method of 'gauging' the number of stars in any direction as seen from the earth, and then deducing from the different densities observed the likely shape of this galactic star cluster *as it would be seen looking 'inwards' from another galaxy*. This was a daring mixture of observation and speculation. Herschel's first galactic diagram appeared like a curious oblong box or tilting parallelogram of stars.[181] But his later calculations produced the now-familiar discus shape of the Milky Way, with its characteristic arms spinning out into space, and the slight bulge of stars at its centre.[182] He was never sure where the solar system was located in the galaxy, and at one point observed that its overall shape was relative, depending on the view as seen by 'the inhabitants of the nebulae of the present catalogue . . . according as their situation is *more or less remote from ours*'.[183]

In the second paper, called simply 'On the Construction of the Heavens' (1785), Herschel began to develop these ideas into a startling new 'natural history' of the universe. He opened by arguing that astronomy required a delicate balance of observation and speculation. 'If we indulge a fanciful imagination and build worlds of our own . . . these will vanish like Cartesian vortices.' On the other hand, merely 'adding observation to observation', without attempting to draw conclusions and explore 'conjectural views', would be equally self-defeating.[184]

His own conjecture would be radical. The heavenly 'construction' was not something architecturally fixed by the Creator, but appeared to be constantly changing and even evolving, more like some enormous living organism. His telescopes seemed to show that all gaseous nebulae were actually 'resolvable' into stars. They were not amorphous zones of gas left over from the Creation. They were enormous star clusters scattered far beyond the Milky Way, and were dispersed throughout thc universe as far as his telescopes could penetrate. The nebulae themselves were active. Their function seemed to be that of constantly forming new stars out of condensing gas, in a process of *continuous creation*. They were replacing stars which were lost.

Herschel found a memorable phrase for this astonishing speculation: 'These clusters may be the *Laboratories of the universe*, if I may so express myself, wherein the most salutary remedies for the decay of the whole are prepared.'[185] He also pursued the possibility that some nebulae may be 'island universes' *outside* the Milky Way, thereby hugely increasing the sense of the actual size of the cosmos. Among these was the beautiful nebula in Andromeda, 'faintly red' at the centre. By 1785 his nebulae count had risen to well over 900. They appeared 'equally extensive with that which we inhabit [the Milky Way] ... yet all separate from each other by a very considerable distance'.[186] He picked out at least ten 'compound nebulae' which he considered larger and more developed than the Milky Way, and imagined the star-cluster view of our own galaxies from theirs. 'The inhabitants of the planets that attend the stars that compose them must likewise perceive the same phenomena. For which reason they may also be called Milky Ways by way of distinction.'[187]

As Kant had speculated, the cosmos might be infinite, *whatever that might mean*. Though Herschel's estimates of cosmological distances were much too small by modern calculation, they were outlandishly, even terrifyingly, vast by contemporary standards. Beyond the visible parts of our own Milky Way, he estimated that a huge surrounding 'vacancy' of deep space existed, 'not less than 6 or 8 thousand times the distance of Sirius'. He admitted that these were 'very coarse estimates'. The implications seemed clear, though they were cautiously expressed in his paper: 'This is amply sufficient to make our own nebula a detached one. It is true, that it would not be consistent confidently to affirm that we were an *Island Universe* unless we had actually found ourselves everywhere bounded by the ocean ... A telescope with a much larger aperture than my present one [twelve inches], grasping together a greater quantity of light, and thereby enabling us to see further into space, will be the surest means of completing and establishing the argument.'[188]

The dramatic implications of these ideas were soon picked up by journalists and popularisers. The following year Bonnycastle assessed the situation in the first edition of his *Introduction* to *Astronomy*: 'Mr Herschel is of the opinion that the starry heaven is replete with these nebulae, and that each of them is a distinct and separate system, independent of the rest. The Milky Way he supposes to be that particular nebula in which our sun is placed; and in order to account for the appearance it exhibits, he supposes its figure to be much more extended towards the apparent zone of illumination than in any other direction ... These are certainly grand ideas, and whether true or not, do honour to the mind that conceived them.'[189]

Also contained in Herschel's revolutionary paper of 1785 were the seeds of a new, long-term project. He was planning the building of a monster forty-foot telescope, with a *four-foot mirror*. This would be the biggest and most powerful reflector in the world. With this he believed he could resolve once and for all the problem of the nebulae – whether they were other galaxies far beyond the Milky Way, or merely gas clouds within it. He would also have a better chance of establishing the true distance of the stars, through the measurement of stellar parallax. Above all he believed he would be able to understand how the stars were created, and whether the whole universe was changing or evolving according to some definite law or plan. Finally, he believed he might establish if there were observable signs of extraterrestrial life, a discovery which would have enormous impact on philosophical and even theological beliefs.

There was one other small, but revolutionary, departure in his 1785 paper. For the first time William Herschel carefully credited Caroline in print with a small 'associate nebula' in Andromeda. It was a previously unknown cluster 'which my Sister discovered on August 27 1783 with a Newtonian 2 foot sweeper'. It was not in Messier's annual catalogue *La Connaissance des Temps*, so this was Caroline Herschel's first new addition to the universe.[190]

3

Balloonists in Heaven

1

Herschel's international success was a great encouragement to Joseph Banks at this time. He was still privately suffering from the loss of his friend Solander, and was being harassed by internecine intellectual feuds at the Royal Society (particularly among the unworldly mathematicians). It was with no little relief that in August 1783 he began to receive secret reports in Soho Square of strange rumours from Paris about the possible existence of a French flying machine.

The dream of flight had haunted men – especially poets, satirical writers and impractical fantasists – since the myth of Icarus. European literature was full of unlikely bird machines, flapping chariots, flying horses and aerial galleons. None of them were remotely practicable. But this was something quite different: a giant 'aerostat' powered by 'inflammable air'. What was more, it was being seriously investigated by the French Academy of Sciences, under the redoubtable Marquis de Condorcet.

Banks's most reliable informant was the wily old American Ambassador to France, Benjamin Franklin, a corresponding member of the Royal Society, and now aged seventy-seven, a shrewd judge of both men and machines. After seven years at the Embassy in Paris, Franklin was still a francophile and an enthusiast, and had just delivered a sparkling report on the craze for mesmerism, or 'Animal Magnetism'. He had noted that Anton Mesmer had earned 20,000 louis d'or 'by this pretended new Art of Healing'.[1]

Franklin now wrote to Banks describing something even more fantastic: a series of aerial experiments involving very large paper bags. These bags had apparently been flown in the open air to a considerable height by a paper manufacturer near Lyons in June that year. The paper-maker, whose name was Joseph Montgolfier, was bringing his bags to Versailles to give a public demonstration before King Louis XVI. It was scheduled to be held in the large courtyard in front of the palace that September.

All this sounded fantastical enough. But then Franklin reported that a member of the Académie des Sciences, Dr Alexandre Charles, had stolen a march on Montgolfier by inflating a silk bag with the newly discovered 'inflammable air' and launching it in public from the Champs de Mars on 27 August. What was astonishing was the lifting power of this simple device. The silk bag or 'balloon', although it was only six feet in diameter, had quickly risen so high that it could no longer be seen. It had crossed the Seine and travelled fifteen miles outside Paris before it burst. This was a distance which a horseman could barely cover in a hour at the gallop.

Then Franklin reported that Montgolfier and his brother Étienne had successfully launched their own balloon from Versailles on 11 September. Unlike Charles's gas balloon, it was powered by hot air, it was very big, and it was beautifully decorated with heraldic symbols. Moreover, its lifting power was spectacular. In a wicker cage attached to the neck of the bag it had carried a sheep, a duck and a cockerel (the French national symbol) right over the rooftops of Versailles, and had stayed aloft for seven minutes. All the animals had returned to earth alive and well.

It was clear what would happen next. Either the Montgolfiers or Charles would try to send a man up in a balloon. The prospect was amazing, and nothing else was talked about in France. Franklin thought that balloons might eventually 'pave the way to some discoveries in Natural Philosophy of which at present we have no conception'. He instanced the examples of 'magnetism and electricity, of which the first experiments were mere matters of amusement'.[2]

Initially Banks wrote back sceptically. 'I see an inclination in the more respectable part of the Royal Society to guard against the *Ballomania* until some experiment like to prove beneficial either to society or science is proposed.' Nevertheless, he conceded by mid-September 1783 that with the Montgolfiers' 'Aerostatic Experiment' at Versailles, the French had 'opened a Road in the Air', and this might mark a new 'Epoch'. If further experiments proved successful, then 'The immediate Effect it will have upon the Concerns of Mankind [will be] greater than anything since the invention of Shipping.'[3]

Paradoxically enough, Banks's first conception of balloon transportation was a thoroughly earthbound one. He saw the balloon as 'a counterpoise to Absolute Gravity': that is, as a flotation device to be attached to traditional forms of coach or cart, making them lighter and easier to move over the ground. So 'a broad-wheeled wagon' normally requiring eight horses to pull it might need only two horses with a Montgolfier attached. This aptly suggests how difficult it was, even for a trained scientific mind like Banks's, to imagine the true possibilities of flight in these early days.[4]

2

Banks was very conscious that the discovery of a lighter-than-air gas had actually been achieved by two English chemists, Henry Cavendish and Joseph Priestley. They had called it 'inflammable air' because of its lightness and explosive properties. Priestley's *Experiments on Different Kinds of Air* had been translated in France in 1768. All the experiments had then been repeated and refined by their rival, the great French chemist Antoine Lavoisier, in Paris. He had measured the buoyancy of this '*gaz*' (a word not yet coined in English) more accurately, and renamed it 'hydrogen'. But no one had manufactured it on a large scale, or realised its dramatic practical applications.

The Montgolfier brothers were commercial paper manufacturers from Annonay, near Lyons, in the Ardèche. They were an effective business team. Joseph was the shrewd entrepreneur, and Étienne was the madcap inventor. They were interested in chemistry for commercial reasons. They had followed Priestley's and Lavoisier's work, and had speculated about putting lighter-than-air gas into paper containers. As early as 1782, Joseph had humorously suggested the theoretical possibility of flying an entire French army into Gibraltar, and seizing it from the English. The troops would fly suspended beneath hundreds of huge paper bags.[5]

Lavoisier's 'hydrogen' was produced by passing sulphuric acid over iron filings. It was one-thirteenth of the weight of common air, and consequently could produce a powerful lift if sufficiently concentrated in a light container (Cavendish had used soap bubbles). But it was slow and dangerous to produce, potentially explosive, and easily escaped from containers made of silk or animal bladders. Hot air, on the other hand, was easily produced by any kind of controlled fire, and could be temporarily contained in inflated silk or paper. It produced a short-term lift, as heat agitates the air molecules, making them move apart and become more buoyant than the surrounding cooler, denser air (and at best about half its weight). This lift was however less powerful than that of hydrogen, was easily dissipated, and consequently required much bigger balloons to sustain the same power of ascent, or carry the same payload.

Joseph Montgolfier later said he had tried Lavoisier's '*gaz*' unsuccessfully, but discovered the principle of hot air by watching his wife's chemise inflating when she hung it over the hearth to dry.[6] He made several small experimental 'aerostats', finally adopting a pear-shaped balloon, with a wide neck that could be lowered over a fire. The Montgolfiers described it memorably as 'putting a cloud in a paper bag'.

On 5 June 1783, they successfully launched their first large paper balloon in open country outside Annonay. It was probably intended as a piece of advertising for their paper business, and it was a dazzling sight. When inflated, their balloon stood thirty foot high, 110 feet in circumference, and took eight men to hold it down. It was crudely constructed of painted silk sections backed with coarse paper and simply buttoned together. In fact it contained no hydrogen gas at all, but simply 22,000 cubic feet of hot air collected from braziers burning straw and damp wool. French hot air proved to have enormous lifting power. When released it rose gracefully to an estimated 6,000 feet, barely visible, and remained aloft for ten minutes.[7]

Perhaps most significant of all, it drew an enormous crowd of onlookers. This ability of the balloon to attract attention and pull large numbers of people has always remained part of its mystique, and an important part of its history. Montgolfier had discovered a scientific principle quite as interesting as that of aerial buoyancy. With ballooning, science had found a powerful new formula: chemistry plus showmanship equalled crowds plus *wonder* plus money. Reports of the flight travelled throughout France, and the Montgolfiers were soon invited to give official demonstrations, first at Versailles and then in Paris. The Marquis de Condorcet, the head of the Académie des Sciences, appointed a committee to investigate the invention and consider sponsoring its development. It assembled France's leading men of science, including Lavoisier and Claude Berthollet.[8]

Now there was the feeling of urgency, even of a race. People began contacting the Montgolfiers, applying to the Académie, or publicly volunteering to be 'the first aerial traveller in the world'. One was a young inventor from Normandy, Jean-Pierre Blanchard, who had already been experimenting with a number of winged flying machines, most notably a flying tricycle. He announced boldly in the *Journal de Paris*: 'Within a very few days I shall be ready to demonstrate my own aerostatic machine, which will climb and dive on command, and fly in a straight line at a constant altitude. I shall be at the controls myself, and have sufficient confidence in my design to have no fear of repeating the fate of Icarus.'[9]

Another, better-connected but no less enthusiastic candidate, was a young Parisian doctor, Jean-François Pilâtre de Rozier. Pilâtre was a professor of natural philosophy who ran a private science museum and college in the rue Saint-Honoré. He was twenty-nine years old. He had invented a gasmask, a hydrogen blowtorch, and a new theory of thunder – all of which seemed equally relevant to ballooning. A small, neat, energetic figure of infectious charm, he was a considerable ladies' man. He had good contacts within the Académie des Sciences and the Ministry of Finance, and some said particularly with 'Madame' (the Comtesse de Provence, Louis XVI's sister-in-law). He would soon be pursued by a number of intellectual aristocrats, such as Madame de Saint-Hilaire. But with his charm went extraordinary *sangfroid* – and of course a head for heights. Pilâtre proved himself fearless and precise during the most alarming experiments, and soon made himself indispensable to the Montgolfiers. He had, in effect, invented the new profession of test pilot. He had the right stuff.[10]

On 21 November 1783 the first manned Montgolfier balloon was launched from the hill of La Muette. This was a commanding site just above the river Seine at Passy, opposite the Champs de Mars (where the Eiffel Tower now stands). The hot-air balloon was enormous, a monster: seventy feet high, and gloriously decorated in blue, with golden mythological figures. It was powered by a six-foot open brazier burning straw. Its chosen 'aeronauts' – another new French term – were Pilâtre de Rozier and an elegant infantry officer, the Marquis d'Arlandes, a major in the Garde Royale. D'Arlandes was selected for his court connections, his enthusiasm and his wealth; and also simply because the Montgolfiers needed a 'counterweight'. Since Pilâtre was to be carried aloft in a circular gallery slung around the open neck of the balloon (and not in a basket), his weight had to be constantly balanced by a second aeronaut on the opposite side. D'Arlandes became therefore, by default, the first co-pilot as well as the first aerial stoker.

D'Arlandes subsequently published a laconic account of their historic voyage, which took them low over Paris for about twenty-seven minutes. The Montgolfier (as the balloon was now known) initially rose to some 900 feet, drifted across the Seine, and then began a series of slow swoops across the rooftops of Saint-Germain, narrowly missing the towers of Saint-Sulpice, rising again over the wooded parkland of the Luxembourg, and finally sinking rapidly onto the Buttes aux Cailles (near the present Place d'Italie in the 13th arrondissement), narrowly missing two windmills.

Because of the circular structure of the gallery, with the neck of the balloon (and the brazier) in the centre, the aeronauts could barely see each other during the flight. This produced a kind of black comedy which was to become familiar in later ascents. Pilâtre spent much of his time calling to the invisible d'Arlandes to stop admiring the view of Paris and stoke the brazier. 'Let's work, let's work! – If you keep gaping at the Seine, we'll be swimming in it soon.'

In fact d'Arlandes seems to have been increasingly (and not unnaturally) overcome by nerves. First he thought the balloon was on fire, then that the canopy was separating from the gallery, and finally that one after another the balloon cords were breaking. He constantly shouted back at the unseen Pilâtre, 'We must land now! We must land *now!*' When the whole balloon shook with a sudden gust of wind above Les Invalides, d'Arlandes screamed at Pilâtre: 'What are you doing! *Stop dancing!*'

Characteristically, Pilâtre ignored these protests, and calmly went on telling d'Arlandes to work at feeding the brazier. He himself took off his bright green topcoat (put on for the crowd), rolled up his sleeves, and went on throwing on straw till his wooden fork broke. Once, when d'Arlandes was desperately shouting at him, 'We must go down! We must go down!', Pilâtre called back soothingly: 'Look d'Arlandes. *Here we are above Paris.* There's no possible danger for you. *Are you taking this all in?*' Many witnesses later said that they could hear the two men shouting excitedly to each other as they passed overhead. They assumed they were describing the glories of flight.

Nonetheless, it was d'Arlandes who had the courage and honesty to record all these exchanges, and to describe his companion, in a phrase that became celebrated, as '*l'intrépide Pilâtre, qui ne perd point la tête*' – the intrepid Pilâtre, who never loses his head. When they landed, d'Arlandes vaulted out of the circular gallery, expecting the huge collapsing canopy to burst into flames at any moment. As he ran anxiously round the outside of the balloon, he found Pilâtre standing quietly contemplating the great gold and blue dome as it finally settled back to earth. 'We had enough fuel to fly for an hour,' was all he said. Pilâtre was holding their basket of provisions, with his green topcoat neatly folded and placed on top. A few moments later a wild, cheering crowd of *le petit peuple de Paris* (not yet *citoyens*) gathered round them. Pilâtre handed them the basket of provisions to celebrate, but they also seized the green topcoat, and tore it into little pieces as souvenirs.[11]

3

This was all very picturesque, and is the 'first flight' that has gone down in the history books. But in fact the Montgolfier was a crude and virtually uncontrollable monster. A far more significant ascent followed just ten days later, when Dr Alexandre Charles made the first ascent in a true hydrogen balloon.

Charles pioneered a number of technical breakthroughs. They included an elongated wickerwork basket safely suspended on ropes beneath the canopy; an impermeable balloon skin made of silk coated in rubber and enclosed in netting; a controllable gas-valve at the top of the balloon for venting; and, most important of all, a finely tuned system of ballast bags filled with sand which could be jettisoned by the kilo or by the gram, precisely as required by the aeronaut. Dr Charles had in effect invented nearly all the features of the modern gas balloon in a single brilliant design.

He launched from the Tuileries Gardens in Paris on 1 December 1783, with a scientific assistant, M. Robert. They attracted what has been estimated as the biggest crowd in pre-Revolutionary Paris, upwards of 400,000 people, about half the total population of the city.[12] It was a glorious pink-and-yellow, candy-coloured balloon, thirty feet tall, and the crowd loved it. The wickerwork basket, a sort of *chaise longue* for two, was completely festooned with flags and bunting. Dr Charles had a full payload of scientific equipment aboard – mercury barometer (which was used as an early form of altimeter), thermometer, telescope, sandbags and several bottles of champagne. In a nice gesture, he handed the release cord to Joseph Montgolfier: 'Monsieur Montgolfier, it is for you to show us the way to the skies!'

Dr Charles later recalled his feelings as the balloon lifted above the trees of the Tuileries and across the Seine. 'Nothing will ever quite equal that moment of total *hilarity* that filled my whole body at the moment of take-off. I felt we were flying away from the Earth and all its troubles for ever. It was not mere delight. It was a sort of physical ecstasy. My companion Monsieur Robert murmured to me – I'm finished with the Earth. From now on it's the sky for me! Such utter calm. Such immensity!'[13] Benjamin Franklin, American Ambassador in Paris, watched the launch through a telescope from the window of his carriage. Afterwards he remarked: 'Someone asked me – what's the use of a balloon? I replied – *what's the use of a newborn baby?*'

Two hours later they landed twenty-seven miles away at Nesle, skimming across a field and chased by a group of farm workers, 'like children chasing a butterfly'. Once the balloon was secured, in a moment of euphoria Dr Charles asked M. Robert to step out of the basket. Released of his weight, and with Charles alone aboard, the balloon rapidly relaunched and climbed into the sunset, reaching the astonishing height of 10,000 feet in a mere ten minutes. One thousand feet per minute: a truly formidable and terrifying ascent. Dr Charles kept calmly observing his instruments, and making notes until his hand was too cold to grasp the pen. 'I was the first man ever to see the sun set twice in the same day. The cold was intense and dry, but supportable. I had acute pain in my right ear and jaw. But I examined all my sensations calmly. I could *hear myself living*, so to speak.'

He began gently to release the hydrogen gas-valve. Within thirty-five minutes he was safely back on *terra firma* – a term that took on new meaning – alighting a mere three miles from his first landing point. His ascent had been almost vertical. It was the first solo flight in history. 'Never has a man felt so solitary, so sublime,– and so utterly terrified.' Dr Charles never flew again.[14]

Public excitement was huge in France that winter. The Musée de l'Air now at Le Bourget has many display cabinets of balloon memorabilia: plates, cups, clocks, ivory draughts pieces, snuffboxes, bracelets, tobacco pipes, hairclips, tiepins, even a porcelain bidet with a balloon design painted on the interior carrying a flag marked 'adieu'. Many sexually suggestive cartoons soon appeared: the inevitable balloon-breasted girls lifted off their feet, monstrous aeronauts inflated by gas enemas, or 'inflammable' women carrying men off into the clouds.[15]

The science writers Faujas de Saint-Fond and David Bourgeois both published handbooks to the science of flight in 1784. Bourgeois opened ecstatically: 'The idea of taking to the air, of flying through sky, and navigating through the ether, has always appealed so strongly to mankind, that it has appeared in numerous classical legends and folktales from the remotest antiquity. The wings of Saturn, the eagle of Jupiter, the peacocks of Juno, the doves of Venus, the winged horses of the Sun all bear witness ... ' He did not mention Icarus.[16] His list of the innumerable benefits of ballooning included weather prediction, telescopc observation of the stars, geographical exploration ('he will cross burning deserts, inaccessible mountains, impenetrable forests, and raging torrents'), military reconnaissance and heavy cargo carrying.[17]

All sorts of ingenious theories about how a balloon might be steered were also proposed: by enormous oars, by wings, by hand-cranked propellers, spinning '*moulinets*', silk-covered paddles, and even giant bellows.

4

In England, George III formally wrote to the Royal Society asking if research into 'air-globes' should be sponsored by the British Crown, or left to private individuals. An enterprising Swiss chemist, Aimé Argand, had released an eighteen-inch hydrogen balloon from the terrace at Windsor Castle on 26 November 1783, first getting the King himself to hold the string and feel the tug. Intrigued, George offered to put up money from his own funds to finance some early experiments.[18] He received a cautious reply from Sir Joseph Banks, who still felt that there was inadequate experimental evidence for balloons' utility. The French, he seemed to imply, were always inclined to mistake novelty for real science.[19] This reaction was very unlike that of the French Académie des Sciences, who were determined to sponsor Pilâtre de Rozier in further ascents and larger balloons, seeing all sorts of possibilities, both commercial and military.

In fact Banks could see the revolutionary nature of the science, but still doubted the technological application. A week after he had received reports of Dr Charles's spectacular demonstration of the first hydrogen balloon, he wrote privately to Franklin in Paris. 'Dr Charles's experiment *seems* decisive ... Practical Flying wc must allow to our rivals. Theoretical Flying we claim ourselves ... Mr Cavendish when he blew soap bubbles of his Inflammable air, evidently performed the [same] experiment which carried Dr Charles on [his] memorable flight.' Banks thought that when the French – 'our Friends on your side of the water' – had 'cooled a little' in their naïve enthusiasm for ballooning, they would realise what advances the English were making in the penetration of the skies through another method – astronomy. Astronomy promised a far greater knowledge of 'the repositories of stars and meteors'. Franklin – 'the old fox' – must have thought this an oddly evasive response; but then he did not know Herschel's plans for the giant forty-foot telescope, whereas Banks did.[20]

Banks said the Royal Society would keep a watching brief, while remaining closely informed of developments in 'the new Art of Flying' by its corresponding Fellows such as Franklin and the English Ambassador to Paris, the Duke of Dorset.[21] Yet Banks himself, still the Romantic explorer, was secretly intrigued and excited. He alerted Henry Cavendish and commissioned his confidant and Secretary, Charles Blagden – a decided francophile – to keep a close eye on developments. Banks also noticed that his sister Sophia had begun to keep an album of balloon cuttings. It included an early street ballad, 'The Ballooniad', which eloquently complained: 'Ye Men of Science! How ye stood aloof/Nor gave of all your Knowledge one kind proof.'[22]

English opinion was generally divided about ballomania. Samuel Johnson had written a 'Dissertation on the Art of Flying' in Chapter 6 of *Rasselas* in 1759. His approach was satirical – his Flying Artist flaps his wings and falls off a cliff into a lake – but he recognised the power of flight over the human imagination: 'How easily shall we trace the Nile through all its passages; pass over to distant regions and examine the face of Nature, from one extremity of the Earth to the other.'[23] Yet when Johnson was asked his opinion by a female correspondent, he at first described the balloon in as deflating a manner as he could muster.

> Happy are you, Madam, that have ease and leisure to want intelligence of air balloons. Their existence is, I believe, indubitable, but I know not that they can possibly be of any use. The construction is this. The chemical philosophers have discovered a body (which I have forgotten, but will enquire) which dissolved by an acid emits a vapour lighter than the atmospherical air. This vapour is caught, among other means, by tying a bladder compressed upon the bottle in which the dissolution is performed; the vapour rising swells the bladder, and fills it. The bladder is then tied and removed, and another applied, till as much of this light

air is collected, as is wanted. Then a large spherical case is made (and very large it must be) of the lightest matter that can be found, secured by some method like that of oiling silk against all passage of air. Into this are emptied all the bladders of light air, and if there be light air enough, it mounts into the clouds, upon the same principle as a bottle filled with water, will sink in water, but a bottle filled with aether would float. It rises till it comes to air of equal tenuity with its own, if wind or water does not spoil it on the way. Such, Madam, is an air balloon.[24]

William Herschel's friend William Watson witnessed one of Pilâtre's preparatory unmanned test flights at Versailles in October 1783. Even though the great wallowing Montgolfier canopy had got caught in some nearby trees, Watson was thrilled by the prospect of regular manned flight, and wrote enthusiastically to the earthbound astronomer, proposing a joint ascent as soon as possible. 'Don't you expect to fly soon? I expect to make many a pleasant flight to Datchet. I forgot to say the machine was 70 foot high and 46 wide.' Herschel immediately thought of the possible use of balloons as observation platforms, carrying telescopes into the clear upper air. It was a development which would eventually lead to the launch of the great orbiting Hubble Space Telescope in 1990.[25]

Surprisingly, balloons did not appeal to the gothic novelist Horace Walpole, though perhaps at sixty-six he was a little old for such perilous novelties. He thought balloons might be sinister: 'Well! I hope these new *mechanic meteors* will prove only playthings for the learned and idle, and not be converted into new engines of destruction to the human race – as is so often the case of refinements or discoveries in Science. The wicked wit of man always studies to apply the results of talents to enslaving, destroying, or cheating his fellow creatures. Could we reach the moon, we should think of reducing it to a province of some European kingdom.' It was an ominous prophecy.[26]

Some considered that there might be an arms race in balloon technology. Franklin could see that balloons might easily be adapted for military purposes. Reconnaissance was the obvious one: 'elevating an Engineer to take a view of an Enemy's army, Works etc. or conveying Intelligence into, or out of, a besieged Town'. Much more menacing, however, especially for the British Isles, was the possibility that they could support an airborne invasion army from France. 'Five thousand balloons capable of raising two men each', Franklin calculated, could carry a force of 10,000 troops rapidly into the field, crossing rivers, hills or even seas with speed and impunity. 'They could not cost more than five Ships of the Line ... Ten thousand Men descending from the Clouds might in many places do an infinite deal of mischief, before a [regular] Force could be brought together to repel them.'[27]

Nevertheless, neither Benjamin Franklin, nor Dr Johnson, nor Horace Walpole could prevent the balloon craze reaching England by summer 1784. Small unmanned gas balloons began to sprout everywhere in the summer sky. Herschel saw them over the Thames Valley, Parson Woodford saw them in Suffolk. Gilbert White wrote a beautiful description of seeing an early manned balloon drifting serenely over his beech wood one idyllic October evening at Selborne in Hampshire: 'From the green bank at the S.W. end of my house saw a dark blue speck at a most prodigious height...In a few minutes it was over the maypole; and then over the fox on my great parlour chimney; and in ten minutes behind my great walnut tree. The machine looked mostly of a dark blue colour; but sometimes reflected the rays of the sun, and appeared a bright yellow. With a telescope I could discern the boat, and the ropes that supported it. To my eye this vast balloon appeared no bigger than a tea-urn.'

White's initial moment of excitement and pure wonder soon altered to a more reflective mood, as the great balloon drifted southwards across Hampshire: 'I was wonderfully struck at first with the phenomenon; and, like Milton's "belated peasant", felt my heart rebound with fear and joy at the same time. After a while I surveyed the machine with more composure, without that awe and concern for my two fellow-creatures, lost, in appearance, in the boundless depths of the atmosphere! (for we supposed then that *two* were embarked on this astonishing voyage). At last, seeing with what steady composure they moved, I began to consider them as secure as a group of storks or cranes, intent on the business of emigration.'[28]

Unmanned and then manned ascents took place in almost every large city in the kingdom – London, Oxford, Cambridge, Bristol, Edinburgh. Dr Johnson's friend the musicologist Dr Charles Burney, father of the novelist Fanny, had a typical reaction: 'I tell my grandchildren they will live to see a regular Balloon Stage [coach] established to all parts of the Universe that have ever been heard of.'[29]

The *Morning Herald* asked its readers to 'laugh this new French folly out of existence as soon as possible'. But the normally conservative *Gentleman's Magazine* described ballooning as 'the most magnificent and most astonishing discovery made – perhaps since the Creation'.[30] Horace Walpole thought it as puerile as schoolboy kite-flying, but then ordered his servants to alert him whenever a balloon flew by, and rushed out into his garden to cheer and wave. 'How posterity will laugh at us one way or the other! If half a dozen break their necks, and Balloonism is exploded, we shall be called fools for having imagined it could be brought to use. If it should be turned to account, we shall be ridiculed for having doubted.'[31]

5

The man who popularised ballooning in Britain more than any other was a twenty-five-year-old Italian, Vincent Lunardi (1759–1806), a young man on the staff of the Neapolitan Legation in London.

Lacking official sponsorship, Lunardi's first remarkable achievement was to launch a successful public subscription. He had his gorgeous red-and-white-striped balloon put on display for several weeks before the launch, hung from the roof of the Lyceum Theatre near The Strand, and charged an ambitious entrance fee. Two shillings and sixpence would purchase a single visit; one guinea would purchase four visits and a front-row seat during the actual launch. Over 20,000 people were said to have visited it, though after payment for balloon equipment, inflation materials and hire of the Lyceum, Lunardi claimed to be penniless.

As interest grew, ballooning quickly became fashionable, and there was talk of an unofficial British Balloon Club, headed by the Prince of Wales and the ultra-progressive Georgiana, Duchess of Devonshire. Several members of the Royal Society also subscribed, and significantly the subscription was headed by none other than Joseph Banks, though in his private capacity. (A guinea entrance ticket is preserved in the Banks Collection, marked number 34.) His sister, the independent-minded Sophia Banks, also admitted to mild ballomania.[32] She made a collection of balloon prints and letters about ballooning, including several from Franklin and Joseph Priestley.[33]

Lunardi's second achievement was to invent for the English the figure of the Romantic aeronaut. Lunardi was a natural showman. He was foreign, of course, but not French. Small, mercurial, and absurdly handsome in the new, almost feminine style, with a fresh face and long, unpowdered hair, he moved lightly and bubbled with infectious enthusiasm. He was a man for whom the adjective 'intrepid' seemed specially invented. He had his portrait painted with his pet dog and cat, both of which he then took on his flight, a sporting gesture calculated to appeal to the English. (Though not, as it turned out, to Horace Walpole.[34]) He was also an incorrigible flirt and ladies' man, as the English naturally expected of an Italian. He once mildly shocked a salon of supporters by proposing a toast to himself: 'I give you me, Lunardi – *whom all the ladies love.*'

His first historic ascent was made from the Artillery Grounds, Moorfields, London, on 15 September 1784. It captured the nation's imagination almost as completely as had the ascents in France. After delays that almost led to rioting, 150,000 people watched the launch at 2 p.m., just two hours late. Led by the Prince of Wales, the gentlemen in the reserved one-guinea seats rose to their feet, and stood gazing upwards in astonished silence. Then they solemnly doffed their hats.

Lunardi drifted north-westwards across London and into Hertfordshire, eating legs of chicken and drinking champagne, and occasionally trying to 'row' his balloon with a pair of aerial oars. One of the oars broke and dropped overboard, starting a rumour that he had jumped out to his death. It was said that the King broke off a cabinet meeting with his Prime Minister, William Pitt the Younger, to watch 'poor' Lunardi float overhead, while a jury in north London hastily brought in a not-guilty verdict so it could run out of the courthouse to watch.

After some time Lunardi's little cat appeared to be suffering from the cold, and he claimed to have briefly 'paddled' his balloon back to earth at North Mimms (now on the M1 motorway). He gallantly handed the shivering animal to a young woman in a field, before releasing ballast and re-ascending. This is a mysterious claim, as unlike Dr Charles, Lunardi had not designed his first hydrogen balloon with a release valve at the top of the canopy, so he could not descend at will (and certainly not by rowing). He had however designed a system of throwing out handfuls of feathers, to tell if the balloon was rising or sinking, and perhaps he had simply lost gas.

Farm labourers harvesting in the fields recalled him shouting through his silver speaking-trumpet. They answered: 'Lunardi, come down!' He threw out several letters, tied with long streamers, one of which was tactfully addressed to 'Sir Joseph Banks, Soho Square, London'.

After two and a half hours, Lunardi finally descended near Ware in Hertfordshire. He tried to land by securing a grappling anchor, but bumped heavily and inelegantly across the fields. With no release valve, he could not deflate the balloon, and his situation became perilous. He called out to some nearby farm workers to help him secure the balloon. But seeing him bounding over hedges and fences, they shouted out that he was riding 'the Devil's horse', and refused to approach. Then happily he spotted a young woman among the group. Graciously raising his hat, he begged for her assistance. Ignoring her terrified menfolk, she gathered up her skirts and darted forward, seized the edge of the errant basket, and saved both balloon and aeronaut. Lunardi climbed out and embraced her tenderly. She was a strong girl, he recalled: 'Elizabeth Brett, a very pretty milkmaid...So I owed my deliverance to the spirit and generosity of a young female.'[35]

A stone monument was raised at this landing place, at Long Mead (field or farm) in the parish of Standon, just outside Ware. It still exists on what is now the village green.

> Let Posterity know, And knowing be astonished! That on the 15th day of September 1784, Vincent Lunardi of Lucca in Tuscany, the First Aerial Traveller in Britain, Mounting from the Artillery Ground in London, and traversing the Regions of the Air, For two hours and fifteen minutes, on this Spot revisited the Earth. On this rude Monument, for [future] Ages be recorded that Wondrous Enterprize, successfully achieved by the powers of Chymistry and the Fortitude of Man, that improvement in Science which the Great Author of all Knowledge... hath generously permitted.

Immediately on Lunardi's return to London, a curiously modern publicity machine began to roll. He sold exclusive rights to his story, and an in-depth interview, to the *Morning Post.* It was headlined 'Lunardi's Aerial Excursion'.[36] He was guest of honour at the Mansion House, and gave lectures in various public halls. Newspaper articles, popular songs (many ribald) and fashion accessories followed. Cups, snuffboxes and brooches were especially popular, but the Lunardi ladies' garter was the *succès de scandale.* Lunardi was introduced to the King, and invited to dine by the Duchess of Devonshire – he tactfully arrived wearing the duchess's own jockey colours of blue and chocolate,[37] and was soon a favourite in her progressive Whig circle. He was given a watch by the Prince of Wales, and had a bronze medallion struck with his profile on one side and his balloon on the other. The Windsor stagecoach was renamed 'The Lunardi'. A master of publicity, he arranged to have a new and bigger striped balloon hung on display at the Pantheon, London, throughout the winter season of 1784, promising further aerial adventures in 1785. The effect of sudden celebrity was as heady as the actual ascent. Lunardi wrote wildly to his Italian guardian: 'I am the idol of the whole nation ... All the country adores me, every newspaper honours me in prose and verse ... Tomorrow I shall put two thousand crowns in the Bank of England.'[38]

Among the more serious opinion-formers, many like Sir Joshua Reynolds, Charles Burney and the MP William Windham (all members of Dr Johnson's Club) were impressed by Lunardi's achievement. Burney wrote a charmingly over-enthusiastic letter to his son Charles junior on 24 September. 'If I had wit enough, or energy of mind sufficient to be *mad* about anything now it would be about *Balloons*. I think them the most wild, Romantic, pretty playthings for grown Gentlemen that have ever been invented, and that the subject, as well as the thing, lifts one to the Clouds, whenever one talks of it.'[39]

Others were less dazzled. Banks wrote privately that Lunardi was 'a charlatan'. Horace Walpole was wittily underwhelmed by the whole thing: 'I cannot fill my Paper as the [newspapers] do, with air balloons; which, though ranked with the invention of Navigation, appear to me as childish as the flying of kites by schoolboys. I have not stirred a step to see one; consequently, have not paid a guinea for gazing at one, which I might have seen by looking up into the air. An Italian, one Lunardi, is the first Airgonaut that has mounted into the Clouds in this country. So far from respecting him as a Jason, I was very angry with him: he had full right to venture his own neck, but none to risk the poor cat's.'[40]

In the end Dr Johnson himself became strangely fascinated by ballooning, though critical of the surrounding showmanship and the lack of scientific rigour. He wrote several letters on the subject in autumn 1784. Two days before Lunardi's flight, he was advising a friend that it was not worth paying for a place in the launch enclosure at the Artillery Ground (Lunardi was charging a guinea a seat), because 'in less than a minute they who gaze at a mile's distance will see all that can be seen'. But he took a surprising and critical interest in technical matters. He thought (rightly) that Lunardi's aerial oars would prove useless in directing flight or altering altitude. 'About the wings, I am of your mind they cannot at all assist it, nor I think regulate its motion.'[41]

Immediately after Lunardi's flight of 15 September, Johnson was amused, and then irritated, to receive no fewer than three long letters or 'Histories' recounting the details of the 'Flying Man in the great Balloon'. He wrote ironically to Sir Joshua Reynolds that he would have been content with just one. 'Do not write about the Balloon, whatever else you may think proper to say.'[42] He was glad that the British were now doing as well in flying matters as their French neighbours, though he continued to be critical of the unscientific approach. 'Lunardi, I find, forgot his barometer and therefore cannot report to what height he ascended.'[43]

On further reflection, he feared that ballooning would not fulfil its first promise, putting his finger with unerring Johnsonian logic on its two apparent shortcomings: 'In amusement, mere amusement I am afraid it must end, for I do not find that [a balloon's] course can be directed, so as that it should serve any purpose of communication; and it can give no new intelligence of the state of the air at different heights, till they have ascended above the height of mountains, which they seem never likely to do.'[44]

A week later, as his last illness drew upon him (dropsy and heart failure, which made him obese and fearfully breathless), Johnson added wistfully: 'To make new balloons is to repeat the jest again. We now know a method of mounting in the air, and I think, are not likely to know more ... I had rather now find a medicine that can ease an asthma.'[45] But this was not to be his last word on the subject.

6

There were other excitements. The actress Mrs Sage, renowned for her Junoesque figure, left a vivid account of being the 'First Aerial female' after an eventful ascent in Lunardi's balloon in June 1785. The launch was made from Hyde Park, attended by a huge and increasingly raucous crowd. Mrs Sage, in a low-cut silk dress presumably designed to reduce wind resistance, was to be accompanied by Lunardi and the dashing Mr George Biggin, a young and wealthy Old Etonian. The gondola was draped in heavy swags of silk, and had a specially designed lace-up door which allowed its occupants to be seen more clearly, as if they were installed in a luxurious aerial salon.[46] But the combined weight of the fixtures and fittings, and the three passengers, proved too much for the balloon, which began wallowing dangerously on its moorings, to the whistles and suggestive jeers of the crowd.

Lunardi made a rapid, though perhaps surprising, decision. Realising that Mrs Sage was the star attraction, after a hasty conference with Mr Biggin, he himself sprang from the gondola, allowing the balloon to make a safe launch with its reduced payload of two. He apparently had no qualms about leaving the control of the balloon (and Mrs Sage) in Mr Biggin's sole care. Unfortunately, in his haste to depart, Lunardi failed to do up the lacings of the gondola's door. As the balloon sailed away over Piccadilly, the crowd were treated to the provoking sight of the beautiful Mrs Sage on all fours in the open entrance of the gondola. The crowd assumed that she had fainted, and was perhaps receiving some kind of intimate first-aid from Mr Biggin.

In fact she was coolly re-threading the lacings to make the gondola safe again. As she later cheerfully admitted, she felt largely responsible for the launching difficulties, as she had omitted to inform Lunardi that she made up '200 pounds of human weight' (over fourteen stone), and he had been far too gallant to enquire. Finally getting to her feet as the balloon floated over Green Park, Mrs Sage trod on Lunardi's barometer and broke it, thus depriving Mr Biggin of any instrument with which to measure their height. Nevertheless, in due course the two of them were lunching peacefully off sparkling Italian wine and cold chicken, occasionally calling to people below through a speaking-trumpet.[47]

The flight followed the line of the Thames westwards, at one point passing through a snowstorm (surprising for mid-June, remarked Mr Biggin nonchalantly), and landed heavily near Harrow on the Hill, smashing through a hedge and dragging across an unharvested hayfield. The infuriated farmer began threatening Mr Biggin and abusing Mrs Sage – she later described him succinctly as 'a savage'. But the honour of the 'first female aeronaut' was unexpectedly saved by the young gentlemen of Harrow School, who rushed out across the fields to greet her, put together a cash collection to pacify the farmer, and carried her bodily (she had hurt 'a tendon in her foot') and in triumph to the local tavern, where everyone evidently got gloriously drunk. Later there was much speculation at Mr Biggin's London club as to whether he had been the first man to *board* a female aeronaut in flight. Gallantly, Mr Biggin refused to comment. The members of Brooks's Club were said to be laying bets on who should first have 'an amorous encounter' in a balloon.

The cry 'Lunardi, come down!' now became a kind of catchphrase, with a suggestive *double-entendre* implied.[48]

Mrs Sage herself felt she had achieved true celebrity, writing modestly to a friend: 'I suppose when I go out I shall be as much looked at as if a native of the Aerial Regions had come down to pay an earthly visit.' She added that the views were magnificent, and that at no point had she needed to open her bottle of smelling salts.[49]

Clearly such ascents remained hugely popular, and even inspirational. But they were also recklessly dangerous, and without any obvious justification beyond entertainment and novelty. It is not surprising that Lunardi's demonstrations were severely criticised by Tiberius Cavallo, FRS, as scientifically useless.[50]

Mockery took other forms. In 1784 the young writer Elizabeth Inchbald (aged thirty-one) managed to get her first play produced at the Haymarket Theatre. It was entitled *A Mogul Tale, or the Descent of the Balloon*. The same year, William Blake wrote and engraved 'An Island in the Moon', a satirical fragment in prose and verse, mocking ideas of flight and pouring scorn on self-deluded 'philosophers', including 'Inflammable Gas' Joseph Priestley. One of his later illustrations shows a spindly ladder leaning against the face of the moon with the caption, 'I want, I want.'

Lunardi used a Union Jack design on all his later balloons, and attracted increasingly large crowds to his launches. In 1785 he took his displays as far north as Edinburgh. But he often had trouble with crowd control, and rowdy disturbances became an important element in the balloon craze. It was dangerous to delay departure beyond the promised hour, even if the balloon was not sufficiently inflated or the wind was adverse. When the newspapers reported a successful launch, it often simply meant that the balloon had lifted off on time and no one in the crowd had been killed.

Lunardi's reputation was badly damaged the following year, when on 23 August at Newcastle a young man, Ralph Heron, was caught in one of the restraining ropes, lifted some hundred feet into the air, and then fell to his death. The impact drove his legs into a flowerbed as far as his knees, and ruptured his internal organs, which burst out onto the ground. He was due to be married the next day.*

---

* Ian McEwan, in the famous opening scene from his 1997 novel *Enduring Love*, describes a similar horrific balloon death.

7

In 1786 Lunardi published *An Account of Five Aerial Voyages in Britain*, in the form of witty, picaresque, self-vaunting letters to his guardian. He had made ballooning fashionable, and started English people thinking about the possibilities of flight, and the new world above the earth. But many, like Banks, still dismissed him as a charlatan, while others wondered why no home-grown British aeronaut had yet taken to the skies.[51]

There were in fact several eccentric amateurs and exhibitionists, but the first serious English pioneer came from a university city, and was largely supported by students. James Sadler (1753–1828) was a baker and confectioner in Oxford's High Street, popular with undergraduates, and also well known as an amateur chemist and inventor. The back room of his bakery was really a laboratory. Sadler had read the work of Cavendish, and followed the news of the Montgolfiers and the French balloon craze of 1783. In spring 1784 he began launching small unmanned balloons, both hydrogen and hot-air type, from the fields around Oxford. He soon attracted financial backing among the undergraduates, and in July 1784 opened a subscription for 'a large Aerial Machine'. In fact he built two: a large hot-air Montgolfier which stood over fifty feet high, and a smaller hydrogen balloon, now known as a 'Charlier'.

Unlike the daredevil Lunardi, Sadler was a family man – happily married with two sons and two daughters. At thirty-one he was modest, quietly spoken, undemonstrative, and his wife could never explain why the dangerous passion for aerostation had seized upon him. Puzzled, but not a little proud, she referred to him as 'the Phenomenon'.[52]

On 4 October 1784 he made the second ascent in England (after Lunardi's), from Christchurch Meadows in the large Montgolfier, which was reported in *Jackson's Oxford Journal*. This ascent was comparatively brief and uneventful, lasting for some thirty minutes, travelling about six miles northwards in the direction of Woodstock. Sadler made a much more dramatic flight in his second ascent, on 12 November. This time he used his hydrogen balloon, and launched before a large crowd from the Physic Garden, with a number of scientific instruments in his boat-shaped basket. By now the soft autumn weather had turned to more wintry and blustery conditions.

The balloon rose rapidly, and it soon became apparent that its speed over the ground was alarmingly swift. Sadler flew southwards towards

Aylesbury, travelling at nearly sixty miles per hour (ground speed) in a brisk wind. There was no time to take any scientific readings. After seventeen minutes the balloon envelope tore, and Sadler was forced to hurl out all his ballast and most of his instruments to prevent an immediate crash-landing. The balloon came down in ploughed fields, dragging Sadler a considerable distance along the ground, destroying the rest of his equipment and most of the balloon. Sadler returned to Oxford with torn clothes and many bruises. But he immediately announced that the hydrogen balloon was the superior aerostat, and planned to construct a much bigger one, capable of sustained flight for twelve hours. With this new balloon he intended to fly across the Channel before Christmas.[53]

Sadler seems to have rekindled the old and ailing Dr Johnson's interest in ballooning, despite his disappointments with Lunardi. While on his last visit to Oxford in October, Johnson sent his black servant Frank to observe Sadler's first launch and report back, being too ill and breathless to go himself. One of the very last notes written in his own hand, dated 17 November 1784, to his old friend Edmund Hector, recalls this: 'I did not reach Oxford till Friday morning, and then I sent Francis to see the Balloon fly, but could not go myself.' It ends with what may be Johnson's last joke about flying. 'I staid at Oxford till Tuesday, and then came back *in the common vehicle* easily to London.'[54] Less than a month later he was dead.

Yet there remained a striking posthumous gesture. Johnson seems to have heard of the disastrous loss of instruments during Sadler's second Oxford ascent of 12 November. Accordingly he presented (or probably bequeathed) to Sadler an enormously expensive barometer, to be used as a precision altimeter on future flights. It was said to be worth 200 guineas, and though Sadler was often tempted to sell it to raise funds, he kept it for over twenty-five years, and always took it on subsequent ascents. Strangely, Boswell nowhere mentions Johnson's touching act of support and encouragement in his *Life*. It was also surely a symbolic gesture from the dying Johnson, who was struggling with his own huge, dropsical, earthbound body.[55]

8

By the end of 1784, the second year of the great balloon craze, no fewer than 181 manned ascents had been recorded, mostly in France and England. There was no sign that the craze was diminishing. On the contrary, it was now being exploited by an intrepid French aeronaut, Jean-Pierre Blanchard (1753–1809), who had earlier challenged Pilâtre and the Montgolfiers in Paris. Learning from Dr Charles's success with hydrogen, he had abandoned his aerial tricycle and constructed his own balloons, and made a number of successful short flights in France. He quickly grasped that a key question with balloons was whether or not they could be navigated.

Blanchard conceived of the balloon as essentially a form of aerial ship, moving through the medium of air as a ship moves through water. It must therefore be capable of being steered, if not directly against the wind, then through several points of the compass across it. There were only two ways to achieve this: either the aeronaut could exploit the wind currents themselves at different altitudes, hoping to find (and perhaps map) ones which blew regularly in different directions, on the analogy of ocean tides and continental currents; or else by providing the balloon with its own independent steering and propulsive instruments.

Blanchard chose to concentrate on the latter. He began experimenting with various ingenious forms of guiding equipment: aerial oars, aerial rudders, sets of flapping wings made of silk stretched across a wicker frame, and most astutely the *moulinet*. This was an early form of hand-cranked propeller, with eight-foot blades also made of stretched silk. Blanchard's theory was as follows. He believed that he could first stabilise his balloon at a fixed height by balancing ballast against hydrogen; and once having achieved this critical 'point of equilibrium', he could then control both its direction and its altitude mechanically with his hand-operated equipment.

In the autumn of 1784 Blanchard came over to London, believing like many entrepreneurs that he could more easily get private financing there than in France. He built a medium-sized hydrogen balloon and made several successful ascents over the Home Counties (it was one of these that *migrated* over the head of Gilbert White). One of his first backers was a member of the Royal Society, Dr Sheldon, who paid for a flight with scientific instruments. However, when the balloon struggled to rise above the London rooftops, Blanchard abruptly threw most of these expensive items overboard. This was wholly characteristic of his high-handed attitude to the 'art of aerostation'. He was a *prima donna* of the air, brilliant, volatile, temperamental, but also utterly fearless.

Like Lunardi, Blanchard was invited to dine with the Duchess of Devonshire, and arranged for a special ascent of a balloon carrying her colours. He met Joseph Banks and several members of the unofficial British Balloon Club. His most significant encounter, however, was with a wealthy and adventurous American physician, Dr John Jeffries. Jeffries, forty years old, was born in Boston. He had qualified both at Harvard and St Andrews, ran a successful practice in Cavendish Square, and had served as a military surgeon on the British side in the American War of Independence. Anxious to be elected an FRS, he had attended the necessary breakfast with Sir Joseph Banks, and was a keen member of the unofficial British Balloon Club.[56]

Jeffries regarded balloon ascents as potentially a part of a major scientific project to discover the secrets of flight, the nature of the upper air, and the formation of weather. After Dr Charles in Paris, he was the first truly trained scientific mind to risk an actual balloon flight. He set out his scientific aims in a paper for Banks, and undertook to write up his ascents for the Royal Society.

In November 1784 he and Blanchard made their first successful trial flight together, with a payload of measuring equipment, across the Thames. Jeffries took carefully prepared instruments with him: a mercury barometer, a thermometer, a hydrometer and an electrometer, to measure the much-feared electrical charges in clouds. In addition he packed maps, a compass and special note-making equipment. He strapped aboard special air flasks, to sample the upper atmosphere at different altitudes, intended for analysis by the reclusive Cambridge scientist Henry Cavendish.

Jeffries drew up a memorandum for the Royal Society, stating the main scientific objectives of the ascents, to be achieved by 'a variety of experiments' and 'not for mere amusement'. He was quite precise: 'Four points need to be more clearly determined. First, the power of ascending or descending at pleasure, while suspended or floating in the air. Secondly, the effect which oars or wings might be made to produce towards this purpose, and in directing the course of the Balloon. Thirdly, the state and temperature of the atmosphere at different heights above the earth. And fourthly, by observing the varying course of the currents of air, or winds, at certain elevations, to throw some new light on the theory of winds in general.'[57]

On this trip Jeffries made the first truly scientific record of a balloon ascent, recording a mass of data – height, direction, air temperature, electrical charges, appearance of clouds, horizon line – at regular time intervals, and taking atmospheric samples for Cavendish. One of the details which emerged was a 'profile' of the characteristic flight path of a hydrogen balloon: not a single smooth parabola, as had been supposed, but a series of looping ascents and descents as the balloon moved above and below its 'equilibrium point'.

Jeffries also gave the first truly vivid account of the changing appearance of the ground as seen 'from a bird's eye view, as it is called'. This too was constantly surprising. As they took off, there was the white sea of upturned faces in the city squares, swiftly reduced to tiny, unrecognisable points. There was the unearthly silence, the sense of their own motionlessness as the earth seemed to revolve below the basket. Though they did not appear to move, their compass needle steadily turned. Below them, the earth appeared transformed. There was the strange flattening out of hills and buildings, the emergence of previously unsuspected patterns in the foliage of woods, or the cultivation marks in fields, or the branching streets of a town. There was the constantly delusive appearance of clouds, and sudden showers of rain or even snowflakes. (No electrical charges were recorded, much to Jeffries' relief.) The whole world became 'like a beautifully coloured map or carpet'.[58]

After this flight, Jeffries agreed to finance Blanchard in an attempt to fly from Dover to France, for the enormous sum of £700.

9

Crossing the English Channel – or for the French, *La Manche* – was an obvious objective for early balloonists. It would be a trial both of balloon technology and of aeronautical nerve. It also carried the distinct undercurrent of an arms race: which nation could command the new element of the air in the event of an invasion? The challenge quickly became an informal national competition, with attempts from both British and French sides of the water. It was seen simultaneously as a scientific, a diplomatic and a sporting battle.

Three main contenders emerged in the autumn of 1784, in the shape of very different and informal teams. They were led by Jean-Pierre Blanchard (Dover), Jean-François Pilâtre de Rozier (Boulogne) and

James Sadler (Oxford). Each was struggling to get financial backing for a suitable balloon. Sadler's balloon never got off the ground. It was destroyed in transit on a Thames barge from Oxford, when a rainstorm soaked the canopy and caused the folds of rubberised silk to stick together.

Pilâtre, who had completed two further epic Montgolfier ascents, was clearly the favourite. He had a large loan of 40,000 crowns from the French Court and the Académie des Sciences, and a big new balloon which aimed to combine the hot-air and the hydrogen principles: a Charlier mounted on a Montgolfier. He was established in Boulogne by November 1784, having built a special hangar for his equipment on the promontory. But he was held up by contrary winds from the north-east, and made up for lost time by making the conquest of a beautiful local convent girl, Susan Dyer, who happened to be English. Romantic trysts alternated with the launching of small test balloons and attempts at weather forecasting, as Pilâtre sat it out with increasing impatience. Rats began to eat the balloon canopy, and his creditors gnawed at his funds.

Meanwhile, Blanchard and Jeffries could not get down to Dover before January 1785, and when they set up in Dover Castle Blanchard quarrelled violently with his American backer. He announced that it would be a solo attempt, and tried to dismiss Jeffries from the entire project. The subsequent arguments, in which the Governor of Dover Castle was forced to intervene on Jeffries' side, delayed the launch by several days.

At the last moment Blanchard tried to hoodwink Jeffries by constructing a lead-weighted belt which he intended to wear beneath his coat, and then announce that the balloon's lift appeared to be too weak to carry two people. Jeffries, an observant man with a cool scientific temperament, spotted the ruse and calmly asked Blanchard to dispense with his personal ballast. But the concept must have stuck in Jeffries' mind, for it would later save both their lives. Blanchard, in return, absolutely refused to take any of Jeffries' scientific instruments on board, except for a barometer and a mariner's compass. What they did agree to take were cork jackets, in case of a forced landing in the sea. They also carried bags of publicity pamphlets, thirty pounds of sand ballast, and Blanchard's patent aerial oars and *moulinet.* Jeffries wore an expensive beaver flying hat to keep out the cold, and fine chamois leather gloves to improve his grip.

Finally, on 7 January 1785, at one o'clock in the afternoon, Blanchard and Jeffries lifted off from the top of Dover cliff to attempt the first ever

Channel crossing. Jeffries' *Narrative of Two Aerial Voyages with M. Blanchard* describes the perilous two-hour flight which followed, interspersed with moments of intense comic rivalry between the two men. Quite early on, each accidentally managed to drop the other's national flag over the side of the basket, and then profusely apologised. Having cleared Dover and its cheering crowds in fine style, the balloon promptly began an easterly drift up-Channel towards the Goodwin Sands. They were soon staring down grimly at its 'formidable breakers'.

The balloon then swung back, and picking up a gentle southerly airstream began to drift towards Calais, but steadily lost height over the sea. By two-thirds of the way across they had progressively jettisoned all the sand ballast, all their food, and most of their technical equipment, except the precious barometer and one bottle of brandy. But the balloon continued to drop, until it was well below the level of the approaching cliffs of the Pas de Calais. They now began to perform a kind of aerial striptease, as Jeffries recorded in his flight diary. 'When two-thirds from the French coast we were again falling rapidly towards the sea, on which occasion my *noble little captain* gave orders, and set the example, by beginning to strip our aerial car, first of our silk and finery: this not giving us sufficient release, we cast one wing, then the other; after which I was obliged to unscrew and cast away our moulinet; yet still approaching the sea very fast, and the boats being much alarmed for us, we cast away, first one anchor, then the other, after which my *little hero* stripped and threw away his coat (great one). On this I was compelled to follow his example. He next cast away his trowsers. We put on our cork jackets and were, God knows how, as merry as grigs to think how we should splatter in the water. We had a fixed cord, &c to mount into our upper story; and I believe both of us, as though inspired, felt ourselves confident of success in the event.'

With nothing remaining as ballast except the bottle of brandy, they were left standing in their underclothes, wearing only their cork jackets. But this made the crucial difference. Less than 120 yards above the sea, the balloon steadied and then began to rise again. As they caught the onshore wind, their ascent turned into a great triumphant arc, taking them high over the cliffs of Calais and twelve miles inland. Blanchard now revealed that he had concealed a small sack of publicity letters, and these were thrown out, to become the first ever airmail delivery. Jeffries calmly noted how the stream of fluttering paper seemed to race across the fields far below them, and took 'exactly five minutes in reaching the surface of the earth'.[59]

Once clear of the coastal updraught, the balloon began an even faster final descent towards the heavily wooded region of Guines forest. A violent and possibly fatal crash-landing in the trees seemed imminent and inevitable. Jeffries, however, maintained a detached, scientific assessment of the situation. He pointed out to Blanchard that there was still one last way of throwing out personal ballast: 'it was contained within ourselves'. Seizing the leather bladders hung in the balloon's rigging as flotation devices, they carefully urinated into them, and threw the contents over the side. In his *Narrative* Jeffries apologised for introducing this 'trivial and ludicrous detail', but pointed out that it was precisely the sort of information that a scientific writer should record.♣

At all events, this 'evacuation' sufficiently checked the rate of their descent, so that the gondola bounced roughly across the tops of the trees instead of plunging violently through the canopy. Jeffries, who was still wearing his chamois leather flying gloves, was able to seize on passing branches until the balloon's progress was gradually halted. It took twenty-eight minutes to release enough hydrogen for the balloon to become manageable. Together Jeffries and Blanchard then carefully hand-manoeuvred the gondola down through a gap between two trees, until at last it safely reached *terra firma*, with the balloon canopy hanging in the branches overhead, gently deflating. They had achieved a first historic crossing.

Jeffries says they staggered around the wreckage of the gondola for several minutes, too stunned and shaking with cold even to congratulate each other. But soon they were surrounded by a crowd of well-wishers, many of whom had followed their course on horseback (like a new form of fox-hunting), and carried them off in triumph to Calais. There is a monument where they touched down, and their balloon car was

♣ Jeffries subsequently claimed that in this manner they lightened the balloon by 'no less than five or six pounds'. As a pint of water weighs a pound and a quarter, he seems to imply that each of them voided well over two pints of urine, which is more than twice the normal male bladder content. Moreover, cold shrinks the male bladder, and Blanchard at least was a notably small and lightly-built man. The probable solution to the surprising amount of weight released is that they defecated as well. No doubt Jeffries felt that this last detail was too much even for scientific candour.

preserved in the Calais museum until 1966. The local *auberge* owner also put up his own quaint but oddly moving sign to commemorate the great crossing.[60]

Later they were rapturously received in Paris, presented to the King, applauded by the Académie des Sciences, and received a standing ovation at the Opéra. They were personally congratulated by Pilâtre (a particularly generous gesture), and asked to give a lecture at his science museum in the rue Saint-Honoré. At one glittering reception, several young ladies rushed up to the aeronauts and crowned them with bay leaves. In between the celebrations, Jeffries spent several quiet evenings with Benjamin Franklin at Passy discussing the future of flight, and the beauty and intelligence of French women.

The English Ambassador, the Duke of Dorset, promised Jeffries that he would be made a Fellow of the Royal Society ('free of all expenses') on his return to London. Jeffries noted in his journal: 'The Duke told me that he was well pleased that I did not suffer the Frenchman to pass over alone.'[61] Despite all the lionising, Jeffries had sent a summary report of the flight to Banks within a week, in a letter dated 13 January 1785. He regretted that it could not yet be sent by aerial post. His report emphasised their remarkable good luck in getting across at all, and made it clear that Blanchard had by no means solved the central problem of navigating a balloon.

The subsequent careers of the two aeronauts were very different. Jean-Pierre Blanchard was awarded a royal pension, and the freedom of the city of Paris. He went on to make sixty-three flights in total, and became the most famous French aeronaut of the first generation. But many of his accounts are unscientific, almost resembling Baron Munchausen-like mixtures of fact and fiction. He founded a Balloon Academy on the Stockwell Road in Vauxhall, and provided balloon entertainments, flying violinists, female aerial acrobats and parachuting animals. Angry crowds eventually wrecked his equipment, and he departed on a global exhibition tour, making ascents in Germany, Holland, Poland, Czechoslovakia and America.

John Jeffries wrote a longer official report to Joseph Banks at the Royal Society, which was published in the Society's *Transactions* in 1786. He never flew again, and his private diary records several exclamations of 'thank God' that he survived. It also includes an almost mystical experience of 'awful stillness and silence' that may be similar to Dr Charles's.[62] He received no honours (except being made 'Baron of the Cinque Ports'), no prize, no pension and no Copley Medal from the Royal Society, but he was at least elected a Fellow. Two months after the flight he revisited Dover in sober and thoughtful mood. 'At noon visited the cliff and spot of our departure on our late aerial voyage into France. The recollection of it was awfully grand and majestick, and my heart filled, I hope, with sincere and grateful acknowledgments to the kind protections of that day. Oh, Gracious Father, may I be influenced by it as I ought, through my life!'[63]

10

This dramatic crossing was soon followed by ghastly tragedy, when the unflappable Pilâtre de Rozier attempted to fly across the Channel in the opposite direction, from Boulogne to Dover, on 15 June 1785. His intention was to recover 'the glory of France', and also perhaps to prove that England could be invaded from the air.

His huge aerostat was not a single balloon but two harnessed together, one on top of the other. The idea was to combine the stable lifting power of the hydrogen balloon (on the top), with the more dynamic and controllable power of the hot-air balloon (below). In appearance this multiple aerostat was oddly menacing, like a warlike mace or club with a short handle. The handle was formed by a thin, tubular-shaped Montgolfier, and the head by a fat, spherical Charlier. Pilâtre and his new co-pilot Pierre Romain stood beneath this contraption in a circular gallery, feeding fuel into an open brazier that could be lowered or jettisoned for landing or in emergencies. The brazier would of course emit a constant stream of sparks.

In theory this dual design combined the best lifting characteristics of hydrogen and hot air. Pilâtre also believed that by allowing him to climb or descend rapidly and at will, it would enable him quickly to find different air currents at different elevations. So he would finally be able to solve the problem of navigation, not by artificial wings or oars, but by naturally harnessing the winds and staying in air currents blowing in the required direction. In this way he would navigate steadily northwards, and easily conquer the Channel and knit together *La Manche* ('the sleeve').

In practice, of course, he had designed a lethal combination of highly inflammable gas and naked flame. He may have had his own misgivings. There was considerable evidence that, being deeply in debt to his sponsors, and forced to carry 'the honour of France', he launched against his better scientific judgement. Certainly he set out in bad meteorological conditions, and with damaged balloon fabric, and persuaded a third passenger to get out of the basket at the last minute. Pilâtre's English fiancée, Miss Susan Dyer, pleaded desperately with him to postpone the attempt.[64] But Pilâtre responded with a dramatic and highly emotional letter. 'For God's sake don't mention such a thing! It is now too late. Give me encouragement. I would rather pierce my heart with a knife than give up this attempt. Even if I were certain of meeting death.'[65]

At 7 a.m. Miss Dyer watched as the cannons fired, and Pilâtre's double balloon rose splendidly in the dawn light to 5,000 feet. It floated out to sea, then seemed to hesitate, and began to drift back over the French coast. It was losing height, and clearly something was wrong. Accounts differ as to what happened next. Using telescopes, witnesses saw Pilâtre – still apparently calm – repeatedly pulling the rope that operated the hydrogen gas-valve at the top of the balloon. It appeared to have jammed in the open position. At the same time his companion Romain was seen frantically lowering the brazier as far below the gallery as possible. The air around the balloon was full of twinkling sparks.

A small, bright crown of yellow flame now began to appear at the top of the balloon, where the hydrogen gas was venting. For a moment the balloon looked to one observer like a heavenly gas lamp, suspended triumphantly above the French cliffs. Then it folded up upon itself, and began to drop to earth, slowly at first, like a long, smoking shroud. According to some farm workers, Pilâtre shouted a warning to them through his speaking-trumpet to keep back. Then he attempted to leap clear at the last moment, possibly to slow the descent of his companion. Both aeronauts were killed, their bodies so horribly broken and ruptured that they were buried the same evening in the little local church at Wimereux.

These were the first recorded deaths of balloonists, and the event shook the scientific community across Europe, and changed the public perception of manmade flight. It seemed all the more shocking because the ascent was a semi-official one, and Pilâtre was a young and glamorous national hero, his name known throughout the Continent.♣

♣ Recent studies by French aeronautical experts, based on the claim that only the top of the Charlier canopy was burnt in the immediate vicinity of the sprung venting valve, have suggested that Pilâtre's basic double-balloon design was perfectly sound. Against all expectation, it appears that the hydrogen was not ignited by a spark from the Montgolfier brazier. The probable cause of the catastrophe was a spark caused by a build-up of static electricity, as Pilâtre pulled the valving line and it chafed against the balloon silk. Audoin

His fiancée Susan Dyer collapsed, and was taken back to her convent. Records show that she died soon afterwards. It is possible that she was pregnant with Pilâtre's child, and committed suicide. Pilâtre de Rozier's fate was even mourned by the English poet Erasmus Darwin:

> Where were ye, Sylphs! When on the ethereal main
> Young Rozier launch'd, and called your aid in vain?...
> Higher and yet higher his expanding Bubble flies,
> Lights with quick flash, and bursts amidst the skies.
> Headlong he rushes through the affrighted air
> With limbs distorted, and dishevelled hair,
> Whirls round and round, the flying crowd alarms,
> And Death receives him in his sable arms!...
> So erst with melting wax and loosen'd strings
> Sunk hapless Icarus on unfaithful wings![66]

11

Hitherto, ballooning in England had been largely dominated by foreigners, French and Italians. This was partly due to the lack of encouragement from the Royal Society, despite the best efforts of Dr Sheldon and Dr Jeffries. But it was also due to the general feeling that ballooning was not a serious scientific pursuit, and was best left to commercial showmen or wealthy private eccentrics. The death of Pilâtre de Rozier in 1785, and Lunardi's accident at Newcastle in August 1786, which resulted in the death of young Ralph Heron, also discredited ballooning with the British public for a generation. From 1790 virtually any balloon sighted in English skies would be assumed to be French and hostile. The aeronaut would find the ground even more dangerous than the skies.

The much-feared aerial invasion by Napoleon's army never materialised. Nonetheless, the French Revolutionary army experimented with a gas observation balloon at the battle of Fleurus against Austria in 1794,

Dollfus, *Pilâtre de Rozier* (1993, Chapter 7, 'Les Causes du Drâme'). It would be interesting to know what Miss Susan Dyer might have thought of this ingenious explanation. Nevertheless, it is true that the first successful non-stop circumnavigation of the globe was performed by a combined helium and hot-air aerostat, with propane burners, the Breitling Orbiter 3, in March 1999.

and the first Corps d'Aerostation and balloon school was formed at Meudon. Lavoisier came up with a cheap method of producing hydrogen for the military, by passing water over red-hot iron, and two young scientists, Charles Coutelle and Nicolas Conte, were appointed to lead the balloon teams and the school. Gaston Tissandier, in his *Histoire des Ballons et Aeronauts Célèbres* (1890), recounts that the young military balloonists took local girls up with them for joyrides and thrilling aerial love-making over the side of the basket, so the first Mile High Club was also formed.

The Corps d'Aerostation eventually fielded four balloons, complete with special hangar tents, winches, gas-generating vessels and observation equipment. Napoleon took the Corps with him to Egypt in 1798, but their equipment was destroyed by Nelson at the battle of Aboukir Bay in July of the following year. Napoleon disbanded the Corps and school at Meudon in 1799, and the rumours of a French airborne army invading Britain remained confined to the realms of fantasy and propaganda. Military balloons were not used again in any conflict until the American Civil War.

In 1810 James Sadler (as inexplicable as ever to his wife) returned to more carefully planned and extended balloon flights, in a series of ascents from Oxford and Bristol. He planned to use his theory of 'oceanic air currents' to navigate across the Irish Sea, a much longer and greater challenge than the English Channel crossing.[67] He believed that these fixed currents existed at various altitudes, flowing steadily in different directions, and could be mapped and used for navigated flight by altering the height of the balloon and so changing direction. On 24 September Sadler made a preliminary ascent over the Bristol Channel, but was baffled by rising and contrary winds, and forced to ditch in a rough sea just off Combe Martin cliffs, when blown perilously towards them without sufficient height to clear them.[68] He said that the worst thing about the entire flight was that he threw overboard Dr Johnson's barometer, in a last futile attempt to lighten the balloon and clear the cliffs. However, the experience he gained of ditching in the sea may subsequently have saved his life.

In July 1811 Sadler continued his experiments by making an ascent from Trinity College Great Court, Cambridge, and landed in a gale near the little village of Stansted. Another, calmer, ascent, from Hackney on 12 August, was made with a scientific observer on board, Henry Beaufoy. Beaufoy kept a minute-by-minute log of the flight, using an array of instruments and carefully noting his physical sensations and impressions. This ascent was also recorded in a beautiful engraving.[69]

Sadler attempted a crossing from Dublin to Liverpool on 1 October 1812, a distance of over a hundred miles, by far the longest balloon flight yet attempted in the British Isles. After a 200-mile dogleg which took him off course north-easterly almost to the Isle of Man, he was swept back safely southwards over Anglesey, where he could have landed. However, he determined to find the direct easterly current to carry him all the way to Liverpool, threw out ballast and climbed again.

Sadler was now steadily swept back out to sea on a northerly airstream, which would eventually have taken him to the Scottish coast or the Isle of Skye – or even to the North Pole. As dusk was coming on, he decided to 'valve' and ditch in the sea, so he could be picked up by a passing boat. But once he was down in the water, the boat refused to approach him. 'I observed that the sailors seemed fearful of coming too near, lest the Balloon should get entangled in their rigging.'[70]

Undismayed, Sadler performed the extraordinary feat of relaunching from the sea by dropping emergency ballast. He flew on northwards till he found a second boat, and ditched for a second time, now almost in the dark. The ship's captain saved him by cleverly running his bowsprit through the balloon's rigging before it sank. Sadler's extraordinary skill and *sangfroid* in ditching and relaunching from the sea, and then ditching a second time virtually in the dark, demonstrated his exceptional gifts as a natural aeronaut. But he had not achieved the historic Irish crossing, and the challenge was eventually to be passed to his son Windham. James Sadler published a vivid account of his attempt, ending with an appeal for greater public support of scientific ballooning: 'Aerostation is too intimately connected with the entire range of Science, its exhibitions are too brilliant and interesting, not to deserve the patronage which a liberal public always confers on desert ... It is so well calculated to throw light on the obscure science of *Meteorology*, to develop many interesting facts in *Magnetism*, and to assist the progress of *Chemistry and Electricity*, that its success must be regarded as interesting to Philosophy [science] in general.'[71]

One Oxford student who responded was Percy Shelley, now aged twenty, who continued to follow Sadler's career and to pursue his fascination with balloons as symbols of liberty. In the winter of 1812 he sent up a series of silk-covered fire balloons, sewn by his teenage wife

Harriet, from the beach at Lynmouth, Devon. Each carried copies of his revolutionary pamphlet 'A Declaration of Rights'. He also composed a rather good sonnet on the subject:

> Bright ball of flame that thro the gloom of even
> Silently takest thine ethereal way
> And with surpassing glory dimmst each ray
> Twinkling amid the dark blue depth of Heaven;
> Unlike the Fire thou bearest, soon shalt thou
> Fade like a meteor in surrounding gloom,
> Whilst that, unquenchable, is doomed to glow
> A watch-light by the patriot's lonely tomb,
> A ray of courage to the opprest and poor...[72]

James Sadler's son Windham made his first solo flight from Cheltenham in 1813, aged only seventeen. He too revealed himself as a natural aeronaut, and four years later, in 1817, he attempted the first Irish crossing that had eluded his father. The flight was again made from Dublin, but this time was better prepared, with careful meteorological planning and the launching of small pathfinder balloons. Windham made the sixty-mile crossing to Wales on a direct easterly course, a journey of five hours. Having learned from his father's experience, the moment he reached land Windham valved the balloon and came down just south of Holyhead.[73]

Like his father, Windham Sadler championed the scientific value of ballooning, and decried its shameful neglect by English backers in the years that followed: 'Strange as it may appear, England, the seat of Science and Literature, has remained satisfied with gazing on the casual experiments of Foreign Aeronauts ... although Cavendish first discovered and Priestley first suggested the application of that powerful agent, Hydrogen Gas, to the purposes of Aerostation!'[74]

But in 1824, at the age of twenty-seven, Windham had a terrible accident when his balloon grapple-line tangled in a chimney during a high-wind landing in the Pennines. He was thrown out of the basket, and hung suspended upside-down by his legs for several minutes, until he finally fell to his death. Devastated, his father James never set foot in a balloon basket again.

12

The early heroic period of ballooning, between 1783 and 1800, appeared to come to a dead end. Individual balloonists went on flying, but not for long. Dr Jeffries returned to Boston in 1789. Lunardi died poverty-stricken in Lisbon in July 1806. Blanchard collapsed with a heart attack after a forced landing in Holland in 1809, and died a few weeks later while apparently attempting to parachute from his new balloon. A few celebratory ascents continued in Paris, notably by the showman Jacques Garnerin and by Blanchard's eccentric young wife Sophia, who specialised in aerial firework displays. But she too was killed in 1819, when her balloon was ignited by fireworks. It must have looked as if ballooning was, scientifically speaking, a *cul de sac.*

Though the hydrogen balloon or Charlier triumphed (temporarily) over the hot-air balloon or Montgolfier, the inability to navigate either form of aerostat appeared to destroy all hopes of finding any immediate technological applications. Balloons simply remained beautiful, expensive and dangerous toys, although the high ascents achieved by the French chemist Joseph Gay-Lussac and others did promise hoped-for advances in meteorology. Gay-Lussac ascended to 23,000 feet above Paris in 1804, establishing the limit at which human beings can breathe. The mysteries of barometric pressure, the function of clouds, the generation of winds and weather systems, were increasingly fascinating.

Interest in meteorology, a nascent science, grew and produced the beautiful cloud classifications of Luke Howard and the valuable wind-scale system of Francis Beaufort. Howard (1772–1864), a Quaker and the first professional meteorologist, published his great study and classification of atmospheric phenomena, *On the Modification of Clouds,* in 1804. He first proposed the four basic cloud-types, using Latin terms in imitation of Linnaean cataloguing. These were *cumulus* (heaped cloud), *stratus* (layer cloud), *cirrus* (long-hair or high mare's-tail cloud) and *nimbus* (raincloud), with various combinations such as *cumulo-nimbus* (the classic heaped-up rain-carrying clouds of an English summer). All are still in use, with additional combinations such as *cirro-stratus* (high, thin, fine-weather clouds). Howard was elected to the Royal Society in 1821, but did not achieve a consistent theory of atmospheric pressures and gradients (high- and low-pressure systems), upon which all weather forecasting would ultimately be based, although he outlined this in his last work, *Barometrographia* (1847). But he called new attention to the formation and transformation of clouds, their seasonal varieties and characteristics, and above all perhaps to their astonishing beauty.[75]

Ballooning added to this new awareness of the complexity and subtlety of clouds, a growing Romantic preoccupation which can be followed in the paintings of Turner and Constable, the notebooks of Coleridge and the poetry of Shelley. When Shelley refers to 'the *locks* of the approaching storm' in his 'Ode to the West Wind' (1819), he is using Howard's definition *cirrus.* 'The Cloud' (1820) demonstrates a remarkably accurate and scientific understanding of cloud formation and the convection cycle.[76] Goethe wrote a number of essays on clouds, atmospheric pressures and weather, and translated passages of Howard, asking him to compose and send his Autobiography to Germany, and describing him as 'the first to define conceptually the airy and ever-changing forms of clouds, thus delimiting and fixing what had always been ephemeral and intangible, by accurate observation and naming'.*

Clouds became fascinating both as scientific phenomena – the generators of electricity, the mysterious indicators of winds and changing air pressure – and as aesthetic phenomena: the 'moods' of the sky reflecting those of the observer, alterations of light over landscape, symbols of change, destruction, regeneration. It could be argued that the Romantics actually invented the idea of 'the weather' itself, as it now preoccupies us; as well, of course, as 'inner weather'.

The first mapping overview of the earth, with drawings made from the balloon basket, revealed the patterns of towns and countryside, the growth of roads, the meandering of rivers, in a new way. Although maps were also the result of trade, exploration, military campaigning and turnpike-building, the creation of the British Ordnance Survey – the first state mapping programme in the world – was partly inspired by balloons.

---

* In fact Lamarck had published a paper 'On Cloud Forms' in Paris in 1802, but his definitions were less authoritative than Howard's, and he used French terms – such as *attroupés* for *cumulus* – which were less easily accepted internationally at this period. If Napoleon had won the European war, weather forecasting might be more Gallic today; as it is, modern French forecasters still give barometric pressures in '*hectopascals*', and have difficulty distinguishing between drizzle, showers and rain. See Richard Hamblyn, *The Invention of Clouds* (2001).

Ballooning produced a new, and wholly unexpected, vision of the earth. It had been imagined that it would reveal the secrets of the heavens above, but in fact it showed the secrets of the world beneath. The early aeronauts suddenly saw the earth as a giant organism, mysteriously patterned and unfolding, like a living creature. For the first time the impact of man on nature was clearly revealed: the ever-expanding relationship of towns to countryside, roads to rivers, cultivated fields to forests, and the development of industry. It was comparable to the first views of the earth from space by the Apollo astronauts in the 1960s, producing a new concept of a 'single blue planet' with its delicate membrane of atmosphere. The famous photograph 'Earthrise' was taken from lunar orbit in December 1968.

Ballooning proved to have extraordinary theatrical power to attract crowds, embody longing, and mix terror and the sublime with farce. It became showmanship, carnival, pure euphoria. A successful balloon launch, in the hands of one of the early masters like Pilâtre, Lunardi or Blanchard, became a communal expression of hope and wonder, of courage and comedy. The balloon crowd (especially in Paris) foreshadowed another kind of crowd – the revolutionary crowd. It contained elements of prophecy, both political and scientific. It was like a collective gasp of hope and longing.

Curiously, it was not the men of science, so much as the poets and writers, who continued to see ballooning as a symbol of hope and liberation. Erasmus Darwin celebrated the daring of the first balloonists, and the new vision of the world their intrepid flights opened up in the 1780s:

> The calm Philosopher in ether sails,
> Views broader stars and breathes in purer gales;
> Sees like a map in many a waving line,
> Round earth's blue plains her lucid waters shine;
> Sees at his feet the forky lightning glow
> And hears innocuous thunder roar below.[77]

Coleridge wrote in his notebooks of the balloon as an image of powerful but mysterious flight. He compared the appearance of a balloon in the sky to that of a flock of starlings climbing and spinning upon itself. It was ultimately an image of human longing and inspiration, both uplifting and terrifying.[78]

Wordsworth began his poem *Peter Bell* (1798) with a playful image of flying in a sort of dirigible airship, or balloon boat.

> There's something in a flying Horse,
> There's something in a huge Balloon:
> But through the Clouds I'll never float
> Until I have a little Boat
> Shaped like the crescent-Moon...
>
> Away we go! – and what care we
> For treason, tumults, and for wars?
> We are as calm in our Delight
> As is the crescent-Moon so bright
> Among the scattered Stars.[79]

Perhaps Shelley put it best, when he was a young undergraduate at Oxford in 1811, and had just witnessed another of Sadler's balloon ascents one sparkling summer morning from Christchurch Meadows: 'The balloon has not yet received the perfection of which it is surely capable; the art of navigating the air is in its first and most helpless infancy; the aerial mariner still swims on bladders, and has not yet mounted the rude raft... It would seem a mere toy, a feather, in comparison with the splendid anticipations of the philosophical chemist. Yet it ought not to be altogether condemned. It promises prodigious faculties for locomotion, and will allow us to traverse vast tracts with ease and rapidity, and to explore unknown countries without difficulty. Why are we so ignorant of the interior of Africa? – Why do we not despatch intrepid aeronauts to cross it in every direction, and to survey the whole peninsula in a few weeks? The shadow of the first balloon, which a vertical sun would project precisely underneath it, as it glided over that hitherto unhappy country, would virtually emancipate every slave, and would annihilate slavery forever.'[80]

4

Herschel Among the Stars

1

Sir Joseph Banks had predicted that British astronomy would go further than French ballooning. In the summer of 1785 William Herschel embarked on his revolutionary new project to observe and resolve the heavens with a telescope more powerful than ever previously attempted. His first move was to draft a preliminary technical specification for Banks to submit to the King. It was a monumental proposal.

What he intended to build was a telescope 'of the Newtonian form, with an octagon tube 40 foot long and five feet in diameter; the specula [mirrors] of which it would be necessary to have at least two, or perhaps three, should be from 36 or 48 or 50 inches in diameter'.[1] The telescope would have to be mounted in an enormous wooden gantry, capable of being turned safely on its axis by just two workmen, but also susceptible to the finest and most minute fingertip adjustments by the observing astronomer. The mirrors would weigh about half a ton each, and cost between £200 and £500. They would have to be cast in London and shipped by barge up the Thames for polishing.[2] Casting would be a major feat of technology, and twenty workmen would be required to effect a continuous process of polishing with newly designed machinery.

The forty-foot would be higher than a house, extremely susceptible to wind, and very exposed to adverse weather conditions, especially frost, condensation and air-temperature changes, which could 'untune' the mirrors like musical instruments. The astronomer (Herschel was now approaching fifty) would be required to climb a series of ladders to a special viewing platform perched at the mouth of the telescope, from which a fall would almost certainly prove fatal. The assistant (Caroline) would have to be shut in a special booth below to avoid light pollution, where she would have her desk and lamp, celestial clocks, observation journals and coffee flasks. But she would see virtually nothing of the stars themselves.

Astronomer and assistant would be invisible to each other for hours on end, shouting commands and replies, although eventually connected by a metal speaking-tube. It would be rather as if they were the tiny crew of some enormous ship, one up on the bridge, the other below in the chart room, intimately dependent on each other but physically isolated. Perhaps this was the premonition of a new kind of vessel: a spaceship flying through the starry night.[3]

All this would require a new level of funding by the King. The estimate of expenses totalled £1,395, with an annual running cost of £150. This huge sum did not include Herschel's annual salary of £200.[4] When he submitted this enormous research-grant application, Herschel austerely did not promise any immediate results – more planets, more comets, more sightings of extraterrestrial life forms. Instead he tried to reassure Banks in the most sober terms: 'The sole end of the work would be to produce an Instrument that should answer the end of inspecting the Heavens, in order more fully to ascertain their *construction*.'[5]

It was one of Sir Joseph Banks's most dramatic diplomatic coups that he had convinced the King to announce a grant by September 1785. The sum was a generous one: the entire construction costs and four years' running expenses – a total of £2,000. The one implied proviso was that Herschel needed to come up with results by the end of 1789. In November 1785 Banks was already sending tactful enquiries through William Watson: 'Sir Joseph Banks is come to Town, & expressed a wish to know from you what preparations you have made relating to the great Telescope, & how far you have proceeded in the work itself. He said that he was very desirous of knowing, that he might be enabled to give the King a history of your proceedings.'[6]

In fact there was no immediate progress that autumn. To Caroline's dismay, Herschel had decided that his grand project required a new house with larger grounds for constructing and erecting the telescope, and more outbuildings for workshops. A first move from Datchet to Clay Hall, nearer to Windsor, had proved abortive when their new landlady objected to the cutting down of trees, and sought to raise the rent on the ingenious grounds that Herschel's monster telescope, if it were ever built, would count as an 'improvement' to the house. Herschel wondered mildly if each new astronomical discovery would increase the valuation, and thereafter require a corresponding increase in the rent.

On 3 April 1786 they moved again (the Herschels' third move in four years), to 'The Grove', a quite small and rather dilapidated country house on the edge of the tiny village of Slough, three miles north of Windsor. It was owned by a wealthy local family, the Baldwins, whose various relations also owned two local inns, the Dolphin and the Crown, and were extensive landowners in the district. The Crown was the main mail-coach halt on the London to Bath turnpike, now rerouted as the modern A4 through Slough. (The original sleepy road junction has become a pedestrianised section of the busy high street, though still known locally as 'the Crown crossroads'.)

The youngest member of the family, Mary Baldwin, was a considerable heiress, and had married a retired London merchant, John Pitt, who was some twenty years older than herself. They decided to lead an easy-going country life, and had their own large, comfortable house less than a mile away from The Grove at the little village of Upton. They proved hospitable and friendly neighbours, and soon got to know the Herschels socially. John was in 'a declining state of health', so William would walk over at weekends, and sit talking to him in his well-appointed library. Caroline seems to have got on rather well with the Pitts' son Paul, an only child who had just started at nearby Eton College.[7]

The Grove stood in secluded grounds, 200 yards south of the Crown Inn, on the east side of the road to Windsor. Though it was surrounded by trees, the ground dropped away sharply to the south, offering a good observational platform. It was also ideal for rapid communications with London and Maskelyne's observatory at Greenwich, as well as remaining close to the King's residence. Indeed the turrets of Windsor Castle could be seen from the terraced walk on the south of the property, a constant reminder of the expectations of the royal patron.[8] The house itself was not large: four bedrooms and a servants' attic. But it had extensive sheds and stables which were gradually converted into workshops and laboratories, and a wash house that became a forge. Above the stables were a series of haylofts which could be converted into a separate apartment. Caroline claimed these for her own. She had them roughly whitewashed, and put into use as a bedroom and a writing room, with a small outside staircase leading up to a flat roof from which she hoped to carry out her comet 'sweeps' in security and independently. This became her 'cottage' and occasional residence, a first step towards domestic independence at the age of thirty-six.

The Grove also had a large flat area of rough gardens in front of it, ideal for levelling and laying the extensive circular brick foundations of the wooden gantry for the forty-foot telescope. The brickwork was capped with Portland stone, though later this was cracked by frost and had to be sheathed in oak.[9] As work progressed, Herschel had all the surrounding trees cut down, including a magnificent row of ancient elms, 'to the grief of everyone who knew that sweet spot', as one neighbour observed. Characteristically, Herschel took no notice of their objections.[10] This scattered collection of buildings would later become known as 'Observatory House', and Herschel's telescope would be marked on the first edition of the Ordnance Survey map for Berkshire in 1830.[11]

The launch of the new project changed the quiet rhythm of the Herschels' lives. The spring of 1786 was 'a perfect Chaos of business', as Caroline put it with a certain relish: 'If it had not been sometimes for the intervention of a cloudy or a moonlight night [bad for stellar observation], I know not when my Brother (or I neither) should have got any sleep; for with the morning came also the workpeople of which there were no less than between 30 or 40 at work for upwards of 3 months together, some employed with felling and rooting out trees, some digging and preparing the ground for the Bricklayers who were laying the foundation for the telescope, and the carpenter in Slough with all his men.'[12]

News of the proposed monster telescope brought a steady stream of visitors to Slough: men of science, academics from the universities, foreign tourists, and too many dignitaries from the Court. Caroline would grow increasingly impatient at their tendency to interrupt Herschel's work. She developed her own laconic way of registering this impatience: 'Professor Sniadecky often saw some objects through the 20 foot Telescope, among others the Georgian satellites. He had taken lodgings in Slough for the purpose of seeing and hearing my Brother whenever he could find him at leisure. Himself was a very silent man.'[13] She was always happy, however, to welcome old friends like William Watson and Nevil Maskelyne, and new supporters from the Royal Society like Charles Burney (who was also in favour of hot-air balloons). Americans were notably well-received.

Sometime in the summer of 1786 the fifty-year-old John Adams, graduate of Harvard University, man of science and future second

President of the United States, turned up one morning uninvited at The Grove. He was shown round all Herschel's new telescopes, and they embarked on an impassioned discussion of the possibility of extraterrestrial life, and the moral implications of there being a 'plurality of worlds'. This was the sort of metaphysical debate that Herschel had once had with his brother Jacob, touching on the speculations of European authors like Fontenelle and Huygens, but which he tended to avoid with his English contemporaries. Neither Herschel nor Caroline recorded exactly what was said, but it is clear from his own diaries that Adams would have put lively and unorthodox views: 'Astronomers tell us that not only all the Planets and Satellites in our Solar system, but all the unnumbered Worlds that revolve round the fixt Stars are inhabited... If this is the case all Mankind are no more in comparison [with] the whole rational Creation of God, than a point in the orbit of Saturn.'

Like the poet Shelley a generation later, Adams liked to press this argument one stage further. If astronomy discovered extraterrestrial civilisations, then surely the earth-based doctrines of Christian redemption became absurd, or at least mighty inconvenient for the Lord. 'I ask a Calvinist, whether he will subscribe to this alternative: EITHER God Almighty must assume the respective shapes of all these different Species, and suffer the penalties of their Crimes, in their stead; OR ELSE all these Beings must be consigned to everlasting Perdition?'[14] *

Amidst all the bustle of visitors and workmen, Herschel was despatched by royal command in July 1786 to deliver and erect one of his ten-foot telescopes, as King George's special gift to the University of

* Adams never forgot this spirited meeting with Herschel. Years later, in 1825, he wrote to Thomas Jefferson, his successor as President, complaining of the orthodox Christian beliefs of most British scientists, and advising Jefferson not to hire them to teach at the University of Virginia, where he was Chancellor. Adams contrasted these scientists' attitudes with Herschel's untrammelled vision: 'They all believe that great Principle which has produced this boundless universe, Newton's universe and Herschel's universe, came down to this little ball [planet earth], to be spit upon by the Jews. And until this awful blasphemy is got rid of, there never will be any liberal science in the world.' This argument would presumably have been satisfactorily concluded the following year, when both Adams and Jefferson died and went to meet the Great Principle. See Michael J. Crowe, *The Extraterrestrial Life Debate* (1986).

Göttingen, which was fast becoming the centre of scientific studies in Germany. His brother Alexander was to accompany him as business manager. This was both a great honour and a great inconvenience, and for the first time in her life Caroline was left wholly in charge of both the construction work on the forty-foot and the continuing observation programme of nebulae and double stars. Currently they had completed 572 sweeps, identified 1,567 nebulae, and found two tiny new moons orbiting Georgium Sidus (a discovery that particularly amused the King).[15]

Caroline's first response was thoroughly domestic. She started a new day book, neatly headed it 'Book of Work Done', drew a careful set of parallel columns, and recorded her inventory of tasks.

> July 3rd 1786. My Brothers William and Alex left Slough to begin their journey to Germany… By way of not suffering too much by sadness, I began with bustling work. I cleaned the brass work for the 7 and 10 feet telescopes and put curtains before the shelves to hinder the dust from settling upon it again. I cleaned and put the [mirror] polishing room in order and made the gardener clean the work-yard, put everything in safety and mend the Fences.

She would not tolerate idling among the workmen, who evidently caused difficulties. The gardener was reprimanded for lolling about the lawns: 'he gave me the name of "Stingy —" in the village, because I objected to his being there when not wanted'.[16] It would be interesting to know what the '—' stood for: the fact that Caroline was female, foreign, diminutive, unmarried, disciplined, or brilliantly gifted perhaps?

That afternoon she did needlework and went shopping in Windsor; when she got back she was mortified to find 'there had been four foreign Gentlemen looking at the instruments in the garden, but did not leave their names'. Later unannounced visitors that month included Nevil Maskelyne and his wife, three members of the great Dollond telescope family, the Duke of Saxe-Gotha, Tiberius Cavallo (the balloon expert from the Royal Society), her friend Dr James Lind, the Prince Resonico and the Plumian Professor of Astronomy at Cambridge, Dr Antony Shepherd. The question of visitors became more awkward as July progressed, and for the first time made Caroline aware of the social anomaly of her position. 'I was often put into great perplexity by such self-inviting

Joseph Banks
An exuberant portrait painted shortly after Banks's triumphant return from the Pacific. Portrait by Sir Joshua Reynolds, 1771–73.

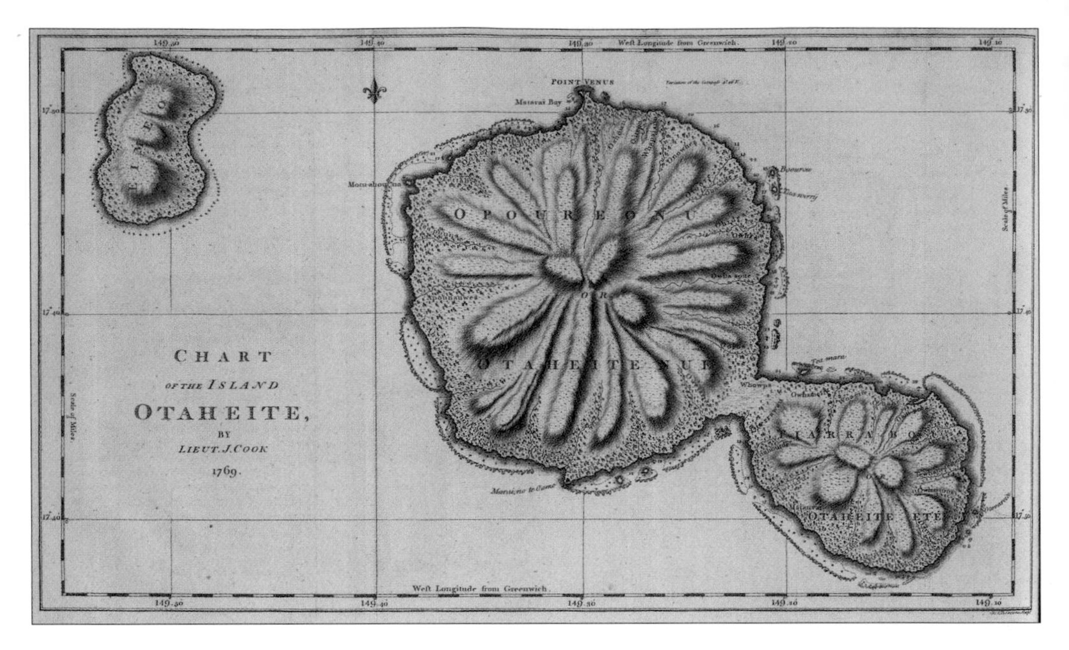

Chart of the island Otaheite, by Lieut. J. Cook, 1769
Matavi bay and Point Venus on the northern shore; and beyond it the island of Eimeo (Moorea), where Banks observed the transit of Venus.

Above *Sketch of Sydney Parkinson, from the frontispiece to his* Journal *(1773)*
Parkinson was only nineteen when he died on the voyage home from Tahiti.

Left *A Woman and a Boy, Natives of Otaheite in the Dress of the Country*
Engraving after Parkinson by T. Chambers, from Sydney Parkinson, *Journal of a Voyage in the South Seas* (1773).

Omai, Banks, and Solander (seated)
Honoured guest or valuable human specimen? Painting by William Parry, c.1775–76.

Dorothea Hugessen, Lady Banks. By Joseph Collyer the Younger, after John Russell, c.1790.
Captain James Cook. Portrait by John Webber, 1776, shortly before Cook's last voyage.

Young William Herschel (locket)
A locket given to Caroline Herschel c.1760, when her beloved brother William was twenty-two, and about to settle in England. With the kind permission of John Herschel-Shorland.

Above *Young Caroline Herschel (silhouette)*
Silhouette made c.1768, when Caroline was eighteen, shortly before she joined William in England.

Sir William Herschel
Herschel, now knighted, powdered and famous after the discovery of Uranus in 1781. Portrait by Lemuel Francis Abbott, 1785.

William and Caroline Herschel
A genteel Victorian image of the Herschels at work, with Caroline administering a sustaining cup of tea. Coloured lithograph, 1890.

Engraved frontispiece to John Bonnycastle's *Introduction to Astronomy* (1811). Urania, the Muse of Astronomy, shows her pupil the new planet.

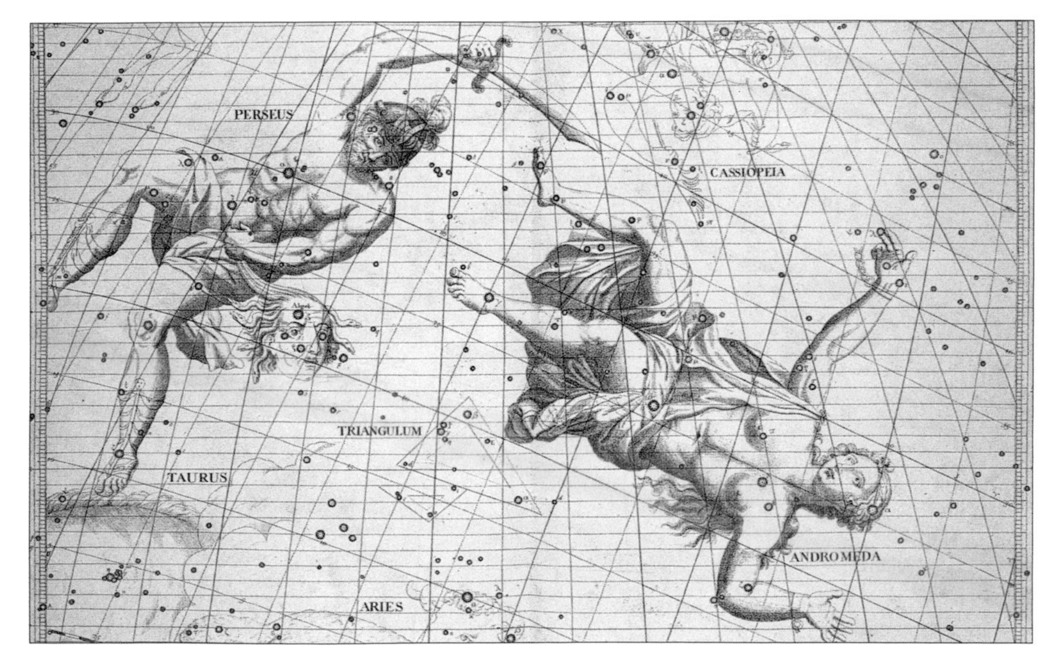

The constellations of Perseus and Andromeda
From John Flamsteed's *Celestial Atlas* (1729). The Andromeda nebula or galaxy is located at the goddess's right thigh.

Above *Close-up of Herschel's seven-foot reflector telescope.* The brass tube on the top is the wide-field 'sighting' scope, and the viewing aperture on the side can take lenses of different magnifications. Whipple Museum, Cambridge. Photograph by Richard Holmes.

Left *Herschel's seven-foot reflector telescope*
The telescope with which Herschel discovered Uranus in 1781. Main reflector mirror in the base, viewing aperture on the side at the top. Note the ingenious portable and adjustable support frame. Royal Astronomical Society. Drawing by Sir William Watson.

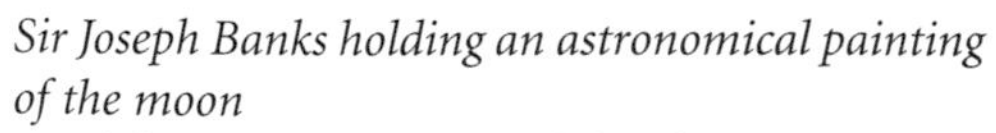

Sir Joseph Banks holding an astronomical painting of the moon
A celebration of the friendship between Banks and Herschel, who had advised the artist John Russell on details of the lunar features.

An astronomical moon globe
Such a beautiful scientific instrument could also serve as decorative furniture. Selenographia Moon Globe by John Russell, London, 1797.

Then felt I like some Watcher of the Skies
When a new Planet swims into his ken,
Or like stout Cortez, when with wond'ring eyes
He star'd at the Pacific, and all his Men
Look'd at each other with a wild surmise—
Silent upon a Peak in Darien—

Detail from the original manuscript of John Keats's sonnet 'On First Looking into Chapman's Homer' (1816)
'... like some Watcher of the Skies/when a new Planet swims into his ken...' Keats later changed 'wond'ring eyes' to 'eagle eyes'.

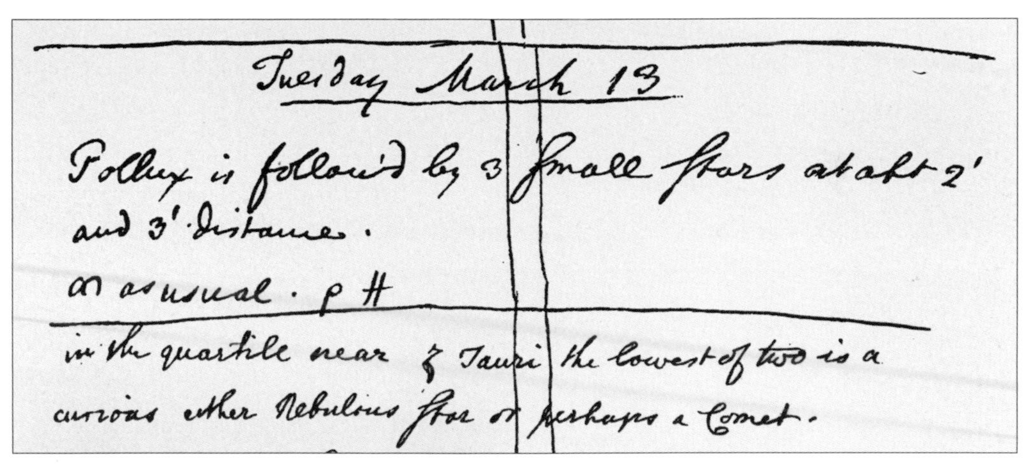
Tuesday March 13

Pollux is follow'd by 3 small stars at abt 2' and 3' distance.

α as usual. ρ H

in the quartile near ζ Tauri the lowest of two is a curious either Nebulous Star or perhaps a Comet.

Detail from Herschel's Astronomical Observation Journal for Tuesday, 13 March 1781
The 'curious either Nebulous Star or perhaps a Comet' was in fact the new planet Uranus.

The planet Uranus
Hubble Space Telescope image of Uranus, showing its ring system and six of its moons. Photograph taken in August 2003.

William Herschel's forty-foot reflector telescope
The engraving was first published as an illustration to Herschel's celebrated paper, 'Description of a Forty Foot Reflecting Telescope', in *Philosophical Transactions* (1795), the journal of the Royal Society.

Sir William Herschel
Herschel the Romantic sage of science under the starry sky he had changed forever. Stipple engraving by James Godby, after Friedrich Rehberg, 1814.

visitors; for I could only look upon myself as an individual who was neither Mistress of her Brother's house, nor of her Time, and for that reason neither could, nor would, ever give invitations.' She also found the endless 'gossipings' of Alexander's pretty but 'foolish' new wife, who came over from Bath, intolerable.[17]

By the end of July Caroline had decided that the only way to remedy this situation was to insist on her own quite separate regime. She would be an astronomer, not a housekeeper. She would check over the calculations of William's nebulae by day, and make her own sweeps up on the roof by night. She would go to bed late (often just before dawn light, around 4 a.m.) and get up late (but always in time to pay the workmen after breakfast). She even wrote William a little imaginary letter about this in her 'Book of Work Done'. In this case it was to be work she would *not* do. 'I find I cannot go fast enough with the registering of sweeps to be serviceable to the Catalogue of Nebulua. Therefore I will begin immediately to recalculate them, and hope to finish them before you return. Besides I think the consequences will be bad of registering the sweeps backwards.'[18] Thus liberated from the nightly duty of William's sweeps, her 'Book of Work Done' began to fill with her own astronomical observations. Symbolically she recorded on 30 August winding up the 'Sidereal Time Piece', the big brass chronometer used to fix stellar positions.

Three summers previously William had built Caroline a special two-foot Newtonian reflector, mounted within an ingenious wooden box-frame. Because of its large aperture, its tube appeared much fatter, heavier and stubbier than normal reflectors of this type: a rotund, almost jovial presence, but not in the least awkward to handle. Suspended from a pivot at the top of the box-frame, the telescope could be precisely raised or lowered by a system of pulleys operated by a large brass winding handle at the bottom. These adjustments were easy to make, and extremely fine. The whole 'contrivance' was set on a solid portable wooden stand, constructed like a three-legged stool, and exactly carpentered to bring the Newtonian viewing lens precisely to the level of Caroline's eye. It also allowed a workman (or Caroline herself) to carry the telescope and stand in two sections, and position it wherever required, downstairs in the garden or upstairs on the flat roof.[19]

This beautiful instrument was designed specifically for its huge light-gathering power and its wide angle of vision. The mirror was 4.2 inches in diameter (the size more usually placed in the seven-foot reflectors), with a large observational field of over two degrees. The magnification was comparatively low at twenty-four times. As with modern binoculars, this combination of low power with a large viewing field allowed the observer to see faint stellar objects very brightly, while placing them within a comparatively wide context of surrounding stars. In effect, Herschel had constructed for Caroline *a hunter's telescope.*

It was a deliberate challenge. The instrument was not suitable for deep space, but it was perfectly designed to spot any strange or unknown object moving through the familiar field of 'fixed stars'. It was designed to find *wanderers* and *messengers* coming into the solar system. In other words, to catch new planets or new comets. It eventually became famous as 'Miss Herschel's small sweeper', and would be joined within two years by 'Miss Herschel's large sweeper'.[20]

On 1 August 1786, only two nights after starting her new sweeps, Caroline thought she had spotted an unknown stellar object moving through Ursa Major (the Great Bear constellation). It appeared to be descending, but barely perceptibly, towards a triangulation of stars in the beautifully named Coma Berenices ('Berenice's Hair', as celebrated in Pope's poem 'The Rape of the Lock'). To find something so quickly, and in such a familiar place (the Great Bear or Big Dipper being the first stop of every amateur stargazer wanting to locate the Pole Star), seemed wildly unlikely. Caroline's Observation Book conveys meticulous caution, but also remarkable certainty.

Unable to calculate the mathematical coordinates of the object, she accompanied her observations with a series of three neat drawings or 'figures', over an eighty-minute time lapse. These showed the circular viewing field of her telescope, with an asterisk shape very slightly changing position relative to three known fixed stars.

> August 1st 1786. 9 hours 50 mins. I saw the object in the center of fig.1 like a star out of focus while the others were perfectly clear. The sec. star is very faint but the weather is hazy, and in a clearer night undoubtedly some more will be visible. ... 11 hours 10 mins. I think the situation is now like in Fig.3 but it is so hazy that I could only imagine I saw the second star & the preceding I could not see at all. The comet is about half way between 53 and 54 Ursa maj. and some stars which I found after looking over the map at leisure to be 14, 15, and 16 Coma Berenices ...[21]

Caroline does not remark that her comet was moving from a male to a female constellation, a fact which might have well have struck her as peculiarly appropriate. But the account written into the 'Book of Work Done' catches something of her growing excitement. The drudgery of daytime calculation in her study was overtaken by the tantalising expectations and frustrations of the nights up on the flat roof.

> August 1st. I have calculated 100 nebulae today, and this evening I saw an object which I believe will prove tomorrow night to be a Comet. August 2nd. Today I calculated 150 nebulae. I fear it will not be clear tonight, it has been raining throughout the whole day, but seems now to clear up a little... 1 o'clock. the object of last night IS A COMET. August 3rd. I did not go to rest till I had written to Dr Blagden [at the Royal Society] and Mr Aubert to announce the Comet. After a few hours sleep I went in the afternoon to Dr Lind, who with Mr Cavallo accompanied me to Slough with the intention of seeing the Comet; but it was cloudy and remained so all night. August 4th. I wrote today to Hanover, booked my observations and made a fair copy of 3 letters... The night is cloudy. August 5th. I calculated nebulae all day, paid the smith... The night was tolerably fine and I SAW THE COMET.[22]

Both Aristotle and Galileo had thought comets were low-level atmospheric phenomena, perhaps lower than the moon. The study of comets was improved by the sixteenth-century Danish astronomer Tycho Brahe, but transformed in 1682 when Edmund Halley famously calculated that the Great Comet of that year, subsequently named after him, would reappear in 1759. It was then finally accepted that comets were outer-space objects that moved in extreme elliptical orbits round the sun, and swung far beyond the known planets. Yet they were still mysterious: of unknown origin and composition, various in their appearance, irregular and alarming in their habits. A reassuring popular view, that they were celestial table-waiters, supplying the planets with moisture and the sun with fire, was expressed by James Thomson in his poem *The Seasons* (1726–30).

> From his huge vapouring train perhaps to shake
> Reviving moisture on the numerous Orbs,
> Thro which his long elipsis winds; perhaps
> To lend new fuel to declining Suns,
> To light up Worlds, and feed the etherial Fire.[23]

By the mid-eighteenth century only about thirty comets had been identified and recorded in the annual French catalogue *La Connaissance des Temps.* The greatest comet-hunter of the age, Charles Messier, had personally found about half of these, and so comet-hunting was generally regarded as a French speciality. Caroline's discovery – even if it had been her only one – would have been an important contribution internationally. Comets (meaning 'hairy stars') were significant because they were the only celestial objects which came in from beyond the known solar system, and therefore carried possible information about conditions further out in space.

The fact that the elliptical path of periodic comets could be calculated according to Newton's laws, and their returns predicted scientifically, seemed to prove that their traditional role as portents of events on earth (usually of sudden disasters) was a meaningless superstition. So the comet that appears in the Bayeux Tapestry turned out to be Halley's on a previous periodic visit; it reappeared without disaster in 1986, and is next scheduled in 2061. However, new comets such as that of 1811 still caused a great popular stir. Adam Smith noted in his *Philosophical Enquiries* (1795): 'the rarity and inconstancy of their appearance, seemed to separate them entirely from the constant, regular, and uniform objects in the Heavens'.[24] ♣

It is revealing that Caroline was too excited to sleep, and that in the absence of Herschel, almost her first reaction was to contact her friend and confidant Dr James Lind, who had spoken up for her over the treatment of her wounded leg. The note dashed off to Alexander Aubert is disarming in its modesty, but hints at her sense of obstacles overcome. 'I hope, Sir, you will excuse the trouble I give you, with my *wag* [vague] description, which is owing to my being a bad (or what is better) no observer at all. For, for the last three years past I have not had an

♣ In modern times the passage of Hale-Bopp (1997) inspired a mass suicide by the Heaven's Gate cult, though that was in California. Even today, great uncertainty surrounds comets. Little more than 1,000 periodic comets have been identified, although several have been visited by space probes. Just as James Thomson suspected, they partly consist of frozen water, and have been described less romantically as 'dirty snowballs' of ice and rock. But current geophysical speculation that comets, as well as volcanoes, may have caused sudden catastrophic climate changes on planet earth in the past curiously brings back their role as portents of disaster. See the chapter 'Geology' in Natalie Angier's exuberant study of current scientific thinking, *The Canon: The Beautiful Basics of Science* (2007).

opportunity to look as many *hours* in the telescope. Lastly I beg you Sir, if this Comet should not have been seen before to take it under your protection.'[25]

Privately she still had grave doubts about her own observation skills, and wrote a frankly unscientific 'Memorandum' in her Observation Book, admitting that the comet seemed to have a mind of its own, and was not behaving at all as it should. 'I am at a loss what to think of the path which this Comet may have, by the figures [drawings] of last night it seemed to move *downwards* but tonights figures show just the contrary. In my letter to Mr Aubert I avoided taking notice of this circumstance ... for my wish was only to say what was just necessary by way of delivering it into better hands.'[26]

Her letter to Charles Blagden, Secretary to the Royal Society and Banks's right-hand man, produced a dramatic reply by return of post: 'I believe the comet has not yet been seen by anyone in England but yourself. Yesterday the Visitation of the Royal Observatory at Greenwich was held, where most of the principal astronomers in and near London attended, which afforded an opportunity of spreading the news of your discovery, and I doubt not but that many of them will verify it the next clear night. I also mentioned it in a letter to Paris, and in another I had occasion to write to Munich.'[27]

The verification of Caroline's comet was achieved much more rapidly than Herschel's planet had been. Its movement through Coma Berenices was relatively easy to ascertain, and its fine hazy tail or coma was unmistakeable. Its cometary status was quickly confirmed by Nevil Maskelyne, and on the following evening, 6 August, an impromptu top-level deputation rode down to Slough. Caroline was astonished to receive Blagden himself, Sir Joseph Banks and the MP Lord Palmerston, demanding to see *her* comet through *her* special sweeper telescope. Gratefully, she recorded that the evening was 'very fine', and everyone was able to get a glorious view of the new visitor, both with her small sweeper and the higher-powered seven-foot telescope.[28]

Banks was in one of his triumphal moods, and announced that her historic letter would be immediately published in the *Philosophical Transactions*, where it duly appeared – though after the usual bureaucratic delay – on 9 November, as 'An Account of a New Comet. In a Letter from Miss Caroline Herschel'. This was her first ever publication by the Royal Society, and an almost unheard-of rarity for a female correspondent.[29]

Maskelyne was also full of praise, patriotically recruiting Caroline into the new ranks of British astronomy at once. 'I hope that we shall, by our united endeavours, get this branch of astronomical business from the French, by seeing comets sooner and observing them later.'[30] Alexander Aubert, realising the personal significance of the find for her, struck a more intimate note: 'I wish you joy most sincerely of the discovery. I am more pleased than you can well conceive that *you* have made it – and I think that your *wonderfully clever and wonderfully amiable* Brother, upon the news of it, shed a tear of joy. You have immortalized your name.'[31]

The idea of a female astronomer intrigued people. When William returned from Germany ten days later, on 16 August, he found that Caroline had become something of a celebrity. In September he was summoned to Windsor specifically 'to exhibit to His Majesty and the royal family the new comet lately discovered by his sister, Miss Herschel'.[32] Fanny Burney the novelist, then a lady in waiting to Queen Charlotte, had evinced little previous interest in the stars. But she now suddenly discovered a lively fascination with astronomy, and leaped at the chance to abandon a game of royal piquet and join the viewing party on the Windsor terrace.

To Fanny's disappointment, Caroline herself was not there (she avoided the Court whenever possible). But the session was interesting 'for all sorts of reasons', the glimpse of the comet-catcher's brother being as fascinating as the comet. 'We found [Herschel] at his telescope. The comet was very small, and had nothing grand or striking in its appearance; but it is the *first lady's comet*, and I was very desirous to see it. Mr Herchel then showed me some of his newly discovered universes, with all the good humour with which he would have taken the same trouble for a brother or a sister astronomer; there is no possibility of admiring his genius more than his gentleness.'[33] Fanny was struck above all by Herschel's total lack of arrogance: 'he is perfectly unassuming . . . yet openly happy in the success of his studies'. But she wondered about his relationship with his reclusive sister.

Intrigued, she soon after persuaded her father to take her on a private visit to Herschel's observatory at The Grove on 30 December 1786. The 'great and extraordinary man' received them in his genial manner with open arms, showed them over the unfinished forty-foot telescope in the garden, and talked unguardedly over tea about 'the new views of the heavenly bodies and their motions' which it would reveal. Fanny was entranced. She exclaimed excitedly: 'he has discovered fifteen hundred universes! How many more he may find who can conjecture?' Charles Burney was also inspired by this visit, and began to compose an extensive 'Ode to Astronomy' in Herschel's honour, which he threatened to read out loud at future convivial suppers.[34]

By contrast, Caroline Herschel was rather silent, and much more of a puzzle. Fanny Burney evidently tried hard, but failed to get on terms with her. 'She is very little, very gentle, very modest, very ingenuous; and her manners are those of a person unhackneyed and unawed by the world, yet desirous to meet and return its smiles.' Those shy smiles seemed to be the extent of their communication. Equally, Caroline did not mention Fanny at all in her day book.[35]

Other visitors to The Grove had better luck. The German novelist Sophie von La Roche gushingly introduced herself to 'the great man's sister, who accompanies him on his path to immortality'. Perhaps Caroline found her fellow-countrywoman easier to placate than Fanny Burney, and made the inspired gesture of picking a bunch of daisies growing in the grass at the foot of the twenty-foot, and presenting them to her as a scientific keepsake. No doubt Sophie was intended to compare them to a star cluster beyond the Milky Way.[36]

Surprisingly, it was Nevil Maskelyne who began to take Caroline's technical prowess most seriously. A correspondence sprang up between them, and slowly blossomed over the next decade. He later wrote a detailed description of her 'large' Newtonian sweeper and her method of working with it. This telescope, built in 1791, was a five-foot reflector with an even bigger aperture of 9.2 inches, but the same low magnification of twenty-five to thirty times, designed for still more effective comet-hunting. Its field of view, being slightly narrower than that of the two-foot sweeper at 1.49 degrees, required even greater familiarity with smaller patterns of the surrounding stars.[37] Maskelyne remarked in passing that, like her brother, Caroline knew all the nebulae in the *Connaissance des Temps* instantly, and sight-read the night sky.[38]

During these same years Caroline was intensely involved in the final stages of setting up the great forty-foot telescope, intended as the climax of Herschel's observation work on the nebulae. While continuing her regular night work as assistant on the twenty-foot, she was also helping to organise a vast team of workmen during the day, overseeing the accounts, and trying to bring some order to Herschel's ever-increasing stream of distinguished and demanding visitors. In autumn 1787 these included the great French astronomer Pierre Méchain, director of the Royal Observatory in Paris and influential editor of the *Connaissance des Temps*. Praising Herschel for his preparatory work on the forty-foot, he also referred gallantly to 'Miss Caroline, your worthy sister, whose celebrity will shine down through the ages'.[39] When she discovered a second new comet in December 1788, any question of beginner's luck melted away even in England.[40] Her reputation continued to grow, especially in France and Germany.

Caroline remembered 1786–88 as the most intense and exciting years of her and William's lives. They were both in their prime: in 1786 he was a vigorous forty-seven, she an animated and increasingly self-confident thirty-six. Their teamwork had never been closer. Thanks to Caroline, Herschel published over a dozen new papers with the Royal Society. ('Very seldom could I get a paper out of his hands in time enough for finishing the copy against the appointed day for its being taken to Town.'[41]) Their great catalogue of nebulae had long since overtaken Flamsteed, and now stood at over 2,000 clusters, her own reputation as 'comet-hunter' gave her an independent scientific standing, and above all the great forty-foot telescope held out the promise of immense new discoveries. Sir Joseph Banks, the Astronomer Royal and the King himself all supported them. Sir William Watson commissioned a bust of Herschel for the Royal Society. Perhaps they would find more planets, new life elsewhere in the solar system, or even new civilisations among the galaxies. By 1789 they would certainly better understand how the universe had been created than at any previous time in history. This moment of scientific optimism coincided with the political optimism in Britain and France. In 1789 the Bastille would fall, and the Rights of Man would be declared.

Caroline's picture of her brother at this period is heroic, but also unintentionally disturbing in its impression of his single-mindedness. The gentle, humorous, sociable man that Fanny Burney had observed is very little in evidence. Here instead was the man who cut down trees. He was in the grip of his dreams, ruled by a new kind of scientific obsession, intensely focused, workaholic, self-denying. He was driving, driven, indefatigable, omnipresent: 'The garden and workrooms were swarming with labourers and workmen, smiths and carpenters going to and fro between the forge and the forty-foot machinery, and I ought not to forget that there was not one screw-bolt about the whole apparatus but what was fixed under the immediate eye of my brother. I have seen him lie stretched many hours in a burning sun, across the top beam whilst the iron work for the various motions was being fixed. At one time no less than twenty-four men (twelve and twelve relieving each other) kept polishing [the mirrors] day and night; my brother, of course, never leaving them all the while, taking his food without allowing himself time to sit down to table.'[42]

The subliminal image of Herschel almost crucified along the top beam of his telescope frame could not have been deliberate. Yet amidst all the bustle and excitement, Caroline slowly became aware of a growing financial crisis, which threatened to bring the entire project to a halt and wreck their fortunes. Over £500 had been wasted on the casting of a first, faulty mirror, a setback so severe that Alexander had urged the mirror's 'secret destruction' because it called into question the whole viability of their casting techniques.[43] Herschel had also seriously underestimated the costs of constructing the revolving gantry and paying the workmen for polishing the mirrors. Despite the sales of telescopes, they were threatened by bankruptcy. The whole glorious project could collapse in disaster and humiliation. By the summer of 1787 Herschel had to consider the delicate business of a new application to the King.

It was once again Sir Joseph Banks, the master of scientific diplomacy, who came to his aid. Although the huge half-ton mirrors were not yet finished, there was still a lot to see at The Grove: the great wooden gantry was partly installed on its turntable, now over seventy feet high, zone clocks and micrometers were assembled, and above all the enormous metal tube of the telescope was lying on its side, slumbering on the grass supported by wooden chocks, and ready to be winched into position. It was the moment, urged Banks, to give a Royal Telescope Garden Party.

Accordingly, on 17 August 1787 an impressive cortège of royal carriages rattled down from Windsor Castle, and Herschel and Caroline played host for the afternoon to a glittering party of dignitaries. The company included King George III and Queen Charlotte, the Duke of York, the Princess Royal, the Princess Augusta, the Duke of Queensberry, the Archbishop of Canterbury, many lords and ladies in waiting, a number of foreign visitors, and several distinguished Fellows of the Royal Society, although Banks himself seems to have remained tactfully absent. It was an impressive display which had the subtle effect, just as Banks would have foreseen, of further publicly committing the King to the scheme for which he was the acknowledged benefactor.

It also provided the occasion for another of the royal witticisms, a great additional gain. Caroline remembered it well even fifty years later. 'One anecdote of the old tube . . . Before the optical parts were finished, many visitors had the curiosity to walk through it, among the rest King George III and the Archbishop of Canterbury: following the King, and finding it difficult to proceed, the King gave him his hand, saying, "Come, my Lord Bishop, *I will show you the way to heaven*." '[44]

Now was the precise psychological moment to apply for the royal top-up grant. Herschel drafted a long letter to Banks for submission to the King, explaining the financial shortfall, the replacement of the faulty first mirror, the technical requirements of the gantry (now to be eighty feet high), and the fact that he expected no immediate profits except purely scientific ones. In an elegant formula probably devised by William Watson (if not by Banks himself), Herschel stated that his sole aims were 'the advancement of astronomy, the honour of a liberal Monarch, and the glory of a nation which stands foremost in the cultivation of arts and sciences'. All details were then costed. The new sum required was huge: £950. But there were also, of course, the continuing running costs, which he estimated could (with careful economies) be kept at £200 per annum. If this increased grant was again assumed to cover operations until 1789, then the total requested (though not specifically stated) was in the region of £1,400 – a very large amount indeed.[45]

Amazingly, Herschel did not stop there. He also raised through Banks something entirely new: the question of a separate royal stipend for Caroline as his official 'astronomical assistant'. No British monarch had ever granted a woman a salary, or even a pension, for scientific work before. The very idea that Caroline might be eligible was as novel as that she might be elected to a Fellowship of the Royal Society, or to a professorship at Oxford or Cambridge or Edinburgh. The one concession Herschel made to convention (again probably advised by Banks himself) was that this stipend might come officially from Queen Charlotte. His phrasing was a fine mixture of reason, *politesse* and provocation. It also contained the interesting claim that the idea for the request had originally come from Caroline herself. She was after all *the ladies' comet hunter*.

> You know, Sir, that observations with this great telescope [the forty-foot] cannot be made without four persons: the Astronomer, the Assistant, and two workmen for the motions. Now, my good industrious Sister has hitherto supplied the place of Assistant, and intends to continue to do that work. She does it indeed much better, to my liking, than any other person I could have, that I should be very sorry ever to lose her from that office.
>
> Perhaps our gracious Queen, by way of encouraging a female astronomer might be enduced to allow her a small annual bounty, such as 50 or 60 pounds, which would make her easy for life; so that if anything should happen to me she would not have the anxiety upon her mind of being left unprovided for.
>
> She has often formed a wish but never had the resolution of causing an application to be made to her Majesty for this purpose. Nor could I have been prevailed upon to mention it now, were it not for her evident use in the observations that are to be made with the 40 foot reflector, and the unavoidable increase of the annual expense which, *if my Sister were to decline*, that office would probably amount to nearly one hundred pounds more for an assistant.[46]

Herschel made no bones about the fact that a female assistant, even his sister, would cost half as much as a male. It is possible to be indignant about this, but contemporary standards must be taken into account. Female domestic servants were paid £10 per annum, while a highly trained governess like Mary Wollstonecraft was paid £40 per annum by Lord Kingsborough in 1787. In fact a £60 stipend would have been handsome, exactly one-fifth of that paid to the Astronomer Royal. In Europe women who wanted to pursue science, like Voltaire's beautiful mathematician Madame du Châtelet, or later Marie-Anne Paulze (Madame Lavoisier), simply had to have supportive or (even better) dead husbands, or private incomes. In Britain they had to be schoolteachers or children's textbook writers, preferably both: like Margaret Bryan (astronomy), Priscilla Wakefield (botany) or Jane Marcet (chemistry). Only in the next generation was it possible to have a career like the physicist Mary Somerville, and (eventually) have an Oxford college named after you. But then, Caroline did live long enough to exchange letters with Mary Somerville drily remarking on this situation.[47]

Six days after the memorable Telescope Garden Party, on 23 August, the King summoned Banks to the palace. His Majesty informed him that he would renew the grant at nearly double Herschel's requested sums, for a total of £2,000 – with an additional £50 per annum for Caroline for life. Here was true royal largesse. It also marked a social revolution: the first professional salary ever paid to a woman scientist in Britain.

But the gift came with a royal sting. The Telescope Garden Party had backfired. The King told Banks that he was annoyed at being placed in such a compromising position by the Herschels. His generosity had been taken advantage of, he had anyway expected quicker results, and in no circumstances would he ever provide a penny more towards the telescope. There were no royal witticisms.

Even Banks was shaken by this royal outburst, later described as a 'Storm'. It may perhaps have taken the alarming form of an early temper tantrum, since George III's madness would declare itself the following year. Banks privately summoned Herschel to Soho Square as early as possible the following morning. He was uncharacteristically tight-lipped: 'I have this moment seen the King, who has granted all you ask but upon certain conditions which I must explain to you.'[48]

The exact nature of these conditions was never put into writing, but probably referred to accounting of expenses, and the timing of future payments. Certainly the Herschels' account books now became minutely detailed, and included such things as the cost of the workmen's beer at lunchtime, and of 'four or five' individual candles burned each night.[49] Caroline later called the stipulations 'ungracious', and said they came with a blunt message 'that more must never be asked for'. They were sufficiently severe for Herschel to consider actually refusing the whole grant, as his old friend William Watson immediately wrote offering to send him 'one or two hundred pounds' instead. Herschel was dismayed at the unexpected turn of events; and for a few days gloomily considered abandoning the whole project. Caroline, despite – or perhaps because of – the success of her own application, was positively indignant. 'Oh! How degraded I felt even for myself, whenever I thought of it!'[50]

Wiser counsels eventually prevailed. As Banks must have pointed out, the grant was, after all, spectacularly generous; and the future of astronomy in Britain was at stake. He may also have warned Herschel, in confidence, about the King's fragile mental state. Watson sent a long, soothing letter on 17 September, urging a larger perspective, a wider field of view: 'I most sincerely sympathise with you, & feel in some measure as you must feel at the unworthy treatment you (& I may add Science) has received. But I sincerely hope by the latter part of your letter that the Storm is passed ... Let me hope, my dear Sir, that this affair has ceased to give you inquietude, & has not lessened your zeal for Science. Remember you have much cause for comfort & even of exultation. By your great discoveries ... you have gained a high and universal reputation.'[51]

Herschel's anger at the King's peremptory attitude gradually faded, as the circumstances of his illness became known. Caroline found it less easy to forgive. She eventually blamed George's courtiers. 'I must say a few words of apology for the good King, and ascribe the close bargains which were made between him and my Brother to the *shabby, mean-spirited advisors* who were undoubtedly consulted on such occasions.' By contrast, Sir Joseph Banks remained 'a sincere and well-meaning friend to the last'.[52]

Relations between Slough and Windsor never quite recovered their initial warmth, and it was only with the advent of the Prince Regent that further grants and honours were to be resumed, over twenty years later. Herschel was to be awarded a late knighthood in 1816, but his £200 stipend as the King's Astronomer at Windsor remained unchanged for the next three decades, by which time wartime inflation had virtually halved its value.[53]

Gradually exultation and the hectic regime were resumed. 'From this time on the utmost activity prevailed to forward the completion of the 40 foot ... and several 7 foot telescopes were finished and sent off.' A new optical workman was hired to oversee the polishing, and an enormous second mirror was successfully cast, much thicker than the first, and weighing in at nearly a ton. The quarterly instalment of Caroline's royal stipend was promptly delivered in October 1787, precisely £12.10*s*. It was her first ever professional payment: as she proudly noted, 'my *Salary*'. The 'astronomical assistant', for all her protests about royal behaviour, was evidently thrilled. It was 'the first money I ever in all my life thought myself to be at liberty to spend to my own liking. A great uneasiness was by this means removed from my mind ... For nothing but Bankruptcy had all the while been running through my silly head.'[54]

In November Pierre Méchain and Jacques Cassini came from Paris to inspect the preparations for the forty-foot, news of which was spreading across Europe. They also observed many of Herschel's 'new universes' through the twenty-foot, and went away thoughtful and deeply impressed.[55]

At about this time, before the forty-foot was finally mounted, the Herschels gave a celebratory banquet which spilled out from the house, across the lawn and finished up with a kind of musical crocodile, dancing into one end of the tube and out the other. It must have been an extraordinary moment, and Caroline was in the highest spirits: 'God Save the King was sung in it by the whole company, who got up from dinner and went into the tube, among the rest two Miss Stowes, the one a famous pianoforte player.' Friends picked up oboes, 'or any other instrument they could get hold of', and accompanied the singing and dancing. 'I, you will easily imagine,' recalled Caroline fondly, 'was one of the nimblest and foremost to get in and out of the tube.'[56]

2

It was a moment of hilarity and the highest spirits. Yet perhaps Caroline's nimble dancing 'in and out of the tube' disguised a certain anxiety. Towards the end of 1787 her emergence as a serious astronomer in her own right was threatened by a crisis in her relations with William that she must have long dreaded. The great brother-sister team was threatened by Herschel's growing friendship with an attractive neighbour. This woman was none other than the thirty-six-year-old Mrs Mary Pitt of Upton.[57]

Herschel and Caroline had often walked over to Upton village on summer afternoons to take tea with the Pitts in their handsome brick-floored parlour, before beginning their long nights of stellar observation. The little path was less than a mile long, eastwards along the escarpment and between large, scented hayfields.[58] It made an idyllic walk, especially when they were coming back in the early evening, with Venus setting in the western sky.

Mary's husband, John Pitt, was frail, and he died in September 1786. That winter the Herschels' teatime visits had become more regular, as a sharp-eyed neighbour, Mrs Papendiek, noticed. 'Widow Pitt, poor woman, complained much of the dullness of her life, and we did our best to cheer her, as did also Dr Herschel, who often walked over to her house with his sister of an evening, and as often induced her to join his snug dinner at Slough. Among friends it was soon discovered that an earthly star attracted the attentions of Dr Herschel.' The 'star' innuendo was an obvious one to make, but the true world of English provincial gossip was revealed in that little word 'snug'.[59]

Gossip did not concern Mary Pitt. She was a large, plain, kindly woman, whose friends described her as 'sensible, good-humoured and unpretending'. An oval miniature portrait shows her in simple country clothes, with her hair caught up in a knotted scarf, as if about to go on a long country walk. But she is also wearing a good, expensive necklace, and her large eyes suggest a certain thoughtful and determined air.[60] She was a woman of independent means, but with few social ambitions, and no wish to live in town. Altogether she had a calm, pleasant, down-to-earth quality that might well have appealed to a distracted astronomer, increasingly driven by his work and his celebrity. Now she was vulnerable, and perhaps that made her doubly attractive to a man like Herschel. Her only son, Paul, was often away from home at Eton; and her elderly mother, the wealthy Mrs Baldwin, was widowed, invalid and demanding. Mary Pitt was lonely, and William Herschel, in his own way, was lonely too.[61]

By early spring 1787 there began to be talk of marriage. Caroline, for whom the evening walks to Upton with her brother had seemed so innocent, was evidently unprepared for this, and shaken once she realised what was afoot. She wrote nothing in her journal, but there are tiny indications of increasingly erratic emotional behaviour. In February, when Alexander's wife died in Bath, she reacted with quite uncharacteristic violence. Her sister-in-law's death was not unexpected, as she had been ill for some time, and anyway Caroline had never been close to her, regarding her as a bore and a gossip. But according to Herschel, who responded phlegmatically enough himself, Caroline was almost hysterical with grief. As he told Alexander: 'Having been up all night Carolina was still in bed when your letter came. Poor Girl, she has hardly had a dry eye today; however our late sister's health had been so very bad we cannot say she died unexpectedly and therefore we should not grieve too much ... Carolina is not well enough to write today but will either tomorrow or next day endeavour to take up the pen. Last week I went to London to cast a 40 foot speculum, much thicker and stronger than my present one.'[62]

Herschel himself was evidently looking towards the future. He foresaw no financial problem in marriage. On the contrary, Mary Pitt turned out to be much wealthier than he had supposed. She had inherited a life interest in her husband's estate (her son Paul was left a handsome £2,000 to be going on with), and her difficult mother had promised to bequeath her all her properties (including the Crown Inn). It was calculated that this alone would bring rents worth at least £10,000 per annum.[63] At the very least, the future financing of the forty-foot would not be put at risk. Though Herschel had never been a fortune-hunter (and would never cease manufacturing telescopes), this must have been a vital consideration, especially after his contretemps with the King.

But what of Caroline? The new situation posed delicate questions of social roles, domestic powers and emotional loyalties, which Herschel tried hard to negotiate. His initial proposal was that he would continue his working establishment, with Caroline as mistress of The Grove, while his married home would become Upton, with Mary. In effect he was proposing a double life: as a husband in one place, and as a man of science in the other. This seemed eminently reasonable to him, and was perhaps not unsatisfactory to Caroline.[64]

It would take one of Jane Austen's unwritten novels to do justice to the social and emotional complications of this unfolding situation. The path between Upton House and The Grove must have been the scene of much drama. Herschel himself was evidently divided between attraction to Mary Pitt, loyalty to his sister and dedication to his science, none of which he hoped need be exclusive. Caroline had much to fear, and very little power of decision in her hands (though more than she first thought). Mary Pitt too was faced with real dilemmas, not least the risk of committing her fortune to a man with divided loyalties, and to coming between such a long-established brother-and-sister collaboration, all the more intense for being largely unexpressed.

Not surprisingly, Herschel's first proposal was promptly turned down by Mary Pitt. The wealthy widow was perhaps not so vulnerable as she appeared. Herschel politely withdrew his offer, as Mrs Papendiek, all agog, soon learned. 'Dr Herschel expressed his disappointment, but said that his [astronomical] pursuit he would not relinquish; that he must have a constant Assistant and that he had trained his Sister to be a most efficient one. She was indefatigable, and from her affection for him would make any sacrifice to promote his happiness.'[65] Caroline was safe.

But only temporarily. After some months the delicate negotiations were reopened, and a different compromise was agreed upon. Mary Pitt as Mrs William Herschel would become undisputed mistress of both establishments, at The Grove and at Upton. She would keep her own servants in both houses, preside at William's table in both, and oversee all his business accounts, including his scientific expenses (which of course she would be underwriting). Two of her maidservants would be kept at both houses, and finally, there would be a footman whose sole job was to take messages along the path between Upton and The Grove.

What was left for Caroline? She would remain at The Grove, but no longer as its mistress and housekeeper. She would move out permanently to the apartment above the stables, next to the observatory buildings. Here she would remain purely as William's 'astronomical assistant', though she could continue as an astronomer in her own right with her sweeper telescopes on the flat roof.

Perhaps a greater blow to her pride was that Caroline would have no further control of the business accounts. In fact she would be offered a quarterly salary of £10, just like a regular employee. This was the same sum that William had once given her just to buy a dress for her singing performances. Caroline pointedly refused this financial proposal, though it is obvious that William would have much preferred her to accept, as no doubt it would have greatly eased his conscience.

But Caroline's increasingly prickly sense of independence would not allow it. Indeed she later came to believe, or at least to claim, that she had arranged the royal salary precisely to avoid having to accept the fraternal one. 'I refused my dear Brother's proposal (at the time he resolved to enter the married state) of making me independent, and desired him to ask the King for a small salary to enable me to continue his assistant. £50 were granted to me, with which I was resolved to live without the assistance of my Brother.' In fact of course the royal stipend had begun eighteen months before the marriage. Caroline would not feel able to accept her brother's offer for another fifteen years.[66]

Whatever Herschel's feelings for Mary Pitt, he was evidently uncertain about the whole arrangement, and was asking advice from the faithful William Watson as late as March 1788, a mere six weeks before the wedding. Watson 'collected the general opinion' of Herschel's friends, and found that all were in favour of the union, 'excepting some little fears with respect to Astronomy'. It was thought that Herschel might 'relax somewhat' the intensity of his night observations. Personally Watson thought this would be a good thing for his health – 'I fear your endeavours are too fatiguing, both to your mind and body' – and that it would 'likewise turn out to the advantage of Science upon the whole'. No one apparently said a word about Caroline, unless she was included under the rubric of 'Astronomy'.[67]

William Herschel and Mary Pitt were married on 8 May 1788, at the tiny parish church of Upton. Sir Joseph Banks rode down from London to be best man. In an inclusive gesture, Caroline Herschel was asked to be one of the two formal witnesses, and when William Watson volunteered to be the other one, Caroline gallantly agreed.[68]

Caroline's last entry in her journal before the wedding is poignant in its determinedly matter-of-fact tone: 'The observations on the Georgian satellites furnished a paper which was delivered to the Royal Society in May. And the 8th of that month being fixed on for my Brother's marriage; it may easily be supposed that I must have been fully employed (besides minding the Heavens) to prepare everything as well as I could, against the time I was to give up the place of a Housekeeper which was the 8th of May, 1788.'[69]

There is no emotional outburst, no tears or recriminations. The only clue to the strength of Caroline's feelings is the fact that she unconsciously repeats the date of William's wedding in the same sentence, and suddenly invents that wonderfully imaginative phrase, 'minding the Heavens'. It is such a tender and ironic description of her entire career: she is the housekeeper to the heavens. But she then seems to hurry it away – the career and the phrase – into a bracket. Caroline did eventually give one further indication of her feelings. It was an entirely silent one, yet was the most dramatic personal gesture she ever made. She completely destroyed her personal journals covering the whole of the next decade of her life. Her records do not begin again until October 1798.*

The only person who may have glimpsed these journals before they were destroyed, or known something of their contents, was another woman, Caroline's future editor Margaret Herschel, the wife of her

* Although there are many dissimilarities, not least that of age, it is interesting to compare Caroline's situation with that described in Dorothy Wordsworth's journal on the day her beloved brother William married their friend Mary Hutchinson in October 1802 at Grasmere. 'At a little after 8 o'clock I saw them go down the avenue to the Church. William had parted from me upstairs. I gave him the wedding ring – with how deep a blessing! I took it from my forefinger where I had worn it the whole of the night before – he slipped

nephew John. Though restrained by strong family loyalties, Margaret left one circumspect but highly sympathetic comment about the journals in print: 'It is not to be supposed that a nature so strong and a heart so affectionate should accept the new state of things without much and bitter suffering. To resign the supreme place by her brother's side which she had filled for sixteen years with such hearty devotion could not be otherwise than painful ... One who could both feel and express herself so strongly was not likely to fall into her new place without some outward expression of what it cost her – *tradition confirms the assumption* – and it is easy to understand how this long significant silence is due to the light of later wisdom and calmer judgement, which counselled the destruction of all record of what was likely to be painful to survivors.'[70]

Over the years Caroline gave various quite different reasons for taking this extreme step. Mostly she passed it off by saying her journals were too dull to be of interest; or would not be understood; or else showed her lack of scientific achievement: 'These books I thought it best to destroy; excepting some fragments which I some 4 or 5 years since sent to my Nephew as waste paper. For, in consequence of my employment at the Clocks and writing Desk, when my Brother was observing I had no other opportunity for looking out for Comets, but when he was absent from home, but this happened so seldom and my sweeps were so broken and unconnected that I could not bear the thoughts of their rising in judgement against me; and besides they contained nothing new but the discovery of 8 Comets and a few Nebs. & clusters of stars.'[71]

At the very least Caroline must have felt that a highly successful scientific partnership was being endangered, one that was now increasingly recognised in the international community of astronomers. But perhaps she felt more, much more. Caroline cannot have forgotten that ten years previously she had given up her own future as a concert singer, when she rejected the offer of a solo appointment after performing arias from Handel's *Messiah* in 1778.[72]

---

it again onto my finger and blessed me fervently. When they were absent ... I could stand it no longer, and threw myself on the bed, where I lay in stillness, neither hearing or seeing anything.' Unlike Caroline, Dorothy contented herself with deleting only a single sentence of her journal, the one about wearing William's wedding ring (*Grasmere Journal*, October 1802). For a subtle and tender account see Frances Wilson, *The Ballad of Dorothy Wordsworth* (2008).

It is hard to believe that she did not feel deeply hurt, and even in some obscure way emotionally rejected, by her brother. But it is difficult to gauge the exact nature of these deeper feelings, and she may not have analysed them too closely herself. More immediately evident was her sudden loss of social status within Herschel's household. During this period in England, and even more so in Germany, previously dependent women – and notably unmarried younger sisters – would expect to be incorporated and remain happily within the newly married household. Caroline's new quarters above the workshops at The Grove were an acceptable adaptation, but her loss of managerial and social responsibility must have felt humiliating. This eventually led her to take the extreme step of abandoning her apartment altogether, and taking rooms in Slough village with the wife of Herschel's head workman, Mr Sprat.[73]

Yet outwardly things went on smoothly. Fanny Burney saw them all at a reception at Windsor later that summer, and thought the situation more amusing than tragic. 'Dr Herschel was there, and accompanied them [the Miss Stowes] very sweetly on the violin; his new-married wife was with him and his sister. His wife seems good-natured; she was rich too! And astronomers are as able as other men to discern that gold can glitter as well as stars.'[74]

When the French astronomer Jérôme Lalande visited The Grove observatory in autumn 1788 he was evidently charmed by Herschel's whole circle, and wrote to thank him with characteristic exuberance: '*Je n'ai jamais passé de nuit plus agréable, sans en excepter celles de l'amour.*' Caroline may have considered that an odd turn of phrase, in the circumstances. Lalande also reported that he had had an audience with King George III, who announced that he was immensely proud of the Herschels and pointedly remarked, while walking on the terrace at Windsor, 'that it was better to spend money on building telescopes than on killing men'.[75]

Caroline's spirits were lifted a little just before Christmas 1788, when on 21 December she discovered her second comet. This time it was moving through the constellation of Lyra, the Harp or Lyre. Although it eventually turned out that it had already been spotted by Charles Messier, this discovery produced much more correspondence than the first, and letters of congratulation – mostly still addressed to William, but sometimes sent directly to her – came crowding in from all sides: from Alexander Aubert, Sir Harry Englefield, Nevil Maskelyne and Jérôme Lalande in Paris. Thereafter Lalande became one of her most faithful, witty and faintly flirtatious correspondents, happily conforming, as he himself pointed out, to the archetype of a Parisian professor. He sent 'a thousand tender respects to *la savante Miss*, of whom I frequently speak with enthusiasm'. But then Lalande liked a little Gallic hyperbole, as he also addressed William on the envelope as: '*Monsieur Herschel, le plus célèbre astronome de l'univers, Windsor, Angleterre*'.[76]

Sir Harry Englefield, a stalwart of scientific committees, adopted a bluffer, but no less satisfactory manner, writing to Herschel on Christmas Day: 'I beg you to make my compliments to Miss Herschel on her discovery. She will soon be the Great Comet Finder, and bear away the prize from Messier and Méchain.'[77]

The most significant, and perhaps unexpected, of these correspondents was the Astronomer Royal, Nevil Maskelyne. Writing directly to Caroline from Greenwich Observatory on 27 December, he began a regular and increasingly confidential exchange of letters. Although he wrote formally to congratulate her, he then added a long, teasing speculation about the interesting possibilities of a close physical encounter with her new comet. He wondered whether Caroline would ever be tempted to ride off on it into space. Any true astronomer like her, he suggested, would consider '*without horror* the thought of our being involved in its immense tail'. However, he hoped she would not be tempted: 'I would not affirm there may not exist *some* astronomers so enthusiastic that they would not dislike to be whisked away from this low terrestrial spot into the higher regions of the heavens by the tail of a comet, and exchange our narrow uniform orbit for one vastly more extended and varied. *But I hope you, dear Miss Caroline*, for the benefit of terrestrial astronomy, will not think of taking such a flight, at least till your friends are ready to accompany you.'

Then he added, formally enough: 'Mrs Maskelyne joins me in best compliments to yourself and Dr and Mrs Herschel.' Yet these light-hearted urgings that Caroline should refrain from departing into outer space perhaps disguised Maskelyne's deeper concern about her unsecured position at Slough. Maskelyne was a family man himself, with an only daughter, Margaret, whom he doted on. Perhaps he understood Caroline's anxieties better than many in the scientific world.[78]

3

When the great forty-foot was at last put into operation in spring 1789, Herschel's first discovery was Mimas, one of the tiny innermost moons of Saturn, with a diameter of only 250 miles. This was a remarkable piece of astronomical observation, and promised well for the powers of the new monster instrument. Mimas is dominated by a single huge crater, eighty miles across and six miles deep, which was much later photographed and named 'Herschel', but only after the *Voyager* flyby of 1980.

Herschel gave a detailed description of the way he managed the forty-foot in a series of papers delivered to the Royal Society, illustrated by careful drawings.[79] He also described Caroline's wooden shed, situated some fifty feet beneath his own platform, equipped with masked candles, star atlases, warning bells and zone clocks.[80]

The completion of the telescope had finally been achieved with grants totalling £4,000 from King George III, an unprecedented amount for the sovereign to spend on a single scientific project of this kind. It was in fact exactly the same sum that the Royal Society had invested in 1768 in the entire scientific team (excluding Banks) for Cook's first three-year expedition to the South Seas. Like King George's Library (presented to the British Library by his son), the forty-foot telescope at Slough became one of the glories of his reign. It quickly became a tourist attraction, and was eventually featured in a popular Victorian magazine as one of 'The Wonders of the World', comparable to the Colossus of Rhodes.[81]

The doctor and writer Oliver Wendell Holmes included it in his tour of famous sites outside London. In his book *The Poet at the Breakfast Table* (1872) he described how he had previously seen an engraving of the great telescope in a child's encyclopaedia back at home in America. So when he rode down the London-to-Bath turnpike road its huge outline reared up over the trees at Slough 'like a reminiscence rather than a revelation'. It seemed a strange, unworldly shape. 'It was a mighty bewilderment of slanted masts, spars and ladders and ropes, from the midst of which a vast tube, looking as if it might be a piece of ordinance such as the revolted Angels battered the wall of Heaven with, according to Milton, lifted its mighty muzzle defiantly towards the sky.'[82]

But Herschel found the huge barrel of the forty-foot telescope unexpectedly difficult to prepare and manoeuvre in anything but perfect weather conditions. The vast surface of the metal speculum was far more susceptible to misting, oxidisation and distortion than those in his smaller telescopes. The one-ton mirrors were also alarmingly cumbersome to change, and Caroline remembered how both William and Alexander 'had many hair-breadth escapes from being crushed' when taking them in and out of the base of the telescope, even with the help of their workmen.[83] By the end of 1789 it was evident that the forty-foot was going to take years, rather than months, to prove its worth.

Through the 1790s Herschel had an ever-increasing sense that he must justify his project, and that the forty-foot was becoming something of a liability. He recorded that in the five years between 1788 and 1793 he managed only seventeen nights of ideal observations, a disastrous statistic.[84] Ironically, the elegant twenty-foot (much preferred by Caroline) continued to be better for deep-space stellar observations, being both more manoeuvrable and more stable. After Mimas, Herschel's best discoveries with the forty-foot remained inside the solar system: he added two new moons to Saturn, five already being known. The forty-foot fared much better as a national scientific showpiece, attracting a large number of European visitors, among them the head of the Paris Observatory and astronomy professors from Berlin, Cracow and Moscow.[85]

The annual need to repolish the huge three-foot mirrors became a growing burden, and in September 1807 Herschel was nearly killed when the one-ton speculum slipped from its harness while being removed from the tube. Many years later, in 1815, he quietly published a paper entitled 'A Series of Observations on the Georgian Planet', in which he admitted the insoluble problems of condensation, manoeuvring and servicing which the forty-foot had brought him.[86]

Yet Herschel's theoretical work now blossomed in an extraordinary and daring way. In 1789, the year of the Fall of the Bastille, he published a paper carefully dated 'Slough near Windsor May 1 1789', and gave it the deliberately anodyne title 'Catalogue of Second Thousand Nebulae with Remarks on the Construction of the Heavens'. This developed his revolutionary 1785 paper 'On the Construction of the Heavens', and extended it with a striking analogy between the botanical cycle as observed on earth, and an organic or 'vegetative' cycle which appeared to be operating throughout the entire universe.

The paper completely overturned any residual idea of a stable, overarching, temple-like universe, created once and for all by the great Celestial Architect and decoratively 'fretted with golden fire', as Hamlet once mentioned. On the contrary, Herschel suggested, the whole universe was subject to enormous fluid movements and changes, over vast periods of time, and these could be observed in the degree of 'compression' or 'condensation' of nebulae, and the 'comparative variety' of size and structure of deep-space star clusters. Herschel's crucial observation was that some galaxies were evidently older, and more evolved, than others. 'We are enabled to judge of the relative age, maturity, or climax, of a sidereal system, from the disposition of its component parts.' Nebulae and star clusters were in effect like 'species of plants', at various stages of growth and decay.

He explained this in his usual quiet, patient manner. 'Youth and age are comparative expressions; and an oak of a certain age may be called young, while a contemporary shrub is already on the verge of its decay.' The fundamental force at work was gravity, gradually over time compressing nebulous gas into huge, bright galactic systems, and eventually condensing into individual stars, 'So that, for instance, a cluster or nebula which is very gradually *more compressed* and bright towards the middle, may be in the perfection of its growth.' While another type of cluster, showing a more equal compression or distribution of individual stars, might be looked upon as 'very aged, and drawing towards a period of change, or dissolution'.

This method of viewing the galaxies ('to continue the simile I have borrowed from the vegetable kingdom') presented the entire universe in a new kind of light, with the most radical implications. 'The heavens are now seen to resemble a luxuriant garden which contains the greatest variety of productions, in different flourishing beds ... and we can extend the range of our experience [of them] to an immense duration.' In a garden we may live 'successively to witness the germination, blooming, foliage, fecundity, fading, withering and corruption of a plant'. Just so, the universe presented 'a vast number of specimens, selected from every stage through which the plant passes in the course of its existence', but brought 'at once to our eyes', and viewed at one particular moment from the earth.[87]

In this paper, astronomy changed decisively from a mathematical science concerned primarily (for practical purposes) with navigation, to a cosmological science concerned with the evolution of the stars and the origins of the universe. The implications were slowly absorbed, most notably by the French astronomer Pierre Laplace, who published his first paper on what he called 'the nebular hypothesis' in 1796.[88] But the revolutionary analogy, which made astronomy a life science with huge philosophical implications, was soon to be celebrated by Erasmus Darwin in the final book of *The Botanic Garden* (1791).

There were other, more personal forms of evolution. The year 1792 saw a decisive change in Herschel's family life. At the age of fifty-three he rejoiced in the birth of his first and only child, his son John. The regime at The Grove became steadily more domestic and sociable. Annual summer holidays, previously unheard of, began, with trips to Cornwall, the south coast and Scotland. Although Caroline rarely participated in these, the arrival of this little child would eventually affect her life too, as much even as her comets.

For the moment, though lonely and isolated, Caroline was doing the best observational work of her career. Pierre Méchain wrote admiringly to William on 25 October 1789, 'her renown will be held in honour throughout the ages'.[89] She continued to find new comets. In 1790 she found her third and fourth, in December 1791 a fifth, and a sixth in October 1793. She herself reported this sixth comet directly to the Royal Society, and her reputation continued to grow fast in astronomical circles. Articles appeared about her work in a number of women's journals, and a faintly scurrilous cartoon was published entitled 'The Female Philosopher Smelling out a Comet'. The comet is depicted as a small child hurtling through the night sky, driven by a fart, while the female astronomer, peering through her telescope and clutching her hands in delight, remarks enthusiastically on the 'strong sulpherous scent' of the comet's coma. But the depiction of Caroline, with her characteristic mass of curly hair, is surprisingly handsome.[90]

Her friendship with Maskelyne, the Astronomer Royal, continued to deepen, and he invited her to stay with him and his family at Greenwich, though she did not immediately accept. With his approval she had begun an updated *Star Catalogue*, which would eventually correct and supersede Flamsteed's, and receive the signal honour of being published at the Royal Society's own expense.

In November 1795 she 'shared' a comet with the German astronomer Johann Encke. Then in August 1797 she found her seventh. She was so excited by this last one that she did something unprecedented for her. After only one hour's sleep, she had a horse saddled for her in Slough, and rode the twenty-odd miles into London at dawn, then crossed the

Thames bridge, and appeared at Maskelyne's observatory at Greenwich for a late breakfast. She gave him a precise memorandum of the comet's position, which he confirmed that night.

At Maskelyne's urging, Caroline wrote to Sir Joseph Banks in Soho Square pointing out that it was a truly historic day, since she had never previously ridden more than two miles beyond Slough. This letter, dated 17 August 1797 from Greenwich, has a light-headed, almost flirtatious tone, which was again quite new for Caroline.

> Sir – This is not a letter from an astronomer to the President of the Royal Society announcing a comet, but only a few lines from Caroline Herschel to a friend of her brother's, by way of apology for not sending intelligence of that kind immediately where it was due... Dr Maskelyne was so kind as to take some pains to persuade me to go this morning 'to pay my respects to Sir Joseph', but I thought a woman who knows so little of the world ought not to aim at such an honour; but go home, where she ought to be, as soon as possible.[91]

It would appear that Caroline stayed on with Maskelyne's family for at least two days. This gesture of independence was shortly followed by a radical change in her lodging arrangements. In October 1797 she moved out of her apartment at The Grove, and into lodgings up the road in Slough village. She also began a new 'day book', in which the first entry read: '1797, in October I went to lodge with one of my brother's workmen (Sprat), whose wife was to attend on me. My telescopes on the roof, to which I was to have occasional access, as also to the room with the sweeping and observing apparatus, remained in its former order [at The Grove], where I most days spent some hours in preparing work to go on with at my lodging.'[92]

The exact significance of this move remains puzzling. To take lodgings with her brother's chief workman clearly sounds like a gesture of defiance against Mary Herschel. The reference to being allowed only 'occasional access' to her telescopes almost suggests that Caroline was being excluded from The Grove against her will. Yet she continued to work successfully on her enlarged and corrected edition of the *Star Catalogue*, which was completed and submitted to the Royal Society the following spring, on 8 March 1798. Its subsequent adoption and publication by the Society was a recognition of outstanding professional merit. Significantly, this was partly achieved through Nevil Maskelyne's good offices.[93]

Caroline wrote to Maskelyne, thanking him for all his support, in terms that begin conventionally enough. 'I thought the pains it had cost me were, and would be, sufficiently rewarded in the use it had already been, and might be in the future, to my brother. But your having thought it worthy of the press has flattered my vanity not a little.' However, she continued in a rather more provoking vein. 'You see, Sir, I do own myself to be vain, because I would not wish to be singular; and was there ever a woman without vanity? . . . . Or a man either? Only with this difference, that among gentlemen the commodity is generally styled ambition.' She leaves the implication of that – does she have any particular gentleman in mind? – hanging in the air, and turns to publication details of her *Star Catalogue.*

She ends her letter on a more intimate note. 'Many times do I think with pleasure and comfort on the friendly invitation Mrs Maskelyne and yourself have given me to spend a few days at Greenwich. I hope yet to have that pleasure next spring or summer. This last has passed away, and I never thought myself well or in spirits enough to venture from home. If the heavens had befriended me, and afforded us a comet, I might, under its convoy, perhaps have *ventured upon an emigration.*' She had not forgotten Maskelyne's original joke about her flying away on the comet of 1788; and perhaps she was also thinking of her first thrilling emigration from Germany with William long ago in 1772.[94]

Caroline's move to lodgings in Slough can be seen as an assertion of professional independence, and even perhaps a recognition of rivalry with her brother. Her day book for the following summer suggests that she had established a steady but solitary routine. In July 1799 she entered: 'My brother went with his family to Bath and Dawlish. I went daily to the Observatory and work-rooms to work, and returned home to my meals, and at night, except in fine weather, I spent some hours on the roof, and was fetched home by Sprat.'[95] In fine weather, of course, she stayed on the roof all night.

But the move must also have reflected an increasing sense of loneliness. She later wrote poignantly about her feelings of solitude and isolation. Such a revelation is extremely rare in her *Memoir*, and it almost seems to have taken Caroline herself by surprise. In the second, revised version of the *Memoir* she writes about her brother Alexander's unhappy love affairs, before he married. Unexpectedly, she adds a footnote: '. . . And I may here remark that I still was and remained almost throughout my long life without a Friend to whom I could have turned for comfort and advice when I was surrounded by trouble and difficulties. This was perhaps in consequence of my very dependent situation, for I never was allowed to form any acquaintance with any other but such as were agreeable to my eldest brother.'[96] This is a surprisingly bitter remark, given Caroline's correspondence with Aubert and Lalande, and above all her blossoming friendship with the Maskelyne family. Indeed, in summer 1799 she went to stay with them in Greenwich for about a fortnight.

When Faujas de Saint-Fond, the science writer and balloon enthusiast, on a long scientific tour of England and Scotland, visited The Grove at this time, he was encouraged to watch Herschel and Caroline working together on night observations. He saw nothing amiss, but on the contrary admired the constancy and 'delightful accord' of the brother-sister team working so closely together on 'this sublime but abstruse science'. Caroline explained to him the unique system of communications between William, high up on the observation platform, and herself below at her desk with its shrouded candlelight, zone clocks and star atlases. To the original method of shouted question and answer – 'Brother, search near the star Gamma Orion' – they had now added a system of coded rope-pulls, hand signals and bells. Later they would add the flexible speaking-tube. Perhaps this was symbolic of their changing relationship.[97]

Herschel himself at last began to mention 'my sister Miss Herschel', or 'my indefatigable assistant, Caroline Herschel', more regularly in his Royal Society papers. She appears in his historic 'The Description of a Forty Foot Reflecting Telescope' (1795), and again in 'A Third Catalogue of the Comparative Brightness of Stars' (1797). But once again his own mind seems to have been elsewhere.

4

In 1791 Herschel had published a highly significant paper with the Royal Society, 'On Nebulous Stars, Properly So-called'. For the first time he had observed an individual star with what he called 'true nebulosity', that is surrounded by a shining cloud of diffuse gas, even though it was already formed as a star. This observation caused him some dismay, because he had previously assumed that gas clouds were simply star clusters too far beyond our galaxy to be 'resolved' by his telescopes (even the forty-foot) into individual stars. The existence of true 'nebulae stars' suggested that perhaps most – or even all – gas clouds without resolvable stars were not distant star clusters after all, but simply nebulae much nearer than he had previously thought. 'Perhaps it has been too hastily surmised that all milky nebulosity, of which there is so much in the heavens, is owing to starlight only.'[98] He began to question his own extra-galactic theories, and wondered if all nebulae actually existed *within* the Milky Way. It therefore seemed possible – though he never stated this – that finally there were no other 'island universes' outside ours. It was a decisive retreat from his most radical thinking about the size and origin of the cosmos.

This theoretical retreat may have partly reflected a growing fear of radical science in England. When Erasmus Darwin published his long scientific poem in two parts, *The Botanic Garden* in 1791, he soon found that he had caused controversy by adopting Herschel's galactic theories without caution or reservations. Drawing on Herschel's earlier two papers on the 'Construction of the Heavens' (1785 and 1789), but ignoring the revisionist 'On Nebulae Stars' (1791), Darwin praised the great astronomer's 'piercing sight' into the cosmos, and his liberating new concept of an evolving universe, with distant nebulae growing and expanding like plants.

In a bravura passage, Darwin also considers Herschel's disturbing suggestion that the entire cosmos may eventually wither back into 'one dark centre'. This implies that the universe not only had a beginning, but will have a physically destructive end, a 'Big Crunch'. There are hints here too of Milton's vision of the falling rebel angels dropping out of the firmament in Book I of Herschel's favourite, *Paradise Lost*. This itself had possible political overtones for a reader in the 1790s, especially after the execution of Louis XVI in 1792.

> So, late descried by Herschel's piercing sight,
> Hang the bright squadrons of the twinkling night...
> Flowers of the sky! Ye to age must yield,
> Frail as your silver sisters of the field!
> Star after star from Heaven's high arch shall rush,
> Sun sink on suns, and systems systems crush,
> Headlong, extinct, to one dark centre fall,
> And death and night and chaos mingle all![99]

Darwin's note to this section calmly remarks: 'From the vacant spaces in some parts of the heavens, and the correspondent clusters of stars in their vicinity, Mr Herschel concludes that the nebulae or constellations of fixed stars are approaching each other, and must finally coalesce in one mass. *Philosophical Transactions* Vol. LXXV.' He adds however the consoling thought that a new universe may arise, phoenix-like, from the collapsed one (which might please contemporary proponents of multiverses). 'The story of the phoenix rising from its own ashes with a twinkling star upon its head, seems to have been an ancient hyroglyphic emblem of the destruction and resuscitation of all things.'[100]

Atheistical ideas were growing among Continental astronomers, and with the declaration of war against France these became even more suspect in Britain. In 1792 Herschel's great friend Jérôme Lalande published a third, enlarged edition of his authoritative *Traité d'Astronomie,* in three volumes, which expressed increasingly sceptical views. Eight years later he wrote an approving Preface to the *Dictionnaire des Athées* (1800). His final view before his death in 1807 was delivered with a flourish: 'I have searched through the heavens, and nowhere have I found a trace of God.'[101]

Pierre Laplace, another avowed atheist, now drew on Herschel's 'nebulae hypothesis' of star formation, and applied it to the formation of the solar system. He expanded this in the first volume of his classic *Mécanique Céleste* (1799). In effect he reasoned that the sun had slowly condensed out of a nebulous cloud of stardust, and then spun off our entire planetary system, just as in a thousand other star systems. There was no special act of Creation. In this way he was able to give a purely materialist account of the creation of the earth, the moon and all the planets. No divine intervention or Genesis was required, nor was it visible anywhere else in the universe.[102] Years later, Herschel's son John would argue that the nebula theory did not apply to the solar system, which was a special case, a 'singularity'.

Laplace's cool confidence in avowing atheistical sentiments was legendary. The story was told that after Napoleon had inspected a copy of Laplace's *Système du Monde,* he challenged the astronomer about his beliefs. 'Monsieur Laplace! Newton has frequently spoken of God in his book. I have already gone over yours, and I have not found His name mentioned a single time.' To this Laplace made the magnificent and disdainful reply: 'Citizen First Consul, I have no need of *that hypothesis.*'[103]

Herschel however was still interested in extraterrestrial life, and in 1795 published one of his most extraordinary papers, 'On the Nature and Construction of the Sun', with the Royal Society, suggesting that the sun had a cool, solid interior and was inhabited by intelligent beings. He reiterated his original claim that the moon was inhabited, and added that by analogy 'numberless globes' among the stars must support 'living creatures'. However, he disapproved of God-hunting within the galaxy, and attacked the 'fanciful poets' who had suggested that the sun was 'a fit place for the punishment of the wicked', viz. a fiery hell constructed for divine vengeance.[104]

Unlike Joseph Priestley, whose library was burnt down by a Birmingham mob in 1791, Herschel managed to avoid any public reputation for heterodox opinions. Visits to his observatory were regarded as uplifting, even religious experiences. Joseph Haydn claimed that his visit to Herschel at Slough in 1792 had helped him compose his oratorio *The Creation*. No scepticism undermines Haydn's joyful celebration of a universe abounding in benevolence, and still safely in the hands of the omnipotent God of Genesis. The dramatic moment of declaration occurs when Chaos, suggested by the key of C-minor, gives way to D-flat major, then to C-major, with the thunderous Scriptural proclamation, 'Let there be light!'[105]

In 1800 Herschel's continued interest in the sun led him to return to the problem of the prismatic distribution of solar light. While making direct observations of the sun (an extremely hazardous operation), he noticed that there were some indications of heat just outside the visible spectrum. In a series of experiments with thermometers mounted along a marked bar, he succeeded in measuring raised temperatures above the visible spectrum of solar light. Though he did not name it, he had discovered the presence of infra-red light. Once again he had broken a boundary of conventional knowledge.

News of this discovery spread rapidly through the scientific community. Henry Cavendish came over from Cambridge to see the experiment, and Benjamin Thompson, a founding member of the newly formed Royal Institution, came down from London. Sir Joseph Banks, delighted to offset Herschel's slow progress on the forty-foot, wrote that he considered that this would ultimately prove a more important discovery even than that of Uranus.[106]

On 3 July 1800 a young Cornishman named Humphry Davy wrote excitedly to his friend Davies Giddy: 'You have undoubtedly heard of Herschel's discovery concerning the production of heat by invisible rays emitted from the sun. By placing one thermometer within the red rays, separated by a prism, and another beyond them, he found the temperature of the outside thermometer raised by more than that of the inside.'[107] This marked a decisive advance on Newton's famous optical experiments with the prism, and implied a hitherto wholly unsuspected power in nature. It would also eventually lead to a decisive breakthrough in stellar astronomy in the twentieth century.

Herschel's public reputation as an astronomer steadily increased. In September 1799 he had been secretly commissioned by the War Office to provide a hundred-guinea spy telescope to be mounted on the walls of Walmer Castle, on the extreme south-east coastal point of Kent, to give early warning of a possible French invasion fleet. It was thought that the telescope could also spot any signs of the suspected aerial invasion by troop-carrying Montgolfiers.[108]

In 1801 Herschel was included in the first volume of the new biographical series of *Public Characters*, alongside Nelson, Pitt, Charles James Fox, Erasmus Darwin, the artists James Northcote and John Opie, Priestley, the anthropologist Lord Monboddo, the actress Sarah Siddons and the Bishop of Llandaff (who was appointed Professor of Chemistry at Edinburgh and promptly blew up his entire laboratory). Besides astronomy, Herschel's entry remarks on his gift for languages, his interest in metaphysics and (perhaps rather out of date) his habit of breaking off a concert to run outside and observe the stars. It also mentions in passing the remarkable talents of his sister Caroline.[109]

In July 1802 Herschel and his wife undertook a visit to Paris, during the short-lived Peace of Amiens. They were greeted as guests of honour by the Institut de France, and chaperoned by their old friend Lalande. They were introduced to the great mathematician Laplace, and were granted an audience with Napoleon, an interview that was chiefly memorable for the ice creams they were given to eat. They initially met the future Emperor in the garden of the Malmaison Palace, where he was supervising the irrigation of some newly planted flowerbeds. He was small and animated, appearing to be expertly informed on whatever subject arose (for example the construction of canals). Then, making a display of extreme informality, Napoleon bustled the party through some french windows into a drawing room, and flung himself down upon an upholstered chair. Herschel pointedly refused to sit in his presence, but carefully answered 'a few questions on Astronomy and the construction of the Heavens'. After further rather stilted conversation, Napoleon became sententious and announced to the assembled company that astronomy 'gave proof of an Almighty Wisdom'. Given the pronounced atheistical views of Laplace, his Chief Scientific Advisor (who was also present), Herschel thought Napoleon was being hypocritical, and actually believed nothing of the sort.

This rather froze the atmosphere, until the conversation turned to English racehorses (admirable, thought Napoleon), the English police system (slack), and English newspapers (unlicensed and amazingly outspoken). Napoleon then had the delicious ice creams served, in several different fruit flavours, while he observed that it was singularly hot, the temperature in the Malmaison garden being precisely 38 degrees in the shade. Herschel noted that the First Consul pointedly used the new centigrade system, and made the quick mental calculation that this meant it was 100.4 degrees Fahrenheit.

Suddenly Napoleon rose from his chair, made a brisk farewell, and without more ado swept out through a side door, pursued by several anxious aides and officers. Herschel only relaxed when he was returning to their hotel in a carriage with Laplace, discussing the rotations of double stars. He suggested that three stars might orbit round a common centre of gravity; but Laplace with an ironic smile contended that as many as six was possible, if not advisable. The First Consul crowned himself sole Emperor four years later.[110]

During this diplomatic episode little John, now aged ten, was left in the care of an elderly Polish Count, who showed him the animals in the Jardin des Plantes, which looked as lonely as himself. Aunt Caroline did not go on this Paris trip, but was left behind at the observatory to look after the telescopes and the visitors. She must particularly have missed meeting Lalande, who always included her in his letters, and would still send 'a thousand tender respects to *la savante Miss*'.[111] ♣

♣ Lalande had published a popular guide to astronomy for women, *Astronomie des Dames* (1795), in which he gave the history of women astronomers, beginning with the legendary Hypatia of Alexandria (also to be praised by Humphry Davy in his lectures) and continuing with Voltaire's mistress Émilie du Châtelet, who translated Newton into French. Caroline Herschel is described as the 'great comet hunter', renowned throughout Europe for her 'proficiency'. The book was subsequently translated into English with the anodyne title of *Astronomy for Ladies* (1815). See Claire Brock, *Comet Sweeper: Caroline Herschel's Astronomical Ambition* (2007).

Yet the trip was to prove important to her in one way. On the family's return from Paris to 'dear old England', the excitement of reunion began a new bond with her nephew John. The little boy had fallen ill on the return journey at Ramsgate, and it was Caroline who nursed him back to health, and listened to his tales of Continental adventures, and how he had sadly missed out on those delicious French ice creams. She had always felt tenderly towards the child, and after her move out to Slough in 1799 she noted: 'my dear nephew was only in his sixth year when I came to be detached from the family, but this did not hinder John and I from remaining the most affectionate friends'. Small herself, she loved sitting beside him on the carpet, 'listening to his prattle'. From the age of eight he would bring scraps of poetry to her, written 'in a most shocking handwriting'.[112]

The solitary, rather solemn little boy came to adore his aunt, and it was she, as much as his father, who inspired in him an early passion for science and astronomy. The shy and diminutive Caroline was able to play with him, and enter deeply into his childhood world, in a way that his father, now about to enter his sixties, was unable, or simply too distracted, to do. She arranged games for him in the garden, and messy experiments on the floor of her lodgings. 'Many a half or whole holiday he was allowed to spend with me … dedicated to making experiments in chemistry, where generally all boxes, tops of tea-cannisters, pepper-boxes, teacups etc served for the necessary vessels, and the sand-tub furnished the matter to be analysed. I only had to take care to exclude water, which would have produced havoc on my carpet.'[113]

When John was found climbing in the scaffolding of the forty-foot, or secretly having tea with the workmen, or cutting geometrical shapes in the panelling of the drawing room with a chisel, it was Caroline who always leaped to his defence.[114] It was also she who gave him several workshop tools for his birthdays, including the small wood-plane, proudly incised with the name 'John' on the handle, which he kept for the rest of his life.[115]

When John was sent to Eton at the age of eight, it was Caroline who saw how unhappy he was there, and tried to persuade William and Mary to choose a different mode of education from John's extrovert step-brother Paul, who had flourished at the school. Mary was reluctant to make the change until she saw John knocked down in a boxing match with an older boy, after which she summarily withdrew him and employed a private tutor, much to Caroline's delight.[116] A portrait of John at this time shows a small, delicate, wide-eyed boy, wistfully holding a wooden hoop, with the towers of Windsor Castle and Eton distantly on the horizon.

In an extraordinary way the relationship between Caroline and her young nephew began to heal whatever suppressed strains and rivalries there were within the Herschel household. Caroline and Mary were increasingly united in their concern for John's welfare, while Caroline knew how to interpret emotionally – as well as scientifically – between father and son. Later, this mentoring relationship would take on unusual importance.*

5

As he grew older, Herschel was becoming a remoter figure in the household. His mind was ranging through the universe. His later papers for the Royal Society had begun to show an increasing awareness of the philosophical significance of astronomy. This was something urged upon him by his old supporter William Watson, who looked forward to conversations 'on Kant's metaphysics', and wished to know how far Herschel agreed with the 'ground and sources' of Kant's philosophy of knowledge.[117]

Already in a paper of 1802 Herschel considered the idea that 'deep space' must also imply 'deep time'. He wrote in his Preface: 'A telescope with a power of penetrating into space, like my 40 foot one, has also, as it may be called, a power of penetrating into time past ... [from a remote nebula] the rays of light which convey its image to the eye, must have been more than 19 hundred and 10 thousand – that is – almost two million years on their way.' The universe was therefore almost unimaginably older than people had previously thought. This idea of deep time was one which required a great deal of explanation to the layman.[118]

---

* Something oddly similar happened in the Wordsworth household, when Dorothy Wordsworth became deeply attached to her brother's first child, her nephew John Wordsworth, who was born in 1803, with Dorothy and Coleridge as his godparents. Dorothy nursed and played with John, who always remained Aunt Dorothy's favourite, while Wordsworth doted on his beautiful daughter Dora (much to her discomfort in later adulthood). Dorothy even acted for several years as John's devoted housekeeper, after he grew up and became a rather solemn young clergyman.

Other papers were unsettling in different ways. 'Observations tending to investigate the Nature of the Sun' (1801) proposed that sun-spot activity could be related to the price of wheat, because it affected the mildness or severity of terrestrial seasons, and hence the fertility of global harvests. Thus the sun, rather than the stars or comets, could bring about political revolutions on earth.[119] Another paper, 'On the Proper Motion of the Solar System', showed that not only did the planets revolve round the sun, but that the entire solar system itself moved through stellar space, orbiting round an unidentified centre in the Milky Way, which was itself moving relative to other galaxies.[120]

Herschel continued to reach carefully towards the idea of an *evolving* universe, a concept as radical in its eventual implications as Erasmus Darwin's notion of evolution within plants and animals. In a late paper published in 1811, 'Astronomical Observations relating to the Construction of the Heavens', Herschel further developed the idea, already explored in 'On the Construction of the Heavens'(1785) and 'Catalogue of Second Thousand Nebulae with Remarks on the Construction of the Heavens' (1789), that all nebulae and large star clusters were at particular points in their sidereal life-cycles, which could be visually identified and catalogued almost in a Linnaean manner. Their characteristic shapes suggested distinct moments of youth, maturing and ageing.

Herschel accompanied this paper with numerous drawings of the nebulae he had observed over thirty years in these different phases: some globular, some spiral, some flattened, some mere blots of incoherent light or chaotic milky spillages. Many, such as the beautiful and characteristic whorl of Andromeda, are now instantly recognisable because of the modern Hubble photographs. These shapes, Herschel argued, were not different because they had been *created* differently, like different species. They were different simply because their stages of development in what he called '*sidereal time*' (meaning stellar time) had reached different points. He was suggesting the inescapable idea of evolutionary youth and age in the universe.[121]

This presentation was radically different from anything seen in any previous astronomical papers anywhere in Europe, except in the broadest philosophical speculations of Kant, the French cosmologist the Comte de Buffon, or Laplace. It presented the universe as a living, growing, organic entity, with all nebulae belonging to one enormous extended family: 'There is not so much difference between them, if I may use the comparison, as there would be in an annual description of the human figure, were it given from the birth of a child till he comes to be a man in his prime.' This comparison is an intriguing premonition of 'time-lapse' photography, now one of the most powerful illustrative tools of modern natural history.[122]

Above all, Herschel's studies of nebulae and the general 'construction of the Heavens' demonstrated how Copernicus' rejection of an earth-centred universe had long been superseded by contemporary science. Not only a sun-centred galaxy, but even a cosmos centred on the Milky Way itself, had to be rejected. This implied an enormous psychological, even spiritual, shift in outlook: seeing our entire solar system as something very small, very far out, on the very edge of the edge of things. As Herschel had written: 'We inhabit the planet of a star belonging to a Compound Nebula [the Milky Way] of the third form ...'[123] ♣

Over the next decade Herschel's work began to be widely known by the younger generation of Romantic writers. Byron visited him at Slough in 1811, and viewed the stars through his telescope, which gave him an alarmingly religious experience: 'The Night is also a religious concern; and even more so, when I viewed the Moon and Stars through Herschel's telescope, and *saw that they were worlds*.'[124] Later Byron defended himself against accusations of atheism. 'I did not expect that, because I doubted the immortality of Man, I could be charged with denying the existence of a God. It was the comparative insignificance of ourselves and *our world*, when placed in competition with the *mighty whole*, of which it is an atom, that first led me to imagine that our pretensions to eternity might be ... *over-rated*.'[125]

♣ Coleridge seized on this idea in a late essay: 'Kepler and Newton, by substituting the idea of the Infinite – for the idea of a finite and determined world assumed in the Ptolemaic Astronomy – superseded and drove out the notion of one central point or body of the Universe. Finding a centre in every point of matter and an absolute circumference nowhere, they explained at once the unity and the distinction that co-exist throughout the Creation by focal instead of central bodies. The attractive and restraining power of the sun or focal orb, in each particular system, supposing and resulting from an actual power, present in all and over all, throughout an indeterminable multitude of systems. And this, demonstrated as it has been by science, and verified by observation, we rightly name the true system of the heavens' – *Church and State* (1830). Hubble put this simply and beautifully: 'Our stellar system is a swarm of stars isolated in space. It drifts through the universe as a swarm of bees drifts through the summer air' – Edwin Hubble, *The Realm of the Nebulae* (1936).

John Bonnycastle's highly successful *Introduction to Astronomy in Letters to his Pupil* was reissued in an enlarged edition in 1811, with an expanded chapter dedicated to Herschel's work and other 'new discoveries'. This was the edition given to John Keats at his Enfield school, and later taken to his lodgings near Guy's Hospital. Bonnycastle continued to include passages of poetry with his scientific explanations, and his work encouraged reflection on the imaginative as much as the philosophical impact of the new astronomy. In theological matters, however, Bonnycastle remained strictly orthodox.

Bonnycastle became apologetic about including poetry, too. In the Preface to his 1811 edition he warned his readers: 'The frequent allusions to the Poets, and the various quotations interspersed throughout the work, were intended as an agreeable relief to minds accustomed to the regular deduction of facts, by mathematical reasoning... Poetical descriptions, though they may not be strictly conformable to the rigid principles of the Science they are meant to elucidate, generally leave a stronger impression on the mind, and are far more captivating than simple unadorned language.'[126]

Keats wrote his sonnet 'On First Looking into Chapman's Homer' very early one autumn morning in October 1816. It celebrates a deeply Romantic idea of exploration and discovery. Without actually naming Herschel, it picks out the finding of Uranus, thirty-five years before, as one of the defining moments of the age. Although combining many sources of inspiration (it is possible that Keats may have attended Charles Babbage's 1815 'Lectures on Astronomy' at the Royal Institution), the poem itself was written in less than four hours.

Keats was twenty, and attending a full-time medical course at Guy's Hospital. He had stayed out all night with his friend and mentor Charles Cowden Clarke at his house in Clerkenwell, drinking and discussing poetry. Clarke had acquired an old 1616 folio edition of Chapman's verse translation of Homer's *Iliad*, and they had taken turns to recite passages aloud. At particular passages Keats 'sometimes shouted' with delight. A favourite was the gloriously extended simile of shining light from Book Five. This compares the golden glow of the Greek warrior Diomed's helmet to the glow of the planet Jupiter rising above the sea in autumn.

> Like rich Autumnus' golden lampe, whose brightness men admire,
> Past all the other host of Starres, when with his cheerful face,
> Fresh washt in lofty Ocean waves, he doth his Skies enchase.

With such images in his head, Keats left Clerkenwell at 6 a.m., shortly before autumn sunrise. The stars were still out as he crossed London Bridge making for his student lodgings at 8 Dean Street, Southwark, near Guy's. He noticed the planet Jupiter, very bright, setting over the Thames. The moment he got to his lodgings, he sat down and began to write, starting with the inspired line, 'Much have I travelled in the Realms of Gold ...'. This perfectly introduced two linked ideas, of thrilling exploration and gleaming brightness, which orchestrate the whole poem.

Keats wrote so quickly that he was able to send a clean copy of the poem straight round to Cowden Clarke that same morning. Clarke remembered opening it at his breakfast table in Clerkenwell by 10 a.m. (a credit also to the postal system). He noticed the historical error – it was Balboa, not Cortez, who reached the Pacific – but was thrilled by the beauty and originality of the sonnet. Among other things, Keats had combined science and poetry in a new and intensely exciting way.[127]

Keats likens his own discovery of Homer's poetry to the experience of the great astronomer and the great explorer finding new worlds.

> ... Then felt I like some watcher of the skies
> When a new planet swims into his ken;
> Or like stout Cortez when with wond'ring eyes
> He stared at the Pacific – and all his men
> Looked at each other with a wild surmise –
> Silent upon a peak in Darien.

Both comparisons turn on moments of physical vision – watching, staring, looking with 'wondering eyes'. (This was the original manuscript reading, although Keats later changed it to the more conventional 'eagle eyes'.) Physical vision – one might say scientific vision – brings about a metaphysical shift in the observer's view of reality as a whole. The geography of the earth, or the structure of the solar system, are in an instant utterly changed, and forever. The explorer, the scientific observer, the literary reader, experience the Sublime: a moment of revelation into the idea of the unbounded, the infinite.

In the case of Herschel's sighting of Uranus, Keats's word 'swims' is brilliantly evocative, because of its sense of new life and movement. The planet is like some unknown, luminous creature being born out of a mysterious ocean of stars. Keats may also have realised that convection currents in the atmosphere, or in the tube of the telescope itself, can give objects the appearance of being seen through a rippling water surface.[128]

Keats's vivid idea of the Eureka moment of instant, astonished recognition celebrates the Romantic notion of scientific discovery. It is appropriate that this is expressed in the oddly anachronistic phrase 'into his ken' (grasp, knowledge), even though it may also be there for the rhyme. The efforts of Maskelyne, Messier and Lexell certainly took weeks, if not months, to confirm the identification of Herschel's 'comet' in 1781. Yet it is also true that Herschel too, despite the evidence of his own Observation Journal, gradually convinced himself that precisely such a moment of instant, sublime discovery had occurred in the garden at New King Street. Herschel in the end may have remembered that night exactly as Keats imagined it.[129]

6

Journalists began to assess the current scientific views of such phenomena, ranging from the 'deep time' geological theories of James Hutton to the 'deep space' nebula theories of Herschel. An essayist in the *Monthly Review* for April 1816 noted archly that there was 'a studious avoidance of any reference to God' in any of these brave new theories. He presented them with a certain scepticism. 'A long dissertation is allotted to the Huttonian Theory of the Earth; but all these speculations, we suppose, must now give way to the "discovery" of Dr Herschel, that planets began their being in the form of nebulous matter, and consist at first of *a vast egg of gas!*'[130]

Meanwhile Herschel was quietly publishing more extraordinary late papers of cosmological speculation, notably 'Astronomical Observations Relating to the Sidereal Part of the Heavens and the Connection with the Nebular Part: Arranged for the Purpose of a Central Examination', a characteristically low-key title for a dramatic paper, dated 24 February 1814.[131] Its final section was headed 'The Breaking Up of the Milky Way'. In it he proposed that there was 'a clustering power' observable in many nebulae, that was producing a 'progressive approaching' within each star group. 'We may be certain that from mere clustering stars they will be gradually compressed through successive stages of accumulation – till they come to what may be called the *ripening period* of the globular, and total insulation.' So, as each star group was increasingly contracting in on itself, it would also be moving away from all the others into a state of greater and greater cosmic isolation.

From these vast movements within the observable universe, Herschel concluded that as some galaxies were being born, others were withering and dying. He had previously touched on this general idea. But he now put forward the apocalyptic proposition that our own galaxy was on the wane, and that there would inevitably be 'a gradual dissolution of the Milky Way'. The progress of this observable dissolution would provide 'a kind of chronometer that may be used to measure the time of its past and future existence'.

At all events, it was clear that the Milky Way 'cannot last forever'; and equally that 'its past duration cannot be admitted to be infinite'. It followed that neither the earth, nor even the solar system, was a separate creation, but merely an infinitesimal part of a galactic evolution. Our galaxy had a physical beginning, and would have a physical conclusion. Our solar system, our planet, and hence our whole civilisation would have an ultimate and unavoidable end.[132]

In 1817 Thomas Chalmers would publish his best-selling *Discourses on Astronomy*. He reflected on the atheistical implications of Herschel's new cosmology, piously attempted to re-establish God's role in the creation, but also raised intriguing questions about extraterrestrial life in the further planets of the solar system, and beyond. The book touched a public nerve, and sold 20,000 copies in its first year, although it was sceptically reviewed by William Hazlitt, one of his rare forays into the physical sciences. But the idea of a dramatically enlarged universe, which surely contained other civilisations apart from our own, was widely accepted among progressive thinkers of the next generation. William Whewell, the future Master of Trinity College, Cambridge, for exampled published an uplifting monograph, *On the Plurality of Worlds*, in 1850.[133]

Despite these apocalyptic conclusions, Herschel was beginning to be regarded as a kindly, silver-haired old sage. For all his remote speculations, there was said to be something disarmingly boyish about him. The poet Thomas Campbell was surprised to find him, with his son John, on holiday in Brighton in September 1813. John, incidentally, was 'a prodigy in science and fond of poetry, but very unassuming'. Campbell was completely captured by the 'great, simple, good old man', as he called Herschel: 'Now for the old astronomer himself: his simplicity, his kindness, his anecdotes, his readiness to explain – and make perfectly perspicuous too – his own sublime conceptions of the universe are indescribably charming. He is 76 [actually Herschel was then seventy-four], but fresh and stout, smiling at a joke ... Anything you ask, he labours with a sort of boyish earnestness to explain ... I asked him if he thought the system of Laplace to be quite certain, with regard to the total security of the planetary system, from the effects of gravity losing its present balance? He said, "No".

Campbell overlooked the startling bluntness of this reply, and its implication that the solar system could easily fly apart (or else implode). Instead he went on to record amiable chat about the newly discovered 'asteroid' belt between Mars and Jupiter. Herschel had actually named 'asteroids' himself in an earlier letter to Banks,[134] and murmured, 'remember there will be thousands more – *perhaps 30 thousand* – not yet discovered'. Herschel also mentioned applying Newton's theories for measuring the speed of solar light to 'inconceivably distant bodies' in the stellar system, with unimaginable results. 'Then speaking of himself, he said with a modesty of manner that quite overcame me, when taken with the greatness of the assertion: "I have looked further into space than ever human being did before me. I have observed stars of which the light, it can be proved, must have taken millions of years to reach the earth." '

Campbell recalled that he felt he had been 'conversing with a supernatural intelligence'. Finally, Herschel completely perplexed the poet by remarking that many distant stars had probably 'ceased to exist' millions of years ago, and that looking up into the night sky we were seeing a stellar landscape that was not really there at all. The sky was full of ghosts. 'The light did travel after the body was gone.' After leaving Herschel, Campbell walked onto the shingle of Brighton beach, gazing out to sea, feeling 'elevated and overcome'.[135] He was reminded of Newton's observation that he was just a child picking up shells on the seashore, while the great ocean of truth lay all before him.[136]

5

Mungo Park in Africa

1

In 1803 Joseph Banks wrote to a friend: 'I am aware that Mr Park's expedition is one of the most hazardous a man can undertake; but I cannot agree with those who think it is too hazardous to be attempted: it is by similar hazards of human life alone that we can hope to penetrate the obscurity of the internal face of Africa.'[1]

Throughout the late 1790s Banks had been increasingly tied down to his presidential chair in London. Physically he was marooned by his gout, and intellectually by the continuous administrative claims of the Royal Society. Yet despite this enforced immobility, and perhaps because of it, Banks's huge imaginative interest in geographical exploration had continued to expand.

From Soho Square his gaze swept steadily round the globe like some vast, enquiring lighthouse beam. The fine, free anthropological adventures in the South Seas of earlier years were a thing of the past, of his lithe youth. But perhaps he could find others to undertake them. He followed the adventures of contemporary travellers with passionate interest. James Boswell gave a mutually flattering account of Banks reading his *Tour of the Hebrides* when it was first published in October 1785: 'The President of the Royal Society clasped his hands together and remained for some time in an attitude of silent admiration.'[2]

Banks determined to support and encourage travel and exploration, both for its scientific value and increasingly for the national interest. In 1779 he had first given evidence before a special committee of the House of Commons recommending 'Botany Bay, on the Coast of New Holland' as the place for colonial settlement and a penal colony. For the next twenty years he kept in close touch with the governors of New South Wales, arranged for a continuous supply of botanical specimens to be shipped back to Kew, and sponsored several expeditions to explore the continent further, such as Matthew Flinders' heroic circumnavigation in 1802–03, and his travels in the mountain ranges of Victoria.

In June 1788 Banks had also become a founder member of the Association for Promoting the Discovery of the Inland Districts of Africa, attending its first historic meeting at the St Alban's tavern in The Strand.[3] Its Secretary was Bryan Edwards, a close friend of Banks's, whose writing on the West Indies and indigenous folklore and witchcraft was later to inspire ballad poems by both Southey and Coleridge (notably Coleridge's 'The Three Graves' in 1798). This pioneering body, which came to be known simply as the Africa Association (and much later, in 1831, was to be merged with the Royal Geographical Society), was soon sponsoring small but highly adventurous expeditions into Egypt and the horn of Africa. Its motives at this stage were scientific and commercial, with no missionary or colonial intentions. Its primary aim was discovery, not conquest. This would change once Banks was appointed to the Privy Council in 1797, and became ever more closely involved in prosecuting the war against Napoleon Bonaparte. From then all exploration took on a more political and frankly imperial significance. Yet Africa and Australia always fascinated Banks for their own sake.

All of the early sponsored African expeditions ended in mystery. John Ledyard was sent out to explore westwards from Cairo in 1788, Major Daniel Houghton to cross the Sahara in 1791, and Friedrich Hornemann to explore southwards from Tripoli in 1799. Various reports and rumours drifted back to Banks and the Africa Association, but none of these early heroic travellers returned alive.[4]

The great prize was to reach the semi-legendary city of Timbuctoo, somewhere south of the Sahara. Here, it was said, lay a great West African metropolis, packed with treasures and glittering with towers and palaces roofed with gold. It was strategically situated astride the fabled river Niger, at the confluence of the Arabic and African trade routes. Beyond Timbuctoo, it was thought that the mysterious Niger might flow due eastwards, providing a trade route across the entire African continent, and eventually meeting up with the Nile in Egypt. But to the Europeans nothing was known for certain, though many speculative maps had been drawn by military cartographers, such as Major John Rennell's 'Sketch of the Northern Parts of Africa', presented to the Association in 1790.

Banks remained optimistically on the lookout for young men of promise and daring. Perhaps he was searching for versions of his earlier self, the fearless young anthropologist and botanist in Tahiti. The fact that his marriage to Lady Banks had produced no children may well have given him a special, personal interest in the careers of these young protégés.

In 1792 he was introduced to a lanky, sandy-haired young doctor from Scotland. Mungo Park had been named by his mother after the Gaelic martyr St Mungo. He struck Banks as a tall, very largely silent, but strangely impressive young man with that promising shine of adventure in his eyes. He was twenty-one years old, unmarried, and announced that he was desperate to travel. Banks immediately sensed a likely candidate, with a suitable physique and a tough, unpretentious background.

He learned that Park had been born into a large, hard-working family at Foulsheils, near Selkirk, in 1771. He had had a happy but Spartan upbringing on a lowland farm, growing up in the valley of the Yarrow river. He was physically hardy and resilient, but also well-read and thoughtful. His background was not unlike Robert Burns's, but his temperament was quite different. Sober, reserved, intensely private almost to the point of withdrawal, Park was a natural loner. But he also had stoic, unshakeable determination, probably influenced by his mother's Calvinism. His faraway eyes had a blue, impassive glitter. If he was a dreamer, he was not afraid of nightmares. Not, at least, to begin with.

At fourteen Park went to live with his uncle, Thomas Anderson, a surgeon in Edinburgh. Here he learned medicine, and made the closest – perhaps the only – friend of his life, his cousin Alexander Anderson. He also admired Alexander's pretty little sister Allison, but she was only eight. Park took his medical degree at Edinburgh University, but could not settle down to domestic doctoring. He wrote poetry, studied astronomy and botany, climbed Ben Nevis, and read travel writers. He was tall, bony, handsome, and deeply uncommunicative. 'His friendship was not easily acquired, for he was ever of a shy, retired, though not suspicious temper,' wrote a later biographer. 'To strangers his calm reserved manner had something of the appearance of apathy and total want of feeling ... Even his dearest friends ... were sometimes ignorant of the designs that lay nearest to his heart, and formed the subjects of his secret meditations.'[5]

In autumn 1792, at the age of twenty-one, he went to London to seek his fortune and find wider horizons. He had been given the introduction to Sir Joseph Banks through his brother-in-law, James Dickson, a botanist who worked at the British Museum gardens. After a breakfast interview at Soho Square, Banks arranged for Park to join a naval expedition to Sumatra in the East Indies, as Assistant Surgeon. He also gave him the run of his library, to prepare himself with reading and study. After his own experiences at Batavia twenty-two years before, Banks must have known that this voyage would be a demanding – perhaps fatal – trial of both the young man's physical constitution and his morale.

On this first expedition to Sumatra, Park quickly discovered his love of travel and his extraordinary sense of self-sufficiency. When he returned eighteen months later in May 1794, tanned and fit, Banks recognised his remarkable qualities, and suggested to the Africa Association that they should send Park to explore the Niger. Speaking in his quiet, lowland accent, Park confessed to Banks that he had 'a passionate desire' to discover the unknown Africa, and 'to become experimentally acquainted with the modes of life and character of the natives'. If he should 'perish in the journey', he was willing that his hopes and expectations should perish with him. He required no promise of 'future reward', and he had no missionary intent. This romantic attitude deeply appealed to Banks, as also to the accountants of the Africa Association.[6]

In the event the Association supplied Park with basic kit for his expedition, and a salary of seven shillings and sixpence a day, or just over £11 a month. They also booked him a passage on a merchant ship bound for the Gold Coast (oddly, like Banks's, it was named the *Endeavour*), and supplied a £200 letter of credit to buy supplies and trading goods at Pisania, the last white outpost on the river Gambia. Park's kit was indeed very basic: it included two shotguns, two compasses, a sextant, a thermometer, a small medicine chest (the regular use of quinine as a prophylactic against malaria had not yet been adopted), a wide-brimmed hat and the indispensable British umbrella. There were also two vital objects of sartorial formality: a blue dress coat with brass buttons, and a malacca cane with a silver top.[7]

'My instructions were very plain and concise,' Park later wrote in his characteristic style. 'I was directed on my arrival in Africa to pass on to the river Niger, either by the way of Bambouka, or by such other route as should be found most convenient. That I should ascertain the course, and, if possible, the rise and termination of that river. That I should use my utmost exertions to visit the principal towns or cities in its neighbourhood, particularly Timbuctoo and Houssa; and that I should be afterwards at liberty to return to Europe, either by the way of the Gambia or by such other route as ... should appear to me advisable.'[8]

2

Mungo Park's ship took a little more than four weeks to reach the Gold Coast, and by 5 July 1794 he was installed in Pisania, a tiny, remote outpost a hundred miles up the river Gambia. It was occupied by only three other white men, each living in a small compound: a doctor, and two white traders whose main business was gold, ivory and slaves. Park kept his views on slavery to himself, took lodgings with Dr Laidley, and was made welcome. As the rainy season set in, he learned the local language, Mandingo, read and botanised, practised navigation by the stars with his sextant, and (after spending too long observing an eclipse of the moon) endured a month-long bout of malarial fever, which 'seasoned' him, in the local terminology, and probably saved his life later on.[9]

Dr Laidley nursed him with great kindness and care, and inspired the first of many vivid evocations of the African experience in Park's *Travels*: 'His company and conversation beguiled the tedious hours during that gloomy season, when the rain falls in torrents; when suffocating heats oppress by day, and when the night is spent by the terrified traveller in listening to the croaking of frogs (of which the numbers are beyond imagination), the shrill cry of the jackal, and the deep howling of the hyena: a dismal concert interrupted only by the roar of such tremendous thunder as no person can form a conception of, but those who have heard it.'[10]

Park laid in a modest £16-worth of trading items – amber, tobacco, beads and Indian silks. These items were carefully chosen, not for profit, but to pay his way in diplomatic gifts and formal permissions to cross tribal territories. He bought a horse and two mules, and hired two servants to accompany him. The first was Johnson, an African guide and interpreter, a calm, stately man who had seen many things: he had been a slave in Jamaica, and then a freed man in service in England, where he married and then returned to Africa. Characteristically, Park paid half Johnson's salary to his wife. The second was Demba, a young African slave boy, 'sprightly', charming and quick-witted, to whom Park promised to purchase his freedom on their safe return.[11] These preparations, and Park's slow recovery from the fever, took five months.

Anxious for Park's safety, Dr Laidley tried to persuade them to leave in the company of a slave caravan, but Park refused, a rejection that was later seen as symbolic. The little expedition left Pisania for the interior on 2 December 1795. 'I believe they secretly thought they should never see me afterwards,' wrote Park.[12] Shortly after, a cheerful letter arrived from Joseph Banks, wondering if Park had returned from Timbuctoo already: 'By the time you receive this you will no doubt have returned from a perilous Journey if you have accomplished the business of seeing Tambookta you will deserve from the Association every thing they can do for you as I have no doubt you will be able to give a good Account of what you have seen.'[13]

In the event, the journey took two years to accomplish. Speculative maps had been drawn of this region, based on the stories of slave traders, but it was virtually unknown territory to any European. It was not even clear where the fabled river Niger rose, or in which direction it flowed. Park had to depend on luck, endurance, local hospitality and his sextant. But he had a Romantic belief in his own destiny, and a strange inner tranquillity, which could accept even the most disastrous turn of events with equanimity.

Park first followed the course of the river Gambia eastwards. He made good progress, but had most of his gifts and valuables claimed or forcibly removed by tribal chiefs in the first few weeks. On 18 February 1796 he reached the point where Major Houghton had written his last note. Here he turned northwards into the region of Ludmar, controlled by a powerful Moorish chieftain, Ali, whose protection Park intended to claim.

But in Ludmar the hoped-for Moorish hospitality imperceptibly changed into captivity, and polite interrogation degenerated into deliberate humiliation. Park had all his remaining goods seized, his interpreter Johnson taken away, and his boy servant Demba abducted. By 12 March he was effectively a solitary prisoner at Ali's camp.[14] He was confined to a hut, and subjected to an intrusive physical examination by Ali's wife Fatima and her entourage of Moorish women. 'A party of them came into my hut, and gave me plainly to understand that the object of their visit was to ascertain by actual inspection, whether the rite of circumcision extended to the Nazarenes [Christians], as well as the followers of Mahomet ... I thought it best to treat the business jocularly.'[15]

Park eventually escaped, and on 20 July 1796 caught his first sight of the river Niger at Sego, some 300 miles inland. It was known locally as the 'Jolliba', or Great Water, and it struck him like a sacred vision.[16] He described this in a striking passage, a mixture of the dreamlike and the familiar. 'Looking forwards, I saw with infinite pleasure the great object of my mission – the long sought for majestic Niger, glittering in the morning sun, as broad as the Thames at Westminster, and flowing *to the eastward.* I hastened to the brink, and having drunk of the water, lifted up my fervent thanks in prayer to the Great Ruler of all things, for having thus far crowned my endeavour with success.'[17] Eastwards, noted Park gravely, exactly as predicted by Herodotus.

Shortly after, the cruelty of the Moors was strangely set aside by an act of unexpected kindness and hospitality. At dusk Park was greeted by a Negro woman who had been labouring in the fields near the river. She invited him back to her hut, lit a lamp, spread a mat and made him supper of fish baked over a charcoal fire. Evidently Park half-expected some kind of sexual overture. But instead the woman invited into the hut various female members of her family, and they all quietly sat round him in the firelight, spinning cotton and singing him to sleep. Park suddenly realised the song was extempore, and the subject was himself. He was amazed when he began to understand the words: 'It was sung by one of the young women, the rest joining in a sort of chorus. The air was sweet and plaintive, and the words literally translated, were these: – "The winds roared, and the rain fell. The poor white man, faint and weary, came and sat under our tree. He has no mother to bring him milk; no wife to grind his corn. *Chorus*: Let us pity the poor white man, no mother has he . . . " '[18]

The women reversed all Park's assumptions about his travels in Africa. He realised that it was he – the heroic white man – who was in reality the lonely, ignorant, pitiable, motherless and unloved outcast. It was he who came and sat under *their* tree, and drank at *their* river. He found it hard to sleep that night, and in the morning he gave the woman four brass buttons from his coat before he left, a genuinely precious gift.

This incident had a huge impact when Mungo Park's *Travels* were eventually published in Britain, and one can imagine what memories it stirred in Banks of his Tahiti nights so many years before. It was however also easy to sentimentalise. The glamorous and well-intentioned Georgiana, Duchess of Devonshire, rewrote the women's song and had it set to music by the Italian composer Giorgio Ferrari, and circulated among the London salons. The first stanza of her version, 'A Negro Song', is remarkably close to the original wording, and retains its strange tenderness:

> The loud wind roar'd, the rain fell fast;
> The White Man yielded to the blast:
> He sat him down, beneath our tree;
> For weary, sad and faint was he;
> And ah! no wife or mother's care,
> For him, the milk or corn prepare.

But Georgiana could not forbear to add a second stanza, which makes the situation far more conventional, and puts the white explorer back in command of his fate. She also added a plangent chorus, which in three lines subtly transformed the African women into pious, domestic supplicants.

> The storm is o'er; the tempest past;
> And Mercy's voice has hush'd the blast.
> The wind is heard in whispers low;
> The White Man far away must go;–
> But ever in his heart will bear
> Remembrance of the Negro's care.
>
> *Chorus:*
> Go, White Man, go! – but with thee bear
> The Negro's wish, the Negro's prayer;
> Remembrance of the Negro's care.[19]

Park travelled on down the river as far as Silla, where, exhausted, he decided to turn back short of Timbuctoo on 25 August 1796. On the return journey he was robbed and stripped by Moorish *banditti* in 'a dark wood' before he reached Kalamia. They took everything – his horse, his compass, his hat, all his clothes except his trousers and his battered boots ('the sole of one of them was tied onto my foot with a broken bridle rein'). They had evidently intended to kill him, but saw him as a feeble white man beneath contempt. They did however throw his hat back to him – not realising that it contained the papers of his travel journal folded up in the band. In what became another famous passage, Park described sitting down in utter despair, believing that the end had come.

'After they were gone, I sat for sometime looking round me with amazement and terror ... I saw myself in a vast wilderness in the depth of the rainy season, naked and alone; surrounded by savage animals, and men still more savage. I was 500 miles from the nearest European settlement. All these circumstances crowded at once on my recollection; and I confess my spirits began to fail me. I considered my fate as certain, and that I had no alternative, but to lie down and perish.'[20]

Park's thoughts turned helplessly towards prayer, and 'the protecting eye of Providence'. But then something curious happened. As he hung his head in utter exhaustion and misery, his gaze began listlessly wandering over the bare ground at his feet. He noticed a tiny piece of flowering moss pushing up through the stony earth beside his boot. In a flash, his scientific interest was aroused, and leaning forward to examine the minute plant, for one moment he forgot his terrible situation. He carefully described this movement out of paralysing despair: 'At this moment, painful as my reflections were, the extraordinary beauty of a small moss in fructification, irresistibly caught my eye. I mention this to show from what trifling circumstances the mind will sometimes derive consolation; for though the whole plant was not larger than the top of one of my fingers, I could not contemplate the delicate conformation of its roots, leaves, and capsula, without admiration.'

In that moment of pure scientific wonder, Park's thoughts and outlook were transformed: 'Can the Being (thought I) who planted, watered, and brought to perfection, in this obscure part of the world, a thing which appears of so small importance, look with unconcern upon the situation and suffering of creatures formed after his own image? – surely not! Reflections like these would not allow me to despair. I started up, and disregarding both hunger and fatigue, travelled forwards, assured that relief was at hand; and I was not disappointed.'

He soon fell in with two friendly shepherds, and continued on his way westwards, towards the sea and the long journey home. Miraculously, he found he could pay his passage by writing phrases from the Koran on loose scraps of paper, saved from his journal, and selling these as religious charms.[21]

Although it was Park's scientific curiosity that saved him – the precise botanical term 'capsula' carries significant weight – a theologian might convincingly describe this moment as an example of the power of the Argument by Design. Coleridge's Ancient Mariner has a similar vision when, alone and becalmed in the Pacific, and dying of thirst, he sees the beautiful, phosphorescent sea creatures playing round the ship's hull, and in a moment of redeeming selflessness he is saved.

> O happy living things! No tongue
> Their beauty might declare:
> A spring of love gushed from my heart,
> And I blessed them unaware:
> Sure my kind saint took pity on me
> And I blessed them unaware.[22]

At this moment the albatross of despair falls from his neck.

Park's moment of revelation fascinated the young Joseph Conrad. He wrote in an essay, 'Geography' (1924), of his inspiring boyhood image of Mungo Park: 'In the world of mentality and imagination which I was entering, it was they, the explorers, and not the characters of famous fiction who were my first friends. Of some of them I had soon formed for myself an image indissolubly connected with certain parts of the world. For instance, the western Sudan, of which I could draw the rivers and principal features from memory even now, mean for me an episode in Mungo Park's life. It means for me the vision of a young, emaciated, fair-haired man, clad simply in a tattered shirt and worn out breeches sitting under a tree.'

It is interesting that Conrad imagined Park in the Sudan, as if he had indeed successfully crossed the whole of Africa from west to east, via Lake Chad.[23]

3

Park slipped back into London just before Christmas 1797. He went quietly into the British Museum gardens to greet his brother-in-law James Dickson, who saw a tall, tanned figure walking up unannounced between the potted plants. Then Park went to Soho Square to receive a thunderous greeting from Banks, who had given him up for lost. In the last week of January 1798 the *True Briton* and *The Times* hailed his return with long articles, though claiming somewhat optimistically that he had glimpsed Timbuctoo and also found the great city of Houssa, a huge, magical metropolis twice as big as London.

Banks wrote delightedly about Park, his 'Missionary from Africa' ('missionary' was still an entirely secular term), to his old crony Sir William Hamilton in Naples. For this sort of despatch Banks adopted a kind of breathless telegraphese. Park, wrote Banks, 'has made most interesting discoveries he has penetrated Africa by way of the Gambia near a thousand miles in a strait line from Cape Verde ... He has discovered a river traced for more than 300 miles till it was larger than the Thames at London. His adventures are interesting in a degree he will publish them soon & I will send you the book he was soon robbed of all his property and proceeded as a beggar sometimes gaining a little by the sale of Charms which he could easily manufacture as they are sentences of the Koran written in Arabic ... hunger and thirst he frequently & patiently Endured & is come Home in good health.'[24]

Banks also announced the success of the expedition to the pioneering German anthropologist Johann Blumenbach, who wrote back from Göttingen: 'how ardently I long to see once Mr Park's own extensive Account of his wonderful & highly interesting Travels'. Blumenbach added a characteristic enquiry: 'I wonder if he has not met with any *white* negroes [Albinos] similar to those you saw at Otaheite ... ?'[25] Banks was not able to help with this, and left Park to spend over a year writing up his original journal. Park began with the editorial help of Bryan Edwards, of the Africa Association, but soon found he had become master of a new form of travel narrative, and continued without further assistance, working away quietly back in Scotland. When the manuscript was at last delivered to Soho Square, Banks was delighted and deeply moved by what he read.[26]

The book revealed Park as the essential Romantic explorer. His heart was a *terra incognita* quite as mysterious as the interior of Africa, about which he wrote with quiet humour and unflinching observation. The manuscript was published, with revised maps by Rennell, in the spring of 1799 as *Travels in the Interior of Africa*, and instantly became a best-seller, enabling Park to marry his childhood sweetheart, Allison Anderson of Selkirk.

Allison was a willowy, beautiful, cheerful young woman who bore Park two sons and a daughter, and encouraged him to settle down as a physician in Peebles. He proved an excellent doctor, quiet and sympathetic, and his fame brought him plenty of distinguished patients, including the young Walter Scott, who lived nearby at Melrose. But Park's wanderlust was not appeased. He began to consider all sorts of exotic places his family might emigrate to, not least Australia or even China. Allison knew he was restless when in 1803 he employed an Arab doctor to teach him Arabic. Scott remembered how he rode over one day to visit Park, but found he was not at home, a more and more frequent occurrence, according to Allison. Scott finally discovered him wandering along the banks of the river Yarrow, solitary and distracted, skimming stones across the water. He explained to Scott how he used to throw stones to gauge the depth of the Niger before attempting a crossing. Then he broke out that he 'would rather brave Africa and all its horrors' than wear out his life as a country doctor, especially in such a cold climate, surrounded by 'lonely heaths and gloomy hills'. Scott guessed that a new journey was being secretly planned.[27]

4

Park's second expedition to West Africa (1805) had a very different complexion to the first. He was financed by the Colonial Office, and given troops and funds to buy his way through the various tribal lands along the Niger. He was offered a salary of £4,000 if he returned, and the same payment to his widow Allison if he did not. He was allowed to take along his best friend, his wife's brother Dr Alexander Anderson, as a companion, and a young Edinburgh draughtsman, George Scott, as the expedition's official artist.

Banks had spent many months trying to organise this expedition, but as war with France continued, its *raison d'être* had clearly altered. It was now transformed from a geographical survey to that of an armed trading caravan, its main purpose to seek to establish a commercial trade route down the Niger. Banks had secretly sent the outline of a grand imperial 'project' to the President of the Board of Trade, the Earl of Liverpool, as early as June 1799. The Niger expedition would form just one small element in this strategy. 'Should the undertaking be fully resolved upon, the first step of Government must be to secure to the British Throne, either by conquest or by Treaty, the whole of the Coast of Africa from Arguin to Sierra Leone...'

For a moment Banks had a heady vision of a vast, benign commercial empire stretching over the dark continent and bringing light and happiness in its wake: 'I have little doubt that in a very few years a trading

Company might be established under the immediate control of Government, who ... would govern the Negroes far more mildly, and make them far more happy than they now are under the Tyranny of their arbitrary Princes ... by converting them to the Christian Religion ... and by effecting the greatest practicable diminution of the Slavery of mankind, upon the Principles of natural Justice and commercial Benefit.'

Banks added that 'the whole Tenor of Mr Park's book' showed that such a strategy was possible, and that the grand civilising mission should include 'the more intelligible doctrines' of the Scriptures and the more useful branches of 'European mechanics'. But then he checked himself, and concluded that he had been 'led away too far by this Idea'. It is not clear how much of this imperial dream he ever vouchsafed to Park himself.[28]

One indication of the changed plan was that Park and Anderson were appointed to the military ranks of captain and lieutenant, in an attempt to give them authority over their troops. Park was uneasy about this, as appears in a letter from Lord Camden at the Colonial Office to the Prime Minister, William Pitt, dated 24 September 1804: 'Mr Park has just been with me. He is inclined to attempt the expedition proposed for the sum I mentioned ... It is therefore to be determined in what manner a Journey of Discovery and of Enquiry for commercial purposes can best be attempted. Mr Park seems to think that he shall be able to travel with less suspicion and therefore with more effect, if he was only accompanied by 2 or 3 persons on whom he could depend.'[29] But in the end he was supplied with forty soldiers.

After a delayed departure from England because of confused expeditionary orders and financing, Park arrived at the island of Goree, off the West African coast, on 28 March 1805. This was barely six weeks before the onset of the rainy season, and was the hottest time of the year for travelling. Nearly a month was spent organising the detachment of forty volunteer troops, commanded by twenty-two-year-old Captain John Martyn, and packing up supplies from the coastal fort. Park finally left the Gambia on 27 April, having written letters to Lord Camden, to Joseph Banks, and to his wife Allison. For the first time, he also made a Will.[30]

The arrival of the rains, long before they reached the Niger, had a catastrophic effect on both their progress and their health. They were ravaged by malarial fever and dysentery, and men dropped behind one by one. They were attacked frequently by wild dogs, by crocodiles, and once by a party of lions. They were continuously soaked by the torrential rains, which fell implacably day and night. Their donkeys' packs (containing gifts of amber beads, pistols, cloth) were split open and looted by tribesmen.

Park was indefatigable in caring for his troops and donkeys, paying natives for help, and arranging staging camps for those left behind. But the death toll was terrible all along the 500-mile march inland from Bamako to join the Niger at Sego. By the time they reached the river on 19 August, only twelve Europeans from the original party were still alive.

The exhausted expedition made camp and began tortuous negotiations with the local leader, chief Mansong. Mansong finally agreed to send them sufficient canoes to embark the remaining men and baggage. These cost Park 'very handsome presents', but the relief of taking to the water was immense. 'The velocity was such as to make me sigh,' he wrote of their swift journey downstream. Although suffering from dysentery and crippling headaches, Park delighted in the elephants, and a passing hippo which blew 'exactly like a whale'.[31]

At Sansanding four more white troops died, and young George Scott. Park dosed himself with mercury calomel to cure a potentially lethal attack of dysentery, and recorded in his journal that with the burning in his mouth and stomach he 'could not speak nor sleep for six days'. It is notable that he somehow managed to keep the knowledge of this illness from his remaining troops, who believed that he was in good health and completely adapted to the terrible conditions. His steady bearing never altered, as catastrophe followed catastrophe, and their surroundings grew steadily more hostile. When Private William Garland died, animals carried away his body from the hut during the night. The Moors urged Mansong to kill the beleaguered white men and seize their goods. 'They alleged that my object was to kill Mansong and his sons by means of charms, that the White People might come and seize on the country. Mansong, much to his honour, rejected the proposal, though it was seconded by two-thirds of the people of Sego, and almost all Sansanding.'[32]

With nine remaining men, including his beloved brother-in-law Anderson, three white troopers, his military friend Captain Martyn, two black slaves (promised their freedom) and his Arabic guide Amadi, Park constructed a forty-foot wooden 'schooner' from the shells of two native canoes roughly carpentered together. It was narrow – just six feet wide – but its shallow one-foot draught made it excellent for negotiating rapids.

He built a small cabin on the stern, armoured the deck with bullock hides and rigged and stocked the craft for a non-stop descent of the river, which he was now convinced (rightly) turned southwards after Timbuctoo and reached the Atlantic in the bay of Benin. He expected opposition, and supplied each remaining man with fifteen muskets apiece and a huge supply of ammunition.

The atmosphere among the surviving members of the expedition is caught in a letter which the cheery, hardbitten Captain John Martyn wrote on 1 November 1805 to a fellow officer, Ensign Megan, safely back at the military station of Goree on the coast. 'Dear Megan – Thunder, Death and Lightning – the Devil to pay! Lost by disease Mr Scott, two sailors, four carpenters and thirty one of the Royal African Corps, which reduces our numbers to seven, out of which Dr Anderson and two soldiers are quite useless ... Captain Park has not been unwell since we left Goree; I was one of the first taken sick with fever and ague ...'

Martyn goes on to describe Park's quiet efficiency, the building of the schooner, and the continued motivation of the expedition to pursue the course of the Niger. 'Captain Park has made every enquiry concerning the River Niger, and from what we learn there remains no doubt that it is the Congo. We hope to get there in about three months or less ... Captain Park is this day fixing the Mast – schooner rigged – 40 feet long – All in the clear. Excellent living since we came here (August 22), the Beef and Mutton as good as ever was eat. Whitbreads Beer is nothing to what we get here ...'

Finally he added a scrawled note on the stained outer flap of his letter, dated 4 November. It captures a soldier's-eye view of the British imperial mission. 'PS Dr Anderson and Mills dead since writing the within – my head a little sore this morning – was up late last night drinking Ale with a Moor who has been at Gibraltar and speaks English – got a little tipsy – finished the scene by giving the Moor a damn'd good thrashing.'[33]

For Park the loss of his close friend and brother-in-law was the most terrible blow, an event that put something like despair for the second time in his heart. He wrote in his journal: 'At a quarter past five o'clock in the morning, my dear friend Mr Alexander Anderson died, after a sickness of four months. I feel much inclined to speak of his merits but ... I will rather cherish his memory in silence, and imitate his cool and steady conduct, than weary friends with a panegyric in which they cannot be supposed to join. I shall only observe that no event which took place during the journey ever threw the smallest gloom over my mind till I laid

Mr Anderson in the grave. I then felt myself *as if left a second time*, lonely and friendless, amidst the wilds of Africa.'[34]

Before setting out from Sansanding, Park wrote three farewell letters: to his sponsor Lord Camden at the Colonial Office, to Sir Joseph Banks, and to his beloved wife Allie. In each he stated that he was in good spirits and determined to press on, and hoped to be back in England the following summer. But he also sent back to Goree by Arabic messenger his journals written up to that date, as if this would be the last chance.

His letters appear to be an extraordinary mixture of dogged courage and feverish delusions. To Lord Camden he wrote with quite uncharacteristic bravado: 'I shall set sail for the east with the fixed resolution to discover the termination of the Niger or perish in the attempt though all the Europeans who are with me should die, and though I were myself half dead, I would still persevere, and if I could not succeed in the object of my journey, I would at least die on the Niger.'[35]

To his wife, carefully dating his letter 'Sansanding 19 November 1805', he wrote more reassuringly and calmly. 'I am afraid that, impressed with a woman's fears and the anxieties of a wife, you may be led to consider my situation as a great deal worse than it really is ... I am in good health. The rains are completely over, and the healthy season has commenced, so that there is no danger of sickness, and I have still a sufficient force to protect me from any insult in sailing down the river to the sea ... I think it not unlikely but I shall be in England before you receive this. You may be sure that I feel happy at turning my face towards home ... the sails are now hoisting for our departure for the coast.'[36]

But to Joseph Banks he wrote with almost visionary detachment, making no mention of hardships or dangers, but as one explorer speaking quietly to another, over a last cigar: 'My dear Friend ... It is my intention to keep to the middle of the River and make the best use I can of Winds and Currents till I reach the termination of this Mysterious Stream ... I have purchased some fresh Shea Nuts which I intend taking with me to the West Indies as we will likely have to go there on our way home ... I expect we will reach the sea in three months from this, and if we are lucky enough to find a vessel, we shall lose no time on the Coast.'[37]

From this point, there is no further direct evidence from Park, as no later letters or journals survive. His last known note records that he was departing, his party reduced to 'three soldiers (one deranged in his mind), Lieutenant Martyn, and myself'.

5

Casting off from Sansanding on about 21 November 1805, Park paddled downriver, keeping well clear of the banks until he hove to outside Timbuctoo, hoping to trade. But apparently he did not dare to disembark because of the threat from hostile Tuareg tribesmen. So finally Mungo Park never entered the city of his dreams.

This dream of 'Timbuctoo' would continue to haunt English writers and explorers for another thirty years. The young Alfred Tennyson submitted a 300-line blank-verse poem entitled 'Timbucto' for the Chancellor's Medal at Cambridge University in 1827. He headed it with an epigraph drawn from Chapman's Homer: 'Deep in that lion-haunted Island lies/A mystic City, goal of high emprise!' Young Tennyson asked dreamily:

> Wide Afric, doth thy Sun
> Lighten, thy hill enfold a City as fair
> As those which starr'd the night o' the elder World?
> Or is the rumour of thy Timbucto
> A dream as frail as those of ancient Time?...

His poem concludes prophetically with a new fear, one which would become frequent in both English and French travel-writing of the mid-nineteenth century (especially in Gérard de Nerval's 1851 *Voyage en Orient*), that the actual discovery of the legendary city would reduce its seductive image to something mundane. Tennyson's private, tantalising mirage of 'tremulous' domes, abundant gardens and 'Pagodas hung with music of sweet bells' would resolve itself into the bleak reality of a few primitive mud huts.

> ... The time is well-nigh come
> When I must render up this glorious home
> To keen Discovery: soon your brilliant towers
> Shall darken with the waving of her wand;
> Darken, and shrink and shiver into huts,
> Black specks amid a waste of dreary sand,
> Low-built, mud-wall'd, Barbarian settlements:
> How changed from this fair City!

Alfred Tennyson won the Chancellor's Medal, but he never went to Africa.[38]

As he proceeded downriver, Park inexplicably refused to pay any tribute to the local chiefs, considering that he had made all necessary payment to Mansong. This was a fatal mistake which the young Mungo Park would never have made. After this failure to render these customary gifts (in effect a river-tax or toll), the boat was attacked from the river-bank almost continuously by infuriated tribesmen. These attacks became more severe when they entered the territory of the Houssa, and their Arabic guide Amadi left them by agreement. On one occasion they were pursued by a flotilla of sixty canoes, and they were constantly subjected to showers of arrows, spears and clubs.♣

Reports agree that the boat was eventually ambushed by Tuareg tribesmen at the rapids of Boussa, some 500 miles downstream from Timbuctoo, and with only another 300 miles to go. Here it seems to have run aground in a narrow, shallow, rocky defile. A witness later found by Amadi described a day-long battle, during which Park threw all his valuables overboard, hoping either to lighten the boat and shoot the rapids, or to placate the tribesmen. If that is true, he achieved neither. At the last, with all their men either killed or wounded, Park and Martyn threw themselves into the river. Their bodies were never recovered. They were either drowned, or killed when they came ashore, or – haunting possibility – disappeared into captivity.

One black slave remained alive on board the *Joliba*. He surrendered, was spared, and was finally released by the local Tuareg chieftain. He was the witness that Amadi eventually tracked down. His account includes one particularly haunting detail: that when Park jumped into the river he held one of the other white men in his arms. There is no explanation for this. Perhaps he was still trying to save one of his wounded soldiers, or was making some sort of last stand with young Martyn.

♣ This final crazed descent of the river in HMS *Joliba*, as Park's vessel was named, can be considered as the first enactment of a journey that was to be repeated many times in subsequent fiction and film. First perhaps in Conrad's *Heart of Darkness* (1899, set in the Congo), and then in such films as *Apocalypse Now* (1979, adapted from *Heart of Darkness*, but set in North Vietnam) and *Aguirre, the Wrath of God* (1972, set in South America). It is made more haunting and resonant precisely by the fact that Park's own journal of these final weeks did not survive. Everything known is reported at second or third hand, and the truth can only – in the end – be imagined.

Nothing else survived – no journals, letters or personal effects of any kind – except for an annotated copy of an astronomical almanac (thought, correctly, to be a sacred book) and a single swordbelt. Amadi was able to buy back the almanac at great expense, but the swordbelt was retained by the local tribal chief as a ceremonial horse's bridle. Park was aged thirty-four at the time of his death (reckoned to be about February 1806), and his widow Allison was paid the compensation of £4,000 by the Africa Association. She died in Selkirk in 1840. Park's *Journal of a Second Voyage* was published in 1815 with a brief, anonymous Memoir; but rumours of his survival persisted for many years in Britain.[39]

The legend of Mungo Park surviving somewhere beyond Timbuctoo – either the prisoner of some tribal king, or else 'gone native' (itself an idea that began to trouble nineteenth-century colonialists) and living as a great chieftain himself – became increasingly haunting. A biography of Park was published by 'H.B.' in 1835, but theories about his disappearance would continue into the twentieth century. In June 1827, the same year as Tennyson's 'Timbucto' poem, Park's eldest son Thomas, obsessed by tales of his father, set out to find him.

Thomas Park had studied science at Edinburgh University, and was now a lieutenant in the Royal Navy. Taking a year's leave of absence, he sailed to Accra on the Gold Coast, where he taught himself the Ashanti language, and made a preliminary sortie into the African interior. From a single surviving letter, sent from Accra to his mother in Scotland, and dated September 1827, it emerges that Thomas had set out on his quixotic mission without warning his family. His jaunty optimism strangely mirrors that of his father's last letters to his wife: 'My dearest Mother, I was in hopes I should have been back before you were aware of my absense. I went off – *now that the murder is out* – entirely from fear of hurting your feelings. I did not write to you lest you should not be satisfied. Depend upon it my dearest mother, I shall return safe. You know what a *curious* fellow I am, therefore don't be afraid for me. Besides, it is my duty – my filial duty – to go, *and I shall yet raise the name of Park.* You ought rather to rejoice that I took it into my head …'

He went on to send love to his siblings, especially his sister, and to mention a possible plan to take his own boat down the Niger. But he gives no other details, no address or means by which he might be communicated with in Accra, and says nothing about companions, preparations or equipment. He signed off in the quiet, resolute Mungo

Park style: 'I shall be back in three years at the most – perhaps in one. God bless you, my dearest mother, and believe me to be, your most affectionate and dutiful son, Thomas Park.'[40]

Thomas embarked on a full-scale expedition in October 1827, marching 140 miles inland to Yansong. It was rumoured that he travelled not like ordinary white men, but in a native style adapted from his father's first expedition. He had taken 'no precaution with regard to the preservation of his health, but, adopting the habits of the people with whom he mingled, anointed his head and body with clay and oil, ate unreservedly the food of the natives and exposed himself with scarcely any clothing to the heat of the sun by day and the influence of the pernicious dews by night'.[41]

Having reached Yansong, Thomas started to make enquiries about his father, but was almost immediately overcome by malarial fever. One account has him lying beneath a sacred tree (like Mungo), awaiting deliverance. Another has him climbing the tree to watch a native festival, drinking too much palm wine in the hot sun, and falling out of its branches. Whatever brought about his death, Thomas Park never returned, and his body was never found. A month later, in November 1827, a clean white shirt, pressed and labelled 'T Park', turned up in a basket of laundry delivered to the explorer Richard Lander at Sokoto, a hundred miles away on the western coast.

6

There are many abiding mysteries about both of Mungo Park's expeditions. In the first, of 1794, there was his extraordinary physical courage combined with a patience amounting to almost suicidal passivity. He refused on principle to engage in personal confrontation, or stand on European 'superiority'. His apparent acceptance of extreme moral and physical humiliation at the hands of native tribesman was exceptional. His reliance on poor villagers, fishermen and native women, rather than on tribal leaders and chieftains, perhaps reflected something of his Scottish upbringing. His dogged determination and adaptability were oddly combined with a strange ineptness and imprudence. His scientific fascination with local wildlife – bees, lions, hippos and birds – seemed instinctive and inexhaustible. His real motives for undertaking the first Niger expedition, beyond a desire for adventure, remain wonderfully enigmatic.

His attitude to slavery is not clear. But his role as an essentially solitary traveller, a lonely wanderer among men and communities, came to seem intensely Romantic.

The second expedition of 1805 was entirely different in both manner and motivation from the first. Britain was now at war with France throughout the globe, and competitive exploration easily became colonial ambition. Mungo Park was ten years older, very conscious of family duties, and interested in financial reward. But equally, his intensely romantic attachment to his wife Allison did not prevent him from returning to the Niger, and the high likelihood of death. His agreement to lead an armed expedition, to accept a military commission and payment (and in effect a form of life insurance) from the Colonial Office, suggests a quite new kind of professionalism. So too does his acceptance of a commercial mission, to search for a 'new trade route into the Sudan', as well as his decision to learn Arabic before he set out. On his first trip he traded mostly in amber and cloth; on his second, in guns and gunpowder.

Whether all this means that Mungo Park had consciously undertaken an 'imperial' mission in his second expedition remains ambiguous. At least up to the last boat journey from Sansanding, he was respectful of all native customs, modest in his behaviour, and humane and honourable in his treatment of anyone he met (including his own troops). The contrast with a soldier like John Martyn (who seems already to have been rehearsing his part for a Rudyard Kipling story) could not be more great.

The dauntless tone of Park's journals, even in the final desperate weeks at Sansanding, may disguise his character as much as it reveals it. The impenetrable optimism of his last letters in November 1805, not only to Lord Camden, but also to Sir Joseph Banks and to his wife, remains enigmatic. So too do the contradictory reports of the circumstances of his death. The tragic obsession of his son Thomas to solve the mystery of his father's disappearance suggests that something far more personal than imperial ambitions was always engaged. Thomas's parting declaration – that he would 'raise the name of Park' – has a curious resonance, and may be said to have been eventually fulfilled by the brass plate that was mounted by Victorian admirers on a monument overlooking the vast and shadowy delta of the river Niger, and dedicated 'To Mungo Park, 1795, and Richard Lander, 1830, who traced the course of the Niger from near its source to the sea. Both died in Africa for Africa'.

Mungo Park's career clearly fits into the wider pattern of great Romantic exploration during this period. His own patron Sir Joseph Banks had established the British tradition, and the few letters they exchanged show a special mutual understanding of the explorer's mixture of endurance and delight. Other figures who actually made it home, like Bryan Edwards (from the West Indies), Charles Waterton (from South America) and William Parry (from the Arctic), would give it an increasingly literary dimension. At the very time that Park died (if he did die) in 1806, Alexander von Humboldt was just publishing the story of his South American wanderings in *A Personal Narrative.*♣

Mungo Park's story inspired a number of poets. Wordsworth included a passage about Park 'alone and in the heart of Africa' in an early version of *The Prelude.* He picked out another moment of crisis, when Park had collapsed in the desert, expecting to die from sunstroke, but later wakened to find

> His horse in quiet standing at his side
> His arm within the bridle, and the sun
> Setting upon the desert.

Wordsworth subsequently withdrew this passage, probably because Robert Southey had used Park's experiences at greater length in his adventure epic *Thalaba the Destroyer* (1801). Southey's fictitious hero is compared to Mungo Park in a long historical prose Note to the poem: 'Perhaps no traveller but Mr Park ever survived to relate similar sufferings.' But this is a case where the historical fact is more powerful than the fiction based upon it. Park's quiet, fresh, limpid prose has easily outlasted Southey's gaudy, melodramatic poem.

♣ Inspired by Cook and Banks, Alexander von Humboldt (1769–1859) had returned from South America in 1804 with 60,000 botanical and zoological specimens, preserved in forty-five enormous packing cases. But unlike Banks, he proceeded to publish his findings in thirty volumes over the next two decades, and later summarised his view of the world in an all-embracing, visionary work, *Cosmos* (1845), which attempted to unite all the contemporary scientific disciplines, from astronomy to biology. He studied volcanoes and oceanic currents, invented isobars, mapped the changes in the earth's magnetic field from pole to equator, and first proposed the science of climate change.

Keats's two Nile sonnets (1816) owe much of their décor to Park and Friedrich Hornemann. But Shelley's epic about his wandering alter-ego, the poet in *Alastor, or The Spirit of Solitude* (1815), deeply reflects the spiritual loneliness of the desert traveller who pursues a perilous river, and knows he will probably never return. Shelley's wilderness, while it includes 'dark Aethiopia in her desert hills', is geographically vague, though it moves more towards India and an imaginary East. But he catches something of Mungo Park's enigmatic wanderlust, and transforms it into an unearthly Miltonic quest for the strange and magnificent limits of the known world:

> The Poet, wandering on, through Arabie
> And Persia, and the wild Carmanian waste,
> And o'er the aerial mountains which pour down
> Indus and Oxus from their icy caves,
> In joy and exultation held his way;
> Till in the vale of Cashmire, far within
> Its loneliest dell, where odorous plants entwine
> Beneath the hollow rocks a natural bower,
> Beside a sparkling rivulet he stretched
> His languid limbs ...[42]

Later, his friend Thomas Love Peacock would remember Shelley stretching his languid limbs on the banks of the Thames, imagining vast and endless expeditions up the Niger, the Amazon, the Nile, though by now these trips would be taken aboard small steamships: 'Mr Philpot would lie listening to the gurgling of the water around the prow, and would occasionally edify the company with speculations on the great changes that would be effected in the world by the steam navigation of rivers: sketching the course of a steamboat up and down some mighty stream which civilisation had either never visited, or long since deserted; the Missouri and the Columbia, the Oronoko and the Amazon, the Nile and the Niger, the Euphrates and the Tigris ... under the over canopying forests of the new, or the long-silent ruins of the ancient world; through the shapeless mounds of Babylon, or the gigantic temples of Thebes.'[43]

Park's *Travels* were widely used (by both sides) in the intense discussions surrounding the abolition of the slave trade in 1807. Ten years later the radical surgeon William Lawrence would refer to Park's observations on African racial types, and particularly the difference between 'Negro and Moor'. John Martin's epic painting *Sadak in Search of the Waters of Oblivion* (1812), showing a bereft and solitary figure painfully pulling himself over desert rocks towards a distant river, may have been partly inspired by Mungo Park and the other explorers who never came back.[44]

Then there was the young explorer Joseph Ritchie, to whom Keats gave a copy of his newly published poem *Endymion*, with instructions to place it in his travel pack, read it on his journey, and then 'throw it into the heart of the Sahara Desert' as a gesture of high romance. Keats received a letter from Ritchie, dated from near Cairo in December 1818. 'Endymion has arrived thus far on his way to the Desart, and when you are sitting over your Christmas fire will be jogging (in all probability) on a camel's back o'er those African Sands immeasurable.'[45] After this there was silence. Joseph Ritchie never returned.

6

Davy on the Gas

1

During the late 1790s Joseph Banks started to get controversial reports of chemical experiments being carried out at a so-called 'Pneumatic Institute' in the Hotwells district of Bristol. They were being organised by Dr Thomas Beddoes, a one-time lecturer from Oxford, who had frequently applied to the Royal Society for subsidy. Despite recommendations from the Duchess of Devonshire and James Watt of the Lunar Society, Banks reluctantly turned down these requests, partly on the grounds that these experiments involved human patients breathing various kinds of gas, in ways which were too controversial to support. But he was also influenced by Dr Beddoes's known radical sympathies.

However, by 1800 Banks had become greatly interested in one of Beddoes's young assistants, a chemist from Cornwall, Humphry Davy. Although only twenty-one years old, Davy had already published several papers on chemistry, a book of *Researches*, and made numerous contributions to *Nicholson's Journal*. He was said to have an outstandingly brilliant and original mind. He even wrote poetry. When Davy came to London in February 1801 to be interviewed for a possible new post at the Royal Institution in Albemarle Street, Banks summoned him to one of his breakfasts in Soho Square.[1]

Sir Joseph met a very unusual young man. Davy was small, volatile, bright-eyed, and bursting with energy and talk. He had measured the cubic capacity of his own lungs, which was vast, especially since he was only five foot five tall, with a chest, which he had also measured, of a mere twenty-nine inches. He spoke English with a Cornish accent, and French with a Breton accent. Though he had never been abroad, he was completely up to date with the latest French chemistry, and was gratifyingly critical of Lavoisier's work on oxygen and 'caloric' in the *Traité Élémentaire de Chimie* of 1789.

Although Davy had attended a grammar school in Truro, and was briefly apprenticed to a physician in Penzance, he was very largely self-taught. He had never been to university, though he told Banks that he always intended to take a medical degree at Oxford. He never had the background in mathematics that shaped the scientific thinking of Newton and Cavendish. Like his contemporaries John Dalton (from Manchester), William Wordsworth (from Cumberland) and Samuel Taylor Coleridge (from Somerset), he always retained strong regional roots.

Banks would come to know a man who loved the high society of London, but was never at home in it, and was often mocked behind his back as a provincial. Davy's patterns of thought, and methods of work, remained highly original and individualistic. He was impulsive, charming and arrogant. Though physically small, he had huge intellectual ambitions. He was a solitary man who was also an incorrigible flirt. He believed passionately in his own 'genius' – a word he used constantly – and in the future of English science. Banks sometimes concluded that Humphry Davy thought these two things were identical; and that very possibly he was right.

Davy was born in Penzance, Cornwall, on 17 December 1778. Penzance was then a tiny seaside town, remote and isolated in the extreme south-west of rural England, dependent largely on fishing and the business of the local tin mines. Its population was below 3,000. Both fish and slabs of stamped tin were sold in the high street, known as Market Jew Street. There were several churches and Wesleyan chapels, many small dark taverns and a local school; but no theatre or learned institutions, except a small subscription library. (This would eventually become the famous Morrab Library.) The front rooms of many of the houses were still floored with beaten earth or sand. On windy days the sound of the breaking sea and the rattle of rigging could be heard in every street.[2]

Eighteenth-century Cornwall was still regarded as being as remote and barbaric as the Scottish Highlands: it was renowned for its fishermen, adventurers, smugglers and (most recently) mining engineers, and also for its rich, creamy, incomprehensible accent. The coach journey to London, which skirted Dartmoor and passed through Exeter and Bristol, covered nearly three hundred miles and took at least three days – and then only if the weather was good. Bad weather could easily cut Penzance off from the rest of the country. Sometimes connections were better by boat, eastwards up the Channel to Plymouth and Southampton; or southwards across the Channel, to northern France and Brittany.

Davy's father Robert was a Cornish craftsman, who had trained in London as a wood carver and gilder. His grandfather was a builder. His uncle Sampson was a watch- and clock-maker with an original touch: he once made a grandmother clock with eyes that winked open and shut each time it ticked. His mother Grace came from an old mining family from nearby St Just. Though originally from Norfolk, the Davy clan had lived for generations in Penzance and its surrounding hamlets, and their modest tombs were crowded into one corner of Ludgvan churchyard, three miles to the east along the coast.*

Robert was a small, genial and unworldly figure, rather disapproved of by the other Davys. He had a reputation as a dreamer and a drinker, 'thriftless and lax in his habits'.[3] In 1782 he unexpectedly inherited a seventy-nine-acre estate of woodlands and marshes called Varfell. It lay immediately south of the village of Ludgvan, with dramatic views of St Michael's Mount, the rocky island in the bay with its ancient abbey and fortress. Robert decided to build a house there, and to raise his family in the depth of the countryside. He continued his wood carving, invested in a small business, and took on local commissions. One of his carvings, a chimneypiece with two sportive griffins, was made for nearby Ludgvan Rectory. Another, illustrating Aesop's fable of the fox and the stork, was sold in London and finished up in the Victoria and Albert Museum.[4]

The wild Varfell estate (its very name full of Celtic echoes) gave Humphry Davy's childhood an extraordinary freedom and independence. He never forgot it, and would always try to recreate it in adult life, especially in his final years. As a boy he was small for his age, but daring and

* These days there are several large, white Davy tombstones leant against the south-east corner of the church, shrouded in nettles and deadly nightshade. They commemorate in deeply-cut but blurred lettering earlier members of the family, dating from the seventeenth or even sixteenth century. Plain and monolithic, one reads simply: 'Sac. Mem. D. Davy et G. Davy', without dates. The Ludgvan parish register records a Davy in 1588. The churchyard looks out over a tiny pub, a wood, and then a wild, bleak landscape of gorse and stones and broken fields, tumbling down to the sea. It reminds one that the Cornish sculptor Barbara Hepworth used to speak of 'the pagan triangle of landscape' between St Ives, St Just and Penzance (Hepworth Museum, St Ives).

mischievous. He was soon running wild through the Varfell woods and down to the adjacent Marazion marshes, with their golden bulrushes and plentiful wildfowl. He wandered through the gorse along the seashore opposite St Michael's Mount, and up into the remote hills behind Penzance. His easy-going father gave him a fishing rod, and then a gun.

He was also allowed to have a dog, Chloe, and eventually a pony called Derby. His taste for country sports, especially shooting and fishing, his intuitive feeling for nature, his love of running water and lyric poetry, were all formed here, and were never forgotten. On one of his birthdays he was allowed to plant an apple tree in the Varfell garden, possibly a local type called the Borlase Pippin, and in honour of Newton and the falling gravity-apple.

Humphry's mother, Grace, whose family came from St Just, was overwhelmingly important to him. She was one of three Millet sisters who had been adopted by a local Penzance surgeon, John Tonkin, on the sudden death of both of their parents from a fever in June 1757, when she was only seven. Tonkin (1719–1801) was one of the leading figures in the town, an old-fashioned philanthropist who was several times elected mayor.[5] Grace remained happily in his household for nearly twenty years, and came to regard him as a father. She married Robert Davy comparatively late, at the age of twenty-six, in 1776. She was the strong, reliable personality who held the family together. Davy was deeply attached to his mother, wrote to her regularly all his life, and kept her informed of all his scientific hopes and triumphs. She in turn, though she never left Penzance, took huge pride in his achievements. Davy eventually had four younger siblings: three sisters (Kitty, Grace and Betsy) and a baby brother, John. The family always remained closely knit. John, twelve years his junior, hero-worshipped him, followed him into medicine, and later became his editor and biographer.

John Tonkin, still acting as the family benefactor, in a way that perhaps suggests Robert's ineffectiveness as a father, paid for young Davy to attend Penzance Grammar School when he was ten. It was agreed that the boy should lodge in Tonkin's large house at the top of Market Jew Street, opposite the White Hart Inn, during term-time.[6] Hoping he might become a physician, Tonkin also encouraged his interest in every form of natural history: fossils, wild birds and animals, botany, chemical experiments. Years later, engravings of their open-air expeditions, the young

Davy skittering beside the aged Tonkin in his large black Quaker hat, became a favourite subject for Davy's Victorian biographers.

Davy was not remembered as an outstanding pupil. There is some suggestion that he had rebelled against his easy-going father, and that Grace had called in Tonkin to provide much-needed discipline. But he was articulate and adventurous, and became famed for 'spouting' stories and poems. His sister Kitty recalled his vivid story-telling, and his staging pantomimes on the back of the carts which were parked at the side of Market Jew Street, known as the Terrace. On summer evenings he would sometimes stand on the porch of the White Hart Inn and deliver 'speeches'. Later he would secretly make fireworks, and let them off in the street. He was a particular favourite of his maternal grandmother, who had an endless fund of Cornish legends and ghost stories. Indeed, she told him, she had herself lived for many years in a haunted house in St Just.

Curiously enough, Davy would later relate his love of science to this fascination with story-telling. What he always wanted to do was to hold an audience spellbound with wonders: 'to gratify the passions of my youthful auditors', as he put it. 'After reading a few books, I was seized with the desire to *narrate*... I gradually began to invent, and form stories of my own. Perhaps this passion has produced all my originality. I never loved to imitate, but always to invent: this has been the case in all the sciences I have studied.' Then he added: '*Hence many of my errors.*'[7]

His holidays were spent at Varfell, rambling, fishing, cliff-climbing and shooting wildfowl all round Mount's Bay. He was small, quick, witty, impetuous, with something mercurial and secretive about him.

In 1793, when he was fourteen, Davy was sent away to Truro Grammar School, also paid for by John Tonkin (then in his seventies). This was perhaps another attempt to discipline him, but also to give a potentially bright boy a better education. He was taught Latin and some Greek, but no science.

Everything changed in December 1794, with the sudden death of his father Robert, aged only forty-eight. It was caused by 'an apoplexy', or a stroke, as Davy would never forget. It was the same year in which the greatest chemist of his generation, Antoine Lavoisier, was guillotined in Paris. Robert left the family in debt to the tune of £1,300, a considerable sum, and the Varfell estate had to be sold. Grace Davy moved the family back to Penzance, started a millinery business with a young refugee Frenchwoman from the Vendée, and took in lodgers. Humphry had to give up his horse, Derby, and shortly after was withdrawn from Truro Grammar School.

Robert was buried in Ludgvan churchyard, and this became a place of secret pilgrimage for Davy. He would walk up the little lane out of Penzance, through Gulval village (where he once stopped to paint the view of St Michael's Mount), past the low stone farmhouse of Varfell, and climb through the trees to the flint church on its small but commanding eminence. Here he would sit with his back to the tombstones, and look out southwards, beyond the roof of Dr Borlase's rectory, far across the fields to the cold blue Cornish sea.

One of Davy's most striking and mysterious poems is set here, in Ludgvan churchyard. It clearly records one such grieving moment, although it may have been written much later, even possibly towards the end of his life. Though grieving, it admits not the least word of Christian comfort. It suggests a purely material philosophy, in which the atoms of his dead ancestors 'dance in the light of suns', but no spirit or soul survives. 'Their spirit gave me no germ/of kindling energy,' as one fragment says. In fact the unfinished poem feels oddly pagan, 'primitive', with a harsh physicality often associated in later Cornish art with the worship of stone, flint and sunlight.

It is also rare among Davy's poems in that it does not rhyme. It is formed from a plain list of terse factual statements: a list of precise observations, such as one might find in a shorthand account of an experiment in a laboratory notebook.

My eye is wet with tears
For I see the white stones
That are covered with names
The stones of my forefathers' graves.

No grass grows upon them
For deep in the earth
In darkness and silence the organs of life
To their primitive atoms return.

Through ages the air
Has been moist with their blood
The ages the seeds of the thistle has fed
On what was once motion and form...

Thoughts roll not beneath the dust
No feeling is in the cold grave
They have leaped to other worlds
They are far above the skies.

They kindle in the stars
They dance in the light of suns
Or they live in the comet's white haze.[8]

John Tonkin clearly felt that Davy was drifting. It was time for him to make his way in the world. On 10 February 1795, just turned sixteen, he was indentured for seven years to John Bingham Borlase, the leading surgeon-apothecary in Penzance, an old friend of Tonkin's. Borlase (1752–1813) had his shop at the top of Market Jew Street, next to the White Hart Inn, and was also several times mayor of Penzance. His father had been rector of Ludgvan, a distinguished Cornish antiquary, botanist and Fellow of the Royal Society. One of his many scientific friends was Dr William Oliver, inventor of the Bath Oliver biscuit.*

Davy moved back into his mentor Tonkin's house (just across the street from Borlase's pharmacy) and was given the run of the attic rooms, one of which he turned into a combined painting studio and laboratory. Tonkin, as generous as ever, supplied him with painting materials, chemicals and some elementary laboratory equipment. In the evenings Davy was given French lessons by one of his mother's lodgers. Officially this was an emigrant priest, Monsieur Dugast. But he may also have been seeing the young refugee from La Vendée, who was known as Nancy.[9]

Davy later told his brother John that this was 'a dangerous period of my life', and, in a wonderful periphrasis, that he had '*yielded to the*

* Borlase's chemist shop, now known as the Peasgood Pharmacy, still exists on the same site at the top of Penzance. The old 'Instructions for Apprentices' still hang framed in its back dispensary, indicating working hours from 7 a.m. to 8 p.m., when 'shutters are put up'. Discipline was strict – 'all joking and trifling are forbidden during hours of business in the shop' – and minute: 'In order to avoid using the mouth to moisten labels it is requisite that the gum brush or label-damper be always used. Biting of corks, licking of labels, licking of fingers or thumb for the purpose of taking up paper, are all prohibited as unclean and unbecoming . . . It is extremely important that all counters are kept free from muddles . . . ' Davy was to remember these instructions when he tried to impose order on the Royal Institution assistants twenty years later.

allurements of occasional dissipation'.[10] Certainly there were rumours of an entanglement, some heartbreak and many sonnets – none of which has survived, possibly because they were in French. There was also the Prospectus for a slim volume: it would contain 'eight Odes', four 'Cornish Scenes', and one long romantic verse tale to be called 'The Irish Lady'.

In this ballad, with skilful displacement, Davy invented a beautiful Irish girl who had fled from Ireland (rather than La Vendée) at the time of the seventeenth-century Protestant persecutions, but was shipwrecked on a rock off Land's End. Though she was drowned, Cornish fishermen occasionally glimpsed her during storms, sitting half-naked on her rock with a rose in her mouth, luring them to destruction.[11] This poem may well refer to Mlle Nancy. But it also belongs to a popular myth or Celtic legend, persistent all along the sea coast of Devon and Cornwall, of the beautiful, fatal woman from over the sea who lures young men to madness or death. Wagner's *Tristan and Isolde*, constructed on Brittany materials, and John Fowles's *The French Lieutenant's Woman*, based on local Lyme Regis legend, also belong to this tradition. In Davy's case the Lady is also alluring him away from science, and this becomes significant in his later life. However, lighter influences show in another long, equally wistful piece, 'Unfinished Poem on Mount's Bay', which plaintively describes Davy's beloved spaniel, Chloe.

French was certainly the language of love, and suitable for sonnets. But it was also, and just as alluringly, the language of Enlightenment science: the language of Laplace, Lamarck and Cuvier; the language of the *Encyclopédie* and the *Biographie Universelle*; the language of the Académie des Sciences, the only scientific body which rivalled the Royal Society in London. Above all it was the language of the greatest chemist of the age, Antoine Lavoisier.

The moment he left school, Davy began to read voraciously. He found he had access to a number of libraries: both Tonkin and Borlase gave him the run of their private collections, a notable privilege for a teenager, and there was the Penzance subscription library. He was also introduced to the son of a wealthy local *savant*, with the promising name of Davies Giddy. Giddy had studied at Oxford University, and was now living at Marazion, the village by the sea opposite St Michael's Mount. He had a large scientific library, and on his one afternoon off a week Davy would walk along the shore to borrow books and discuss them avidly.

His reading exploded: classical authors including Homer, Lucretius, Aristotle; English poets including Milton and James Thomson; and French science writers, especially Buffon, Cuvier and Lavoisier. He plunged into William Enfield's recently published two-volume *History of Philosophy* (1791), which was in effect a history of European science to date. He later observed wryly of this time: 'The first step towards the attainment of real discovery was the humiliating confession of ignorance.'[12]

It is evident that the death of his father, and all the subsequent emotional upheavals, profoundly shook the sixteen-year-old Davy, and started an intellectual ferment that never left him. Besides writing poetry, he also started his first diary, set himself reading lists and work timetables, and began a series of essays on religion versus materialism. During 1796 he wrote an essay 'On Mathematics', and another 'On Consciousness', which gleefully explored the implications of materialism. He described the body as 'a fine tuned Machine', and wrote a syllogistic proof that the 'soul' could not exist, since it was said to be eternal and 'unchangeable', while every known part of the human body, including the brain, was temporary and changed perpetually. 'QED the soul does not exist.'[13]

The experience of 'paralytic strokes' (like his father's), which destroyed 'perception and Memory' as well as physical motion, proved that the physical brain was the single centre of 'all the Mental faculties'. Children were not magically endowed with intelligence and souls at birth. On the contrary: 'A Child is not superior in Intellectual power to a common earthworm. It can scarcely move at will. It has not even that active instinctive capacity for Self-Preservation.' Such speculations gave Davy a sense of growing excitement and freedom. He wrote two supreme declarations of faith on page 61 of his working notebook. The first was: 'Man is capable of an infinite degree of Happiness.' The second was: 'The perfectibility of science is absolutely indefinite.'[14]

When he played billiards with Tonkin, Davy tried to extrapolate the Newtonian laws of motion from the concussion of the balls. He read James Thomson's great poem *The Seasons*, and imitated it in his own poem about energy in nature, 'The Tempest'. A long poem of self-dedication, 'The Sons of Genius', went through innumerable drafts, that can be dated anywhere between 1795 and 1799, when it was first published.

To scan the laws of Nature, to explore
The tranquil reign of mild Philosophy:
Or on Newtonian wings to soar
Through the bright regions of the starry sky!

From these pursuits the Sons of Genius scan
The end of their Creation, hence they know
The fair, sublime, immortal hopes of Man,
From whence alone undying pleasures flow.

Theirs is the glory of a lasting Name,
The meed of Genius, and her living fire!
Theirs is the Laurel of eternal flame,
And theirs the sweetness of the Muses' lyre.[15]

2

In 1797 Davy quite suddenly became fascinated by chemistry. The subject, closely linked to radical ideas about the nature of material reality, was going through its own revolution. At this time it was becoming the Romantic science *par excellence.* The last of the old alchemy was being replaced by true experiments, accurate measuring and weighing, and a new understanding of the fundamental processes of combustion, respiration and chemical bonding.

It was into this exciting new world that Davy was drawn. He could read both the English and the French accounts, which he found were often in contention, and this brought him an added sense of drama and immediacy. His English text was William Nicholson's *Dictionary of Chemistry* (1795), a full, solid explanatory work which laid out the current state of the science, how it had emerged from alchemy, and what the future challenges and theories were. His French text was Antoine Lavoisier's short, elegant and epoch-making *Traité Élémentaire de Chimie*, originally published in 1789. This developed new theories of 'oxygen' and 'caloric', a new table of elements, and proposed a whole new system of 'chemical nomenclature'. Both books came from Davy's mentor John Tonkin's library.

Davy saw that the time of alchemists was over, and a great new and revolutionary age of chemical experiment was opening. With intense excitement, he scrawled in his Penzance notebook a ringing declaration of intellectual freedom, which he would later include in his first published essay. 'Chemistry, which arose from the ruins of alchemy, to be bound with the fetters of phlogiston, has been liberated, and adorned with a beautiful philosophic theory. The numerous discoveries of Priestley, Black, Lavoisier, and other European philosophers in this branch of science, afford splendid proofs of the increasing energies of the human mind.'[16]

The traditional notion of the 'four elements' as the fundamental and unchanging building blocks of the material world, which went back to the Greeks and Aristotle, was being overturned. Earth, Air, Fire and Water were not what they seemed. To start with, it had been suspected since 1780 that the most basic of all elements – common water – was actually a subtle composition. It was finally 'decomposed', and shown to be an elastic compound of hydrogen and oxygen (H_2O), in a classic public experiment by Lavoisier in his laboratory at the Paris Arsenal on 28 February 1785.[17] This was later repeated, in an electrical experiment by William Nicholson and Anthony Carlisle reported in *Nicholson's Journal* in 1800. It would not be lost on Davy that this simple but spectacular result was achieved through the use of the newly invented voltaic battery.

Fire, long supposed by Joseph Priestley and others to depend on a single mysterious and volatile substance known as 'phlogiston', had again been analysed quite differently by Lavoisier. He proposed that fire was the rapid combination of carbon with oxygen, a process known as combustion, by which things actually grew heavier than before, not lighter. Despite all appearances, escaping 'phlogiston' (still championed by Priestley) did not exist. Nonetheless, Lavoisier still thought that heat itself was a substance, which he now proposed to call 'caloric'.

As for supposedly common air, the new science of pneumatics was to show analogous things. In reality air was an elastic mixture of oxygen and nitrogen ('azote', in Lavoisier's nomenclature), with traces of several other gases. In animal respiration it was utterly changed: oxygen was extracted by the lungs and passed into the bloodstream, while carbon dioxide was exhaled. Both Priestley and Lavoisier agreed on that.

Exactly the reverse happened with plants: vegetation 'restored air corrupted by combustion or respiration'. Plants absorbed carbon dioxide through photosynthesis, and returned oxygen to the economy of nature. This was demonstrated not by Lavoisier, but by Priestley, in another classic series of experiments, using air pumps and vacuum flasks, published in his *Experiments on Different Kinds of Air* (1774–77).[18]

These findings had inspired both the painter Joseph Wright of Derby and the young poet Anna Barbauld, who often visited Priestley's laboratory in Bowood House, Wiltshire. Yet the global significance of this crucial equilibrium between plant and animal life was not yet apparent.♣

Finally, with the 'element' of earth, it was suspected in a similar way that alkaline substances found in the common earth, such as potash and soda, hid compound secrets, if a way could be found to unlock them.

♣ Priestley gave a vivid account of an experiment in 1775, in which a mouse struggled for life in his air pump for over half an hour, thus proving that de-phlogistated air (oxygen) was a supporter of animal respiration. See Jenny Uglow, *The Lunar Men* (2002). Perhaps it also proved something else, as deduced by Anna Barbauld. She wrote about this experiment from the mouse's point of view, in a touching poem in which the 'freeborn mouse', cruelly imprisoned in its laboratory cage, appeals for its right to life, perhaps the first animal-rights manifesto ever written.

For here forlorn and sad I sit,
Within the wiry grate,
And tremble at the approaching morn
Which brings impending fate . . .

The cheerful light, the vital air,
Are blessings widely given;
Let Nature's commoners enjoy
The common gifts of Heaven.

The well-taught philosophic mind
To all compassion gives;
Casts round the world an equal eye,
And feels for all that lives.

'The Mouse's Petition to Dr Priestley, Found in the Trap where he had been Confined all Night' (1773).

The disappearance of the traditional world of the 'four elements' was revolutionary. It was as radical in the world of chemistry as Copernicus's proof that the earth was not the centre of the solar system; or (some said) as Robespierre's claim that the people, not the king, embodied sovereignty. Moreover, it was *counter-intuitive*: it went against common sense and common appearances. Surely water and air were primary, simple elements? Not at all: chemical experiment and scientific instruments could prove that they were not what they seemed to human senses – just as Newton, with his optical experiments with the prism, had shown that white sunlight was not what it seemed to the human eye, but a composite rainbow or spectrum of coloured light. Goethe had mused on the counter-intuitive nature of science: 'When we try to recognise the idea inherent in a phenomenon we are confused by the fact that it frequently – even normally – contradicts our senses. The Copernican system is based on an idea which was hard to grasp; even now it contradicts our senses every day [that the sun rises] ... The metamorphosis of plants contradicts our senses in this way.'[19]

It was characteristic of young Davy that he saw chemistry primarily as an expression of growing mental power, of creative hope. Yet he also relished the precise technical challenge it now presented. The first task lay in the decomposing or analysing of chemical substances into their true compounds, and precisely weighing, measuring and recording the process. Some twelve primary 'elements' were now established – beginning with hydrogen, carbon, oxygen and nitrogen – and many more were expected. These would later form the basis of the Periodic Table, first suggested by John Dalton as a 'Table of 20 Elements' in 1808 (and organised by the Russian chemist Dmitri Mendeleyev in 1869, using the card game of patience as a model).

Then, much more needed to be discovered about the three processes of transformation as defined by Priestley and Lavoisier: combustion, respiration, oxidation. Finally, chemistry needed to be applied to the human condition itself: the workings of the human body and mind, medicine, the cure of diseases, and what Davy called 'the laws of organic existence'. Together, these would provide the key to life on earth itself. The whole field was wide open to a new generation, and the time for a truly great chemist to emerge was ripe. No one was more aware of this than Joseph Banks at the Royal Society.

Years later, in his Geological Lectures of 1811, Davy would nonetheless praise the contribution of early alchemists like Paracelsus and Albertus Magnus, and particularly the first woman chemist, the legendary Hypatia of Alexandria, who worked in the fourth century, 'a single bright star in a night of clouds and obscurity', as he called her with a characteristic flourish. He would return to this theme, one that also fascinated Mary Shelley, in his last book, *Consolations in Travel, or The Last Days of a Philosopher* (published posthumously in 1830).[20]

Antoine Lavoisier had been the leading chemist in Europe. Elected to the Académie des Sciences in 1768 at the early age of twenty-five, he had established at the Arsenal in Paris the finest chemical laboratory of its time. Earning vast sums from his official post at the royal tax-collecting agency, the Fermiers-Général, Lavoisier poured his wealth into scientific research. His laboratory was equipped with the most sophisticated and expensive instruments available, such as the precision pair of scales made by Nicholas Fortin and said to be worth 600 livres. He also had a beautiful and highly intelligent wife, Marie-Anne Paulze, whom he trained up as a full-time scientific colleague.

Only thirteen when she married Lavoisier, Marie-Anne learned English and translated all the scientific papers by Priestley and Cavendish as soon as they appeared. She also acted as Lavoisier's laboratory assistant, wrote up his scientific journals, and drew all the illustrations for his *Traité*. Lavoisier's execution by order of the Revolutionary Convention in 1794 (he was accused of embezzling tax funds) was a disaster for French science. It also nearly overtook Marie-Anne: her beloved father was guillotined on the same day as Lavoisier, the next man to climb the scaffold (in the present Place de la Concorde) after his gifted son-in-law.[21]

Lavoisier had written an influential seven-page Preface to his *Traité Élémentaire*, defining his scientific method. This declaration seized young Davy's imagination. Writing with great simplicity and clarity, Lavoisier championed the idea of precise experiment, close observation and accurate measurement. Above all, the man of science was humble and observant before nature. 'When we begin the study of any science, we are in the situation, respecting that science, similar to that of children ... We ought to form no idea but what is a necessary consequence, and immediate effect, of an experiment or observation ... *We should proceed from the known facts to the unknown.*'[22]

Lavoisier was not of course the first to champion scientific observation and precision.* He criticised Descartes' speculative theories, and quoted the philosopher Condillac – 'instead of applying observation to the things we wished to know, we have chosen to imagine them' – who in turn quoted Bacon and the early members of the Royal Society in London: Newton, Halley, Hooke. Lavoisier was a great anglophile. He praised Bacon's philosophy of discovery, and set out the aims and ideals of experimental science as a great Romantic adventure of the mind. Davy never lost this vision, and it remained with him until the very last of his writings, set down in an essay to be called 'The Chemical Philosopher'.[23]

Now for the first time there are accounts of Davy's own experiments, as recalled by his brother John: 'His apparatus consisted chiefly of phials, wine-glasses, teacups, tobacco pipes, and earthen crucibles; and his materials were chiefly the mineral acids and alkalis in common use in medicine. He began his experimental trials in his bedroom in Mr Tonkin's house.'[24] On the cover of one notebook Davy carefully drew in ink an olive wreath encircling the flame of a lamp: the bays of poetry surrounding the light of science. Characteristically he headed another notebook 'Newton and Davy'.

* Lavoisier's 1789 Preface makes one wonder when people really did first begin to look at objects in nature carefully, *for their own sake*. The idea of exquisite, close observation of natural phenomena has its own literary history. Robert Hooke's *Micrographia* of 1664, with its exquisite drawings of fleas and other tiny creatures, championed the idea of minute observation at scales smaller than normal human vision. But the precise, even reverent contemplation of nature is clearly associated with the Romantics, and can be seen arriving in private journals and letters from the 1760s onwards. The journals of Joseph Banks and his colleagues in the South Seas, of Gilbert White in Hampshire, of Coleridge in Somerset, of Dorothy Wordsworth in the Lake District, all demonstrate this (almost sacred) attention to things simply and precisely observed. William Herschel wrote a brilliant paper on the nature of objective observation, and its particular problems in astronomy. Goethe wrote another in 1798 on the general problems of subjectivity, 'Empirical Observation and Science'. In 1788 Edward Jenner published a chilling paper in the Royal Society's *Philosophical Transactions* on observing the murderous activity of a baby cuckoo in a sparrow's nest. Jenner's quiet, meticulous description of the baby cuckoo (while still blind) relentlessly wheelbarrowing its smaller 'rival' sparrow chick backwards, between its half-formed wings, up the side of the nest until it was thrown out, has all the power of a moral allegory, but remains completely objective in tone. 'Observations on the Natural History of the Cuckoo, in a Letter to John Hunter FRA' (1788). See Tim Fulford (ed.), *Romanticism and Science* (2002), vol 4.

For a dizzy moment he believed he had disproved one of Lavoisier's basic claims, the existence of heat as a separate element called 'caloric'. By rubbing together two large lumps of ice in a vacuum, Davy produced heat by simple friction (motion), which steadily melted the ice, though no 'caloric' element had been separately introduced, and nothing had been allowed to escape. He thereby believed he had demonstrated that 'caloric' could not be a chemical entity in itself, and that the most famous French chemist must have been wrong. In fact the heating effect of friction had already been demonstrated by Count Rumford in Munich (by boring metal cannons), and Davy had partly misunderstood Lavoisier's terminology. Nevertheless, hugely excited, he began to compose a series of scientific papers, part experimental and part speculative, which he entitled 'Essays on Heat and Light'.

In summer 1797 a new lodger came to stay with Grace Davy, arranged through the ever-solicitous Tonkin. Gregory Watt was the prodigal son of the great Scottish engineer James Watt. At twenty-five he was the youngest member of the Lunar Society, brilliantly clever but physically frail – probably consumptive – and emotionally unstable.[25] He had graduated in geological sciences from Glasgow University, and had been sent to Cornwall to convalesce from what was termed a 'nervous illness'.

After initial suspicions, Davy formed a close friendship with Watt, and took him on madcap field expeditions to explore the local slate and tin mines, plunging fearlessly into the nearby Wherry mine, which ran out deep under the sea. They gathered a huge range of mineral specimens, and went drinking in the evenings. Watt – fully six years older – teased Davy as 'my dear Alchemist', and announced that he would be Davy's 'mystagogue in his initiation into the orgies of the mirth-inspiring Bacchus', by which one may understand that they drank a little French wine together in honour of Lavoisier – and possibly of Mlle Nancy.[26] Years later, in his Geology Lectures, Davy would fondly recall these expeditions. Gregory passed on Davy's name to his father, who in turn wrote about the young prodigy to his friend Dr Thomas Beddoes of Bristol.

Thomas Beddoes was regarded indulgently as a sort of secular saint by the Watt family: a holy fool of science. A gifted physician and lecturer, he had been forced to resign from his Fellowship at Oxford for his staunchly (and tactlessly) held republican and atheist views. He was a friend of Erasmus Darwin, and was much liked by the whole group of Lunar Men based around Birmingham, but especially by the Watts. At Oxford one of his best students had been the wealthy young Cornishman Davies Giddy, who was already lending Davy books from his extensive scientific library.

Beddoes, who lived in Rodney Place, Clifton, had a wide knowledge of European science, and had probably the most up-to-date scientific library in the west of England. He praised Lavoisier for his 'study of impalpable substances ... bringing within the sphere of the senses ... fire, electricity, and magnetism'.[27] In 1798, although it was wartime, he planned to open a new kind of democratic clinic, the Bristol Pneumatic Medical Institute, at 6–7 Dowry Square, Hotwells, on a hillside above the river Avon. Beddoes was now thirty-eight years old, and he felt it was the moment to try out his big idea: a radical centre for free public medicine, and research into inhalable gases, drugs and diets.

The Pneumatic Institute had been on his mind since 1794. Using the Bristol publisher Joseph Cottle, Beddoes had issued a number of idealistic pamphlets and questionnaires, to drum up financial and medical support. He wrote: 'The Institution will be conducted with the utmost publicity so that all mankind may reap the benefit of it. The expense is estimated at 3 or 4 thousand pounds.' Reassuringly for subscribers, his bankers would be Coutts & Co. of London.[28]

Beddoes had already tried treating various diseases (notably consumption, palsy and strokes) with drug regimes, using opium and digitalis, and experimental diets. His new idea was based on the recently discovered chemistry of respiration. His concept was that inhaled gases, 'factitious airs', by entering the bloodstream via the lungs, could alter and improve the whole constitution, and thereby cure major diseases. On 31 October 1794 he wrote to Davies Giddy: 'Incontestable proof has been given that the application of airs or gases to the cure of diseases is both practicable and promising. There is for instance the best reason to hope that Cancer, the most dreadful of human maladies may by some of these substances be disarmed of its terror and its danger too.'[29]

He planned to house up to a dozen in-patients, and to treat up to 300 out-patients a week, most of them without charge.[30] But financing philanthropy was always difficult. Eventually Beddoes thought that sufficient income could be generated by the sale of portable gas-inhaling equipment to local aristocrats, who he assumed were always more or less ill, and in need of gas treatment. But he needed initial capital: he asked Giddy for a gift of £350, got financing from James Watt, applied publicly to Joseph Banks at the Royal Society, and privately to the Duchess of

Devonshire. Knowing perhaps that the duchess was not averse to a flutter, Beddoes put his proposal in terms of a wager, promising he could 'cure gout for 500 guineas with a new specific', but was happily prepared 'to forfeit 5,000 guineas' if he failed. Five thousand guineas was also the sum he had hoped for from the Royal Society.[31] In the event it was another liberal aristocrat, William Henry Lambton, who supplied most of the funds, in return for having Beddoes tutor his sons.

Beddoes's republican sentiments were always closely tied up with his view of public medicine. The following spring, while informing Giddy that he was treating a young woman for stomach ulcers, he observed that 'the quondam Patriot William Pitt was almost done for', and ironically enclosed a brown silk hat-ribbon printed in gold letters with the patriotic slogan: 'Licensed to Wear Hair Powder. Pitt for Ever!'[32] ♣

It was a philanthropic project, typical of an age that also produced in Bristol at exactly this time Coleridge and Southey's Pantisocratic scheme to start a self-governing commune on the banks of the Susquehanna in America.[33] Beddoes was now looking for a young, enthusiastic assistant to promote this quixotic scheme. But he also wanted to appear realistic. He wrote carefully to Davies Giddy in July 1798: 'I can open for [Davy] a more fruitful field for investigation than anybody else. Is it not also the most direct road to fortune? ... He must devote his time [here] for two or three years ... It will be considered as part of his medical education ... He does not undertake to discover cures for this or that disease; he may acquire just applause by bringing out clear, though negative results ... I would gladly place [these] at the head of my first volume.'

In effect Beddoes was offering Davy the chance of his first scientific publication, as well as a salaried research post.[34]

Davy now began his own lively correspondence with Beddoes in Bristol, describing his 'new theories' of combustion and respiration. He announced that he had a whole series of other papers on gases, electricity, heat and – most intriguingly – the universal energy transmitted by starlight. Beddoes read these eagerly, and, encouraged by James Watt, invited Davy – not yet twenty – to join the Institute as an assistant.

♣ In 1795 Pitt had levied a tax on hair powder, to help raise funds for military campaigns abroad. The ribbon fell out of Beddoes's letter as I unfolded it in the Truro archive, and I let out a republican whoop! that almost led to my ejection.

It is significant that Davy (and his mentor Tonkin) clearly saw this as a step forward to a career in medicine, not in chemistry or the physical sciences. Such a career – that of the professional research scientist – did not yet exist. (Neither of course did the term 'scientist' itself, as will emerge.) Davy would continue to think of a career in medicine, even of taking a medical degree at Oxford, until he was thirty. But what he was pioneering was the role of the public man of science in British society, and this was to be one of his greatest and most fruitful inventions.[35]

On 1 October 1798, Davy was formally released from his indentures in Penzance, and appointed Superintendent of the Pneumatic Medical Institution in Bristol. It was a momentous move, both in terms of geography and career. His old supporter Tonkin approved, but his mother was acutely anxious and wept at his parting, while his little brother John was inconsolable. Davy set out on the long journey eastwards, round Dartmoor and through Exeter, riding on the top of the coach for economy. All along the route he passed through villages hung with flags and bunting. On enquiring, he was told that the whole nation was celebrating the news of Nelson's victory over the French at the battle of the Nile. But it was almost as if they were cheering *him* – Humphry Davy from Penzance, a son of genius.

At Bristol Hotwells he found Dr Beddoes to be 'uncommonly short and fat', a kindly but distracted host, and 'extremely silent' unless holding forth on some scientific topic or theory. Beddoes suffered from asthma, hated all physical exercise, but was passionately committed to his idea of public medicine, especially to benefit the poor. Rather surprisingly, he had married into the Edgeworth family, a clan of gifted Dublin doctors and intellectuals. His Irish father-in-law once described Beddoes as 'a little fat Democrat of considerable abilities, a great name in the Scientific world as Naturalist and Chemist – good humoured, good natured, a man of honour & virtue [though] his manners are not polite'.[36]

Davy was prepared for eccentricity in a scientific genius. The real surprise was Beddoes's young wife, twenty-four-year-old Anna, the younger half-sister of the novelist Maria Edgeworth. She was the precise opposite of the doctor (a good proof of 'polarities', Davy later thought): thin, energetic, talkative – and dazzlingly pretty – and not at all a bluestocking. A miniature of 1787 shows her with long blonde hair cut in a fringe, wide provoking eyes, and a tender voluptuous mouth. Anna was vivacious and alarmingly direct, with a love of the countryside inherited from her Irish roots. Davy reported back innocently to his mother in Penzance that Mrs Beddoes was 'the reverse of the Doctor, extremely cheerful, gay, witty; she is one of the most pleasing women I ever met with ... we are already very great friends'.[37] Soon they were going for long walks together along the banks of the Avon, and Davy was half in love with her. Several years later he would recall these walks in one of his best poems, 'Glenarm by Moonlight', describing the 'hours of confidence' they shared.

That winter Beddoes published Davy's earliest speculative essays on the chemistry of heat and starlight, which followed Lavoisier's ideas on 'oxygen' but also challenged his concept of what Davy called briskly 'the imaginary fluid caloric'. They appeared in Beddoes's annual anthology, published by Joseph Cottle, *Contributions to Physical and Medical Knowledge, principally in the West of England*, which was intended to give publicity to the Institute and encourage donations. Cottle had also, as it happened, just brought out that autumn an anonymous little book of poems entitled *Lyrical Ballads.*

Davy's two main essays were far the most ambitious contribution to the anthology, and announced his intellectual arrival in Bristol. He set out to champion chemistry, and speculate about its future, on the grandest metaphysical scale. In a Penzance notebook he had exclaimed: 'What we mean by Nature is a series of *visible images*: but these are constituted by light. Hence the worshipper of Nature is a worshipper of light.'[38] In his Essay 1, 'On Heat, Light and the Combinations of Light', he developed this into an entire cosmological vision, in which the whole universe was powered by starlight as well as Newtonian gravity, and would eventually be understood as a single unified idea. 'We may consider the sun and the fixed stars (the suns of other worlds) as immense reservoirs of light, destined by the great Organizer to diffuse over the Universe organization and animation. And thus will the law of Gravitation, as well as the Chemical laws, be considered as one great end – PERCEPTION. Reasoning thus it will not appear improbable that one law alone may govern and act upon matter, – an Energy of Mutation impressed by the will of the Deity – a law which may be called the law of Animation.'

He added confidently that 'the further we investigate the phenomenon of Nature, the more we discover simplicity and unity of design'.[39]

Even more radical was his suggestion that all human consciousness depended directly on physiological processes and 'corpuscular' changes.

'Perception, ideas, pleasures and pains, are the effect of these changes ... The laws of mind then, probably, are not different from the laws of corpuscular motion.'[40] As a result the chemistry of the human body would provide a key to human well-being in the broadest sense. 'We cannot entertain a doubt that every change in our sensations and ideas must be accompanied by some corresponding change in the organic matter of the body. These changes experimental investigation may enable us to determine. By discovering them we should be informed of the laws of our existence ... Thus would chemistry, in its connection with the laws of Life, become the most sublime and important of all sciences.'[41]

Davy was making an almost metaphysical claim that chemistry might prove to be the path to ultimate knowledge. In an unpublished essay from this time, 'An Essay to Prove that Thinking Powers Depend on the Organization of the Body', he went much further towards a materialist position. He played with the idea that all mental powers were produced by 'the peculiar action of fluids upon solids', that is, that there was a defining neurochemistry of the human brain. The 'soul' itself might ultimately be, or depend upon, a material entity. He argued that it was scientifically incorrect to believe that 'God is unable to make matter think'. All mental problems – including pain and unhappiness – might be cured by the chemistry of drugs and gases.[42]

In Essay 2, 'On the Generation of Phosoxygen', Davy developed Lavoisier's theory that all plants, when acted upon by sunlight, decomposed 'carbonic acid gas' (carbon dioxide) and released oxygen into the atmosphere. He also claimed to show experimentally that aquatic plants, when exposed to sunlight, oxygenated the surrounding water. Since all animal life did the reverse – absorbing oxygen in respiration and releasing carbonic gas – there was an essential equilibrium or harmony within nature. Davy had in effect described what is now known as the 'carbon cycle'.[43]

3

Davy began his regular work at the Institute, seeing patients and administering gas and drugs according to Dr Beddoes's instructions. These treatments were based on the 'Brunonian system', the theoretical work of Scottish physician John Brown (1735–88), hotly debated in the Edinburgh medical schools, which divided all medicines into stimulants and depressants. In fact this had very little basis in trials or experiment, as Davy gradually came to realise (and as Banks at the Royal Society had long suspected). Beddoes also introduced him to his Bristol publisher Joseph Cottle, and sent him to visit the Institute's most influential supporters: the powerful Wedgwood family at Cote House, and James Watt and the Lunar Society in Birmingham. Davy made an excellent impression on everyone he met, and his circle of acquaintances rapidly expanded.

Initially Davy boarded with the Beddoes family in their large house at 3 Rodney Place, Clifton. Later he moved down the hill to live directly above the Institute and its laboratories and garden, in a corner of Dowry Square, Hotwells. As its name implied, the Hotwells district had a long tradition of thermal baths and healing spa establishments. But the small, reclusive Georgian square, tucked away into the hillside below Clifton village and wood, seemed an odd location for an experimental medical practice, with its daily stream of poverty-stricken patients, and its pungent aroma of chemicals and gases.

Hitherto the square had been an elegant *cul de sac*, with only its southern end opening onto the main Hotwells coaching road into Bristol city and docks. Until the arrival of Dr Beddoes's Institute, it had evidently been a haven of tranquillity and respectability. The fine new brick and sandstone houses, with their tall sash windows and pillared porticoes, quietly enclosed a private garden on three sides. The Institute occupied two adjacent buildings, Nos 6 and 7, on an L-shaped site in what had previously been the north-west corner of the square, the quietest and furthest from the road. Beddoes chose the elegant No. 7 to house the main reception rooms and infirmary, while No. 6, more of a rabbit warren, contained the laboratories and staff quarters, and opened directly out onto a steep garden at the back. A separate outbuilding in that garden was used for the manufacture of gases and the storage of chemical compounds. No. 6 also had a wide tradesmen's entrance, where medical supplies could be delivered in bulk by cart, and bodies (usually of small animals) could be removed.[44] ♣

♣ The present square is subdued and beautiful, if distinctly run-down. No. 7 has no plaque behind its wisteria, and No. 6 has become the offices of a builders' merchant. Nevertheless the distant memory of its great, subversive medical tradition still haunts the square: one worn old stone plate reads: 'Robert Young, Surgeon', while another ancient brass proclaims 'Clifton Dispensary'.

As part of his policy of progressive public medicine, Beddoes advertised free pneumatic treatments for people suffering from consumption, asthma, palsy and scrofula. Untreatable or anti-social diseases, such as venereal infections, were also included. For more wealthy patients, the Institute offered inhaling kits that could be purchased and used in the home. This was one aspect of the Institute that Banks had objected to, as he felt it was open to quackery.[45]

For the first few months Davy, though delighted by his quarters, found himself acting largely as a medical superintendent. There were a number of assistants under his command, including two ancient bottle-washers, Dwyer and Clayfield, and young Dr Kinglake, whom he quickly dominated. But gradually the working rooms were fitted out, and for the first time in his life Davy was in charge of a well-equipped chemical laboratory.

By the spring of 1799 Beddoes agreed to Davy setting up a monitored series of gas-inhaling experiments, to see if any real scientific data could be gathered on the healing power of gases. In fact he intended to use the new empirical chemistry of Priestley and Lavoisier to test, and if necessary challenge, the Brunonian system of medicine by controlled experiment. He wrote to James Watt, an outstanding engineer, for designs of gas-inhaling equipment, including a silken face-mask with a wooden mouthpiece. The masks and gas bags were based on balloon technology.[46]

In April 1799 Davy began his analysis of common air, and the workings of human respiration within the lung. He spread his initial experiments over various compounds of 'factitious airs', including hydrogen, carbon dioxide and carbon monoxide, and several combinations of nitrous gas. Before trying anything out on his patients, he tested everything on himself, often at grave risk. Fainting fits, nausea and stunning migraines frequently overcame him. But he was undaunted.

One early unguarded experiment with carbon monoxide (a lethal gas, still much favoured by garage suicides) almost killed him.[47] At two in the afternoon he began to inhale four quarts of 'pure hydrocarbonate' in the presence of his assistant Patrick Dwyer and a new laboratory recruit, James Tobin. On inhaling the third quart he collapsed. 'I seemed sinking into annihilation, and had just power enough to drop the mouthpiece from my unclosed lips ... I faintly articulated, "I do not think I shall die." ' Davy still had the presence of mind to take his own pulse – 'threadlike and beating with excessive quickness' – then staggered out of the laboratory into the garden of No. 6 Dowry Square.

Here he collapsed on the lawn, trembling and seized with agonising chest pains. He was semi-conscious for some minutes, and was given oxygen by the terrified Dwyer. After half an hour he thought he was recovered, but he became giddy again and was helped to a bed. He lay there for the rest of the day, suffering from 'nausea, loss of memory, and deficient sensation'. He vomited, and was then overcome by 'excruciating pain' between the eyes. Finally by ten o'clock at night his symptoms began to ease, and he fell into an exhausted sleep.

Davy had nearly recovered his strength by the next evening, that is some thirty hours later. He concluded calmly that if he had taken 'four or five [quart] inspirations instead of three', he would have 'destroyed life immediately without producing any painful sensations'. A week later he was trying to inhale 'carbonic acid' (perhaps vaporised phenol), which so burnt his epiglottis that he choked.[48]

It is remarkable that these effects did not frighten or deter him, and these early experiments give a first glimpse of the reckless courage and impetuosity that always drove Davy in the laboratory. Nonetheless, it is also notable that he had previously prepared a bladder of oxygen in case of emergency, and Dwyer was instructed to apply it. The publisher Joseph Cottle, who was convinced of Davy's genius and hoped eventually to print the results of his experiments (in the unlikely event that he survived them), recalled melodramatically: 'No personal danger restrained him from determining facts, as the data for his reasoning... He seemed to act as if in case of sacrificing one life, he had two or three others in reserve, on which he could fall back in case of necessity... Occasionally I half despaired of seeing him alive the next morning.'[49]

Finally Davy decided that the properties of nitrous oxide (N_2O, or laughing gas) made it the safest and the most promising for trials. He set himself his first experimental research programme, to test different concentrations of the gas: first on himself, then on animals, and finally on other human volunteers. Initially he was especially interested to analyse the *exhaled* air of the lungs, to discover what quantities of the gas were absorbed into the human bloodstream. He devised ingenious equipment to measure and control both inhalations and exhalations: various silk bags and bladders, glass vacuum flasks, a mercurial 'air-holder and breathing machine' of cast iron, made by his assistant Clayfield, wooden and metal mouthpieces, corked tubes which could be placed in the nostrils, face-masks and hand pumps, and finally (after nine months) a complete portable gas chamber with entrance and exit valves.[50]

At first Davy was largely concerned with the process of respiration, and possible therapeutic benefits. Later, with his human subjects, he became more interested in the physiological reactions of the whole body; and effects of pleasure and pain. Finally he became fascinated by purely psychological responses. He wrote proudly to his mother in Penzance: 'We are going on gloriously. Our palsied patients are getting better; and, to be a little conceited, I am making discoveries every day.'[51] He also boasted that he had been invited to contribute poetry to the leading Bristol literary magazine the *Annual Anthology*. It was published by Cottle, and edited by the young poet and one-time Pantisocrat Robert Southey, recently returned from Spain.

4

Nitrous oxide was not without risks. It was considered a lethal gas by both Priestley and the American chemist Dr Samuel Mitchill.[52] But Davy went ahead anyway. He heated crystals of ammonium nitrate, collected the gas released in a green oiled-silk bag, passed it through water vapour to remove impurities, then inhaled it through a mouthpiece while his assistant Dr Kinglake monitored his pulse rate.[53] The immediate obvious danger was that the ammonium nitrate would explode at a temperature above 400 degrees; the other was that the first inhalations would kill him or permanently damage the linings of his lungs.

But Davy's first experiment went superbly. After inhaling four quarts of gas, he experienced 'highly pleasurable thrilling, particularly in the chest and extremities. The objects around me became dazzling, and my hearing more acute.' The next day the entire experience appeared dream-like, he could not recall his sensations, and only by rereading his laboratory notes was he convinced that the experiment had taken place at all.[54]

Davy frankly admitted the extraordinary first effects of nitrous oxide. He experienced strange 'thrillings', increased bodily heat in his extremities, giddiness, raised pulse rate and (carefully observed in a mirror) a facial flush or suffusion of blood so 'my cheeks became purple'. He noted: 'Sometimes I manifested my pleasure by stamping or laughing only; at other times, by dancing round the room and vociferating.[55] He sent his first account to his Cornish supporter Davies Giddy in a letter of

10 April 1799. 'This gas raised my pulse upwards of twenty strokes, made me dance about the laboratory as a madman, and has kept my spirits in a glow ever since.'[56] Shortly afterwards he sent three rather more sober reports to the leading scientific magazine of the day, *Nicholson's Journal.*

These earliest experiments he also recorded in verse, partly to see how far his linguistic skills were affected, and also to explore whether the experience could be imaginatively described. In this case the poetry was itself a form of scientific data. The result was very bad verse, but surprisingly precise physiological information. He headed it 'On Breathing Nitrous Oxide'.

> Not in the ideal dreams of wild desire
> Have I beheld a rapture-waking form;
> My bosom burns with no unhallowed fire:
> Yet is my cheek with rosy blushes warm
> Yet are my eyes with sparkling lustre filled
> Yet is my mouth replete with murmuring sound
> Yet are my limbs with inward transport thrilled
> And clad with newborn mightiness around.[57]

Davy suggests in the opening three lines that his physiological state could be compared to that of spontaneous sexual arousal produced by an erotic dream ('dreams of wild desire . . . a rapture-waking form'). Yet in fact this is *not* the cause (although Anna Beddoes may have provoked the comparison). He then goes on in the next four lines carefully to define his physical sensations, the facial blushing and so on, and ends with the overwhelming delusion of physical power ('newborn mightiness'). Davy's verse is normally very clear, so the uncharacteristic confusion of grammar and syntax here, opening with the emphatic but barely coherent 'Not', and tailing off into the repeated 'Yet', is itself interesting empirical evidence of his mental state.

His discovery of the gas's potential filled him with excitement and ambition. One of his earliest laboratory notes records: 'This evening April 27th [1799] I have felt a more high degree of pleasure from breathing nitrous oxide than I ever felt from any cause whatever – a thrilling all over me most exquisitely pleasurable, I said to myself I was born to benefit the world by my great talents.'[58]

Nitrous oxide inhalations now became a regular part of his laboratory routine. He recorded in his notes: 'Between April and June I constantly breathed the gas sometimes three or four times a day for a week, at other times, four or five times a week only. The general effect of it I can describe with great difficulty, nor can I well discriminate between its agency and that of other physical and moral causes. I slept less than usual, I thought more in bed, I had a constant desire of action.' He did not define what this 'action' might be, but he thought he had 'increased sensibility of touch', and the tips of his fingers were 'pained' by rough things, even by paper.[59] He felt he was 'more irritable than usual', though that might have had other 'moral' causes. In retrospect it appears that these 'moral causes' might have been connected with Anna Beddoes.[60]

Normally the gas was taken in laboratory conditions, with Dr Kinglake as his assistant, while rigorous notes were taken. But sometimes Davy returned in the evenings, apparently alone. These sessions seem to have been particularly intense. 'I have often felt very great pleasure when breathing it alone, in darkness and silence, occupied only by ideal existence.'[61] On the evening of 5 May 1799 Davy prepared to subject himself to a special session, to see what the psychological effects might be of a deliberately excessive dose. He prepared for this by going for a long, solitary, moonlit walk along the banks of the Avon, clearing his mind and tuning his feelings to the beauties of nature. 'After eating a supper, drinking two glasses of brandy and water and sitting for some time on the top of the wall reading Condorcet's *Life of Voltaire*', he returned to Dowry Square, and immediately inhaled six quarts of pure nitrous oxide. The experiment was monitored by Kinglake. The results were striking, but oddly disappointing, because they produced no further mental or spiritual revelations: 'The pleasurable sensation was at first local, and perceived in the lips and the cheeks. It gradually, however, diffused itself over the whole body, and in the middle of the experiment was for a moment so intense and pure as to absorb existence. At this moment, and not before, I lost consciousness; it was, however, quickly restored, and I endeavoured to make a bystander acquainted with the pleasures I experienced by laughing and stamping. I had no vivid ideas.'[62]

There was no extension of his earlier visionary glimpses to record in his notes. Though he had 'vivid and agreeable dreams' later that night, nothing further emerged from his unconscious mind. The large dose of nitrous oxide brought him no spiritual revelation, no deeper contact with 'the universe'. Perhaps it was Davy's disappointment with this that made him overlook at first the significant physiological fact that the gas could be used in an entirely different way: to banish consciousness altogether, and then safely and quickly 'restore' it. A gas that could blot out feelings – *ana-thesia* – and then bring them back.

Yet in retrospect Davy clearly grasped what had occurred, as he returned to this experience some twenty years later in his book *Salmonia*. While discussing the nature of pain as felt by animals and even fish, and the sense of impending death, he recalled the process of 'losing consciousness' by inhaling both nitrous oxide and (the lethal) carbon monoxide gas.[63]

He continued experimenting on himself until July, usually three or four times daily, and sometimes in the evening after drinking wine or brandy. Yet he remained meticulous in all his scientific records. For example, he repeatedly measured his lung capacity, finally refining his figures down to these: 254 cubic inches when he forcefully inhaled; 135 cubic inches when he naturally inhaled; and, perhaps most interesting, forty cubic inches of residual air. He also analysed the lung content of natural air without nitrous oxide: 71.9 per cent nitrogen, 15.2 per cent oxygen, and 12.8 per cent carbon dioxide, which is startlingly close to a modern chemical analysis.[64]

In May 1799 he tentatively began nitrous oxide trials with the clinic's regular patients. The results could be unpredictable. Some reported delicious sensations of bodily heat and stimulation. Others recorded alarming muscular spasms or mental confusion. Yet others were merely made dizzy or sleepy.

Davy now pioneered the 'blind' experimental method. He deliberately did not tell his subjects what concentration of nitrous oxide they were breathing, or whether they were in fact inhaling ordinary air (which they sometimes were). He carefully recorded pulse, muscular reactions, visual distortions, blushing and sexual stimulation, and any episodes of mental confusion or hysteria. He also asked his subjects to describe in detail their own subjective sensations.

His laboratory journals show that he was increasingly fascinated by the hallucinogenic properties of the gas, and its effects on human consciousness and perceptions. Slowly he became aware of its power to alter moods, stimulate the body's energy and deaden pain. Then he hovered again around the revolutionary notion of *anaesthesia*, and one further decisive step: the idea that controlled doses of nitrous oxide might be used in a surgical context.[65]

Davy initially used voluntary patients who were already attending the clinic, or members of Beddoes's immediate circle of family and friends. Most were male, but there were several young women, though they remain tactfully unnamed in his reports. Anna Beddoes certainly inhaled the gas under Davy's supervision, perhaps for pleasure as much as in the interests of science. As Beddoes himself wryly observed: 'Mrs. Beddoes had frequently seemed to be ascending like a balloon up the hill to Clifton.' Robert Southey's young wife Edith also inhaled, though Davy's note records that she was 'very little affected, only rendered giddy'.[66]

But one young woman, described only as 'Miss J', had such a violent reaction that Davy was evidently alarmed. 'Miss J breathed this morning six quarts ... in about a minute she suffered the bag to drop. She then began to sob most violently, then cried and laughed alternately. She used most vivid muscular actions and appeared to be perfectly delirious in about ten minutes.' Davy dragged her over to the window, and attempted to calm her by getting her to take deep breaths of fresh air. 'She relapsed again and continued in the hysterical fit for near two minutes. Her muscular motions were uncommonly violent.'[67]

Joseph Cottle recorded a rumour that another young woman was overcome by hysterical excitement, ran out of the laboratory, and rushed screaming down the street towards the Avon, where she was somewhat bizarrely reported to have 'jumped over a large dog' before she could be restrained and brought back. This case does not appear in Davy's notes, but the idea that women could be made to lose their inhibitions, and might even be sexually aroused by nitrous oxide, persisted.

With his experimental subjects, Davy monitored pulse rates, and required them to undergo certain standard tests, such as gazing at a candle flame and listening to bells. He wanted to record physiological changes, such as distortions of vision and hearing. But gradually he became more and more interested in subjective responses. He asked his Institution patients to put into words exactly what they were feeling. This proved surprisingly difficult, and early responses ranged from 'I don't know how, but very queer'; to 'I felt like the sound of a harp.'[68]

Davy now conceived a new and original line of investigation. He began to enlist perfectly healthy subjects, chosen from his highly articulate circle of Bristol friends, and asked them to describe their sensations as precisely as possible. They included the poet Robert Southey, several members of the Edgeworth family, Gregory Watt and his father James, Tom Wedgwood, the heir to the great Staffordshire pottery company, and a number of young writers and scholars like Peter Roget and John Rickman.

A few, like Cottle, refused to volunteer, either on moral or prudential grounds. But it is striking how many accepted. Eighty pages of these accounts were eventually published in Davy's *Researches.* Many of their strange, gasped phrases included the idea of rebirth: 'I seemed a new being'; 'I seemed a sublime being newly created'; 'I felt as if possessed of new organs.'[69] Initial enthusiasm was naïve and unbounded. Gregory Watt spoke of 'heavenly inhalations'. Robert Southey wrote to his brother: 'O, Tom! Such a gas has Davy discovered, the gaseous oxide. Oh, Tom! I have had some; it made me laugh and tingle in every toe and finger-tip. Davy has actually invented a new pleasure, for which language has no name. Oh, Tom! I am going for more this evening! It makes one strong and happy! So gloriously happy!'[70]

Maria Edgeworth visited Clifton at this euphoric time, and gave a novelistic spin to her sister Anna's glowing accounts of Davy's experiments. The handsome young Cornishman 'enthusiastically expects wonders will be performed by the use of certain gases, which inebriate in the most delightful manner, having the oblivious effects of Lethe, and at the same time giving the rapturous sensations of the Nectar of the Gods!' Maria also noted with an amused eye that Anna seemed particularly full of 'grace, genius, vivacity, and kindness' that spring, implying that this had not always been the case before Mr Davy's arrival.[71]

But some accounts were more prosaic. Southey and several others recorded the simple terror they felt on first putting the wooden mouthpiece between their lips, and attempting to breathe the gas. This was followed by sensations of relief, giddiness or weightlessness, falling over, and finally helpless laughter: 'The laughter was involuntary but highly pleasurable, accompanied by a thrill all through me.' It is noticeable that Southey's accounts (given both to Beddoes and to Davy in the laboratory) are much more restrained than those in his private letters to his brother Tom. This marks an interesting problem in the gathering of supposedly objective, scientific evidence.

Dr Peter Mark Roget, then a young medical student from Edinburgh, and the future compiler of *Roget's Thesaurus* (1852), found, ironically enough, great difficulty in choosing the words to describe his feelings aptly. 'I felt myself totally incapable of speaking ... My ideas succeeded one another with extreme rapidity, thoughts rushed like a torrent through my mind.' He felt forced to try analogies, and towards the end managed an accurate description of fainting: 'I suddenly lost sight of all the objects round me, they being apparently obscured by clouds, in which were many luminous points, similar to what is experienced on rising suddenly and stretching out the arms.'[72]

Mr Coates primly observed 'a degree of hilarity altogether new to me'; while Miss Rylands was circumspectly 'deprived of the power of speaking, but not of recollection'. Beddoes's brother-in-law, the jolly Mr Lovell Edgeworth, 'burst into a violent fit of laughter, and capered about the room without having the power of restraining myself'. He remarked wonderingly that he almost bit through the wooden mouthpiece.

What Davy began to see was that reactions reflected personal temperament, as much as simple physiological changes. So what the musician Mr Wansey reported was an experience like 'some of the grand choruses of the Messiah' which he had heard played by 700 instruments in Westminster Abbey five years previously. While Southey's great friend, the down-to-earth radical tanner of Nether Stowey, Tom Poole, was reminded of climbing mountains in Glamorganshire.[73]

5

The friendship which now formed between Southey and Davy was one of the most important of his Bristol years. They spent many evenings together at Dowry Square in the spring of 1799, discussing politics, science, literature and medicine, as well as inhaling nitrous oxide. They sometimes walked out together to Tom Poole's house in Nether Stowey, or to Southey's lodgings in Wiltshire. 'When I went to the Pneumatic Institute,' Southey recalled, 'he had to tell me of some new experiment or discovery and the views which it opened for him, and when he came to Westbury there was a fresh bit of *Madoc* for his hearing.'[74] Southey was hugely impressed by Davy's energy and idealism. He wrote to his friend William Wynn: 'Humphry Davy possesses the most miraculous talents I ever met with or heard of, and will I think do more for medicine than any person who has ever gone before him.'[75]

The rival claims of poetry and science became a passionate topic between them. Davy tentatively showed Southey his poems, scattered through his laboratory notebooks, many still in draft, and Southey promised to select the best and publish them. On 4 May 1799 he wrote to Davy constructively criticising his poem 'Mount's Bay' and urging him to continue writing.[76] Poetry, he argued, would be good for Davy's science. He also promised to introduce his extraordinary friend Coleridge, who was due (indeed *overdue* as usual) to return from Germany, full of Blumenbach's scientific lectures at Göttingen University and wild tales of the witches of Walpurgisnacht in the Hartz mountains. Coleridge was proposing to translate Blumenbach's *Manual of Natural History*, though he should also have been finishing his poem 'Christabel'. Among other wonders Southey also described the strange Valley of the Rocks near Lynmouth, and asked Davy whether a scientific explanation (sea erosion) or a mythological one (giants' abandoned castle) would be more satisfactory.

In this way Southey gently encouraged Davy not to abandon his vision as a poet, amidst all the excitement of the gas experiments. 'I must not press the subject of poetry upon you, only do not lose the feeling and habit of seeing all things with a poet's eye: at Bristol you have a good society, but not a man who knows anything of poetry. Dr Beddoes's taste is very pessimism.'[77]

Southey wrote again in August, proposing that he and Davy collaborate on an epic poem set in Peru. (The only other writer with whom Southey had previously collaborated was Coleridge.) This would develop Davy's poet's eye, and be 'a relaxation from more important studies'. The 1799 *Annual Anthology*, which included Davy's five poems, was nearly ready for publication. 'You still, I suppose, go on working with your gaseous oxide, which according to my notions of celestial enjoyment, must certainly constitute the atmosphere of the highest of all possible heavens. I wish I was at the Pneumatic Institution, something to gratify my appetite for that delectable air, and something for the sake of seeing you.'[78]

Nevertheless, when Davy proudly sent a copy of the *Annual Anthology* to his mother in Penzance, he felt the need to reassure her. 'Do not suppose I am turned poet. Philosophy, Chemistry and Medicine are my profession. I had often described Mount's Bay to my friends here. They desired me to describe it poetically.'[79]

It was to be a memorable autumn. On 11 October Gregory Watt wrote to Davy: 'get an air holder of gas prepared for I am determined to ascend the heavens'.[80] A new round of nitrous oxide experiments had begun, and these brought Davy's first meeting with Samuel Taylor Coleridge, still full of his trip to Germany and just celebrating his twenty-seventh birthday. It took place at Dowry Square on 22 October 1799. Coleridge was only in Bristol for a fortnight before hurrying off to join Wordsworth and Dorothy in the Lake District. Nevertheless he spent several evenings talking excitedly with Davy, and had repeated inhalation sessions at the Dowry Square laboratory. He must have compared the gas with his already extensive and overpowering experience of opium.[81]

In fact Coleridge's accounts of his reactions to the gas seem oddly prosaic. His heart thumped 'violently', he involuntarily 'beat the ground' with his feet, and he watched some trees in the garden becoming 'dimmer and dimmer', as if seen through tears. Nitrous oxide seemed strangely reassuring to Coleridge, even homely: 'an highly pleasurable sensation of warmth over my whole frame, resembling what I remember once to have experienced after returning from a walk in the snow into a warm room'.[82] He used only one descriptive phrase which is reminiscent of a line from his great opium poem of 1797, 'Kubla Khan': he spoke of 'more *unmingled pleasure* than I had ever before experienced'.

Yet Coleridge was evidently intrigued by the whole phenomenon of nitrous oxide and its '*psychosomatic*' (a word he coined) implications, and would return to it in a dazzling series of letters written to Davy the following year. Altogether Coleridge was much struck by the young chemist ('an admirable young man'), but he still hurried away to visit Wordsworth in the Lake District. It was only when Davy travelled up to London for the first time in his life, in late November 1799, that the friendship was truly formed.

Coleridge was now living with Charles and Mary Lamb in the Middle Temple, translating Schiller's play *Wallenstein* (not Blumenbach after all) and writing articles for the *Morning Post.* He saw Davy frequently over a ten-day period, and took him to dine with the anarchist philosopher William Godwin. They were joined by Lamb, the poet Charlotte Smith and the portrait painter James Northcote. This was a memorable dinner, and Davy talked brilliantly about the future of science to his artistic listeners. Godwin, at the height of his philosophical fame, having just published his notorious *Memoir* of his wife Mary Wollstonecraft, was hugely impressed with Davy, although he thought he would 'degrade his vast Talents' by limiting them to chemistry. Yet everyone was agreed: Davy was 'extraordinary'.

Coleridge immediately began to fantasise about setting up a 'little colony' with Davy and Wordsworth (though they had not even met) – 'Precious stuff for Dreams'. After Davy returned to Bristol, Coleridge wrote a long letter in January 1800, opening with a characteristic suggestion: 'I wish in your researches that you and Beddoes would give a compact compressed History of the Human Mind for the last century . . .' As to Godwin's criticism of chemistry, Coleridge described his robust defence to Davy. 'Why, quoth I, "how Godwin! Can you thus talk of science, of which neither you nor I understand an iota" etc, and I defended Chemistry as knowingly at least as Godwin attacked it – affirmed that it united the opposite advantages of immaterialising the mind without destroying the definiteness of the Ideas – nay even while it gave clearness to them.'

Here Coleridge was defending the intellectual discipline of science as a force for clarity and good. He then added one of his most inspired perceptions. He thought that science, as a human activity, 'being necessarily performed with *the passion of Hope*, it was poetical'. Science, like poetry, was not merely 'progressive'. It directed a particular kind of moral energy and imaginative longing into the future. It enshrined the implicit belief that mankind could achieve a better, happier world. This is what Davy believed too, and 'Hope' became one of his watchwords.[83]

6

Throughout the momentous year 1799 Davy continued to fill his notebooks with visionary essays and poems. But he did not forget Cornwall. In October, to Grace Davy's delight, the prodigal son had returned home to Penzance for a month. He brought her fashionable jewellery from Bristol, and an impressively large case of chemical apparatus. Davy later described how an entire portable chemical laboratory, including air pump, electrical apparatus and 'a small forge', could be fitted into a single trunk. He visited Davies Giddy and other old friends, walked up to his father's tomb at Ludgvan church, went fishing, shooting and geologising, and wrote some dreamy half-rhyming poems, distinctly inspired by Coleridge's 'conversation poems' in the *Lyrical Ballads*.

> Many days have passed
> Beloved scene, since last I saw
> The moonbeams gild thy whitely-foaming waves . . .
> The dew of labour has oppressed my brow,
> On which the rose of pleasure never glowed;

For I have tasted of that sacred stream
Of science, whose delicious water flows
From Nature's bosom ...[84]

In December, as promised, his five poems appeared in Southey's *Annual Anthology*, including 'The Sons of Genius', 'Saint Michael's Mount' and 'The Tempest'.

It was in this same month that Davy first used a portable gas chamber especially designed by James Watt. This device allowed a much longer total exposure to nitrous oxide, and also psychologically isolated the subject from his laboratory surroundings. It was a narrow, dark, boxed chamber 'like a sedan chair', about five feet high, completely sealed with stretched canvas and pasted paper to make it airtight. Air was pumped out from a two-inch vent above the subject's head, while gas was introduced by another 'about the height of the knee'. The subject was supplied with 'a feather fan' to mix the gas around him. 'On each side and in front should be a pane of glass about twelve by eighteen inches, that you may see the patient during his confinement.' The hermetic seal on the chamber allowed gas to be introduced under slightly higher than normal atmospheric pressure.[85]

There is a vivid account of Davy's first use of this faintly sinister machine, on 26 December 1799. Naturally he tried it himself first. He stripped to the waist, placed a large mercury thermometer under his armpit, took a stopwatch to time his pulse, and had himself sealed into the chamber by Kinglake. Over a precisely agreed time of seventy-five minutes, Kinglake pumped in ('threw in') exactly eighty quarts of nitrous oxide. Davy's pulse rose to 124, his temperature to 106, and his cheeks went bright purple. But, amazingly, he remained conscious. Kinglake then released him, and gave him as planned a final twenty quarts of pure gas to inhale through a mouthpiece. It had been agreed that Davy (if he could still speak) would then try to describe his sensations as accurately as he could to Kinglake.

This is the published version: 'By degrees as the pleasurable sensations increased, I lost all connection with external things; trains of vivid visible Images rapidly passed through my mind and were connected with words in such a manner, as to produce perceptions perfectly novel. I existed in a world of *newly connected and newly modified ideas*. I theorized; I imagined that I had made discoveries. When I was awakened from this semi-delirious trance by Dr Kinglake, who took the gas-bag from my mouth, Indignation and pride were my first feelings ... My emotions were enthusiastic and sublime; and for a minute I walked round the room perfectly regardless of what was said to me ... With the most intense belief and prophetic manner, I exclaimed to Dr Kinglake, – "Nothing exists, but Thoughts! – the Universe is composed of impressions, ideas, pleasures and pains!" '[86]

The unpublished verbatim version from Davy's 1799 laboratory notebook is rather more colourful: 'I was now almost completely intoxicated ... The sensations were superior to any I ever experienced. Inconceivably pleasurable ... Theories passed rapidly thro the mind, believed I may say intensely, at the same time that every thing going on in the room was perceived. I seemed to be a sublime being, newly created and superior to other mortals, I was indignant at what they said of me and stalked majestically out of the laboratory to inform Dr Kinglake privately that nothing existed but thoughts.'[87]

Other extreme experiments included the combination of nitrous oxide with alcohol. On one December night in the laboratory, Davy drank an entire bottle of wine 'as fast as possible that it might produce its full effects'. The gas did not prevent 'complete intoxication' in less than an hour, but it helped his hangover the next morning. The experiment was repeated, with the same results. He noted: 'On December 23 I breathed after a terrible drunken fit a larger quantity of gas, 2 bags and two bags of oxygen, it made me sick.' The next day, 'no headache came on, and my appetite was almost canine'.[88] He wondered if his experiments were getting out of hand.

With the completion of the intensive work on *Researches*, Davy gave himself up to further intellectual speculations. In his notebooks he scribbled extensive essays, often unfinished, on subjects such as the 'Formation of the Intellect' (starting in the womb, before birth); 'The History of Passion', 'On Genius' ('what is this generating faculty of man, which acts through the immensity of ages?'); and 'On Dreaming'.[89]

Now besides the poetry, there were fictional fragments, erotic fantasies, and some unusual sections of self-analysis. Some clearly show Coleridge's continuing influence, and touch again on the difference between the scientific and the poetic imagination: 'Today, for the first time in my life, I have had a distinct sympathy with nature. I was lying on the top of a rock to leeward; the wind was high, and everything in motion ... everything was alive, and myself part of the series of visible impressions; I should have felt pain in tearing a leaf from one of the trees ... Deeply and intimately connected are all our ideas of motion and life, and this, probably, from very early association. How different is the idea of life in a physiologist and a poet!'[90]

In June 1800 Davy published his first individual work, *Researches Chemical and Philosophical chiefly Concerning Nitrous Oxide or Dephlogisticated Nitrous Air, and its Respiration.* It was issued in London by Joseph Johnson, the radical publisher of William Godwin, Mary Wollstonecraft, Coleridge and Wordsworth. Davy thus joined an author list which was largely literary and philosophical in character, and with a strong radical reputation.[91]

'Ten months of incessant labour were employed in making them,' he wrote of the *Researches*, 'and three months in detailing them.' Crisply divided into four 'Research' sections, the historic monograph described the entire range of his gas experiments, presenting the previous history of gases (Research 1); his own chemical analysis and decompositions of nitrous oxide (Research 2); his examination of the whole phenomenon of respiration (Research 3); and finally eighty pages of detailed accounts of the individual inhalation sessions (Researches 4). It was these that caused the sensation among general readers.

Humans were not Davy's only subjects. The section 'Research 3' contains multiple gas experiments (nitrous oxide, hydrogen and carbon monoxide) on live animals, including dogs, cats, birds and rabbits. He also immersed fish in de-oxygenated water; and butterflies, bees and house flies in mixtures of the gases. Many of these subjects died in convulsions, and were calmly dissected. None of his scientific reviewers remarked on the problematic nature of this research, but Davy became more and more uneasy at the pain he was causing, and Coleridge would later call his attention to pain as a phenomenon in itself.

Davy's style is plain, discursive and never sensational. He presents himself throughout as the objective narrator of each experiment, the calm man of science who can observe an animal asphyxiating without emotion, and can still take his own pulse when he thinks he himself is dying. In this sense the book invents a scientific persona, the unflinching teller of true tales. The *Researches* were dedicated to Dr Thomas Beddoes, as 'pledges of more important labours'. Yet Davy carefully avoided drawing any general conclusions, and in particular made no medical claims at all about the therapeutic value of gas treatments at the Pneumatic Institute.

In an important Preface Davy described how his new, empirical approach to scientific investigation had altered over the last eighteen months.[92] As a result of the critical reception of his highly speculative early essay 'On Light and Heat', he had begun an extensive reappraisal of his own scientific methods at Bristol. In his notebooks of late 1799 there are humiliating confessions critical of his own premature 'pursuit of speculations and theories', and of the 'dangers of false generalization'. He now believed, not entirely convincingly, that 'the true philosopher' avoided 'theories' altogether. He upbraided himself fiercely: 'It is more laborious to accumulate facts than to reason concerning them; but one good experiment is of more value than the ingenuity of a brain like Newton's.'[93]

In the published Preface, these self-criticisms are only slightly modulated. Self-criticism itself was now becoming an effective part of Davy's scientific style: 'I have endeavoured to guard against sources of error; but I cannot flatter myself that I have altogether avoided them. The physical sciences are almost wholly dependent on the minute observation and comparison of properties of things not immediately obvious to the senses ... I have seldom entered into theoretical discussion, particularly concerning light, heat and other agents ... Early experience has taught me the folly of hasty generalization. We are ignorant of the laws of corpuscular motion ... Chemistry in its present state, is simply a partial history of phenomena, consisting of many series more or less extensive of accurately connected facts.'[94]

Here was, apparently, the sobered scientific empiricist who would appeal to Banks and the Committee of the Royal Institution.

It was vital to Davy how his first publication would be received. In July 1800, anxiously awaiting news, he took himself off on a long summer walking tour into Wales with the painter Thomas Underwood, a rich and bohemian young man who had scientific interests. Underwood also happened to be one of the proprietors of the newly formed Royal Institution in London, but they restricted their conversation to the benefits of sunlight, starlight and fishing.

First reactions to the book were very mixed. Preliminary accounts of the experiments had already been unguardedly described by Beddoes in a pamphlet, *Observations made at the Medical Pneumatic Institution*, published at the end of 1799. Though circulated in Bristol, this unwittingly prepared the way for scandal in London.[95] The polemicist Richard Polwhele quickly published a nimble poem, 'The Pneumatic Revellers' (1800), a satirical attack on the nitrous oxide experiments which used suggestive dialogue and innuendo to imply that the Bristol laboratory witnessed scenes of intoxication, hysteria and even sexual debauchery.[96] The experimenters and their subjects were mocked with glee:

> And they cried, everyone, 'twas a pleasure ecstatic
> To drink deeper drafts of the mighty Pneumatic!

Davy and Beddoes were also attacked, as Banks had feared, in an anonymous pamphlet, *The Sceptic* (1800). They were described as a pair of 'Bladder conjurors and newfangled Doctors pimping for Caloric'. In a word, they had used gas to seduce their female subjects when unconscious. By putting female subjects 'in a state of gas', they insidiously 'gained admittance to their lovely persons'. In a fantasy sequence, the pamphlet describes how the dastardly Dr Caloric 'warms their snowy bosoms; blows up the latent spark of soft desire; explores each hidden source of human bliss; and unsuspected riots in their Charms!' One 'fair patient' was even rumoured to have been made pregnant under nitrous oxide.[97]

Such attacks were still continuing four years later, when Robert Harrington published his polemic essay entitled 'The Death Warrant of the French Theory of Chemistry' (1804). The new chemistry was dismissed as charlatanism, and linked to the craze for ballooning. 'This is supposed to be the age of *airial* philosophy; I wish it were the age of common-sense for at present it has taken an airial flight; and unfortunately, candour and justice have flown away with it!' Beddoes and Davy were described as 'aerial flying chemists' pursuing 'ecstatic, lunatic and Laputatic sensations'.[98] The *Anti-Jacobin* magazine made a more general link between radical politics, inhaling gas, flying balloons and mesmerism. But eventually these attacks were to prove far more damaging to Beddoes in Bristol than to Davy once he was established professionally in London.[99]

Davy was already growing restless with Beddoes's regime of gas treatments. Secretly, he believed he had come to a dead end. He was becoming more and more interested in galvanism, and the experimental possibilities of the new electrical pile or 'battery' invented by Alessandro Volta of the University of Como. This had been described in a paper published by Banks in the Royal Society's *Philosophical Transactions* that summer of 1800. The voltaic battery could produce an electrical charge by purely chemical means, and hold it for many hours.

The accounts of Davy's first electrical experiments appear in a notebook headed 'Clifton 1800, from August to November'.[100] He had read a paper about the crucial experiment by Nicholson and Carlisle, who used a voltaic pile to 'decompose' water, and wrote breathlessly to Davies Giddy in Penzance: 'an immense field of investigation seems opened by this discovery: may it be pursued so as to acquaint us with the laws of life!'[101] To Beddoes's dismay, troughs of voltaic batteries, with their rows of square metal plates and pungent smell of oxidising acids, began to replace the glass gas tanks and silken bags in the Institute's laboratory.

Davy wrote to Coleridge in November: 'I have made some important galvanic discoveries which seem to lead to the door of the temple of life.'[102] Extensive correspondence continued between them about the 'hopeful' and progressive nature of science, the theory of chemistry, and the physiology of pleasure and pain, throughout the rest of the year.[103] Coleridge followed Davy's publications eagerly, and wrote with delight when he saw the title of one of his new galvanic essays advertised in the *Morning Post*: 'Upon my soul, I believe there is not a letter in those words, round which a world of imagery does not *circumvolve*: your room, the garden, the cold bath, the Moonlit Rocks . . . and dreams of wonderful Things attached to your name!'[104]

Coleridge's own notebooks began to show a new, scientific precision in the observation of plants, water and weather at this time. 'River Greta near its fall into the Tees – Shootings of water threads down the slope of the huge green stone. – The white Eddy-rose that blossom'd up against the stream in the scollop, by fits and starts, obstinate in resurrection. – It *is the Life* that we live. Black round spots from 5 to 18 in the decaying leaf of the Sycamore.'[105] He felt that the new poetry and the new science were so closely entwined that they must somehow merge, and invited Davy to move north and establish a chemistry laboratory in the Lake District. Coleridge announced: 'I shall attack Chemistry, like a Shark.'[106]

But could they really combine? Southey was one of the first of the Romantic poets to suggest that there might be a profound difference between the scientific and the artistic temperament. This was a subject he would pursue with Coleridge, who did not entirely agree. In February

1800 Southey was already writing to his friend William Taylor: 'Davy is proceeding in his chemical career with the same giant strides as at its outset . . . Chemistry, I clearly see, will possess him wholly and too exclusively: he allows himself no time for acquiring other knowledge. In poetry he will do nothing more: he talks of it, and that is all; nor can I urge him to perform promises which are perhaps better broken than kept. In his own science he will be first, and the high places in poetry have long been occupied.'[107]

Despite Southey's doubts about Davy's literary interests, Davy did see through the press both the second edition of the *Lyrical Ballads*, and Southey's *Thalaba*, in 1800, and agreed to help edit a third volume of the *Annual Anthology*. He also privately continued writing poetry about Anna Beddoes, and his own memories and visions. Eighteen months later, in August 1801, Southey was confidently informing Coleridge: 'I wish it were not true, but it unfortunately is, that experimental philosophy always deadens the feelings; and these men who "botanize upon their mothers' graves", may retort and say, that cherished feelings deaden our usefulness; – and so we are all well in our way.' Here Southey was quoting from a poem by Wordsworth, 'The Tables Turned'. But Coleridge still had different ideas on the matter.[108]

Not all Davy's Bristol friends agreed that a great poet had been lost. Gregory Watt was glad Davy had not gone on contributing to the *Annual Anthology*. He later mocked poetry as an 'exquisitely insidious' form of delusion, and described most poets as 'sporters with the feelings of the world', whose effusions deserved to be burnt by the public hangman. 'You, my dear philosopher,' he reassured Davy, 'have wisely relinquished the stormy Parnassus, where transient sunshine only contrasts the cloudy sky, for the mild and unvarying temperature of the central grotto of science.' Then, serious for a moment, he urged Davy to remain in his calm laboratory and be 'guided by the light of your own creation'.[109]

But Davy had not relinquished Parnassus, though he chose never to publish his poems after 1800. For the rest of his life he filled his laboratory notebooks with drafts and fragments of poetry, which were afterwards faithfully collected by his brother John, and scattered posthumously throughout his *Memoirs of Sir Humphry Davy*. Most of these would be travel pieces ('Fontainbleau', 'Mont Blanc', 'Athens', 'Canigou'), loose forms of descriptive verse-diary, which show great sensitivity to seasons and landscape – especially rivers and mountains. They are exactly what you would expect of a meditative fisherman who had read Coleridge and Wordsworth, and also Izaak Walton. Yet they are surprisingly conventional in language and feeling.

However, there are a number of striking confessional pieces, of much greater intensity, in which Davy tried to work through some of his strange metaphysical ideas about death, fame and hope. The style is plain, often rather awkward, but here the thought is often highly original.* It is difficult to imagine what other writer of the period (except perhaps Caroline Herschel) would have imagined the dead Lord Byron touring the universe on a comet, saluted by extraterrestrial beings, and accelerating towards the speed of light.

> Of some great comet he might well have been
> The habitant, that thro' the mighty space
> Of kindling ether rolls; now visiting
> Our glorious sun, by wondering myriads seen
> Of planetary beings; then in a race
> Vying with light in swiftness, like a king
> Of void and chaos, rising up on high
> Above the stars in awful majesty.[110]

Davy also referred frequently in his later lectures to comparisons between the poetic and the scientific imagination. In 1807 he wrote in terms that would be echoed both by Coleridge and by Keats: 'The perception of truth is almost as simple a feeling as the perception of beauty; and the genius of Newton, of Shakespeare, of Michael Angelo, and of Handel, are not very remote in character from each other. *Imagination, as well as the reason, is necessary to perfection in the philosophic mind. A rapidity of combination, a power of perceiving analogies, and of comparing them by facts, is the creative source of discovery.* Discrimination and delicacy of sensation, so important in physical research, are other words for taste; and love of nature is the same passion, as the love of the magnificent, the sublime, and the beautiful.'[111]

---

* These confessional poems include 'Written After Recovery from a Dangerous Illness' (1808, *Memoirs*, pp.114–16); 'The Massy Pillars of the Earth' (1812, *Memoirs*, p.234); 'The Fireflies' (1819, *Memoirs*, pp.251–4); 'The Eagles'(1821, *Memoirs*, p.279 and *Salmonia*, pp.98–100); 'On the Death of Lord Byron' (1824, *Memoirs*, p.285); 'Ulswater' (1825, *Memoirs*, pp.320–2); 'Thoughts' (1827, *Memoirs*, p.334); 'The Waterfall of the Traun' (1827, *Memoirs*, p.360). They will all appear later in this narrative.

7

By the end of 1800 Davy's *Researches*, and his early papers on galvanism in *Nicholson's Journal*, were rousing serious interest in London. He began receiving unofficial approaches from Sir Joseph Banks and Benjamin Thompson, and there was talk of a professorship in chemistry. In February 1801 he again visited London, and was officially interviewed by the Committee of the Royal Institution – Banks, Thompson and Henry Cavendish – who had been considering offering him an initial post as Assistant Chemical Lecturer, with a possible professorship to follow. He then had the decisive Soho breakfast with Banks, who quickly determined to poach him from Beddoes, and capture him for the Institution. His first shrewd move was to send him on to have informal drinks with Benjamin Thompson, an altogether different kind of patron.[112]

Thompson (1753–1814) was a strange and remarkable man, with none of Sir Joseph's diplomatic bonhomie, but with equal energy and an even more ruthless drive. A Fellow of the Royal Society, he was an American citizen from Boston, but had been knighted by the British government and then appointed Count Rumford by the Elector of Bavaria, an unusual combination of honours. In the course of his extraordinary, picaresque life, Rumford was variously a professional soldier, an inventor and man of science, a Minister of State, a philanthropist and a philanderer. His tall, thin, imposing figure, permanently stooped, combined with large, bright, attentive eyes and a spectacular Roman nose, gave him the appearance of some powerfully beaked and faintly sinister bird of prey about to pounce – an appearance much loved by cartoonists such as James Gillray. As the inventor of various heating and lighting appliances, and propounder of a correct theory of heat (proving Lavoisier's 'caloric' to be a product of friction), Thompson instantly recognised young Davy's potential – and duly pounced.

Attractive terms of employment were mooted (though not agreed), and Davy was already writing to Davies Giddy on 8 March 1801 about the wonderful prospects held out by Banks and Thompson, his imminent move to London, and the promise of fresh funding for his work on galvanism. He acknowledged Beddoes's plans at Clifton for 'a great popular physiological work' on therapeutic gases, but this, he had to admit, was a work on which he would not be collaborating after all. It might, in fact, be a dead end. At all events, science now called him elsewhere.

8

There may also have been non-scientific reasons for Davy's departure from Bristol. Amidst the youthful and enthusiastic circle of the Pneumatic Institute, it was his friendship with Anna Beddoes that had developed the most unexpectedly.[113] He had gradually discovered that her sunny directness disguised deep unhappiness within the marriage.[114] Thomas Beddoes was not a tender or communicative husband, and suffered from bouts of deep depression, what he himself called his 'Hamlet complaint'. In June 1801 he would become so depressed about his debilitating asthma attacks that he allegedly asked Anna's permission to commit suicide, or so she later told Davy.[115]

In consequence, Anna was a young woman secretly desperate for affection, and the early romantic walks along the Avon with Davy had soon turned into tearful confessions and declarations. Both had begun sending each other poems. One of Davy's started unguardedly:

> Anna thou art lovely ever
> Lovely in tears
> In tears of sorrow bright
> Brighter in joy ...

In return she sent him unsigned verses addressed to 'Mr Davy, Pneumatic Institution, Dowry Square, Hotwells'. She apologised for troubling him with her frustrations and miseries, but suggested moments of precious intimacy:

> When to thy trembling hand I silent gave
> My bloodless arm, impatient for the grave ...

Davy's poems are simple and fragmentary, sometimes trying out his elementary Greek in short epigrams:

> The beautiful girl
> Is not mine,
> Not mine the beautiful ...

He also covered his notebook with repeated pencil drawings of Anna's profile and blonde, windswept hair.[116]

Anna's poems, some copied into Davy's notebook, are longer and more melodramatic than his. They seem to play with the idea of suicide, and with the guilty sense that she is making impossible emotional demands on Davy. In one she imagines (or perhaps repeats) his angry refusal of these poignant advances. Here she appears to write accusingly about herself in Davy's own voice, or perhaps quoting him:

> Am I then called to minister relief
> To her who rudely plunged me into grief,
> Who dropped my infant errors into day
> And tore the veil of secrecy away . . . ?[117]

It is not clear what Davy's 'infant errors' might be, unless memories of Mademoiselle Nancy at Penzance. Davy also kept other poems signed 'Fidelissima' ('your most faithful lady') from a slightly later date, which may be a continuation of Anna's, or from another woman entirely.[118] Perhaps there was some more tangled history of deception and betrayal at Bristol, among the young ladies who appear only as initials in his published *Researches.* Certainly Davy later wrote to his confidant Dr John King of indulging in 'physical sympathies' at Bristol of which he was subsequently ashamed. A little more light is thrown on this flirtation or frustrated love affair (or whatever it was) with Anna by several fragments of later correspondence.

Shortly after Davy left Bristol, his friend Dr King sent him news that Anna and Dr Beddoes were expecting a baby. Davy was clearly pleased, and wrote back on 4 November 1801. He thought motherhood would greatly improve what he now regarded as Anna's unstable state of mind. 'How delightful a thing it would be to see that woman of Genius, of feeling, of candour, of idleness, of caprice, instructing and nourishing an infant, losing in one deep sympathy many trifling hopes and many trifling fears.'[119]

In spring 1802, Anna Beddoes gave birth to a baby girl, christened Anna Maria, to whom Davy addressed several poems over the next few years, calling her 'Nature's fairest child/A flower of springtime, rude and wild'. In 1804 he wrote the child a nine-stanza birthday poem, dedicated 'To A.B – 2 years old' and again describing her fondly as 'a sweet blossom of the early spring of life'.[120]

At the same time he continued sending occasional poems to Anna herself, carefully annotating one, 'Written in the coach December 25 1803, passing from Bath to Bristol'. In these simple, sentimental lyrics he looked back nostalgically to what he described as 'Life's golden morn', and the 'anguish and the joys' of the early Bristol days.

> Its love was wild its friendship free
> Its passion changeful as the light
> That on an April day you see
> Changeful and yet ever bright.[121]

In the event, Anna was not in fact made any happier or more fulfilled by motherhood. Soon after her daughter's birth she abandoned her husband and ran off for several months with Davy's wealthy Penzance friend and patron Davies Giddy.[122] Yet this did not break the marriage to the long-suffering Dr Beddoes, nor the confidential friendship with Davy. Five years later, in 1806, she and Davy met again in London, and she congratulated him on his lectures and asked him to send her his portrait, a characteristically provocative request. Davy sent it with a copy of his poem 'Glenarm by Moonlight', with further nostalgic memories of their Avon riverside walks together.

> Think not that I forget the days,
> When first, through rough unhaunted ways,
> We moved along the mountain side,
> Where Avon meets the Severn tide;
> When in the spring of youthful thought
> The hours of confidence we caught . . .[123]

Perhaps he did not realise what emotions these lines would stir up. But either then, or shortly after, he gently tried to disentangle himself from the correspondence, and asked Anna to forget him. This produced an agonised, and not entirely coherent reply, written on 26 December 1806. It gives the best glimpse of what might have happened between them at Bristol. 'I suppose from the experience you have had of my conduct – but you cannot tell how much pain your last observation gave me – *that I should forget you, and think no more about you!* Yet I must certainly deserve it for I have in former days treated you with unkindness that you have the generosity to forgive . . . Of all those who know you best I have most reason to value the qualities of your heart, and I believe at this moment you have not a more sincere admirer, no, not even amongst the young and beautiful, than she who has treated you with such ungentleness.'

Anna went on to say that the last time she saw Davy in London, he appeared 'so vividly alive' that it roused the 'almost expiring spark of ambition' in her bosom. Ambition for what, she does not explain. The letter tails off in ill-defined sadness, indeed perhaps exactly the kind of ill-defined sadness that originally captured young Davy's heart and sympathies. 'I know not for what reason it is but I cannot write or think of you without the most melancholy sensations. Adieu ... I am almost ashamed to send this letter ... *destroy it*.'[124]

Clearly some powerful emotions existed between them during these years. A decade later Davy would confide to another woman: 'You can have no idea at all of what [Anna] was ... She possessed a fancy almost *poetical* in the highest sense of the word, great warmth of affection, and disinterestedness of feeling; and under favourable circumstances, she would have been, even in talents, a rival of [her half-sister] Maria.'[125]

If there was attraction, even seduction, it would be hard to tell who initiated it. But it may be wrong to consider Davy as emotionally naïve, or exploited, at Bristol. Coleridge later remarked enigmatically: 'a young poet may do without being in love with a woman – it is enough if he loves – but to a young chemist it would be salvation to be downright romantically in love'.[126] Davy's later bawdy remarks about women to his friend Dr King suggest a certain worldliness, while Tom Wedgwood gave him, as a suitable parting present to take to London, an exquisite porcelain statuette of a naked Venus.[127]

9

Bristol also saw the end of a scientific love affair. After some eighteen months of extensive experiments on nitrous oxide in the Dowry Square laboratory, Davy had been forced to conclude that the gas, remarkable as it was, could not be used for therapeutic purposes. This was his private opinion, although no such explicit statement was published in *Researches*. For Thomas Beddoes this was a crushing disappointment, particularly as it was exactly what Joseph Banks had always predicted. Banks had written to James Watt: 'in the case of Dr Beddoes's project – I do not fully understand it, & ... I do not expect any beneficial consequences will be derived from its being carried into execution.'[128] It looked as if Beddoes's young protégé had inadvertently undermined the entire *raison d'être* of the Pneumatic Institute.

Yet there was one major scientific discovery which hovered tantalisingly close to Davy's grasp. Had he fully seized it, he would have made himself, Beddoes and the Pneumatic Institute famous forever. Though nitrous oxide could not cure physical disease, it could do something just as valuable: it could temporarily suspend physical pain, or at least the sensation of pain. The gas provided the key to an entirely new science, that of *anaesthetics*: literally, 'the negation or blocking of feelings'.

Characteristically, Davy pounced upon the new concept, asserting in his laboratory notebook that the gas could certainly be used for suppressing even 'intense physical pain'.[129] He speculated on the physiological mechanism: 'Sensible pain is not perceived after the powerful action of nitrous oxide because it produces for the time a momentary condition of other parts of the nerve connected with pleasure.'[130] He successfully tried treating his own toothache from impacted wisdom teeth with nitrous oxide. 'The pain diminished after the first four or five inspirations.' But the effect did not last, and he did not take the next logical step of having the offending teeth removed while under the influence of nitrous oxide. The gas was seen as blotting out the consciousness of pain with pleasure, rather than suspending consciousness itself. Yet in many of his extreme experiments Davy had deliberately pushed himself into unconsciousness, and he knew this could be done without harm.

Later, writing up his experiments for *Researches*, he more explicitly stated the gas's surgical potential: 'As nitrous oxide in its extensive operation appears capable of destroying physical pain, it may probably be used with advantage during surgical operations in which no great effusion of blood takes place.'[131] He added the caution about bleeding not to limit the gravity of operations in which anaesthesia might be applied, but because he believed nitrous oxide was only absorbed through venous blood. It might therefore become ineffective during a major operation when extensive haemorrhaging took place.

Part of Davy's originality was simply to conceive of the radical idea of pain-free surgery. He later had long discussions with Coleridge about the nature and significance of human pain. Coleridge wondered, for example, why God might have created a world in which human childbirth, one of the great productive aims of nature, was so painful as well as so dangerous for women.[132] This was a prophetic speculation, as it turned out, for nitrous oxide mixed with oxygen eventually became one of the standard anaesthetic procedures used during labour, especially if difficult or prolonged.

When he read *Researches* in December 1800, sent to him by Longman with the new edition of *Lyrical Ballads*, Coleridge wrote to Davy wondering if he had had further communication with the leading London surgeon Sir Anthony Carlisle, 'concerning *Pain*', as they had all once discussed it during his London visit. 'It is a subject which *exceedingly interests* me – I want to read something by somebody expressly on *Pain*, if only to give arrangement to my own thoughts, though if it were well treated, I have little doubt it would revolutionise them.' He later urged Davy himself to write such a philosophical treatise on pain.[133]

Many standard operations in the early nineteenth century – being cut for kidney stone, having teeth removed, or a wounded limb amputated – were unimaginably painful. The pain also caused shock, which itself could kill. The only known form of painkiller – the soldier's use of alcohol – was largely a method of controlling terror and deadening shock, not true anaesthesia.

But having made his momentous suggestion, Davy failed to pursue it. Characteristically, he rushed impetuously on to other discoveries. Although he published his conclusions, neither he nor Beddoes saw that the immense possibilities of anaesthesia were taken up. The loss to human well-being, in the alleviation of terror and suffering on the operating table for another two generations, was incalculable. Fanny Burney's account of her own mastectomy – having a breast removed, without anaesthetics, by a military surgeon in her Paris apartment in 1811 – is perhaps more shattering than any account of a limb amputated on the battlefield during the Napoleonic Wars.

It would seem that Davy missed the greatest medical opportunity of his early career. As late as 1831, his polemical biographer J.A. Paris dismissed the whole nitrous oxide experiment as absurd. 'It will be admitted that there must have been something singularly ludicrous in the whole exhibition. Imagine a party of grave philosophers, with bags of silk tied to their mouths, stamping, roaring and laughing about the apartment.'[134] Nitrous oxide only began to be tried again experimentally some forty years later. This was in America, when Dr Horace Wells had a tooth extracted under the gas during a demonstration lecture in Connecticut in December 1844. Wells awoke, announced that he had not felt 'a pin-prick', and proclaimed 'a new era in tooth-pulling'.[135]

But it was quite another chemical, ether, that provided the first true anaesthesia for major operations. The American surgeon William Thomas Morton successfully amputated a man's leg under ether at the Massachusetts General Hospital on 16 October 1846. Two months later, on 31 December, a British surgeon, Mr Lansdowne, performed a similar amputation at the Bristol General Hospital. Thereafter anaesthesia by ether was used widely in both the Crimean and the American Civil War. But the final acceptance of anaesthesia in Britain did not really come until Queen Victoria admitted to having taken a whiff of chloroform during the birth of her son Prince Leopold in April 1853.

Yet the historic Bristol operation of December1846 suggests that Davy's speculation about anaesthesia did eventually bear fruit. The chemist supplying the ether, William Herapath, sent a detailed description of the anaesthetic procedure to the *Bristol Mirror* in January 1847. It is clear from his closing remarks that he knew of Davy's *Researches*, still a legend in Bristol, and had been both inspired and cautioned by them: 'I have no doubt the inspiration of nitrous oxide (laughing gas) would have a similar effect upon the nerves of sensation as the vapour of ether, as I have noticed that persons under its influence are totally insensible to pain; but I do not think it would be advisable to use it in surgical cases, from its frequently producing an ungovernable disposition to muscular exertion, which would render the patient unsteady and embarrass the operator.'[136] ♣

♣ To this day there is still much controversy in the medical literature about Davy's and Beddoes's failure to pursue anaesthesia at this time, a debate hosted by the Association of the History of Anaesthesia. Several scholars suggest a 'cultural' as much as a technical inhibition. They argue that the late-eighteenth-century attitude to pain, in a surgical context, did not admit to the concept of a 'pain-free' operation. Pain itself was a natural and intrinsic part of the surgical procedure, and a surgeon's ability to handle a patient's pain – through his imposed psychological authority, his dexterity, and above all his sheer speed of amputation and extraction – was an essential part of his profession. In a word, there was the need for a 'a paradigm shift' to conceive of pain-free surgery. See Stephanie J. Snow, *Operations without Pain* (2005), Dr A.K. Adam, 'The Long Delay: Davy to Morton', in the *Journal of the Royal Society of Medicine*, vol 89, February 1996. For a masterly overview of surgical pain and procedure in the early nineteenth century, see Druin Burch, *Digging Up the Dead* (2007). The whole question will be pursued in my Chapter 7, 'Dr Frankenstein and the Soul'.

Before leaving Bristol, Davy wrote a long letter of thanks to his old benefactor, John Tonkin in Penzance. Besides expressing his gratitude and his determination to do 'something for the public good', he included an interesting *tour d'horizon* of scientific developments as he saw them in January 1801. While public affairs, economic hardship and the war with France filled him with 'confusion' and dismay, the immense possibilities of scientific research had never looked brighter. The cowpox inoculation, pioneered by Dr Edward Jenner, was becoming general 'not in England alone, but over the whole of Europe', and promised to annihilate smallpox. Galvanism held out immense possibilities, and 'promises to unfold some of the laws of our nature'. Even the Pneumatic Institution, 'in spite of the political odium attached to its founder', might yet cure some 'obstinate diseases'. There is no record of what Davy said, or wrote, on these matters to his mentor Dr Beddoes; or to Anna Beddoes.[137]

10

On Monday, 9 March 1801, Humphry Davy left Bristol to take up his post as Assistant Lecturer in Chemistry and Director of the Chemical Laboratory at the Royal Institution, Albemarle Street, London. His salary was £100 per annum, plus 'coals and candles' and a small set of attic rooms. This was his first professional scientific post, and he was to remain associated with the Institution for the rest of his life.

Founded only two years previously in 1799, the Institution had originally been conceived by Count Rumford as a centre for displaying the latest mechanical inventions, and providing workshops and lectures for the poor, modelled on his earlier philanthropic ventures in Munich. But under the influence of Cavendish and Banks (who obtained a Royal Charter in January 1800), an ambitious new lecture theatre and laboratory were installed in the new Albemarle Street premises, and the Institution's emphasis began to move towards original scientific research and 'regular courses of philosophical lectures and experiments'.

But the Royal Institution had yet to make its mark on London intellectual life, or to find its natural audience – or indeed any significant audience at all – as the founders (and their subscribers) were only too aware. Accordingly, Davy seized his chance, and set out to make a splash with his inaugural lecture on 25 April, choosing the challenging and perplexing subject of 'Galvanism'. Contrary to expectation, the crowd that gathered that evening was large and fashionable, with Banks and Count Rumford sitting expectantly and alarmingly in the front row.

Davy bounced onto the dais – small, youthful, glowing and enthusiastic. He spoke directly to his audience without notes. He made a thrilling narrative out of each experiment, performing a series of spectacular galvanic demonstrations – sparks, fulminations, explosions – with all the skill of a conjuror. Yet his scientific explanations were simple, logical and lucid. He also had the highly unusual gift of putting the science in its historical and social context: he spoke of 'the history of galvanism, detailed the successive discoveries', and its possible future.

He received glowing reports in the press. The *Philosophical Magazine* described how 'Mr Davy (late of Bristol)' presented a wholly new branch of philosophy, 'we mean the galvanic phenomena', and held his audience, including many ladies, completely spellbound. 'Mr Davy, who seems to be very young, acquitted himself admirably well. From the sparkling intelligence of his eye, his animated manner, and the *tout ensemble*, we have no doubt of him attaining distinguished excellence.' Banks and Rumford realised they had chosen a winner.[138]

Later that exciting year, Davy succeeded in publishing his first paper in the Royal Society's prestigious journal, the *Philosophical Transactions*. To Banks's satisfaction it had nothing to do with therapeutic gases, but concerned an improved form of voltaic battery, sufficiently powerful and sustained that it could be used to enhance an entirely new form of chemical analysis.[139]

But Davy's departure had a disastrous effect on Thomas Beddoes, who soon after abandoned further research, and returned to more conventional medicine. He converted the Pneumatic Institute into a charitable dispensary, the Preventative Medicine Institution, and published his collected essays as *Hygia: Essays Moral and Medical* (1802). Though as generous and philanthropic as ever, Beddoes came to believe that all his work on gases had been a failure. His private life was also falling apart. Anna twice left him to pursue Davies Giddy to London, in 1804 and 1806. He still practised as a physician, but was overworked and overweight, and discovered he was suffering from heart disease. In *Hygia* he gave a sombre definition of 'the philanthropic doctor', a far cry from his original ideals: 'One who is humane in his conduct not so much from sudden impulses of passion and pity, as from a settled conviction of the misery prevailing among mankind.'[140]

Meanwhile, Davy was going on triumphantly to success in London. He discovered that he was extremely good at lecturing, and what's more, he adored doing it in front of large audiences. In June 1801 he wrote to his confidant John King at Bristol: 'My labours are finished for the season as to public experimenting and public communication. My last lecture was on Saturday evening. Nearly 500 persons attended ... There was Respiration, Nitrous Oxide, and unbounded Applause. Amen!'

He was now being sought out by members of the scientific community from all over London, and he gave private demonstrations in the basement laboratory of the Institution. Regular parties of philosophers met to inhale the 'joy inspiring' gas. 'It has produced a great sensation. Ca ira!' Davy presented nitrous oxide to the members of the Askesian Society, but again there was no suggestion of anaesthetic applications. He added dizzily: 'I dream of greatness and utility – I dream of Science restoring to Nature what Luxury, what Civilization have stolen from her – pure hearts, the forms of angels, bosoms beautiful, and panting with Joy & Hope.'[141] Not the least satisfactory part of lecturing was that Davy – still only twenty-two – found that he had attracted a large number of these 'angels' to his mixed audiences. From this time on he began to receive invitations, *billets-doux*, and especially Valentine poems, from young women, many of them anonymous.

He spent the rest of the summer walking in Wales, looking back at his two momentous years in Bristol. He told his old colleague King: 'I think of you often and my heart often yearns towards the old ideas of Clifton, the Hotwells, and the moral and natural beings that beautify them.' But he felt that in coming to London he had passed through a necessary period of emotional turmoil and 'transition', and that he could never go back. He coded his words to King in chemical jargon: 'The season that I passed in the country was a season of mental reaction – new ambitions had produced in my mind new hopes and new fears. It was necessary that these hopes and fears should sink into consciousness. *Irritability was induced and physical Stimulation was recurred to.* You will understand me, and the explanation will plead as an excuse for me that I sometimes drowned moral sympathy in the vicious & vile physical sympathy.'[142]

The source of physical stimulation was probably nitrous oxide, to which he had become briefly addicted. But what exactly Davy meant by 'vicious & vile physical sympathy' is less obvious. He seems to imply unhappy sexual involvement, yet it is hardly the sort of phrase he would have used in relation to Anna Beddoes. Again the question arises, whether young Davy had become entangled with other young women patients at Bristol. Now, at any rate, in London he had left it all behind. 'I am now a new being. Pardon my egotism. I have moral feelings, deep moral feelings saying to me, "remember your friends". They are connected with vivid dreams of hope, dreams of … their happiness.'[143]

On 21 January 1802 he launched a hugely successful second set of lectures on Agricultural Chemistry. Among the most enthusiastic members of his audience was Coleridge, who regularly attended his lectures, filling sixty pages of notebook with materials. 'I attended Davy's lectures to enlarge my stock of metaphors,' Coleridge wrote afterwards. 'Every subject in Davy's mind has the principle of Vitality. Living thoughts spring up like Turf under his feet.'[144]

These lectures opened with Davy's masterly 'Discourse Introductory to a Course of Lectures on Chemistry', which became famous as a Romantic statement of the progressive role of science in society.[145] He began by claiming a central place for chemistry in the development of scientific knowledge: botany, zoology, medicine, physiology, agriculture, all ultimately depended on knowledge of chemical processes. He even included the 'sublime' science of astronomy, with a salute to Herschel and his superb telescopes and metal alloy mirrors: 'The progress of the astronomer had been in some measure commensurate with that of the chemical artist, who, indeed, by his perfection of materials used for astronomical apparatus, has afforded the investigating philosopher the means of tracing the revolution of the planets, and of penetrating into space, so as to discover the forms and appearances of the distant parts of the universe.'[146]

But Davy wished to make even bigger, philosophical claims for the scientific spirit and imagination. Drawing on his previous exchanges with Coleridge about the 'hopeful' nature of scientific progress, he put before his audience a vision of human civilisation itself, brought into being by the scientific drive to enquire and create. Science had woken and energised mankind from his primal ignorance and 'slumber'. This was in effect Davy's version of the Prometheus myth: 'Man, in what is called a state of nature, is a creature of almost pure sensation. Called into activity only by positive wants, his life is passed either in satisfying the cravings of the common appetites, or in apathy, or in slumber. Living only in moments he calculates little on futurity. *He has no vivid feelings of hope, or thoughts of permanent and powerful actions.* And unable to discover causes, he is either harassed by superstitious dreams, or quietly and passively submissive to the mercy of nature and the elements.'

But once woken by science, man is capable of 'connecting hope with an infinite variety of ideas'. He can provide for his basic needs, and anticipate future enjoyments. Above all science enables him to shape his future, *actively*. 'It has bestowed on him powers which may almost be called creative; which have enabled him to modify and change the beings surrounding him, and by his experiments to interrogate nature with power, not simply as a scholar, passive and seeking only to understand her operations, but rather as a master, active with his own instruments.'[147]

Davy announced to his spellbound audience that they were witnessing the dawn of 'a new science', and it would be wonderful: 'The dim and uncertain twilight of discovery, which gave to objects false or indefinite appearances, has been succeeded by the steady light of truth, which has shown the external world in its distinct forms, and in its true relations to human powers. The composition of the atmosphere, and the properties of gases, have been ascertained; the phenomenon of electricity has been developed; the lightnings have been taken from the clouds; and lastly, a new influence has been discovered, which has enabled man to produce from combinations of dead matter effects which were formerly occasioned only by animal organs.'[148]

Davy was deliberately proposing a revolutionary view of science, and for a moment his audience must have believed that the wild young man from Bristol was going to propose political revolution as well. Banks, and others in the front row of the theatre, held their breath when Davy launched into the following declaration: 'The guardians of civilization and of refinement, the most powerful and respected members of society, are daily growing more attentive to the realities of life; and, giving up many of their unnecessary enjoyments in consequence of the desire to be useful, are becoming the friends and protectors of the labouring part of the community.'[149] What French, insurrectionary sentiment would follow from Beddoes's erstwhile protégé?

Yet Davy knew very well that it would be fatal to raise any suggestion of social revolution, anything that smacked of 'Continental' ideology. With a skilful change of pace and direction, he hastened to dispel any vision of a democratic future. On the contrary, 'the unequal division of property and of labour, the difference of rank and condition amongst mankind, are the sources of power in civilized life, and its moving causes, and even its very soul'.*

The final version of this passage stands out as deliberately aimed at the aristocratic supporters of the Institution, and must have greatly relieved Banks. Nonetheless, Davy found his own way of emphasising the radical nature of scientific progress, and this too Banks must have liked. He returned to the image of dawning light, but he used it with deliberate British understatement. Science did not deal in extravagant republican dreams, utopian nonsense, or dangerous French political abstractions. It was plain, reasonable, empirical, patriotic: 'In this view we do not look to distant ages, or amuse ourselves with brilliant, though delusive dreams concerning the infinite improveability of man, the annihilation of labour, disease, and even death. But we reason by analogy with simple facts. We consider only a state of human progression arising out of its present condition. We look for a time that we may reasonably expect, for a bright day of which we already behold the dawn.'[150]

But this was not quite all. Davy's final claim for science was an extraordinary one, and must have much struck Coleridge. Science was psychologically, even spiritually, therapeutic. 'It may destroy diseases of the imagination, owing to too deep a sensibility; and it may attach the affections to objects, permanent, important, and intimately related to the interests of the human species.' The value of science was, in this sense, universal, 'even to persons of powerful minds', whose primary interests were 'literary, political or moral'. It strengthened the habit of 'minute discrimination', and encouraged a language of 'simple facts'. But perhaps Coleridge would have felt that Davy was on less certain ground when he added that science tended 'to destroy the influence of terms connected only with feeling'.[151]

---

* Davy had tried various versions of this prudential sentence in his notebooks, and it had evidently given him some trouble. He had, for example, first written not 'sources of power', but 'germs of power in civilized life', a much more tentative idea. Also, the uncharacteristic word 'soul' had originally been the more exuberant 'spirit of life'. Most striking of all, he had finally deleted from the text of his *Discourse* any ideal of a cooperative society, in which the wealthy had a duty to fund research, an essential aspect of Beddoes's vision. In his manuscript he had originally written: 'What might we not hope for in a state of society in which the character of the philosopher [scientist] was united with that of the artist, and in which it became the business of men of property and power eminently to patronise the sciences?' (Davy Archive, Royal Institution, Ms Box 13 (c), pp.57–8.)

Among other crucial ideas in the body of the lectures, Davy further explained and popularised the concept of the 'carbon cycle', as originally discovered by Priestley and Lavoisier. He presented it as the key to life on earth, a continuous universal recycling of carbon and oxygen between plants and humans. The metaphysical emphasis he gave to this law of harmony in nature may well have influenced Coleridge's letters and poems on 'the One Life', written during the spring and summer of 1802.

Later that year Coleridge worked with Wordsworth on the Preface to the third edition of *Lyrical Ballads*, describing the function and future of poetry. They collaborated on a famous passage connecting the 'discoveries of the men of science' with the imaginative work of the poet. Davy was clearly the inspiring figure behind this declaration: 'If the labours of Men of science should ever create any material revolution, direct or indirect, in our condition, and in the impressions which we habitually receive, the Poet will sleep no more than at present; he will be ready to follow the steps of the Man of science, not only in those general indirect effects, but he will be at his side, *carrying sensation into the midst of the objects of science itself*. The remotest discoveries of the Chemist, the Botanist, or Mineralogist, will be as proper objects of the Poet's art as any upon which it can be employed.'[152]

Davy's success gave him new confidence. When Anna Beddoes came up to London with her sister Maria Edgeworth, he proudly showed them round the Institution in a way that much struck Maria, her sharp novelist's eye seeing a new character emerging. 'He was much improved since I saw him last – talking sound sense and has left off being the "cosmology" man. After we had seen all the wonders of the Royal Institution, Mr Davy walked with us and got into the depths of metaphysics in the middle of Bond Street. I don't know whether he or the Bond Street Loungers amused me most.'[153]

Davy's 'metaphysics' and charismatic lecture style increased the Institution's annual subscriptions dramatically, and began to stabilise its problematic finances, much to Count Rumford's satisfaction (since he had been the Institution's major private donor). By 1803 Albemarle Street had been designated the first one-way in London, to avoid the traffic jam of carriages on Davy's lecture days. It was also noticed that he was especially popular with young women, and a cartoon by Thomas Rowlandson shows a commanding, romantic figure holding his mixed audience entranced, with a clutch of young women crowded into the left-hand balcony seats. They are gazing down attentively at Davy, while themselves being ogled by an ancient academic. This was the new science as sexual chemistry.

Another cartoon by James Gillray, 'New Discoveries in Pneumaticks!', played on the less romantic but equally popular notion of chemistry as 'stinks', and the idea that laughing gas could produce a truly room-shaking fart. Professor Thomas Garnett, Count Rumford and Sir Joseph Banks are all identifiable in this cartoon, and again nearly half Davy's audiences are women, many of them scribbling notes (or possibly *billets-doux*).

A French tourist, Louis Simond, described Davy's electrifying lecture technique, with its special appeal to young students and women. Though his ideas were so radical, he noted that Davy was careful to make conventional references to the beauties of divine creation. Meanwhile he staged spectacular and often dangerous chemical demonstrations, that produced gasps of amazement and bursts of applause. Though some were critical of these performances as mere showmanship, they were skilfully designed as genuine scientific demonstrations, and Davy believed that surprise and wonder were central to a proper appreciation of science. Banks heartily approved.[154]

Davy's advancement now became rapid, not to say meteoric. In June 1802 he was promoted from Assistant to full Lecturer. The following year, after another brilliant series on Agricultural Chemistry, he was appointed to the Professorship of Chemistry at the Royal Institution, replacing Garnett, who had found himself eclipsed by the young star and quietly resigned. Davy's salary increases were also rapid. In 1803 he received £200 per annum. In November 1804 he was elected Fellow of the Royal Society, and his professorial salary at the Institution was doubled to £400 per annum. He soon began to receive invitations to give summer lectures in other university cities, notably Dublin, receiving payments that quickly trebled his annual income.

Coleridge, marooned in Keswick in the Lake District and suffering increasingly from opium addiction and marital unhappiness, could still take genuine delight in Davy's success. 'I rejoice in Davy's progress. There are three Suns recorded in Scripture – Joshua's, that stood still; Hezekiah's, that went backwards; and David's that went forth and hastened on his course, like a bridegroom from his chamber. May our friend's prove the latter!' Continuing in scriptural mood, he prophesied that Davy in London would have to battle 'two Serpents at the cradle of his genius', Dissipation and Devotees, which might degrade true scientific Ambition into Vanity. 'But the Hercules will strangle both the reptile monsters.' He still felt a powerful belief in Davy's scientific mission. 'I have hoped, and do hope, more proudly of Davy than of any other man... he has been endeared to me more than any other man, by being a Thing of Hope to me (more, far more than myself to my own self in my most genial moments).'[155]

Davy was proud to play that role. The following year, hearing that Coleridge was about to take ship to the Mediterranean, an exile which he secretly believed would be permanent, Davy sent him a magnificent, hyperbolic letter of valediction and encouragement: 'In whatever part of the World you are, you will often live with me, not as a fleeting Idea but as a recollection possessed of Creative energy, as an IMAGINATION winged with fire, inspiriting and rejoicing. You must not live much longer without giving to all men the proof of your Power... You are to be the Historian of the Philosophy of feeling. – Do not in any way dissipate your noble nature. Do not give up your birthright.'[156]

The original West Country group of friends was gradually being scattered. Davy continued to write to Southey, now settled in the Lake District, and to Tom Poole, busy with his tanning in Nether Stowey; but he steadily lost touch with Beddoes and King in Bristol. One of the sharpest breaks with his Penzance past came with the news of Gregory Watt's lingering death from consumption in October 1804, at the age of thirty-two. Watt had delivered a paper on Cornish geology at the Royal Society only the previous spring, and had recently written to Davy, 'full of spirit' about his future scientific work. The loss of his old friend shook Davy strangely. He was shocked into writing one of his most reflective and bleak speculative letters, his feelings giving 'erring wings' to his religious doubts and his rarely expressed scepticism.

For an uncharacteristic moment, individual ambition and achievement seemed meaningless to Davy: 'Poor Watt! – He ought not to have died. I could not persuade myself that he would die... Why is this in the order of Nature, that there is such a difference in the duration and destruction of her works? If the mere stone decays, it is to produce a soil which is capable of nourishing the moss and lichen. When moss and lichen die and decompose, they produce a mold which becomes the bed of life to grass, and to a more exalted species of vegetable... But in man, the faculties and intellect are perfected: he rises, exists for a little while in disease and misery, and then would seem to disappear, without an end, and without producing any effect.'[157]

Nowhere in this bleak and heartfelt letter, written to another Bristol friend, his erstwhile laboratory assistant Clayfield, did Davy mention God, or any conventional notion of heaven. Instead, he tried to tell himself that there was some incomprehensible 'arrangement' by which Watt's unfulfilled gifts would be 'applied' by Nature. He drew on a familiar analogy with the transformation of the butterfly, but even this took him in an odd and unexpected philosophical direction: 'The caterpillar, in being converted into an inert scaly mass, does not appear to be fitting itself for an inhabitant of air, and can have no consciousness of the brilliancy of its future being. We are masters of the earth, but perhaps we are the slaves of some great and unknown beings ... We suppose that we are acquainted with matter, and with all its elements, and yet we cannot even guess at the cause of electricity, or explain the law of the formation of the stones which fall from meteors. There may be beings – thinking beings, near us, surrounding us, which we do not perceive which we can never imagine. We know very little; but, in my opinion, we know enough to hope for the immortality, the *individual immortality of the better part of man*.'[158]

This was the nearest Davy would come to any idea of the soul, or personal immortality. But what is surprising is his notion of mankind, the 'masters of the earth', being themselves subject to other and greater masters, alien powers elsewhere in the universe. He did not conceive of these as gods, but more like the extraterrestrial intelligences of science fiction, 'thinking beings', close by, but invisible, imperceptible, even unimaginable. He would return to this idea in the last book he ever wrote, *Consolations in Travel*.

In 1805 Davy branched out and gave his spring lectures at the Royal Institution 'On Geology'. For this he had read the work of Hutton and Playfair, and grappled with the new controversies about the age of the earth, and whether its rocks were formed by flood or by volcanic action. His demonstrations included a large-scale model of a volcano, mounted on an insulated plate, that innocently emitted smoke, then suddenly burst into flames, and finally erupted in a cloud of 'ash'. He also recalled his expeditions in Cornwall with Gregory Watt, and delivered a heartfelt elegy on this early, lost friend.[159]

Davy spent a relaxed summer in the Lake District, and climbed Helvellyn with Wordsworth, Southey and Walter Scott. They talked of Coleridge, who was still absent somewhere in the Mediterranean, and writing home ever less frequently. Davy hoped nonetheless that 'his genius will call forth some new creations, and that he may bring back to us some garlands of never-dying verse'. He wrote to Coleridge, urging him to return to England and give a course of lectures on Poetry at the Royal Institution.

Davy returned to London to be awarded the Copley Medal by Banks (for some humdrum work on agricultural chemistry), and elected to the Council of the Royal Society. He was also invited to give the important annual series of Bakerian Lectures at the Royal Society, starting the following autumn. There seemed nothing that could now stop his career, and his meteoric rise to fame at the age of twenty-six. But he still needed to achieve a major scientific discovery to secure his name. In October his early hero Horatio Nelson, whose victory at the Nile he always associated with his professional beginnings in science, was killed at the battle of Trafalgar.

11

On 20 November 1806 Humphry Davy gave his first Bakerian Lecture, to a packed theatre at the Royal Society, with Joseph Banks presiding in the chair. It was a prestigious but challenging appointment. The series had been founded in 1775 by Daniel Defoe's son-in-law, Henry Baker, and dedicated to advances in 'Experimental Philosophy'. Some genuinely new discovery had to be demonstrated, and Davy was expected to choose as his subject either gases, or geology, or agricultural chemistry. Instead he announced that he would be 'investigating and elucidating' the nature of electricity, the use of the new voltaic battery, and the possibilities of opening up a wholly new field of 'electro-chemical analysis'. The lecture created an international sensation. It would be followed by four more over the next four years: the second Bakerian on 19 November 1807, the third on 15 December 1808, the fourth on 15 November 1809, and the fifth and last on 15 November 1810.[160]

Davy began his first lecture with a characteristically enticing *tour d'horizon*: 'It will be seen that Volta has presented to us a key which promises to lay open some of the most mysterious recesses of nature. Till this discovery, our means were limited; the field of pneumatic research had been exhausted, and little remained for the experimentalist except minute and laborious processes. *There is now before us a boundless prospect of novelty in science; a country unexplored, but noble and fertile in aspect; a land of promise in philosophy.*'[161]

He first set out to clarify the nature of electricity, which was still not remotely understood. It was popularly regarded as an invisible and volatile fluid stored in glass Leyden jars, ever ready to leap out with a bang. Against all appearances, Davy argued, the electrical charge stored in Leyden jars or produced by voltaic batteries was no different in kind from that produced by a stormcloud, a 'torpedo' or electric eel, or a hand-cranked friction generator, except that it was more manageable and sustained. Moreover, it was energy produced by *chemical* changes. The tingle produced by acid saliva on a metal tooth filling is just such a chemical change. 'A plate of zinc and a plate of silver, brought into contact with each other, and applied to the tongue, produce a strong caustic sensation. This is analogous with ... the experiment of Galvani, on the excitation of the muscles of animals.'[162]

Next, he demonstrated that electricity did not itself 'generate' matter, as most of his contemporaries thought, but was a form of pure energy. It was, he argued, essentially *bi-polar* energy, divided into a negative charge (associated with heating and expansion) and a positive charge (associated with cooling and contraction). Lightning, for example, was generated by negatively charged stormclouds meeting positively charged ones.[163] (Modern physics would explain lightning as caused by a massive discharge of static electricity, generated by agitated electrons within a single stormcloud.)

Davy then set out a meticulous series of experiments, using a number of different salts and alkalis, which proved that there was such a thing as 'chemical affinity' throughout nature. That is, chemical materials were held together, or bonded, by the positive and negative energies of electricity. By using the voltaic battery as an analytical tool, and 'decomposing' various metals and earths by electrolysis (usually over several days), he promised to reveal entirely new elements, hitherto unknown and unnamed. These investigations 'can hardly fail to enlighten our philosophical system of the earth; and may possibly place new powers in our reach'.[164]

Throughout, Davy referred confidently to scientific work going on across Europe, notably by Berzelius and Potin in Stockholm, and

The first balloon crossing of the English Channel, 7 January 1785
A dramatic retrospective image of the hydrogen balloon flown by Blanchard and Jeffries leaving the coast of Dover in Kent and heading southwards into the lowering skies over Calais. Detail from an oil painting by E.W. Cocks, c.1840.

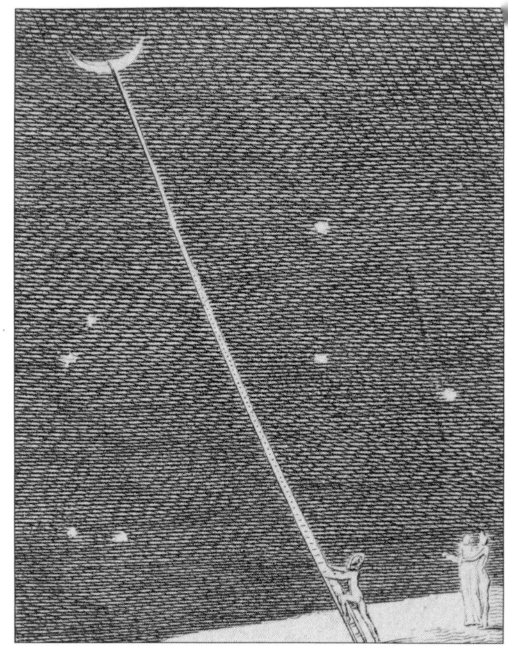

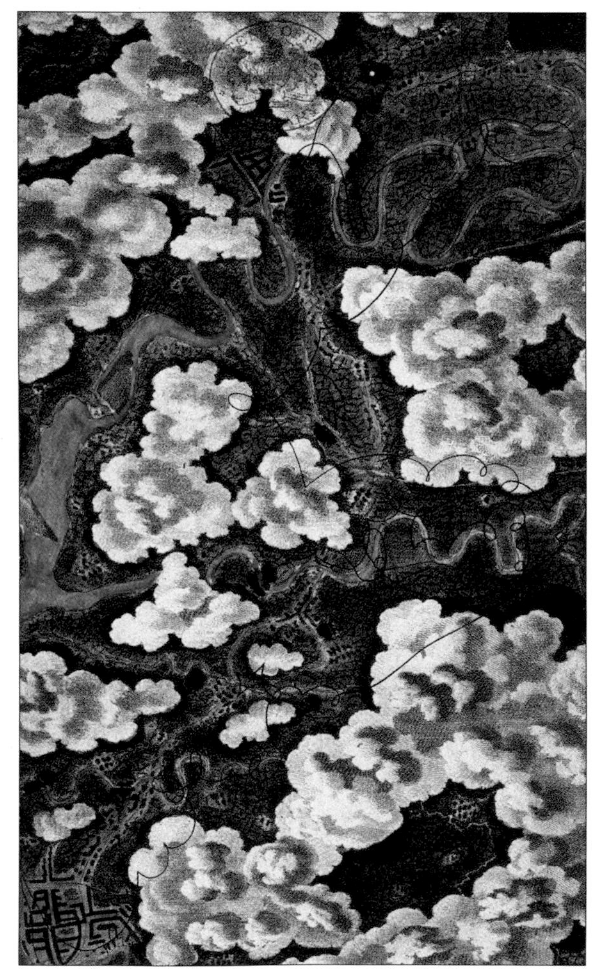

Clockwise from top left

The first manned ascent in a Montgolfier hot-air balloon, Paris, 21 November 1783
Pilâtre de Rozier and the Marquis d'Arlandes stand on either side of the platform, balancing the giant aerostat and preparing to supply fuel to the six-foot brazier suspended (invisibly) within the canopy. Plate taken from *Le Journal.*

'I want, I want'
William Blake's mocking view of scientific endeavour. Line engraving from *For the Sexes: The Gates of Paradise* (1793).

Historic early view of the earth from a balloon, 1786
Beneath the small, puffy shapes of rain-carrying cumulo-nimbus clouds, the meanderings of rivers (depicted in red), the unmetalled roads curving through fields and the grid patterns of village streets clearly emerge. Coloured engraving from an actual sketch by Thomas Baldwin, *Airopaidia* (1786), made at an estimated height of 4–5,000 feet.

The first manned ascent in a hydrogen balloon or Charlier, Paris, 1 December 1783
Coloured print showing Dr Alexandre Charles and M. Robert ascending from the gardens of the Tuileries, with the old Louvre palace on each side and the Champs Élysées beyond. Note the circle of barrels containing acid and iron filings, which have been used to generate the hydrogen. In reality the crowd numbered a quarter of a million people.

Above *Jean-Pierre Blanchard*
A slightly caricatured miniature of the aeronaut Blanchard, the French prima donna of British ballooning. Engraving by J. Newton after an original work by R. Livesay, 1785.

Left *John Jeffries, American balloonist and physician*
Note the chamois-leather flying gloves, beaverskin flying helmet and large, expensive barometer (as altimeter) – Americans already had good kit. Engraving after an original by Tissandier, c.1780s.

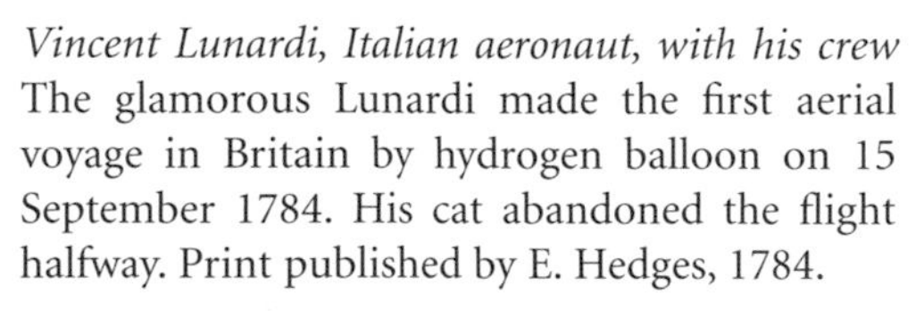

Vincent Lunardi, Italian aeronaut, with his crew
The glamorous Lunardi made the first aerial voyage in Britain by hydrogen balloon on 15 September 1784. His cat abandoned the flight halfway. Print published by E. Hedges, 1784.

James Sadler
The English aeronaut was supported by the students of Oxford University and Dr Johnson. By Edmund Scott, after James Roberts, 1785.

Sadler plaque
A patriotic plaque to James Sadler, 'First English Aeronaut', at Merton Field, Oxford.

Mungo Park
The young Scottish doctor and explorer shortly before setting out for West Africa. Miniature after Henry Edridge, c.1797.

Mungo Park
Cartoon by Thomas Rowlandson, vividly suggesting the sufferings endured during Park's first African travels. Watercolour, c.1805.

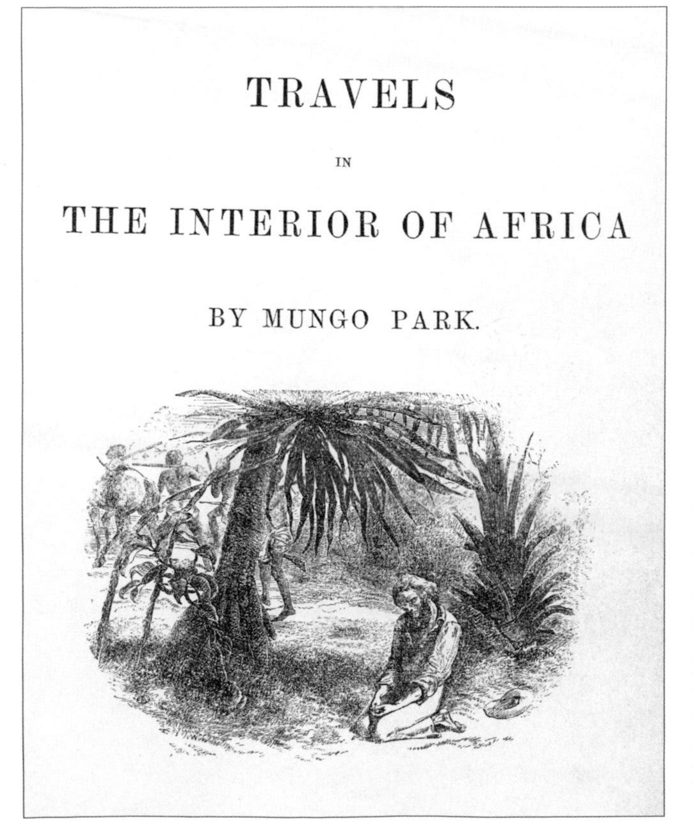

TRAVELS

IN

THE INTERIOR OF AFRICA

BY MUNGO PARK.

Mungo Park's *Travels in the Interior of Africa* (1799)
Title page with vignette showing Park sitting under a tree, having been robbed, stripped and left to die during his first journey into the heart of Africa. Note the precious hat containing all his journals hidden in the crown. From the 1860 edition of his *Travels.*

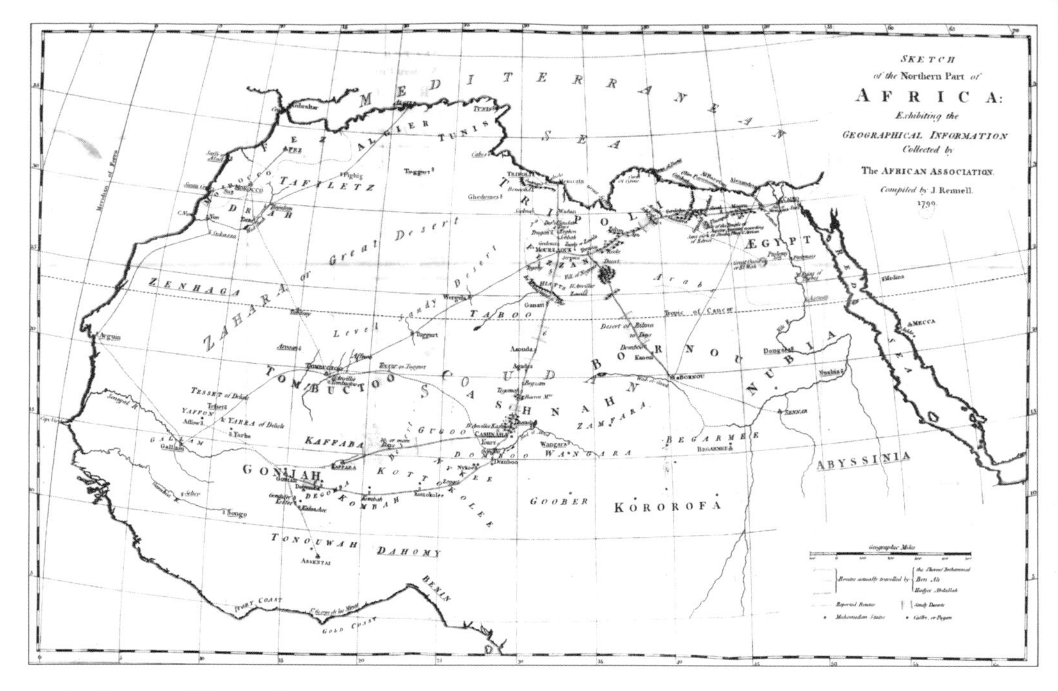

A sketch map of the northern part of Africa

By Major John Rennell, 1790, based on reports brought back by various travellers before Mungo Park's expedition. The Niger is a dotted line rising near Timbuctoo and apparently flowing eastwards across Africa towards Abyssinia.

'The people began to attack him, throwing lances, pikes, arrows and stones'

A Victorian image of the death of Mungo Park on the river Niger, during his second journey, 1805–06. From the 1860 edition of his *Travels.*

Coleridge
Portrait by Peter Vandyke, 1795, shortly before he met Davy at the Pneumatic Institute, Bristol.

Byron
Already famous, Byron was about to be introduced to Banks, Herschel and Davy. Portrait by Richard Westall, 1813.

Keats
Keats at Hampstead, where he studied stars and nightingales, and wrote his poem 'Lamia'. Drawing by Charles Armitage Brown, 1819.

Erasmus Darwin
The polymath poet who championed the encyclopaedic science of the Lunar Society. After Joseph Wright of Derby, 1770.

Shelley
Shelley, atheist and scientific enthusiast, travelled abroad like Davy and Byron. Portrait by Amelia Curran, painted in Rome, 1819.

Blake
Blake mocked the pretensions of most scientific men, and particularly demonised Newton. Portrait by Thomas Phillips, 1807.

Gay-Lussac and Thénard in Paris. But he calmly explained that he had corrected, refined or overtaken all of their experiments. He was in effect claiming that British chemistry, for the first time since Hooke and Boyle in the seventeenth century, led the scientific world. This first Bakerian Lecture was both brilliant and challenging, but so far Davy had merely set out his stall. It was not until the revelations of the second Bakerian Lecture that he hoped the full, revolutionary impact of his work would be recognised.

In the interim, Davy spent much of the summer of 1807 either fly-fishing or trying to persuade Coleridge to undertake a series of literary lectures at the Royal Institution. These turned out to be very similar pursuits. Coleridge had returned from Malta inspired and refreshed by his experiences (among other things, he had fallen in love with a Sicilian *prima donna*), but more hopelessly addicted to opium than ever. Aware of his fragile state, Davy wrote briskly to their mutual friend Tom Poole: 'In the present condition of society, his opinions in matters of taste, literature and metaphysics must have a healthy influence.'

Davy was also optimistic about the state of the war against France. 'Buonaparte seems to have abandoned the idea of invasion, and if our government is active, we have little to dread from a maritime war . . . The wealth of our island may be diminished, but the strength of the people cannot easily pass away; and our literature, our science, and our arts, and the dignity of our nature, depend little upon external relations. When we had fewer colonies than Genoa, we had Bacons and Shakespeares.' In fact, for Davy, science was becoming increasingly patriotic.[165]

On 19 November 1807 Davy gave his second Bakerian Lecture at the Royal Society. He dramatically described how he had just isolated two wholly new elements, potassium and sodium, by 'electrolysis'. By ingenious use of the Institution's voltaic batteries he had over several hours charged and decomposed the common alkalis soda and potash in a vacuum flask, and found the unknown chemicals forming in a crust at the positive and negative poles of the battery. When he extracted the globules of potassium from the crust of potash, they burst spontaneously into an astonishing, bright lilac-coloured flame. Sodium reacted similarly when plunged into water, producing an equally vivid orange flame. Here matter itself seemed to be breaking into life from a previously secret and hidden world, at the chemist's sole command.

Davy had left these experiments perilously late, only a few weeks before the second Bakerian Lecture was due. Almost deliberately, like a journalist working to a deadline, he put himself under extraordinary pressure, and had to work with hectic speed. But that was what he liked. On 6 October 1807 his laboratory notebook records in huge, triumphant letters: 'CAPITAL EXPERIMENT PROVING THE DECOMPOSITION OF POTASH'.

The description of this historic discovery is given dramatically enough in Davy's final lecture text: 'The potash began to fuse at both its points of electrization. There was a violent effervescence at the upper [positive] surface; and at the lower, or negative surface, there was no liberation of elastic fluid; but small globules having a high metallic lustre, and being precisely similar in visible characters to quick silver, appeared, some of which burnt with explosion and bright flame ... These globules, numerous experiments soon showed to be the substance I was in search of.'[166]

But Davy's real excitement and relief is only revealed in his assistant's account of that day. The twenty-eight-year-old Professor of Chemistry became a schoolboy again. 'When he saw the minute globules of potassium burst through the crust of potash, and take fire as they entered the atmosphere, he could not contain his joy – he actually danced about the room in ecstatic delight; some little time was required for him to compose himself to continue the experiment.'[167] Davy had discovered the principle of 'electro-chemical' analysis, a term that he coined, and opened up the vast field for experiment that he had promised.*

The lectures were greeted with universal excitement and praise. Banks was deeply impressed. When he had his official portrait painted in the

* The temporary laboratory assistant in 1807 was Davy's young cousin Edmund Davy. The following year the post was handed over to his brother John, then eighteen, who remained until 1811. The tense, excited, youthful atmosphere of the laboratory was crucial at this stage in Davy's work. It would be recreated with the young Michael Faraday eight years later. The explosive appeal of potassium and sodium to young chemists is wonderfully caught in Oliver Sacks's autobiography, *Uncle Tungsten: Memories of a Chemical Boyhood.* Sacks and two teenage friends first watch as 'a frenzied molten blob' of potassium threatens to set light to his bedroom laboratory, and then go out to throw a three-pound lump of sodium in Highgate Ponds. 'It took fire instantly and sped around and around the surface like a demented meteor, with a huge sheet of yellow flame above it. We all exulted – this was chemistry with a vengeance!' Oliver Sacks, *Uncle Tungsten* (2001), pp.122–3.

Royal Society's presidential chair the following year, wearing all his insignias and honours, he was shown holding a transcript of Davy's Bakerian Lectures. In Bristol, his old patron Beddoes recorded proudly: 'Davy has just solved one of the greatest problems in chemistry by decomposing the fixed alkalis.'[168] The newly founded *Edinburgh Review* ran a series of long articles on the Bakerian Lectures, written by its rising young intellectual star Henry Brougham.[169] Coleridge wrote to praise Davy's 'march of glory'.[170]

Even Davy's acknowledged rivals saluted him. The renowned Scandinavian chemist Jacob Berzelius described it as one of the finest of all modern chemical experiments. The French Académie des Sciences, partly through the good offices of Gay-Lussac, awarded Davy the new Prix Napoléon, worth the enormous sum of 60,000 francs. The Académie issued a formal invitation that Davy come to Paris to collect it. It was a fine and challenging gesture at a time of bitter war between the two nations, but diplomatically there were difficulties from the start. Davy assumed he had won the total sum donated by Napoleon, but the actual prize consisted of the annual interest from that capital, a rather more modest 3,000 francs.[171]

For all his success, Davy was overworked and exhausted. Immediately after his lecture he had undertaken to oversee a ventilation scheme for Newgate Prison, and he fell dangerously ill with a form of jail fever in December 1807. He was near death for several weeks, and an invalid for several months. His patient physician, Dr Thomas Babington, a fellow fly-fisherman, became a friend for life. He did not return to his laboratory until 19 April 1808.

This brush with death at the age of twenty-nine only increased Davy's celebrity, and raised his reputation in scientific circles. The Royal Institution issued daily public reports about his health, and gave a special lecture examining the significance of his work and comparing him to Bacon, Boyle and Cavendish. It also voted funds for a huge new voltaic battery to be constructed for his future use, a trough of '600 double plates of four inches square', said to be four times as powerful as any in England. The following year, after the fourth Bakerian Lecture, a private subscription provided a 2,000-plate battery, now said patriotically to be more powerful than any in Europe – including, of course, France.[172]

From his sickbed, Davy succeeded in organising Coleridge's first set of lectures on 'Poetry and the Imagination', which finally took place – after many dramatic interruptions – at the Royal Institution in spring 1808. Brilliant but intermittent, they have been called Davy's most dangerous experiment, and he wrote his own long private reflections on Coleridge's mixture of genius and 'ruined' sensibility. He adopted the view of the man of science looking down dispassionately on an artist, though in terms so florid that he seemed somehow to entangle himself in Coleridge's own situation: '[Coleridge] has suffered greatly from Excessive Sensibility – the disease of genius. His mind is a wilderness in which the cedar & oak which might aspire to the skies are stunted in their growth by underwood, thorns, briars and parasitical plants. With the most exalted genius, enlarged views, sensitive heart & enlightened mind, he will be the victim of want of order, precision and regularity. I cannot think of him without experiencing mingled feelings of admiration, regard & pity.'[173]

Poetry, genius and eternity were much on Davy's mind during his convalescence. Perhaps the break in his punishing laboratory routine allowed suppressed emotions to surface. As he was slowly recovering, he celebrated by writing a striking, hymn-like poem, 'Lo! O'er the earth the kindling spirits pour'. It was filled with pantheistic visions of change and transformation, of godlike forces energising the whole earth; and with images that seem to refer to Coleridge's view of intelligent Creativity active in the universe, and playing upon matter like an Aeolian harp:

> All speaks of change: the renovated forms
> Of long-forgotten things arise again;
> The light of suns, the breath of angry storms,
> The everlasting motions of the main.
>
> These are but engines of the eternal will,
> The One intelligence, whose potent sway
> Has ever acted, and is acting still
> While stars, and worlds and systems all obey.
>
> Without whose power, the whole of mortal things
> Were dull, inert, an unharmonious band.
> Silent as are the harp's untuned strings
> Without the touches of the poet's hand.[174]

Other, perhaps forbidden, emotions also surfaced. During his fever Davy later said that he had a repeated hallucination of a beautiful, tender, unknown woman who nursed him, held him and had 'intellectual conversations' with him. '[When] I contracted that terrible form of typhus fever known by the name of jail fever ... there was always before me the form of a beautiful woman ... This spirit of my vision had brown hair, blue eyes, and a bright rosy complexion, and was, as far as I can recollect, unlike any of the amatory forms which in early youth had so often haunted my imagination ... Her figure was so distinct in my mind as to form almost a visual image ... [but] as I gained strength the visits of my good angel, for so I called it, became less frequent.'[175]

Davy was fascinated by this hallucinatory experience, which may have recalled some of his earlier nitrous oxide visions. Yet it obviously touched some much deeper chord. He added, revealingly, that he was 'passionately in love at the time'. From the fragmentary notes and Valentine poems he preserved (probably by mistake) in his notebooks, there were a number of young women who may have set their caps at him during his lectures of 1807. But this visionary woman, he insisted, was someone quite different, someone utterly unknown to him.

He was strangely precise on the matter. His vision was not the 'lady with black hair, dark eyes, and pale complexion' who was currently the object of his 'admiration'. She was a younger woman, almost a girl, and of a different physical type: a glowing, youthful brown-haired girl, 'a living angel'. Stranger still, Davy would claim that he actually met this 'visionary female' ten years later, in 1818, 'during my travels in Illyria'. She was then 'a very blooming and graceful maiden of fourteen or fifteen years old'.[176] Finally he would meet her a third time, ten years later still, in 1827–28, when she was in her mid-twenties and he was stoically enduring what turned out to be his last illness.[177]

Later Davy would remark on the mysterious ten-year pattern of this amorous cycle, each recurrence apparently taking place at the beginning of his own new decade: at thirty, at forty, at fifty. If there had been an earlier one, it would have been in 1798, when he turned twenty, and had just met Anna Beddoes. It suggests an area of private emotion and turmoil that would only be hinted at in confidential letters to his brother John, and in his later poetry.[178]

12

On 24 December 1808 Dr Thomas Beddoes died in Bristol, aged only forty-eight. He had been suffering from chronic heart disease, and had been faithfully nursed by his erring wife Anna, who had returned to him in his time of need. It seems he had written a number of letters to friends, including Davy, but received few replies, and felt forgotten. Neither Davy nor Coleridge had been in touch with their old mentor for several years.

Beddoes's last publication was 'A Letter to Sir Joseph Banks, President of the Royal Society, on the Prevailing Discontents, Abuses and Imperfections in Medicine' (1808). In it he advocated a five-year training course for all physicians, financed by public taxation, and a national policy of preventative medicine: the first remarkable glimmerings of a National Health Service. There were also glimpses of the old radical doctor. He suggested that family health would be universally improved if all wives were provided (free of charge) with anatomy lectures, washing machines (steam-powered), fresh vegetables and pressure cookers.[179]

His death caused uneasy stirrings among the scattered circle of the original Bristol Institute, and self-questioning about their own careers. Davy wrote to Coleridge: 'my heart is heavy. I would talk to you of your own plans, which I shall endeavour in every way to promote; I would talk to you of my own labours which have been incessant since I saw you, and not without result; but I am interrupted by very melancholy feelings which, when you see this, I know you will partake of... Very affectionately yours.'[180]

Coleridge replied with a passionate, guilty letter about Beddoes's selfless medical career and quixotic generosity. He said he had wept 'convulsively' at the news of his death. It emerged that Anna had been looking for someone to write Beddoes's biography. She had successively approached Davy, then Southey, then Davies Giddy, and finally Coleridge. But all finally turned her down. A dull *Memoir* was written by Dr John Stock in 1811, but it was Peter Roget who eventually produced a fine article for the *Encyclopaedia Britannica*, though not until 1824. Meanwhile Davies Giddy became the guardian of Anna's children, having first taken the precaution of marrying a Miss Gilbert, and changing his perilous surname accordingly to Davies Gilbert. Anna herself moved to Bath, then Italy, settling in Rome. Her son, the poet Thomas Lovell Beddoes (1803–49), would spend twenty-five years writing a strange, semi-dramatic poem entitled *Death's Jest-Book*, replete with grotesque imagery from medical surgery and his father's laboratory, which he could just remember. When he could write no more, Thomas committed suicide in Basle, at the age of forty-five.[181]

Davy wrote thoughtfully in his journal of Beddoes's shyness, his apparent remoteness in conversation, and his 'wild and active imagination', which he judged was equal to Erasmus Darwin's, but too often hidden. He added wistfully: 'On his death he wrote to me a most affecting letter regretting his scientific aberrations. I remember one expression: "*like one who has scattered abroad the Avena Fatua of knowledge from which neither brand nor blossom nor fruit has resulted. I require the consolation of a friend.*" ' That last phrase would come to haunt Davy himself.[182]

By 1809 Davy's reputation was powerfully in the ascendant. He was included in Volume 7 of the influential series *Public Characters: Biographical Memoirs of Distinguished Subjects*. The long entry praised his modesty and genius, described his 'galvanic experiments' at the Royal Institution, and gave a long summary of the Bakerian Lectures. He was presented as an exemplary figure from the new world of British science, a dedicated researcher 'bent over his retorts', unworldly and ignoring public fame. There was no hint of his future reputation for arrogance, ambition and professional jealousies.[183]

A huge composite portrait, *Eminent Men of Science Living in 1807–8*, was painted to commemorate the historic expansion in British science at this moment. In fact it represents a retrospective view, as the artist William Walker actually painted it around 1820. In it he depicted a group of thirty figures, all male, standing and sitting with grave formality in some ideal clubland smoking room. They include in the front ranks Davy, Herschel, Banks, Dalton, Cavendish and Jenner. Dr Thomas Beddoes is nowhere to be found. He had been forgotten, just as he feared.♣

♣ Beddoes was recovered in the fine biography by Dorothy Stansfield (1984), and through the work of the late, lamented Roy Porter in a number of brilliant essays, notably 'Thomas Beddoes and Biography' in *Telling Lives in Science*, ed. Michael Shortland and Richard Yeo (1996), and finally in Porter's masterwork on the history of medicine, *The Greatest Benefit to Mankind* (2001).

The poet Anna Barbauld, who had previously written about Joseph Priestley's epoch-making experiments, now singled out Davy's scientific lectures as one of the glories of the age. Future historians would

> Point where mute crowds on Davy's lips reposed,
> And Nature's coyest secrets were disclosed.[184]

Her faintly mischievous image did not overlook the sexual frisson surrounding Davy's popular performances, and naturally this aroused some jealousy. Some of Davy's oldest friends worried about the effect on him of success and celebrity. Coleridge feared he was sacrificing himself to London fashion, 'more and more determined to mould himself upon the age, in order to make the age mould itself upon him'.[185] But perhaps this was the inevitable cost of such success.

One consequence of Davy's celebrity was tempting offers to give public lectures outside London. In 1810 he accepted an invitation to Ireland to give his Chemical and Geological lectures for an exceptional fee of a thousand guineas. He lectured to packed theatres in Dublin in spring 1810, and again in spring 1811, when Trinity College conferred on him an honorary doctorate. In these lectures he particularly stressed the importance of scientific knowledge for women's education, and for 'the improvement of the female mind'. Milton was wrong on the subject, and Mary Wollstonecraft was right.[186] Among his attentive and admiring audiences was a strikingly pretty and vivacious Scottish widow called Jane Apreece. At a glittering reception afterwards, Jane Apreece told Humphry Davy that she loved fishing.

7

Dr Frankenstein and the Soul

1

In September 1811 the Herschels' old friend Fanny Burney, by then the married Madame d'Arblay, underwent an agonising operation for breast cancer without anaesthetic. It was carried out by an outstanding French military surgeon, Dominique Larrey, in Paris, and so successfully concluded that she lived for another twenty years. What is even more remarkable, Fanny Burney remained conscious throughout the entire operation, and subsequently wrote a detailed account of this experience, watching parts of the surgical procedure through the thin cambric cloth that had been placed over her face. At the time the surgeon did not realise that the material was semi-transparent. 'I refused to be held; but when, bright through the cambric, I saw the glitter of polished steel – I closed my eyes. I would not trust to convulsive fear the sight of the terrible incision.'

On the subject of pain, and Humphry Davy's failure to pursue anaesthesia, it is worth considering what Fanny Burney wrote about her terror *before* this mastectomy operation: 'All hope of escaping this evil being now at an end, I could only console or employ my mind in considering how to render it less dreadful to [my husband] M. d'Arblay. M. Dubois had pronounced: "you must expect to suffer – I do not wish to mislead you – you will suffer – you will suffer *very much*!" M. Ribe had *charged* me to cry! To withhold or restrain myself might have seriously bad consequences, he said. M. Moreau, in echoing this injunction, enquired whether I had cried or screamed at the birth of Alexandre. Alas, I told him, it had not been possible to do otherwise. "Oh then," he answered, "there is no fear!" – What terrible inferences were here to be drawn!'[1]

Indeed, she screamed throughout the operation. 'When the dreadful steel was plunged into the breast – cutting through veins – arteries – flesh – nerves – I needed no injunction not to restrain my cries. I began a scream that lasted unintermittingly during the whole time of the incision – & I almost marvel that it rings not in my ears still! So excruciating was the agony … All description would be baffled … I felt the Knife *rackling* against the breast bone – scraping it!'

One of Burney's many extraordinary reflections was whether extreme physical pain could not only induce unconsciousness – 'I have two total chasms in my memory of this transaction' – but actually force the soul out of the body. She also found that the act of recollection carried its own pain, and that she had taken three months to complete the account, as a letter of nearly 10,000 words to her sister Esther. She had severe headaches every time she tried to go on with it. Once finished, she could not look back over what she had written. 'I dare not revise, nor read, the recollection is still so painful.' It is an astonishing record of courage, not least in Fanny's determination to protect her husband from the trauma of the operation. But it also recalls what the real conditions of surgery were at this period.

Throughout the Napoleonic Wars, medical science had been spurred on by the immediate bloody demands of the battlefield. It became ever more daring and more ruthless. Larrey, for example, had performed 200 amputations in twenty-four hours after the battle of Borodino, and been awarded the Légion d'Honneur.[2] But as the conflict wound down, it began to turn again to more speculative enquiries. France was still held to lead Europe in medicine and surgery, and its great state hospitals in Paris, notably the Hôtel Dieu and La Salpêtrière on the Left Bank, still pioneered surgical techniques and anatomical theory. Here Xavier Bichat and Baron Cuvier reigned supreme. Nevertheless, the collapse of the French wartime economy, and the overwhelming demands made on the country's medical resources by returning veterans and *mutilés de guerre*, began to hinder its scientific advances.

By contrast, the teaching hospitals of London and Edinburgh were now gaining an international reputation. Ever-vigilant, Joseph Banks was anxious to foster this growing advantage and prestige through elections to the Royal Society. Medicine was fashionable, and the hospitals began to attract a new and gifted generation of medical students and teachers, such as Henry Cline (St Thomas's), John Abernethy (St Bartholomew's), Joseph Henry Green (Guy's) and Astley Cooper (Guy's).

Banks drew them into the Royal Society wherever possible. For example, he ensured that Astley Cooper's pioneering operation of piercing the human *membrane tympani* (eardrum) to relieve potentially fatal inner-ear infection reached a wide audience by being written up for the Society's *Philosophical Transactions* in 1801.[3] Thanks to Banks, Cooper was elected to the Royal Society and awarded the famous Copley Medal for this work, at the age of thirty-three. Cooper's students would include John Keats at Guy's Hospital in 1814. Other medical men made important contacts in the literary world. The radically minded Henry Cline befriended the writers Horne Tooke and John Thelwall, and even acted as a character witness during their trial for treason in 1794. John Abernethy treated Coleridge for opium addiction, and J.H. Green would become Coleridge's amanuensis in 1818. In 1816 Byron chose as his travelling companion in exile the young Dr William Polidori, newly qualified at Edinburgh Hospital. Polidori secretly contracted with the publisher Murray to keep a diary of Milord's exploits (perhaps not quite in accordance with his Hippocratic oath).

Banks kept his all-seeing eye on all these too. Among them, he soon became aware of a talented and unorthodox young surgeon, William Lawrence, working at St Bartholomew's Hospital under John Abernethy. As early as 1802, when Lawrence was a mere medical student aged only nineteen, Banks had spotted him and recommended him to William Clift, the conservator of the Hunterian Collection. 'Sir, I beg leave to introduce to your acquaintance the bearer Mr William Lawrence, a Comparative Anatomist who is able to give – as well as receive – information. He wishes to see the [skeletons of the] Elephant & the Rhinoceros, & will probably find in the Collection many more things that he will desire to look at.'[4] It was Lawrence who would rekindle one of the most disturbing scientific debates of the Romantic period, and stir up the controversy that became known as the Vitalism Debate in 1816–20.

In 1813, Banks had carefully supported the election of Lawrence to the Royal Society, at the strikingly early age of thirty. Two years later, in 1815, Lawrence was given his first academic appointment of real public significance when he was made a Professor of Anatomy at the Royal College of Surgeons. Promotion within the medical profession was still largely in the hands of an oligarchy, and the appointment depended on the good offices of his mentor John Abernethy. As Abernethy himself had long held one of these professorships, this was an expression of great personal confidence. A glittering career now opened before Lawrence under the wing of Abernethy.

John Abernethy was a powerful ally, at the height of his powers and influence. Born in 1764, he had originally trained at the world-famous medical schools at Edinburgh, then came south to study under the great surgeon John Hunter (1728–93), working among the bodies at his grim dissecting rooms in Great Windmill Street. (There is a curious *Ackroydian* historical resonance in the fact that a century later this became the home of the first English burlesque and nude tableaux shows.) Later he pursued a successful career in surgery, built a highly lucrative practice as a consultant physician in Mayfair, and was appointed senior surgeon at Bart's Hospital in 1815. At fifty-one he was at the top of his profession, and also held a Professorship of Anatomy at the Royal College of Surgeons.

A squat, sandy-haired figure, Abernethy was a pious, plain-spoken Scot of 'unconquerable shyness' in society, but famed for his blunt bedside manner with patients, and his brusque diagnoses.[5] Brought up a Calvinist, he had no time for niceties. He made no secret of his opinion that most of the diseases of his richer clients could be cured by cutting back on food and alcohol, and saying their prayers. When consulted by Coleridge in 1812 for a huge range of complex stomach complaints and subtle nervous afflictions (including chronic nightmares), he unhesitatingly diagnosed a simple case of opium addiction, and indirectly helped Coleridge find asylum with the physician James Gillman in Highgate four years later.[6]

Among the medical students at Bart's, Abernethy was one of the most popular lecturers of his time, partly because of his cussedness and eccentricity. So there was great interest when in 1814, as Professor of Anatomy at the Royal College of Surgeons, he began the series of annual public lectures known as the Hunterian Orations. He chose a subject for his discourse inspired by his old anatomy teacher: what he called 'An Enquiry into the Probability and Rationality of Mr Hunter's Theory of Life'.

The topic was unexpected. Hunter's celebrity had been based on his practical – indeed terrifying – skills as a surgeon, and his vast knowledge of comparative anatomy. His characteristic last paper, published posthumously, was 'A Treatise on Blood, Inflammation and Gunshot Wounds' (1794). Hunter had established a sophisticated collection of comparative anatomy specimens. After his death in 1793 they were purchased by the state, and in 1800 entrusted to the newly established Royal

College of Surgeons in Lincoln's Inn Fields. The Hunterian Museum exists there to this day. In a way the collection was an intellectual time-bomb, for, sequentially displayed, the specimens visibly demonstrated, to anybody who cared to examine them, how directly and evidently man's skeletal structures (skull, hands, feet) and internal organs (heart, liver, lungs) had evolved from 'lower' animal forms. They were compelling proof of a certain kind of continuous physiological 'evolution', and they clearly suggested that man had developed directly from the animal kingdom, and was not a unique 'creation'.

But this was not the subject that Abernethy chose. His old master Hunter, like many scientific men towards the end of their lives, had developed certain undefined mystical yearnings. Abernethy had found among his bloodstained and chaotic manuscripts various ill-defined theories of a Life Force or Life Principle, which suddenly seemed of great contemporary interest. Hunter speculated that this Force was somehow associated with spontaneous motions inherent in the human physiology: systolic and diastolic pulses of the heart, the circulation of the blood, healing inflammation, male erection, and female blushing. Above all he believed that blood itself held the secret of Vitality: 'it is the most simple body we know of, endowed with the principle of life'.[7]

Building on Hunter's speculations, Abernethy proposed a theory of human life based on a semi-mystical concept of a universal, physiological life force. Blood itself could not explain life, though it might carry it. This universal 'Vitality' was a 'subtle, mobile, invisible substance, super-added to the evident structure of muscles, or other form of vegetable and animal matter, as magnetism is to iron, and as electricity is to various substances with which it may be connected'. Abernethy further suggested that this theory brought scientific evidence – if not exactly proof – to the theological notion of the soul. If the Life Force was 'super-added', some power outside man must obviously have added it.[8]

In drawing his analogies between Vitality and electricity, Abernethy also called on the authority of Humphry Davy's Bakerian Lectures at the Royal Society. Like many scientific men of the day he was entranced by the potentialities of the voltaic battery, and its possible connections with 'animal magnetism' and human animation. Electricity in a sense became a metaphor for life itself. 'The experiments of Sir Humphry Davy seem to me to form an important link in the connexion of our knowledge of dead and living matter. He has solved the great and long hidden mystery of chemical attraction, by showing that it depends upon the electric properties which the atoms of different species of matter possess... Sir Humphry Davy's experiments also lead us to believe, that it is electricity, extricated and accumulated in ways not clearly understood, which causes those sudden and powerful motions in masses of inert matter, which we occasionally witness with wonder and dismay.'[9]

The lectures excited great interest in the medical profession, but not yet among the general public. This would soon change. In 1816, to his surprise and irritation, Abernethy found his fashionable speculations on the mysteries of the Life Force, and the role of electricity in animating 'inert matter', scathingly attacked and denounced. The critic was none other than his fellow Professor of Anatomy, his youngest and most gifted pupil, the thirty-three-year-old William Lawrence.

It was a wholly unexpected blow, though later Abernethy said that Lawrence had been known 'to decry and scoff' at his views behind his back on the wards at Bart's.[10] He felt bitterly about it, as Lawrence had been Abernethy's assistant since the age of sixteen, and Demonstrator at Bart's from 1803 until 1812, largely under his protection and patronage. He had even lodged with Abernethy for three years, and was considered his protégé.

Clearly, Lawrence felt that none of these claims weighed against scientific truth, and had grown impatient with his old chief. Temperamentally Lawrence was the opposite of his patron. Tall, thin, ambitious, elegant and highly articulate, he regarded medicine as a pure science, with no outside allegiances. European in his outlook, flamboyant and radical in his thinking, he was well read in French and German medical literatures. He knew the writings of the Jena University circle, and had developed strong leanings towards Cuvier and Bichat and the materialist view of human life. If not an avowed atheist, he had little time for conventional pieties.

Unusually for an English medical student, William Lawrence had actually studied anthropology under Banks's old friend Blumenbach at Göttingen in Germany, where he had been noted for his brilliance and theoretical daring.[11] Blumenbach had developed a whole new science of craniology: the collecting, measuring and classing of animal and human skulls. His huge skull collection at Göttingen University was popularly known as 'Dr B's Golgotha', and he frequently wrote to Banks requesting specimens. Craniology was also pursued by Alexander von Humboldt in

South America, and the classification of racial types, a particular Germanic concern, had begun.[12] At twenty-four, while still acting as Abernethy's humble Assistant and Demonstrator at Bart's, Lawrence had translated Blumenbach's seminal work, *Comparative Anatomy*, in 1807.

This book was a battle cry of a theoretical kind. It raised new and intensely controversial questions about human racial types, and the hypothetical link between skull shape, brain size and intelligence. Blumenbach introduced the first classic racial divisions between Caucasian, African, Asiatic and Indian types. But, perhaps most significant of all, he tried to define the physical structure of the human brain, and how it produced 'a mind'. He came within a pace of dismissing the existence of a human 'soul', and suggesting the purely material basis for life itself. But having an extremely restricted and expert medical readership, the book caused little immediate stir in England, and it is doubtful if Abernethy himself ever read it.

While Abernethy was consulted by Coleridge in 1812, Lawrence found a much younger but equally demanding literary figure in his consulting rooms in July 1815. This was the twenty-two-year-old Percy Bysshe Shelley, suffering from a cocktail of nervous diseases including abdominal spasms, nephritic pains, suspected tuberculosis and a writing-block. Lawrence – literate, radically minded and well travelled – quickly gained the poet's confidence. 'My health has been considerably improved under Lawrence's care,' Shelley wrote with some surprise in August, 'and I am so much more free from the continual irritation under which I lived, as to devote myself with more effect and consistency to study.'[13] A month later, in September 1815, Shelley was drafting his long poem about travel and self-searching, *Alastor, or The Spirit of Solitude*, and a series of speculative essays about the nature of life, and also of death, as in his 'Essay on a Future State'.[14]

These medical consultations would continue regularly over the next three years, until Shelley and his young wife Mary departed for Italy in 1818. They took place during the height of the Vitalism debate, and not unnaturally they developed a literary as well as a medical aspect. It was Lawrence who recommended the warm, smiling Italian climate as 'a certain remedy' for all Shelley's diseases. It was also Lawrence, with his unusual knowledge of French and German experimental medicine, who helped turn the Shelleys' joint scientific speculations along a more controversial path.[15]

The natural tendency of most English doctors and surgeons was to avoid too much theory and speculation. This evidently did not apply to Lawrence, or to his intellectual masters on the Continent. The great French naturalist Georges Cuvier approached all animal life as part of a continuous 'successive' development. The celebrated Parisian doctor Professor Xavier Bichat developed a fully materialist theory of the human body and mind in his lectures *Physiological Researches on Life and Death*, translated into English in 1816. Bichat defined life bleakly as 'the sum of the functions by which death is resisted'.

Even more radical were the 'Machine-Man' theories of the French physiologist Julien de la Mettrie. He argued that the theologian, with his 'obscure studies', could say nothing intelligible about the soul, and that only physicians and surgeons were in a position to study the evidence. 'They alone, calmly contemplating our soul, have caught it a thousand times unawares, in its misery and its grandeur, without either despising it in one state or admiring it in the other.'[16]

William Lawrence was only waiting the opportunity to bring such radical ideas to bear. As part of his new professorship he was required to give the series of public lectures at the Royal College, starting in spring 1816. These immediately followed on the series given by Abernethy. It was the custom that one Hunterian Lecturer would preface his remarks with an appropriate salute to the endeavours of the previous incumbent. But on entering the lecture hall, after a few elegant throwaway compliments, Lawrence began roundly to attack Abernethy's theories. He stated bluntly that there was absolutely no such thing as a mysterious Life Principle, and that the human body is merely a complex physical organisation. In a phrase that became notorious, he claimed that the development of this physiological organisation could be observed unbroken, 'from an oyster to a man'.[17]

Lawrence's references to Abernethy became steadily more aggressive and sardonic. 'To make the matter more intelligible, this *vital principle* is compared to magnetism, to electricity, and to galvanism; or it is roundly stated to be oxygen. 'Tis like a camel, or like a whale, or like what you please ...' This last was a contemptuous, and deliberately literary, allusion to Shakespeare's Hamlet mocking the foolish old Polonius. Other smart literary quotations came from the poems of Alexander Pope and John Milton.[18]

Lawrence eventually went on to broaden his attack. Science, he argued, had an autonomous right to express its views fearlessly and objectively, without interference from Church or state. It must avoid 'clouds of fears and hopes, desires and aversions'. It must 'discern objects clearly' and shun 'intellectual mist'. It must dispel myth and dissipate 'absurd fables'.[19] The world of scientific research was wholly independent. 'The theological doctrine of the soul, and its separate existence, has nothing to do with this physiological question ... An immaterial and spiritual being could not have been discovered amid the blood and filth of the dissecting room.'[20]

Finally he attacked the very nature of the religious, mystifying or unscientific philosophy which Abernethy appeared to be promulgating: 'It seems to me that this *hypothesis or fiction* of a subtle invisible matter, animating the visible textures of animal bodies, and directing their motions, is only an example of that propensity in the human mind, which had led men at all times to account for those phenomena, of which the causes are not obvious, by the mysterious aid of higher and imaginary beings.'[21]♣

As the controversy became more public, Lawrence was accused of personal betrayal, ingratitude and atheism. Between 1817 and 1819 he and Abernethy continued to exchange increasingly vitriolic views in their Royal College lectures, and student groups of supporters formed round each. Abernethy was the senior figure, but Lawrence would not back down, and published his lectures in a book that became notorious, his *Natural History of Man* (1819).

It became clear that this was no ordinary academic wrangle, but that the subject in contention was the fundamental nature of human life. The larger implications were clearly social, political and even theological. There was also a strong overtone of imperial controversy: foreign versus British science. Thus Vitalism was the first great scientific issue that widely seized the public imagination in Britain, a premonition of the debate over Darwin's theory of evolution by natural selection, exactly forty years later.

♣ This is a line of argument that has a long scientific footprint, and can be found being used to great rhetorical effect today by Daniel Dennett and Richard Dawkins. Against this, it is interesting to read the defence of the necessary and dynamic notion of 'mystery' by Humphry Davy in his lectures (see my Prologue), or by the great twentieth-century American physicist Richard Feynman in *The Meaning of It All* (posthumously published in 1999). Though not a religious man, Feynman believed that science was driven by a continual dialogue between sceptical enquiry and the sense of inexplicable mystery, and that if either got the upper hand true science would be destroyed. See James Gleick, *Richard Feynman and Modern Physics* (1992).

2

In fact Vitalist ideas had been stirring for over a generation. Ever since the 1790s the new developments in Romantic medical science and theory had begun to raise fundamental questions about the nature of life itself. What distinguishes organic from inorganic ('dead') matter, or vegetable life from animal life? Was there some form of animating power throughout nature, and if so, was it identical to – or analogous with – electricity? These led on, inevitably, to an enquiry about the nature of mind, spirit and the traditional concept of 'the soul': how could this be explained or defined in scientific terms, or should it simply be dismissed?

Such questions, traditionally the province of theologians and philosophers, were now increasingly considered by physicians, science writers, and those who studied what Coleridge called 'the science of mind'.[22] They had already been the subject of ingenious scientific experiments in Europe, which gave rise to increasingly fierce debates surrounding the work of Luigi Galvani in Italy and Franz Anton Mesmer in France. By 1792 Galvani's supposedly 'magnetic' frog experiments were proved to be erroneous by Alessandro Volta: the mysterious 'vital electrical fluid' came not from the animal itself, but from the chemical action of the metal plates to which it was attached during experiments.

Similarly, the French Académie des Sciences had appointed a scientific commission in 1784, headed by Franklin and Lavoisier (both experts in electrical phenomena), to examine the claims of 'animal magnetism'. They set up a series of elegant 'blind' trials, in which mesmerists were asked to identify objects that had been previously filled with 'vital fluid', including trees and flasks of water. They signally failed to do so. The commissioners then went on to examine the supposed curing of 'mesmerised' patients. As with the Montgolfier trials, Franklin wrote in detail to Banks of their findings. With impressive precision, the commission concluded that some 'mesmerised' patients did actually show marked signs of improved health. But this was not because of any 'magnetic' influences or 'vitalising' electrical fluids. It was simply because the patients *believed they would be cured.**

---

* This is possibly the first scientific identification of the famous 'placebo effect', although it would not be properly tested and defined until the 1950s. It has been claimed that over 30 per cent of all patients show a 'placebo' response, most notably in cases of depression, heart disease and chronic muscular pain. This figure has recently been questioned, since

But speculation continued to flourish in Germany, where a group of young writers, gathered at the University of Jena, began to explore the philosophical ideas of Friedrich Schelling and what he called *Naturphilosophie.* This doctrine, perhaps best translated as 'science mysticism', defined the entire natural world as a system of invisible powers and energies, operating like electricity as a series of 'polarities'. According to Schelling's doctrine, the whole world was indeed replete with spiritual energy or soul, and all physical objects 'aspired' to become something higher. There was a 'world-soul' constantly 'evolving' higher life forms and 'levels of consciousness' in all matter, animate or inanimate. All nature had a tendency to move towards a higher state.

So carbon for example 'aspired' to become diamond; plants aspired to become sentient animals; animals aspired to become men; men aspired to become part of the *Zeitgeist* or world spirit. Evolutionary, idealist, electrical and Vitalist ideas were all evidently tangled up in this system, which had an obvious appeal to imaginative writers in the Jena circle like Novalis, Schiller and Goethe, as well as experimental physiologists like Johann Ritter.[23] It had its attractions, not least in its optimism and its sense of reverence for the natural world. But it also constantly teetered on the brink of idiocy. One of its wilder proselytisers, the Scandinavian geologist Henrick Steffens, was said to have stated that 'The diamond is a piece of carbon that has come to its senses'; to which a Scottish geologist, probably John Playfair, made the legendary reply: 'Then a quartz, therefore, must be a diamond run mad.'[24]

These ideas gradually crossed the Channel to Britain, though not of course escaping the sceptical, all-weather eye of Banks. In January 1793

---

the earlier trials may have been methodologically flawed (they lacked a neutral 'control' group of patients); and the definition of 'cure' itself is open to a high degree of subjective distortion. e.g. Who is to say when a depression is cured, or how to measure if a severe pain is reduced to a milder one? This is similar to the problems Davy encountered when trying to describe objectively the effects of nitrous oxide. Nevertheless, the 1784 commission's work indicates why Vitalism raised genuine scientific questions, and also drew attention to that mysterious area which Coleridge (again) would define as 'psychosomatic' – the mind-body interface. There is an elegant passage in his notebooks wondering what causes men to blush, and the female nipple to become erect. Shelley composed an intriguing poem, 'The Magnetic Lady to her Patient' (1822), and Thomas De Quincey wrote a fine reflective essay, 'Animal Magnetism', for *Tait's Magazine* in 1834, investigating this subject, which remains alive in the continuing debate about 'alternative medicine'.

the radical journalist John Thelwall gave a hugely controversial public lecture on 'Animal Vitality' organised by the Physical Society at Guy's Hospital, under the auspices of the surgeon Henry Cline. The topic was so popular among the medical students that discussions were renewed over five subsequent and increasingly rowdy meetings.[25] 'Citizen' Thelwall had become known for his wish to 'demystify' various forms of authority and received opinions, and the following year, in May 1794, he was to be prosecuted for political sedition, a charge which carried the death penalty. It was partly the support of Cline, and the young Astley Cooper, which saved him from the gallows.[26]

Attacking what he saw as the potential mystifications in Hunter's theories, Thelwall proposed an openly materialist thesis that no 'spark of life' was divinely conferred, and that no soul was implanted by some external source. Yet he did not believe, like Hunter, that a 'life principle' could be simply explained by blood passing through the lungs. On the other hand he also maintained that '*Spirit*, however refined must still be material.' But what then was its source, if not blood – and not God?

Here Thelwall raised the Vitalist question which haunted a whole generation of Romantic writers. 'What is this something – this *vivifying principle*? – Is it atmospheric air itself? Certainly not ... It has been proved by experiment, that in the arteries of the living body there is no air. Something, however, it must be, that is contained in the atmosphere, and something of a powerful and exquisitely subtle nature.'[27]

Attempts to define this exquisite but powerful 'something' deeply concerned the young poets Thelwall came to meet after his release from prison, when he fled with his wife to the West Country in 1797. But for them it still seemed more a psychological than a physiological question. Coleridge in his conversation poems was exploring the metaphysical notion of a 'one Life' that unified all living forms; while Wordsworth in 'Tintern Abbey' wrote tentatively and beautifully of 'a sense sublime/Of something far more deeply interfused/Whose dwelling is the light of setting suns'. Both writers, at this most radical point in their lives, were trying to avoid an explicit reference to God, while retaining their intuitions of a 'spiritual' power – whatever that might be – both within man and within the natural universe. It was a balancing act that, perhaps, could only be performed in poetry.

All these Vitalist speculations were dramatically brought back to life, ten years after Thelwall's lecture, by an astonishing and brutal series of public experiments performed in London on 17 January 1803. The perpetrator, as Banks noted grimly, was another Italian, the Professor of Anatomy from Bologna, Giovanni Aldini. Banks had received earlier reports from Charles Blagden in Paris of Aldini 'experimenting on animals' with voltaic batteries the previous year, but remained dubious about his authenticity. There were unconfirmed rumours of Aldini dazzling the Galvanic Society with his 're-animation' exhibitions, but also of what Blagden called his 'excessive puffings and pretensions'.[28] In London, surrounded by eager publicity, Aldini attempted to revive the body of a murderer, one Thomas Forster, by the application of electrical charges six hours after he had been hanged at Newgate.

His demonstrations were graphically and melodramatically reported in the press: 'On the first application of the [electrical] arcs, the jaw began to quiver, the adjoining muscles were horribly contorted, and the left eye actually opened ... The conductors being applied to the ear, and to the rectum, excited muscular contractions much stronger ... The arms alternately rose and fell ... the fists clenched and beat violently the table on which the body lay, natural respiration was artificially established ... A lighted candle placed before the mouth was several times extinguished ... Vitality might have been fully restored, if many ulterior circumstances, had not rendered this – *inappropriate*.'[29]

That small, grotesque detail of the opening eye may well have caught a young novelist's imagination. Later experiments involved oxen's heads, dogs' bodies, and another human corpse which was said to have laughed and walked. The reports eventually caused such a public outcry that the experiments were banned, and Aldini forced to leave the country in 1805.

3

So when Abernethy and Lawrence began to clash in 1816, it was not entirely surprising that their angry exchanges quickly revived the old Vitalism debate in renewed form. For all its misgivings, the Royal College of Surgeons must have been glad to have raised a subject which brought such publicity. The exchanges were now closely followed by such serious literary journals as the *Edinburgh Review* and the *Quarterly Review*. At its base there lay a theological question: whether the 'super-added' force, if it existed, was the same as a spirit or soul, or some 'intermediary' element between body and spirit, or some form of 'vital' electrical fluid? By 1819, and the publication of both Abernethy's original lectures and Lawrence's *Natural History of Man*, the issues had also become heavily politicised. Here was humane, pious English science fighting against cruel, reductive, atheistical French science.

The conservative *Quarterly Review* found a more personal line of attack: 'We at the Quarterly Review, would ask what is it that Mr Lawrence, who is generally in the habit of smiling at the credulity of the world, modestly requires us all to believe? That there is no difference between a man and an oyster, other than that one possesses bodily organs more fully developed than the other! That all the eminent powers of reason, reflexion, imagination, and memory – the powers which distinguish a Milton, a Newton, and a Locke, – are merely the function of a few ounces of organized matter called the brain! ... Mr Lawrence considers that man, in the most important characteristics of his nature, is nothing more than an orang-outang or an ape, with "more ample cerebral hemispheres"! ... Mr Lawrence strives with all his powers to prove that men have no souls! ... Mr Lawrence has the sublime confidence to tell us that it is only "the medullary matter of the brain" that thinks or has spiritual consciousness!'[30]

As such questions caught the public imagination, they also spread among writers and artists. The influential idea that the group of writers first known as the 'Lake Poets' (with the later addition of the 'Cockney School') were particularly opposed to all scientific advance seems to have begun at precisely this time. This gradually hardened into the dogma that the 'Romantic poets' (as they eventually became known) were fundamentally anti-scientific. The myth can be observed forming on one signal occasion at a dinner party hosted in his north London studio in December 1817 by the painter and diarist Benjamin Haydon. This subsequently became known as the 'Immortal Dinner' (though it was in fact an extended, rather drunken luncheon). Indeed, it has been thought to exemplify the permanent, instinctive, deep-seated antagonism between Romantic poetry and science. But the truth seems rather different.

The poets present included Wordsworth, Charles Lamb and John Keats (but not, significantly, Coleridge, Byron or Shelley). Its aim was to celebrate the first stage of Haydon's enormous oil painting *Christ's Entry into Jerusalem*, upon which he had been labouring for three years, and which would take him another three to complete. His subject was the dominion of religion over the arts and sciences. Haydon had laid his dinner table directly beneath the huge rectangular canvas. A triumphant, youthful, bearded Christ rides at evening through the ancient city of Jerusalem, surrounded by a mob of enthusiastic disciples. The whole crowd sweeps downwards towards the viewer. But in one remote corner, set apart at the right of the picture, appear unmistakeable portraits of Wordsworth, Newton and Voltaire. Newton here represents analytic science, Voltaire godless French philosophical scepticism, and Wordsworth natural English piety. Haydon, perhaps provokingly, had dressed his old friend in a kind of monkish robe. There is one other striking figure just behind them. The young John Keats, his mouth wide open with a kind of shout of wonder, appears in animated profile from behind a pillar.

During the increasingly rowdy dinner-table discussion that developed, the painting provoked a debate about the powers of Reason versus the Imagination. The destructive and reductive effects of the scientific outlook were mocked. Warming to the theme, Lamb mischievously described Newton as 'a fellow who believed nothing unless it was as clear as the three sides of a triangle'. Keats joined in, agreeing that Newton had 'destroyed all the poetry of the rainbow, by reducing it to a prism'. Haydon jovially records: 'It was impossible to resist them, and we drank "Newton's health, and confusion to Mathematics."'[31]

Keats was wittily referring to the classic experiment in Newton's *Optics*, already much criticised in an essay by Goethe, in which a shaft of sunlight was passed through a prism, and separated out into the rainbow light of the spectrum. In fact the point of the experiment was that when the separated rainbow colours were *individually* passed through a second prism, they did *not* revert to white sunlight, but remained true colours (that is, in modern terms, they remained at the same wavelength). The rainbow was *not* a mere scientific trick of the glass prism. It genuinely and beautifully existed in nature, through the natural prism of raindrops, although paradoxically it took a human eye to see it, and every human eye saw it differently. It seems unlikely that Keats did not know this; but perhaps he did not wish to admit (in that company) that Newton had actually *increased* the potential 'poetry of the rainbow', by showing it was not merely some supernatural sky-writing, as asserted in Genesis: 'I do set My bow in the cloud, and it shall be for a token of a covenant between Me and the earth.'

This playful and eventually drunken attack on the reductive effects of science was orchestrated and eagerly recorded by Haydon in his diary. Unlike the others, he was a passionate fundamentalist Christian, and believed that most science was inevitably godless, and probably blasphemous. His view is often linked with what Wordsworth wrote in his poem 'The Tables Turned':

> Sweet is the lore which Nature brings:
> Our meddling intellect
> Misshapes the beauteous forms of things: –
> We murder to dissect.

'Murder to dissect' was certainly already a poets' rallying cry. It was exactly this poem that was quoted by Southey in his 1801 letter to Coleridge about Davy's apparent rejection of poetry. Here the attack moved from physics to medicine. Yet there was evidently much popular misunderstanding of what anatomical 'dissection' actually involved, equating it more with Aldini than with Hunter. Although it necessarily began with the opening up of a corpse (an act still surrounded by many unconscious taboos), it was not in fact mainly a procedure of 'cutting up' with scalpels. The important instruments were forceps, rounded metal probes, and the surgeon's own fingers. The essential process was one of separating out tissue and organs, and laying bare the various independent systems, such as the heart, the lactating breast, or the reproductive system, for meticulous study. These were often the object of the most exquisitely refined and delicate anatomical drawings (though these too could arouse horror).

Yet the act of dissection could also be seen as one of profound attention and reverence for nature. This is how John Abernethy described his teacher John Hunter at work: 'He would stand for hours motionless as a statue, except that with a pair of forceps in either hand he was picking asunder the connecting fibres of some structure ... patient and watchful as a prophet, sure that the truth would come: it might be as in a flash, in which, as with inspiration, intellectual darkness became light.'[32]

Wordsworth's brief poem had been written nearly twenty years earlier, and does not really express his considered view of Newton, the heroic, voyaging figure of the later *Prelude*. If either Coleridge or Shelley had been present at the dinner (they were both in London), one imagines the conversation would have taken a rather different tack. Shelley had already baited Haydon for his 'religious superstitions' on an earlier occasion, remarking on 'that most detestable religion, the Christian', and always defended progressive science,[33] while Coleridge had made his own experiments with prisms in the Lake District, and really did understand the formation of the rainbow, both poetically and scientifically. He knew it was a refraction of light through a fleeting curtain of raindrops, but also saw it was a powerful mythological symbol.

Like one of his literary heroes, the great seventeenth-century physician and essayist Sir Thomas Browne, Coleridge did not accept any contradiction between the two modes of vision. He wrote in his *Notebooks*: 'The Steadfast rainbow in the fast-moving, fast-hurrying hail-mist. What a congregation of images and feelings, of fantastic Permanence amidst the rapid change of a Tempest – quietness the Daughter of Storm.' Coleridge accepted that the rainbow was produced by refraction through the 'hail-mist', but also that its paradoxical effect *on the observer* of beautiful steadiness amidst terrifying chaos had a powerful psychological and poetic symbolism. The 'quiet Daughter' is perhaps a reference to Cordelia in Shakespeare's *King Lear*. Cordelia could even be understood as Lear's rainbow during the storm on the wild heath, the steadfast and reassuring symbol of love seen through the prism of tears.*

At Highgate, Coleridge and his doctor and confidant James Gillman had decided to intervene actively in the Vitalism debate, and collaborated on a paper, 'Notes Towards a New Theory of Life', which tried to steer a metaphysical path between the two extreme positions. Coleridge, anxious to reconcile science with a sacred concept of life, argued that the soul existed, but had no analogy with 'electricity'. While denying that life was purely physical organisation, he rejected the idea of some mystical life force with dry humour. 'I must reject fluids and ethers of all kinds, magnetical, electrical, and universal, to whatever quintessential thinness they may be treble-distilled and (as it were) super-substantiated!'[34]

He also discussed the question with his learned friend J.H. Green, who was a member of the Royal College. Green's speciality was eye diseases, and he was a Demonstrator at Guy's Hospital while Keats was training there. It was Green who made the historic introduction between his young student and the ageing Coleridge on one of their walks across Hampstead Heath to Kenwood, 'in the lane that winds by Lord Mansfield's park',

---

* Richard Dawkins has praised this passage from Coleridge as 'good science', in his remarkable study of Science and Romanticism, *Unweaving the Rainbow* (1998, Chapter 3, 'Bar Codes in the Stars'). The whole chapter gives a scientist's lively view of Haydon's dinner party, to which Dawkins, the Professor of the Public Understanding of Science, would clearly have liked to have been invited – as would I.

in spring 1819. One of the many subjects that Keats remembered from this long, oracular perambulation – besides 'Nightingales' – was 'First and Second Consciousness'.[35]

Green later became Coleridge's full-time amanuensis at Highgate, and continued adding to the 'Theory of Life' and discussing the implications of the Vitalism debate over several years, though nothing was ever published in his lifetime. Coleridge's position remained that the 'life principle' certainly did exist, but had nothing to do with physiology. It consisted in an inherent drive towards 'individuation', which moved up the chain of creation, and finally manifested itself in the unique form of human 'self-consciousness', which included the moral conscience and the spiritual identity or 'soul'.

This of course was a metaphysical, not a medical explanation. It was clearly an adaptation of Schelling's *Naturphilosophie*. But it did have the crucial effect of suggesting that the real subject of the Vitalism debate was the mysterious nature of this 'consciousness' itself: how it began, how it grew, to what degree it was shared with animals, and what happened to it when the body died. How exactly the physical brain itself 'generated' this consciousness Coleridge did not presume to say. Green pursued the problem long after Coleridge's death, and after he became President of the Royal College of Surgeons he published some of Coleridge's speculations as *Spiritual Philosophy* (1865).

The nature of 'consciousness' remains a major challenge to modern neuroscience, and one of the great abiding scientific mysteries. The French physiologist Pierre Cabanis, much admired by Lawrence and deprecated by Abernethy, suggested that 'just as the stomach, liver, and other glandular organs produced their typical secretions, so also did the healthy brain secrete moral thought'.[36] Coleridge dismissed such suggestions as mechanistic, and his 'Theory of Life' is also interesting because it represents the nearest he came to suggesting the evolution of human intelligence, an argument that he was otherwise inclined to dismiss as the 'absurd orang-utang theory'.

Indeed, it is still sobering to read Coleridge – one of the outstanding minds of his generation – on the subject of evolution. As he wrote to Wordsworth: 'I understood you would take the Human Race in the concrete, have exploded the absurd notion of Pope's Essay on Man, [Erasmus] Darwin, and all the countless believers – even (strange to say) among Christians – of Man's having progressed from an Ouran Outang state – so contrary to all History, to all Religion, nay, to all Possibility – to have affirmed a Fall in some sense.'[37]

Keats's two years of medical training at Guy's (1816–17) under Astley Cooper and J.H. Green are recorded in his extensive Anatomical and Physiological Notebook, together it must be said with many doodlings of flowers and faces in the margins. So he undoubtedly knew more about medical, chemical and dissection procedures – as well as the Vitalism debate itself – than anyone else at Haydon's dinner party. His witticism at Newton's expense could be seen as the typical knowing humour of a clever medical student.

Yet Keats did appear to make a later attack on the cruel and 'demystifying' aspects of science in his ornate narrative poem 'Lamia', written in 1820. This poem appears to be very different in spirit from his earlier sonnet praising Herschel, the inspired scientific 'watcher of the skies'. It was based on a strange misogynist medical 'case history' he had found in Robert Burton's *Anatomy of Melancholy*, in which a beautiful, seductive woman is revealed by a wise physician-philosopher, one Apollonius, to be a terrifying snake, or 'lamia'. In Burton's version, Apollonius's 'diagnosis' saves Lamia's infatuated young bridegroom Lycius, just in time, on his wedding night, 'although Lamia wept, and desired Apollonius to be silent'.[38]

Keats changed many of the details of Burton's story, not least that his Lycius is so heartbroken at the loss of his lovely Lamia – whether she be a serpent or not – that far from thanking the scientific Apollonius, he retreats to his bed in misery and dies at the end of the poem. Keats prepares for this dénouement in a striking passage in which he refers to the icy touch and 'cold philosophy' of science, which destroys the beautiful mystery of all natural objects, like the rainbow – or, indeed, like the serpent-woman.

> ... Do not all charms fly
> At the mere touch of cold philosophy?
> There was an awful rainbow once in heaven:
> We know her woof, her texture; she is given
> In the dull catalogue of common things.
> Philosophy will clip an Angel's wings ...
> Unweave a rainbow, as it erstwhile made
> The tender-personed Lamia melt into a shade.[39]

But *is* the serpent-woman a natural object? Or is she something artificial and lethal, an alien life force which will prove fatal to man, and particularly to her naïve young bridegroom, who is innocently besotted with her? This is the question that Keats seems to pose by the end of his poem. What is the role of science (represented by the fierce old sage Apollonius) in protecting man from seductive but destructive delusions?

The Lamia poem is the one in which Keats himself said he had made 'more use of my judgement' – meaning his powers of intellectual analysis – than any other. It is in fact full of intellectual provocations – not least about the nature of sexual attraction, and the indiscriminate drive of the life force – and is replete with chemical and surgical imagery. There is another passage, far less known, in which Keats describes the Lamia herself, before she is transformed into a woman. Here she is presented not as some conventional erotic anaconda (as Humboldt might have encountered in the Amazon forest), but as if she were the result of some astonishing new chemical or biological combination, producing a gleaming, seductive but utterly alien new life form.

> She was a gordian shape of dazzling hue,
> Vermilion-spotted, golden, green and blue;
> Striped like a zebra, freckled like a pard,
> Eyed like a peacock, and all crimson barred;
> And full of silver moons, that, as she breathed,
> Dissolved, or brighter shone, or interwreathed
> Their lustres with the gloomier tapestries –
> So rainbow-sided, touched with miseries
> She seemed ...
> Her head was serpent, but, ah bitter-sweet!
> She had a woman's mouth with all its pearls complete.[40]

This extraordinary creation is both sexually alluring and yet clearly menacing and 'demonic'. By using the term 'rainbow-sided' of her body, Keats even seems to be recalling his old Newtonian joke, and inventing his own mysterious *biological* rainbow, a living creature who is both a spectre and a spectrum. There are many other passages which play with medical and scientific imagery in the poem – for example Hunter's theory of 'inflammation' as proof of vitality. When Lycius desperately grasps Lamia's *chilly* hand, 'all the pains/Of an unnatural heat shot to his heart'.[41]

But most memorable and disturbing is the passage in which Lamia the snake changes into Lamia the woman, 'a full-born beauty new and exquisite!' This new birth is described in semi-scientific terms, as if Keats were observing a violent chemical experiment in a laboratory, or a surgical procedure (like Fanny Burney's), or one of Aldini's electrical trials. It is agonising. Lamia's serpentine body begins to convulse, her blood 'in madness' runs through her length; she foams at the mouth, and her saliva 'so sweet and virulent' burns and 'withers' the ground where it spatters. Her eyes 'in torture fixed' become glazed and wide. The 'lid-lashes' are seared, and the pupils flash 'phosphor and sharp sparks'.

> The colours all inflamed throughout her train,
> She writhed about convulsed in scarlet pain:
> A deep volcanian yellow took the place
> Of her milder-moonèd body's grace;
> And as the lava ravishes the mead,
> Spoilt all her silver mail, and golden brede;
> Made gloom of all her frecklings, streaks and bars,
> Eclipsed her crescents, and licked up her stars ...[42]

Keats never lets his reader forget this traumatic birth, and what it has cost the serpent to become a human being. His extraordinary invention, perhaps the most brilliant and thought-provoking of all his narrative poems, engages many of the moral issues surrounding Vitalism, the nature of life, and the notion of human consciousness. Above all, perhaps, it asks if the beautiful Lamia has a soul.

4

But the most singular literary response to the Vitalism debate was Mary Shelley's cult novel *Frankenstein, or The Modern Prometheus* (1818). In this story, originally thought to have been written by a male author – either Walter Scott, William Godwin or Percy Shelley – a sort of human life is physically created, or rather reconstructed. But the soul or spirit is irretrievably damaged.

Mary Shelley's preliminary ideas for the novel can be dated back remarkably early, to the year 1812, when her father William Godwin took her to hear Humphry Davy give his public lectures on chemistry at the Royal Institution. She was then only fourteen. Her young Victor Frankenstein would also begin as an idealistic and dedicated medical student, inspired by the lectures of the visionary Professor Waldman at

Ingolstadt. Mary Shelley would eventually draw directly on the published text of Davy's famous 'Introductory Discourse', in which he spoke of those future experiments in which man would 'interrogate Nature with Power ... as a master, active, with his own instruments'.[43]

Waldman's lecture on chemistry expands Davy's claims, and has an electric effect on the young Victor Frankenstein.

> 'The ancient teachers of this science,' said he, 'promised impossibilities and performed nothing. The modern masters promise very little; they know that metals cannot be transmuted, and that the elixir of life is a chimera. But these philosophers, whose hands seem only to dabble in dirt, and their eyes to pore over the microscope or crucible, have indeed performed miracles. They penetrate into the recesses of Nature, and show how she works in her hiding-places. They ascend into the heavens; they have discovered how the blood circulates, and the nature of the air we breathe. They have acquired new and almost unlimited Powers: they can command the thunders of heaven, mimic the earthquake, and even mock the invisible world with its own shadow.'
>
> Such were the Professor's words – rather let me say such the words of Fate – enounced to destroy me. As he went on I felt as if my soul were grappling with a palpable enemy; one by one the various keys were touched which formed the mechanism of my being. Chord after chord was sounded, and soon my mind was filled with one thought, one conception, one purpose. So much has been done! – exclaimed the soul of Frankenstein: more, far more will I achieve! Treading in the steps already marked, I will pioneer a new way, explore unknown Powers, and unfold to the world the deepest mysteries of Creation.[44]

When Mary eloped with Shelley to France and Switzerland in 1814, their shared journal indicates that they were already discussing notions of creating artificial life. As they returned penniless, by public riverboat down the Rhine, they remarked on the monstrous, inhuman appearance of several of the huge German labourers on board, and noticed that they sailed beneath a lowering *schloss* known as 'Castle Frankenstein'.[45] On their return Shelley began writing the first of his series of speculative and autobiographical essays, mixing scientific ideas with psychology, under such titles as 'On the Science of Mind', 'On a Catalogue of the Phenomenon of Dreams' and 'On Life'. He evidently discussed these disturbing ideas with Mary, for she remembered on one occasion how he broke off from writing one of them, 'overcome by thrilling horror'.[46]

Mary's brilliance was to see that these weighty and often alarming ideas could be given highly suggestive, imaginative and even playful form. In a sense, she would treat male concepts in a female style. She would develop exactly what William Lawrence had dismissed in his lectures as a 'hypothesis or fiction'. Indeed, it was to be an utterly new form of fiction – the science fiction novel. Mary plunged instinctively into the most extreme implications of Vitalism. In effect, she would take up where Aldini had been forced to leave off. She would pursue the controversial – and possibly blasphemous – idea that vitality, like electricity, might be used to reanimate a dead human being. But she would go further, much further. She would imagine an experiment in which an entirely new human being was 'created' from dead matter. *She would imagine a surgical operation, a corpse dissection, in reverse.* She would invent a laboratory in which limbs, organs, assorted body parts were not separated and removed and thrown away, but assembled and sewn together and 'reanimated' by a 'powerful machine', presumably a voltaic battery.[47] Thus they would be given organic life and vitality. But whether they would be given a soul as well was another question.

This extraordinary fiction was begun at the Villa Diodati on Lake Geneva in the summer of 1816, in a holiday atmosphere of dinner parties and late-night talk, but very different from that at Haydon's 'Immortal Dinner'. The talk was quick, clever, sceptical, teasing and flirtatious. Mary Shelley records that she, Shelley and Byron, inspired by Dr Polidori (himself only twenty-two), discussed the galvanic experiments of Aldini, and various speculations about the artificial generation of life by Erasmus Darwin. They then, famously, set themselves a ghost-story-writing competition.

Byron scrawled a fragment about a dying explorer, 'Augustus Darvell' (dated 17 June 1816); Shelley composed his atheist poem 'Mont Blanc'; Polidori dashed off a brief gothic bagatelle, 'The Vampyre', which he later tried to pretend was actually Byron's (so he could sell it), while Mary Shelley wrote – but very slowly, over the next fourteen months – an intricately constructed 90,000-word fiction, which gradually became, draft crafted upon draft, *Frankenstein, or The Modern Prometheus.* She handed in the completed manuscript to Lackington, Allen & Co. in August 1817, just three weeks before her baby Clara was born on 2 September.

The actual writing of Mary's novel can be followed fairly closely from her journal in Switzerland, and then back in England at Great Marlow on the Thames. What is less clear is where she gathered her ideas and materials from, and how she created her two unforgettable protagonists: Dr Frankenstein and his Creature. One is tempted to say that the Creature – who is paradoxically the most articulate person in the whole novel – was a pure invention of Mary's genius. But in Victor Frankenstein of Ingolstadt she had created a composite figure who in many ways was typical of a whole generation of scientific men. The shades of 'inflammable' Priestley, the deeply eccentric Cavendish, the ambitious young Davy, the sinister Aldini and the glamorous, iconoclastic William Lawrence may all have contributed something to the portrait.

Yet Frankenstein is essentially a European figure, a Genevan – perhaps of German Jewish ancestry – studying and working at Ingolstadt in Germany.[48] The importance of the German connection, and the experiments already done there, was pointed out by Percy Shelley in the very first sentence of his anonymous 'Preface' to the original 1818 edition of the novel. 'The event on which this fiction is founded, has been supposed, by Dr Darwin, and some of the physiological writers of Germany, as not impossible of occurrence.'

So who had Mary Shelley been thinking of? The outstanding young German physiologist known in British scientific circles at this time was Johann Wilhelm Ritter (1776–1810). His work at the university of Jena had been reported to Banks regularly at the turn of the century, and his election and move to the Bavarian Academy of Sciences in Munich in 1804, when still only twenty-eight, was closely followed.[49] Banks and Davy kept a particularly keen eye on his work, since Ritter had anticipated Davy's improvements on the voltaic battery, had invented a dry-cell storage battery, and had followed up Herschel's work on infra-red radiation from the sun, by identifying ultraviolet rays in 1803. He was also known for certain undefined 'galvanic' experiments with animals, which were the talk of the Royal Society, although amidst a certain amount of head-shaking.[50] But among his colleagues at Jena he was regarded as a portent. The young poet Novalis (Frederick von Hardenburg, also a mining engineer) exclaimed: 'Ritter is indeed searching for the real *Soul of the World* in Nature! He wants to decipher her visible and tangible language, and explain the emergence of the Higher Spiritual Forces.'[51]

In September 1803 Banks received a confidential report from the chemist Richard Chenevix, a Fellow of the Royal Society and the recipient of the Copley Medal in 1803, who was on a scientific tour of German cities. Writing from Leipzig, Chenevix noted that the 'most interesting' work at Jena was being done by Ritter, who was using a huge voltaic battery to obtain 'most capital results', having 'a very powerful effect upon the animal economy' but without damaging 'the most delicate organs'. Apparently holding back further details for a separate paper, Chenevix added to Banks: 'In communicating these experiments to you who *are at the centre*, they will immediately find their way to other Philosophers of London. Mr Davy I am sure will be particularly interested.'[52]

But by August of the following year, when Ritter had moved to Munich, Chenevix's reports had taken on a rather different tone. 'Ritter the galvanist is the only man of real talent I have met with; and his head and morals are overturned by the new philosophy of Schelling. I have declared open war against these absurdities.'[53] Chenevix's final report, of 7 November 1804, while still praising Ritter, now has an openly sarcastic edge, and ends on a disturbing note, as if he had witnessed something terrible which he cannot quite bring himself to describe: 'You may remember that I mentioned to you Ritter's experiments with a Galvanic pile ... Ritter is experimenter in chief, or as they term him, *Empyrie of the New Transcendent School*. I saw him repeat his experiments; and they appeared most convincing. Whether there was any trick in them or not I cannot pretend to say ... Ritter with a large body of Professors and pupils, is gone from Jena; and Bavaria is now enlightened by their Doctrines. It is impossible to conceive anything so disgusting and humiliating for the human understanding as their dreams.'[54]

That these 'dreams' are related to those of the fictional Dr Frankenstein seems more than possible. Experiments that had been forbidden by the Prussian government in Jena were taken up again when Ritter moved to the traditionally more libertarian atmosphere of Munich. From his desultory and posthumous memoirs, *Fragments of a Young Physicist* (1810), it would seem that in Munich Ritter fell fatally under the influence of one of the wildest of the *Naturphilosophie* practitioners, a certain Franz von Baader. Experiments that began with water divining, 'geoelectrical' mapping and 'metal witching' turned to the revival of dead animals by electrical action, and possibly the 'disgusting and humiliating' revival of dead human beings, although there is no definitive evidence of this. At all events, Ritter's Bavarian colleagues were gradually alienated, his students abandoned him, and his mental stability became increasingly fragile. He neglected his family (he had three children), withdrew into his laboratory, and grew increasingly remote and obsessive. Finally, his promising career was destroyed, and he died penniless and insane in 1810, aged thirty-three. In other circumstances his *Memoirs* might have been those of young Victor Frankenstein.[55]

Ritter's tragic story was clearly known to Banks, to Davy, and very probably to Lawrence after his time in Göttingen with Blumenbach. Whether it was known to Dr Polidori, and whether it was he who told it to the Shelleys in 1816, is speculation. But they clearly knew from some source about 'the physiological writers of Germany'. Moreover, the novel owes something else to Germany. Mary Shelley chose to narrate Frankenstein's act of electrical reanimation, or blasphemous 'creation', in a gothic style that owes nothing to the cool British manner of the Royal Society reports, but everything to German ballads and folk tales.

> It was on a dreary night of November, that I beheld the accomplishment of my toils. With an anxiety that almost amounted to agony, I collected the instruments of life around me, that I might infuse a spark of being into the lifeless thing that lay at my feet. It was already one in the morning; the rain pattered dismally against the panes, and my candle was nearly burnt out, when by the glimmer of the half extinguished light, I saw the dull yellow eye of the creature open. It breathed hard, and a convulsive motion agitated its limbs.
>
> How can I describe my emotions at this catastrophe, or how delineate the Wretch whom with such infinite pains and care I had endeavoured to form? His limbs were in proportion, and I had selected his features as beautiful. Beautiful! – Great God! His yellow skin scarcely covered the work of muscles and arteries beneath. His hair was lustrous black and flowing; his teeth of pearly whiteness. But these luxuriances only formed a more horrid contrast with his watery eyes, that seemed almost of the same colour as his dun-white sockets in which they were set, his shrivelled complexion and straight black lips.[56]

5

As her novel developed, Mary Shelley began to ask in what sense Frankenstein's new 'Creature' would be human. Would it have language, would it have a moral conscience, would it have human feelings and sympathies, *would it have a soul*? (It should not be forgotten that Mary was pregnant with her own baby in 1817.) Many of Lawrence's reflections on the metaphysics of the dissecting room and the theory of brain development seem to be echoed in ideas and even complete phrases used in *Frankenstein*. Here again it seems that Shelley, who was attending medical consultations with Lawrence throughout spring 1817, and may sometimes have been accompanied by Mary, made an opportunity for all three of them to explore these specialist themes.[57]

Mary Shelley's idea of the mind was, like Lawrence's, based on the notion of the strictly physical evolution of the brain. This is how Lawrence was provocatively challenging his fellow members of the Royal College of Surgeons in his lectures of 1817: 'But examine the "mind", the grand prerogative of man! Where is the "mind" of the foetus? Where is that of a child just born? Do we not see it actually built up before our eyes by the actions of the five external senses, and of the gradually developed internal faculties? Do we not trace it advancing by a slow progress from infancy and childhood to the perfect expansion of its faculties in the adult... '[58]

Frankenstein's Creature has been constructed as a fully developed man, from adult body parts, but his mind is that of a totally undeveloped infant. He has no memory, no language, no conscience. He starts life as virtually a wild animal, an orangutan or an ape. Whether he has sexual feelings, or is capable of rape, is not immediately clear. Although galvanised into life by a voltaic spark, the Creature has no 'divine spark' from Heaven. Yet perhaps his life could be called, in a phrase of the medical student John Keats, a 'vale of soul-making'.

Almost his first conscious act of recognition, when he has escaped the laboratory into the wood at night, is his sighting of the moon, an object that fills him with wonder, although he has no name for it: 'I started up and beheld a radiant form rise from among the trees.* I gazed with a kind of wonder. It moved slowly, but it enlightened my path... It was still cold... No distinct ideas occupied my mind; all was confused. I felt light, and hunger, and thirst, and darkness; innumerable sounds rung in my ears and on all sides various scents saluted me.... Sometimes I tried to imitate the pleasant songs of the birds, but was unable. Sometimes I wished to express sensations in my own mode, but the uncouth and

* Mary Shelley adds as her own footnote in the novel: '* the moon'.

inarticulate sounds which broke from me frightened me into silence again ... Yet my mind received, every day, additional ideas.'[59]

From this moment the Creature evolves rapidly through all the primitive stages of man. Mary's account is almost anthropological, reminiscent of Banks's account of the Tahitians. First he learns to use fire, to cook, to read. Then he studies European history and civilisation, through the works of Plutarch, Milton and Goethe. Secretly listening to the cottagers in the woods, he learns conceptual ideas such as warfare, slavery, tyranny. His conscience is aroused, and his sense of justice. But above all, he discovers the need for companionship, sympathy and affection. And this is the one thing he cannot find, because he is so monstrously ugly: 'The cold stars shone in their mockery, and the bare trees waved their branches above me, the sweet voice of a bird burst forth amidst the universal stillness. All, save I, were at rest ... I, like the arch-fiend, bore a hell within me, and finding myself unsympathised with, wished to tear up the trees, spread havoc and destruction around me, and then to have sat down and enjoyed the ruin.'[60]

On the bleak Mer de Glace glacier in the French Alps, the Creature appeals to his creator Frankenstein for sympathy, and for love. 'I am malicious because I am miserable. Am I not shunned and hated by all mankind? You, my creator would not call it murder, if you could precipitate me into one of those ice-rifts ... Oh! My creator, make me happy! Let me feel gratitude towards you for *one* benefit! Let me see that I excite the sympathy of *one* existing thing. Do not deny me my request!'[61]

This terrible corrosive and destructive solitude becomes the central theme of the second part of Mary Shelley's novel. Goaded by his misery, the Creature kills and destroys. Yet he also tries to take stock of his own violent actions and contradictory emotions. He concludes that his one hope of happiness lies in sexual companionship. The scene on the Mer de Glace in which he begs Frankenstein to create a wife for him is central to his search for human identity and happiness. The clear implication is that a fully human 'soul' can only be created through friendship and love: 'If you consent [to make me a wife], neither you nor any other human being shall ever see us again. I will go to the vast wilds of South America. My food is not that of man. I do not destroy the lamb and the kid to glut my appetite; acorns and berries afford me sufficient nourishment. My companion will be of the same nature as myself, and will be content with the same fare. We shall make our bed of dried leaves; the sun will shine on us as on man, and will ripen our food. The picture I present to you is peaceful and human, and you must feel that you could deny it only in the wantonness of power and cruelty.'[62]

The Creature is here offering to go westwards to South America or the Pacific, and to return to that primitive Edenic state glimpsed by Cook and Banks. He and his mate will live as vegetarians, kill nothing, cook nothing, build nothing, and reject everything that European civilisation stands for. They will become, in fact, Noble Savages.

In response, Frankenstein goes to London (rather than Paris) to study the latest surgical techniques. He consults with 'the most distinguished natural philosopher' of the day in this 'wonderful and celebrated city', though this man is not named.[63] He then sets up a second laboratory in Scotland, in the remote Orkney islands, where he plans to create a second Creature, a woman. Her companionship will satisfy the male Creature.

But Frankenstein is overcome with doubts. 'Even if they were to leave Europe, and inhabit the deserts of the New World, yet one of the first results of those sympathies for which the Demon thirsted would be children, and a race of devils would be propagated on earth ... Had I the right, for my own benefit, to inflict this curse upon everlasting generations?'[64] His eventual decision to destroy his handiwork is perhaps the grimmest scene in the novel. The laboratory is revealed as a place of horror and blasphemy: 'I summoned sufficient courage, and unlocked the door of my laboratory. The remains of the half-finished creature, whom I had destroyed, lay scattered on the floor, and I almost felt as if I had mangled the flesh of a human being ... With trembling hand I conveyed the instruments out of the room, cleaned my chemical apparatus, and put the relics of my work into a basket with a great quantity of stones.'[65]

There is something more than deathly about those stones. It is as if Frankenstein is burying scientific hope itself beneath the earth.

In his grief and fury the Creature revenges himself on his creator by destroying Frankenstein's friend Clerval, and then his bride, Elizabeth. From thenceforth both are locked in a pact of mutual destruction, which eventually leads pursuer and pursued to the frozen wastes of the North Pole – the antithesis of the warm, Pacific paradise. In a sense, both have lost their own souls. Drawing on the Miltonic imagery of *Paradise Lost*, both see themselves as fallen angels, doomed to eternal solitude and destruction. The dying Frankenstein remains unrepentant as he gasps: 'All my speculations and hopes are as nothing. Like the archangel who aspired to omnipotence, I am chained in eternal hell ... I have myself been blasted in all my hopes ... *Yet another may succeed.*'[66]

But the Creature has attained a kind of self-knowledge, and even humility: 'When I call over the frightful catalogue of my deeds, I cannot believe that I am he whose thoughts were once filled with sublime and transcendent visions of the beauty of the world. But it is even so. The fallen angel becomes a malignant devil. Yet even that enemy of God and man had friends and associates in his desolation. I am quite alone ... He is dead who called me into being; and when I shall be no more, the very remembrance of us both will speedily vanish ... I shall no longer see the sun or stars, or feel the winds play on my cheeks.'[67]

6

Victor Frankenstein's experiment in soul-making had ended in disaster. The novel itself disappeared into temporary obscurity, and fewer than 500 copies were sold of the first edition. But it was made famous, if not notorious, in the 1820s by no less than five adaptations for the stage. These caused widespread controversy. The first was staged in London in July 1823, at the English Opera House in The Strand. It was entitled portentously *Presumption: or The Fate of Frankenstein*. From the start there was sensational publicity:

> Do not go to the Opera House to see the Monstrous Drama, founded on the improper work called FRANKENSTEIN!!! Do not take your wives, do not take your daughters, do not take your families!!! – The novel itself is of a decidedly immoral tendency; it treats of a subject which in nature cannot occur. This subject is PREGNANT with mischief; and to prevent the ill-consequences which may result from the promulgation of such dangerous Doctrines, a few zealous friends of morality, and promoters of the Posting-bill (and who are ready to meet the consequences thereof) are using their strongest endeavours.[68]

The part of 'The Creature', which was cleverly and sinisterly left *blank* in the programme, made the actor T.P. Cooke famous (despite his terrible gout) – just as it later made Boris Karloff famous. Over the next four years there were fourteen separate productions, mounted in London, Bristol, Paris and New York.

Presumption made several fundamental changes to Mary Shelley's novel, all without her permission. Nor did she receive any copyright fees. Curiously she did not seem to mind, and when she herself went to see the play in September 1823 she loved it. 'But lo & behold! I found myself famous! Frankenstein had prodigious success as a drama at the English Opera House ... Mr. Cooke played the "*blank's*" part extremely well – his trying to grasp at the sounds he heard – all he does was well imagined and executed ... it appears to excite a breathless eagerness in the audience ... in the early performances all the ladies fainted and hubbub ensued! ... They continue to play it even now.'[69]

Yet the changes have influenced almost all subsequent stage and film productions. They altered the scientific and moral themes of the book, and shifted it permanently towards a mixture of gothic melodrama and black farce. Victor Frankenstein is made the archetypal mad and evil scientist. He has stood for this role ever since. But in the original novel he is also a romantic and idealistic figure, obsessive rather than evil, and determined to benefit mankind. His demoniac laboratory becomes the centre of dramatic interest, with fizzing electrical generators, sinister bubbling vats and violent explosions. But no such laboratory is described in the novel: Frankenstein works by candlelight at a surgical table. He is also given a comic German assistant called Fritz, who adds gothic farce to the whole proceedings. There is no such assistant in the novel: Frankenstein's work is essentially solitary and dedicated, like that of an artist.

But the most important change of all is this. Mary Shelley's unnamed Creature is transformed into the 'Monster', and made completely dumb. He is deprived of all words, whereas in the novel he is superbly and even tragically articulate: 'And what was I? Of my creation and my creator I was absolutely ignorant ... Where were my friends and relations? No father had watched my infant days, no mother had blessed me with smiles and caresses; or if they had, all my past life was now a blot, a blind vacancy in which I distinguished nothing ... I was, besides, endued with a figure hideously deformed and loathsome. I was not even of the same nature as man ... When I looked around I saw and heard of none like me. Was I, then, a Monster, a blot upon the earth, from which all men fled and whom all men disowned? I cannot describe the agony that these reflections inflicted upon me ... Oh, that I had forever remained in my native wood, nor known nor felt beyond the sensations of hunger, thirst and heat!'[70]

7

William Lawrence's experiment ended in an altogether different way. At the end of 1819 he withdrew his *Natural History of Man*, yielding to pressure from the Royal College of Surgeons and a number of medical institutions. But he continued to speak out in favour of scientific freedom. 'I take the opportunity of protesting, in the strongest possible terms... against the attempt to stifle impartial enquiry by an outcry of pernicious tendency; and against perverting science and literature, which naturally tend to bring mankind acquainted with each other, to the anti-social purpose of inflaming and prolonging national prejudice and animosity.'[71]

Lawrence allowed the radical publisher Richard Carlile to reissue a pirate edition of the *Natural History* in 1822, which ran to nine editions (Carlile also successfully pirated Shelley's *Queen Mab*). Carlile wrote his own pamphlet, *Address to the Men of Science* (1821), in which he urged Lawrence and others to retain their intellectual independence. When Carlile died, in a final gesture of support, he gave Lawrence his corpse for dissection, an almost unheard-of bequest.[72]

Lawrence was also supported by Thomas Wakley, mercurial editor of the newly founded medical journal the *Lancet*. In scintillating and lively articles, Wakley attacked the old guard of the Royal College, and satirised the attempts of Abernethy and others to bring theology into the surgical theatre. Whenever they dissected some haemorrhaging organ, or pulsating artery, Wakley mocked, they would exclaim 'with uplifted eye, and most reverentially contracted mouth: "*Gintilmen*, behold the *winderful* evidence of *Desin!*" '[73]

But in 1829 William Lawrence stood for the Council of the Royal College of Surgeons, a body famous for its conservatism. Silently renouncing his radical and 'materialist' views, he went to see his old patron and enemy John Abernethy. It was not a meeting on the Mer de Glace. After long discussions, Lawrence received forgiveness and wholehearted support from his old mentor. Lawrence was unanimously elected, and when his old comrade-in-arms Thomas Wakley came to protest on behalf of the *Lancet*, Lawrence helped to physically manhandle him out of the Council chamber. Sir William Lawrence finished his career as Surgeon-General to Queen Victoria, and was created a baronet. But perhaps he had lost his own soul.

8

Davy and the Lamp

1

After the hugely successful Geology Lectures of spring 1811 in Dublin, Humphry Davy returned to the west of England on a summer fishing expedition. Here, while innocently angling along the banks of the river Wye, he himself was hooked by a small, dark and vivacious Scottish beauty, Jane Apreece. For the first time in his life he fell desperately in love, and felt a power that might be greater than science.

Jane, who had heard him lecture in Ireland, was, initially at least, rather cool about him. She wrote to a friend on 4 March 1811: 'Mr Davy is remarkably pleasant, & all the fashion & celebrity of admiration do not injure his unaffectedness. It is said that a more dangerous Power in the sprightly form of an Irish Peeress may probably burn some of his combustible matter & at least singe if not scarifying his heart.'[1]

Jane was thirty-one, a widow and an heiress. She was known in Edinburgh as a wit and a *belle esprit.* She dressed beautifully, and talked flamboyantly: she had a kind of electrical energy about her. Davy loved energy. Jane had travelled widely in Europe, and spoke fluent French and Italian. She could read Latin, and she liked going to lectures. She was clever, self-confident and original.

Apart from Anna Beddoes, Davy's decade of success and glamour at the Royal Institution had brought him various flirtations, as evidenced by the many Valentine poems addressed to him there.[2] But as he wrote to his mother, until he met Jane Apreece he had never seriously considered marriage, and had felt that a scientific career was not compatible with a wife and family. His true bride was Science. However, perhaps his notions of scientific celibacy were changing.

Jane was a romantic figure. She was the daughter of Charles Kerr of Kelso, who had made a fortune in Antigua, and left her a considerable inheritance. She may also have had some West Indian blood in her veins. There was certainly something tropical in her temperament. Her first marriage, at nineteen, to a much older man, a decrepit Welsh baronet languidly named Shuckburgh Ashby Apreece, had been unhappy and childless. Its best aspect, said Jane, was that he had frequently taken her abroad. In Geneva she made friends with Madame de Staël, and later claimed to be the original of the heroine of de Staël's sensational romance *Corinne* (1807), about a lonely woman who finds love in southern climes.

She had other literary connections. She knew Sydney Smith and the waspish novelist Horace Walpole. In London, she once dined with William Blake. She was a subscriber to Coleridge's philosophical magazine *The Friend*. Walter Scott was a distant cousin, and a close friend. In the summer of 1810 they had toured the Highlands and the Hebrides together, and he observed that she was headstrong, inquisitive and not frightened by storms. They got on well, teasing each other as cousins should, but Scott was clearly a little in awe of her. He wrote in his journal that he thought her 'more French than English, and partaking of the Creole vivacity and suppleness'. It is not quite clear what he meant by this last compliment, perhaps that Jane was volatile and sexually provocative. She certainly had social ambitions: 'as a lion-catcher, I would pit her against the world. She flung her lasso over Byron himself.'[3]

But Jane was also clever and independent-minded. After she was widowed in 1809, she established an intellectual salon in Heriot Row, and cut a swathe through the Scottish academics. She was particularly drawn to scientific men. The mathematician Professor John Playfair, who had superbly interpreted Hutton's geology to the world, was said to have once knelt submissively in Princes Street to resolve the complicated stratifications of her laced boots. The wit Sydney Smith – who was also lecturing at the Royal Institution – was enchanted by her, and throughout his life retold endless suggestive anecdotes about her encounters. Everyone agreed that beneath a certain flamboyance and affectation, Jane had 'an excellent heart'.

Evidently Jane Apreece was a vivid personality, and someone who attracted gossip all her life. Yet her story is not well-documented, compared to Davy's, and it is strange for such a beauty that no portrait exists in a public collection.[4] Little of her early correspondence is known to have survived, though it is remarkable that she eventually kept more than ninety of Davy's letters to her.[5] He began writing them in

August 1811, after they had been briefly together in a sailing party on the Wye. He was still fishing at Denham and preparing his autumn lectures, while she had returned to Scotland. From the start his letters were a dreamy combination of science and sensibility: 'The clear and rapid Colne which moves over a green bed, living with beautiful aquatic plants the flowers of which glisten on its surface, is immediately beneath the window at which I am writing . . . I have scarcely a wish beyond the present moment except that I might see you as the Naiad of this stream, but you are now a mountain Nymph & scorn our low and quiet pastoral scenery.'[6] Receiving letters from her gave him 'a higher sensation than even exhilatory gas. I may be permitted a chemical allusion as we are both now pursuing the same science.'[7]

Rather surprisingly, Davy consulted his old flame Anna Beddoes about Jane Apreece. Anna had met Jane socially through the Edgeworth family in Ireland, and Davy innocently passed on her barbed compliments. 'Mrs Beddoes says "I do admire Mrs Apreece, I think her very pleasing, feel her abilities and almost believe if I knew her I should love her – more I suppose than she should love me." '[8]

That autumn, Jane left Edinburgh and moved to London, taking up residence in an elegant house at 16 Berkeley Square, strategically placed within ten minutes' walk of the Royal Institution.[9] Davy began sending her books – Izaak Walton's *Compleat Angler*, of course, but also Anacreon and other classical love poets. Then came copies of his Chemical Lectures 'decyphered' into plain English; and then – his own sonnets. She in turn began to attend his autumn lectures, announced that she was 'of the true faith of the genuine Angler', and gallantly set herself to undertake a private course of 'chemical studies'.

It was now Jane's turn to send verses to Davy, though these have not survived. He responded gravely: 'Your mind is "of poetical frame" for there is no mind in which so much feeling is blended with so much thought.' The man who had once seen the poems of Southey and Wordsworth through the press risked the mild criticism that perhaps her verses were a little artificial by Romantic standards: 'You want only the habit of connecting pictures from natural imagery with moods of human passion to become a genuine poet.'[10] Jane took this well.

Throughout that autumn Davy assiduously introduced Jane to the lions of the scientific world. She was escorted to his lectures by the distinguished chemist Professor Charles Hatchett ('we shall both be proud to be in your train'), and dined in a party with William Herschel, when they discussed the distance of the furthest stars.[11] He was also able to introduce her to Robert Southey, and share literary gossip about the quarrel between Coleridge and Wordsworth.

Davy was now less intimate with Coleridge. In March 1809 Coleridge had come near to quarrelling with him, because Davy would not let the Royal Institution support his scheme to publish *The Friend*: 'Davy's conduct *wounds* me.' Coleridge felt fame had gone to Davy's head, and that his high-handed (or perhaps prudent) behaviour betrayed their friendship, even though they had been 'intimate these nine years or more'. He claimed he had written a long poem – 'the only verses I have made for years' – praising Davy's 'Genius and great Services to mankind'. But now he had no thoughts of publishing it: Davy was too caught up in his own fame, exactly as Coleridge had once prophesied.[12]

But love, not fame, was on Davy's mind. By 1 November he was writing to Jane with increasing intensity, the romantic fisherman now replaced by the romantic man of science. 'There is a law of sensation which may be called the law of continuity & contrast of which you may read in Darwin's *Zoomania* [sic]. An example is – look long on a spot of pink, & close your eyes, the impression will continue for some time & will then be succeeded by a green light. For some days after I quitted you I had the pink light in my eyes & the rosy feelings in my heart, but now the green hue & feelings – not of jealousy – but of regret are come.'[13]

When Davy left for his December lectures in Dublin, absence only deepened his feelings. His lectures were heaped with praise, he was awarded an honorary doctorate by Trinity College, and was 'overpowered' by admirers at receptions and banquets. Yet despite all this, he could think of nothing but Jane Apreece. His courtship became more open and direct. Amidst his public triumph, Davy secretly gave way to the language of love. On 4 December 1811 he wrote from his rooms at the Dublin Society: 'I have the power of dreaming and picture-making as strong as when I was fifteen. I call up the green woods and the gleams of sunshine darting through them, and the upland meadows where we took our long walk. I seem to hear, as then, the delightful sound of the nightingale interrupted by the more delightful sound of your voice. You will perhaps laugh at this visionary mood, and call it romance; but without such feelings life would be of little worth ... Without this, its tones are like those of the Aeolian harp, broken, wild, and uncertain, fickle as the wind that produced them, beginning without order, ending without effect ... To see you is the strongest wish of my heart.'[14] The imagery of several of Coleridge's poems rises opportunely through these letters.

On his return to the Royal Institution, Davy set himself to storm her with more scientific seductions. 'You are my magnet (though you differ from a magnet in having no repulsive points) and direct my course.' By March 1812 he was writing: 'I no longer live for anything but you ... Your felicity will be the pole star of my future course.'[15] But he was intimidated by her aristocratic friends, and perhaps by her money. He may secretly have feared that Jane's sparkling wit and love of socialising might interfere with the necessary routine and self-concentration of his laboratory work. He continued gallantly to insist that they would not; and what is more, so did she.

Jane, in turn, admired Davy's brilliance, his handsome boyish figure, and the intellectual glamour that attached to him as celebrated lecturer. She had many other suitors at this point in her life, but none so intense or determined – or so serious. Perhaps that may have been a problem. In private she may have laughed at Davy's didactic and over-earnest moments, the lecturer overcoming the suitor, as was sometimes revealed in the solemn *longueurs* of his love letters: 'Your moral virtues always improved me & exalted my ideas of human nature.' Jane was not impressed by her own moral virtues.

When she once teased him with being absurdly romantic about her, he was incapable of wittily turning the shaft, as Sydney Smith would certainly have done. Instead of a seductive epigram, he delivered a solemn oration. 'If this be romantic, it is romantic to pursue one's object in science; to attach the feelings strongly to any ideas; it is romantic to love the good, to admire the wise, to quit low and mean things and seek excellence.'[16]

Jane may also have been worried by his Cornish background, though in a way she was a social adventurer just like him. She shrewdly suspected that her only real rival was chemistry. Davy himself once unguardedly admitted that 'the pleasure I derived from your conversation interfered with my scientific pursuits', though hastily adding: 'I have gained much and lost but little.'[17]

On this score, both sets of friends predicted disaster. She was made for society, he was made for the laboratory. Sydney Smith, now clearly jealous, cattily used chemical imagery to beg Jane to reject Davy out of hand. On 29 December he wrote: 'Pray remain single and marry nobody... you will be annihilated the moment you do, and instead of being an exciting alkali or acid, become a neutral salt. You may very likely be happier yourself, but you will be lost to your male friends.'[18]

So Jane Apreece prevaricated in a way that Jane Austen (just writing *Pride and Prejudice*) would have approved of. She twice refused Davy's offers of marriage, took to her bed in Berkeley Square and announced she was ill and incommunicado. But she was astonished by the tender and unguarded declaration this released from Davy: 'For the first time in my life I have wished to be a woman that I might watch by your bedside; I might wish that I had not given up the early pursuit of medicine for then I might have been admitted as a Physician. Though an untoward Beau, you would find no more devoted Nurse.'[19] In the passionate declarations that followed, it seems that each was able to reassure the other. Davy agreed to the momentous step of giving up full-time lecturing at the Royal Institution (a thing he had secretly wanted to do for some time), while Jane assured him that her fortune would allow them to travel, while he continued his chemical researches independently. This was a tantalising prospect for both of them.

Davy had one further scientific seduction to offer. He confided to Jane that the Prince Regent was about to confer a knighthood upon him, for services to chemistry, in the forthcoming Birthday Honours. It would be the first scientific knighthood of the Regency, indeed the first since Sir Isaac Newton. She need no longer feel ashamed of him at the dinner tables of Mayfair. At the third time of asking, Davy's proposal of marriage to Jane Apreece was at last accepted. He reacted with genuine rapture. 'I have passed a night sleepless from excess of happiness. It seems to me as if I began to live only a few hours ago ... The great future object of my life will be your happiness ... My happiness will be entirely in your will.'[20]

Congratulations were now in order, and Sir Joseph Banks was pleased and rather amused that one of his young scientific protégés had made such a fine – and wealthy – match: 'She has fallen in love with Science and marries him in order to obtain a footing in the Academic Groves ... It will give to Science a new kind of *eclat*; we want nothing so much as the countenance of the ladies to increase our popularity.'[21] Banks evidently teased the bridegroom in a worldly way. 'Davy is on the point of being married to the gentle Dame Apreece who has at least £4,000 a year, the half of it her own. He swears he will never desert Science. I tell him she will bring him into Parliament and make a fool of him. We shall see how this matter will end.'

Davy had no such political ambitions, and believed that Jane fully accepted his passionate commitment to science. In fact Banks was quite handsome about Jane, and saw her as a valuable, if probably bossy, addition to the social life of the Royal Society: 'If she is satisfied with being installed as the Queen of Literature we shall all be ready to put ourselves under her Dominion, and I think she is Quick and Clever enough to reign over us and keep us all in very Good Order.' Coming from Banks, who had no high opinion of bluestockings, this was handsome indeed.[22]

Davy wrote delightedly to his mother in Penzance, proudly and rather grandly announcing the news. He said that but for Jane, he had never expected to marry. His letter to his younger brother John of March 1812 has a touching nobility and simplicity.

> My dear John,
>
> Many thanks for your last letter. I have been very miserable. The lady whom I love best of any human being, has been very ill. She is now well and I am happy.
>
> Mrs Apreece has consented to marry me, and when this event takes place, I shall not envy kings, princes, or potentates.
>
> Do not fall in love. It is very dangerous! …
>
> Ever most affectionately yours
>
> H. Davy.[23]

The romantic dénouement was carefully orchestrated. In April Davy delivered what was agreed would be his final lectures at the Royal Institution. He was awarded an emeritus professorship, and granted the continuing use of all research facilities. On 8 April he was knighted by the Prince Regent, and three days later, on 11 April, he married Mrs Apreece, who thereupon became Lady Davy.

Among much expensive jewellery, he gave her a symbolic wedding present. He had collected a decade of his Royal Institution lectures, and now edited and assembled them as his *Elements of Chemical Philosophy*. This book was the cornerstone of his early scientific career, and proclaimed the progressive value of science and its power to 'investigate and master' nature. On 1 June it was published with a formal dedication to Lady Jane Davy.[24]

Chemical Philosophy was too technical to achieve a wide general readership, but it contained a powerful historical 'Introduction', which placed chemistry at the forefront of all contemporary scientific research. By contrast, his more popular *Agricultural Chemistry*, published simultaneously, ran to many editions over the next decade. With these two publications, for which he was paid 1,000 guineas (a sum which compares well with what Walter Scott received for his poems), Davy had made chemistry as popular as astronomy.[25] He himself seemed to symbolise the hopes and ambitions of Romantic science to produce a better world for all mankind. Among others, the young poet Percy Shelley began to incorporate Davy's ideas into his work, beginning with his visionary materialist poem *Queen Mab* of 1812, with its long scientific prose notes.[26]

Shelley's book order for 29 July 1812, when he was beginning the poem at Lynmouth in Devon, included Mary Wollstonecraft's *Vindication of the Rights of Woman*, David Hartley's *Observations on Man* and Davy's *Elements of Chemical Philosophy*: a characteristic mixture of radical politics, sceptical philosophy and the new science. The vogue for attaching explanatory prose notes, both historical and scientific, to epic poems had been popularised by Erasmus Darwin in *The Botanic Garden*, taken up by Southey in *Thalaba* (which Davy had edited for the press), and then admiringly imitated by the twenty-year-old Shelley in *Queen Mab*. Underlying it was the formal problem of how far scientific data could any longer be convincingly expressed in poetry (as Lucretius had done). De Quincey would later suggest that they had to be separated as the 'Literature of Knowledge' and the 'Literature of Power'. Shelley's *Prometheus Unbound* (1820) was to be arguably the last successful attempt to combine the two in a major English poem.[27]

Davy's lectures in *Chemical Philosophy* opened with a brilliant short survey of the entire field: 'An Historical View of the Progress of Chemistry'.[28] Starting with the early Egyptians and the Greeks, continuing with the 'delusions' of medieval alchemists and the revolutionary discoveries of the early Royal Society in seventeenth-century London, he then celebrated the astonishing advance of Enlightenment chemistry throughout Europe over the past thirty years, culminating in the work of Priestley, Cavendish and Lavoisier, and his own discoveries in electrochemistry, which he emphasised were being rapidly developed by many other chemists on the Continent. Davy was notably generous to the

French – Lavoisier, Berthollet and Gay-Lussac – and the Scandinavians; but he drew a convincing picture of an entire scientific community of minds at work across Europe. He made numerous unexpected asides: the importance of Arabic chemistry, the witty suggestion that Cleopatra might have been an 'experimental' chemist with her love potions, the crucial importance of 'new instruments' (such as the voltaic battery) in taking research forward, and the paradoxical fact that Newton's genius in many ways hindered chemistry by turning attention to 'optics, mechanics, and astronomy'.[29]

Most striking is Davy's power to engage the reader in a direct, non-technical way. The essay opens with poetic simplicity: 'The gradual and almost imperceptible decay of the leaves and branches of a fallen tree exposed to the atmosphere, and the rapid combustion of wood in our fires, are both chemical operations. The object of Chemical Philosophy is to ascertain the causes of all phenomena of this kind, and to discover the laws by which they are governed. The ends of this branch of knowledge are the applications of natural substances to new uses, for increasing the comforts and enjoyments of man; and for the demonstrating of the order, harmony, and intelligent design of the system of the earth.'[30]

From this time chemistry joined astronomy and botany as the most popular and accessible forms of modern science for amateurs, and as a new doorway into the 'intelligent design' of the universe. It was a sign of the times that 'Portable Chemical Chests' began to go on sale in Piccadilly, priced between six and twenty guineas.[31] Davy would later emphasise how few pieces of equipment an experimental chemist needed.[32] Chemistry guides and primers, besides Jane Marcet's, began to be widely available. Coleridge reflected in one of his notebooks of this time: 'Whoever attended a first Course of Chemical Lectures, or read for the first time a Compendium of modern Chemistry (Lavoisier, Parkinson, Thomson, or Brande) without experiencing, even as a *sensation*, a sudden *enlargement & emancipation* of his Intellect, when the conviction first flashed upon him that the Flame of the Gaslight, and the River Water were the very same things (= elements) and different only as AB *uniting* with B, and as AB united?'[33]

While annotating the visionary works of the German mystic Jakob Boehme, Coleridge added a further aside on the clarifying intellectual impact of the scientific approach: 'Humphry Davy in his laboratory is probably doing more for the Science of Mind, than all the Metaphysicians have done from Aristotle to Hartley, inclusive.'[34] Later he would fear, wrongly, that Davy was becoming a 'mere Atomist', but his recognition of the significance of Davy's 'chemical revolution' and the 'dynamic' vision of nature that it revealed never faltered, despite their personal estrangement.[35]

In many ways spring 1812 was the climax of Davy's early career. The unknown boy from Penzance had achieved a European reputation in science, an emeritus professorship and a knighthood, and a glamorous society marriage. Yet he was still only thirty-three. A formal visit to Cornwall by the new Lady Davy was promised at this moment of celebration, but never in fact materialised. It seems that Davy was still embarrassed by his humble roots.

A rather unusual honeymoon followed in Scotland that summer. They were joined by Davy's younger brother John, now twenty-two years old, the only member of her husband's family that Jane ever met. He was fresh from medical studies in Edinburgh, and very much on his best behaviour. Jane approved. They embarked on a tour of misty lochs and highland castles. Davy frequently leapt from the open carriage and abruptly disappeared with his fishing tackle to explore a river, while John was left to manage the horses and amuse Jane. This arrangement worked well, and there was much hilarity, some at Davy's expense. Jane would recall it regretfully the following year, while touring in different circumstances.

The official holiday business was salmon-fishing and grouse-shooting. But the Davys were a celebrity party, the most eminent scientific family in the land, and they stayed in succession with the Marquis of Stafford, the Duke of Gordon, the Duke of Atholl and Lord Mansfield. This was delightful and natural for Jane, but also appealed to Davy, who had a growing relish for aristocratic company. Their triumphal progress was finally cut short in September by Jane twisting her foot, and Davy achieving his ambition of shooting a stag.[36]

They were much in love, but there was no talk of children, and no visit to Davy's beloved childhood haunts in Cornwall. Instead, he wrote to his mother praising John's behaviour during the trip, and generously offering to pay his student allowance of £60 per annum. Perhaps on Jane's promptings, he was also tactful towards his younger brother: 'Lest it should injure John's feelings of independence, it may appear to come from you.' As to Jane: 'She desires her kind remembrances to all my family. We are as happy and well-suited as I believe it is possible for people to be; we have nothing to regret in our past lives, and everything to hope for.'[37]

Perhaps there was something wistful in that last phrase. Yet all was well, and the prospects for the glamorous young couple were glowing. Davy was soon happily back at work in the laboratory, and a further tour – perhaps even to Jane's beloved Italy – was planned for the following spring.

2

Davy now began his life as an independent man of science. His first project, with the blessing of the Royal Society, was a patriotic one: he intended to contribute to the British war effort. He began to investigate explosives, using a formula communicated to him by the French physicist André Ampère. This was not at the Royal Institution, but at a secret commercial manufactory at Tunbridge in Kent. The plan was to produce improved high explosives for the Royal Engineers. They were to be used against Napoleon's troops in Spain, for the mining of besieged cities and the blowing up of fortifications during the Peninsular War. The scheme had the unofficial encouragement of Joseph Banks, although it was dangerous work. Ampère warned Davy that one French chemist had lost an eye and a finger.

In November 1812 Davy was nearly blinded by a test tube explosion while mixing chlorine and ammonium nitrate. Slivers of glass punctured his cornea, and cut his cheek. He informed Banks at the Royal Society that he had discovered a substance so powerful that a quantity as small as 'a grain of mustard seed' had caused the damage. He did not remark on the more sinister slant this gave to the principle of scientific progress, or on the paradox that French science was being turned against the French. He tried to hide the seriousness of the 'slight accident' from Jane, telling her on 2 November, 'there is no evil without a good, always excepting the toothache'. But his vision was seriously impaired for many weeks, and he sought for an amanuensis to help him write up his report for the Royal Society.[38]

Some odd rumours went around about this accident. William Ward, the future Lord Dudley and a prize London gossip, wrote speculatively to a friend in December: 'I have been to see Sir Humphry Davy Kt, who has hurt one of his eyes. Some say it happened when he was composing a new fulminating oil, and this I presume is the story that the Royal Society and the Institut Imperial [in Paris] are expected to believe; others that it was occasioned by the blowing up of one of his own powder mills at Tunbridge; others again that Lady D scratched it in a moment of jealousy – and this account is chiefly credited in domestic circles.'[39]

But other gossips gave an idyllic impression of the couple. Henry Crabb Robinson, garrulous friend of the Lake Poets and one-time foreign correspondent for *The Times*, came across them at a London literary dinner given for Wordsworth a few months later, in May 1813. He noted in his diary: 'Sir Humphry and Lady Davy there. She and Sir H seem hardly to have finished their honeymoon. Miss Joanna Baillie [the Scottish playwright] said to Wordsworth, "We have witnessed a picturesque happiness!" '[40] Not the least picturesque thing was the way in which the Davys effortlessly united the worlds of literature, science and high society.

When Davy had returned to London, half-blinded in a patriotic cause, he urgently sought help to continue his experiments. In his absence he found an increasing chaos had overtaken the Royal Institution laboratory. The most basic materials were neglected – pens, ink, towels, soap, the servicing of the huge voltaic battery. 'The laboratory is constantly in a state of dirt and confusion ... I am now writing with a pen and ink such as was never used in any other place.'[41] Although no longer officially on the staff, Davy peremptorily dismissed the drunken laboratory assistant William Payne, and began to search for a replacement. On 1 March he interviewed a young bookbinder for the post of Chemical Assistant at the Royal Institution. The young man's father had been a London blacksmith. His chief recommendations were punctuality, neatness and sobriety. His name was Michael Faraday, aged twenty-one.

Faraday had read Jane Marcet's *Conversations in Chemistry, Mainly intended for Young Females*, which particularly singled out Davy's contribution. Her book was a new kind of science popularisation, aimed at opening 'young minds' to scientific methods and the wonders of the natural universe.[42] The first edition (1806) had cautiously recounted Davy's experiments with 'nitrous oxide or exhilarating gas' ('some people become violent, even outrageous ... I would not run any risk of that kind'). In the new edition (1811) Marcet gave a heroic assessment of Davy's

Bakerian Lectures. 'In the course of two years, by the unparalleled exertions of a single individual, chemical science has assumed a new aspect. Bodies have been brought to light which the human eye never before beheld, and which might have remained eternally concealed under their impenetrable disguise.'[43]

Thus inspired, Faraday had begun attending Davy's lectures in 1812, having been given free tickets. He had taken detailed notes, immaculately written out and illustrated in his neat hand. He then bound them in his spare time at the bookbindery where he worked off Oxford Street. When interviewed by Davy, he submitted the bound book as his *curriculum vitae* and proof of his dedication. He was given lodgings in the attic of the Royal Institution, coal and candles and an evening meal, and a tiny salary of twenty-five shillings a week. Davy described him as 'active and cheerful, and his manner intelligent'.[44]

His new employer immediately departed on a recuperative fishing expedition, writing to Jane from Launceston in Cornwall on 9 March 1813: 'The weather my dear love has been delightful. We have traced one of the Devonshire streams to its rocky source (the Dart). It is a river clear, blue and bright.'[45] If he used this expedition to make a swift visit to his mother at Penzance, he did not mention it.

The following month Davy went off on another fishing trip, this time to Hampshire, where the sight of flies dancing above the river in the evening light, and fish sporting on the surface of the water, was simply 'irresistible'. Writing to Jane from Whitechurch on 14 April, he tried to construct a humorous mythology of his escapes: 'I flirt with the water nymphs, but you are my constant goddess. I make you the personification of the spirit of the woods, and the waters, and the hills, and the clouds ... This is the earliest form of religion.' He slightly jeopardised this sylvan vision, by then competitively informing his wife in a postscript that he had caught five trout, while his friends had only caught one between them all.[46] He wrote again more reassuringly the following day. 'I breathe a sigh upon my paper from the thought of being apart from you for only two days. My dear, dear Love creates a void which no interest or amusement can fill ... The longer I live, the more I shall love you, my dearest Jane.'[47]

For all his protestations, a pattern in Davy's flights to the riverbank was becoming clearer. Jane was often 'indisposed' in London, or else embarked on a vigorous round of tea parties and receptions. Davy, exhausted by laboratory work, became 'irritable' and sought the open air. As he wrote from the banks of the Avon near Fordingbridge, he was soothed by the simple fact of 'the moving water and changing sky'.

Sometimes he hoped Jane would join him. On the riverbank 'one learns as it were to become a part of nature; the world and its cares & business are forgotten, all passions are laid asleep ... We live a life of simplicity and innocence according to the primary laws of nature, losing all trifling and uneasy thoughts, keeping only what constitutes the vitality of our being, the noble affections. Of mine, you know the highest and constant object.'[48] But, understandably, Jane may have had her reservations about the 'primary laws of nature', for they could also mean the sporty masculine life of competitive fly-fishing, interminable tall stories and riverside taverns.

That spring of 1813 in London, Jane in turn tried introducing Davy to a new sport of fashionable lionising. Young Lord Byron had recently returned from the Near East and published the first two cantos of *Childe Harold* (1812) and *The Bride of Abydos* (1813). He was taking the literary salons and ladies by storm, and proved an early capture. Jane wrote airily to a friend back in Edinburgh: 'Lord Byron is still here, but talking of Greece with the feelings of a poet and the intentions of a Wanderer. He is to have a quiet breakfast here with the intention of an Introduction to Miss Edgeworth ... I expect the sense of one, the imagination of the other, with the genius of my own Treasure [Davy] to afford a high intellectual banquet.'[49] Unexpectedly, Byron and Davy hit it off rather well, and later when His Lordship was self-exiled in Italy, Davy remained one of the few Englishmen that he could stand to meet. He would even put him in his poem *Don Juan* (1819–24).

Davy continued to compensate for London socialising with solitary fishing, sometimes slipping as far away as his beloved Cornwall. On 15 April 1813 he wrote Jane a long, tender letter from Bodmin, describing himself alone beside a remote river at sunset: 'About the time you were sitting down to dinner I was standing in the midst of a secluded valley upon a bridge above the fork of the [river] Allan, watching the last purple of the sky dying away upon the rapid water, and by its decay making visible a bright star.' There is perhaps an undertone of reproach in this, yet also a kind of romantic understanding. As often as he could, Davy sent her propitiatory fish wrapped in ice on the overnight mail coach to London: gleaming trout and tender young grayling.[50]

3

In the north of England, far from the London salons, quite other events were unfolding. On 24 May 1812 the great Felling colliery mining disaster had shaken the population of Sunderland. Every miner in the coalpit, all ninety-two of them, was killed under horrific circumstances: some mutilated, some 'scorched dry like mummies', and some blown headless out of the mineshaft 'like bird-shot'. An underground fire raged for many days, and it took more than six weeks before the bodies could be recovered.[51] Hitherto Felling had been a model pit, with a clean accident record. The disaster shook the whole mining community of the North-East. Ever deeper mineshafts were bringing increasing fatalities, and it was calculated that over 300 miners had been killed in the past five years, almost all by explosion of 'fire-damp'. This was a lethal gas released by newly opened coal seams. It was believed to be some form of hydrogen, which when mixed with air could be ignited by a single miner's candle flame.*

A Safety Committee led by the Duke of Northumberland and the Bishop of Durham was formed to find a practical solution. Campaigns were launched by the vicar of Jarrow, the Rev. John Hodgson, and by Dr Robert Gray, a future Bishop of Bristol. Mining experts put forward various ideas, including ventilation schemes and several prototype safety lamps, one manufactured by Dr William Clanny, a Sunderland physician, and another by a local mining engineer, George Stephenson. But none was considered sufficiently effective or reliable, and the Committee dithered. A second explosion shook the Felling colliery in December 1813, and a further twenty-two men died. Matters now became urgent. After several meetings it was decided that a professional scientific opinion should be sought at a national level, and an official approach by Dr Gray to Sir Humphry Davy in London was decided upon. But by the time a

* Detailed accounts of the Felling colliery disaster of 1812 can be found in the remarkable archives of the Durham Mining Museum, Northumberland, and on its website. These include the names and ages of every one of the ninety-two mine-workers killed, of whom it emerges that more than twenty were fourteen or younger – the youngest being eight years old. The names are collected under the heading 'IN MEMORIAM', and their places of burial are also given where known: a tribute to the lasting loyalties and strength of feeling among the mining communities to this day.

letter was sent to the Royal Institution in the winter of 1813, Davy and Jane were already on the Continent.[52]

On 13 October 1813 Sir Humphry and Lady Davy (very conscious of both their rank and their nationality) embarked in their own carriage on an eighteen-month Continental tour. With them went the young Michael Faraday as their travelling companion: awkward, socially naïve, but very anxious to please. He belonged to a rare sect of Biblical fundamentalists, the Sandemanians, puritan and unworldly in outlook, though with a strong sense of public duty and service. Davy treated him easily, as he had done in the laboratory: as a scientific assistant and promising young protégé. Jane, with her happy memories of travelling with John Davy in Scotland, was less patient. Faraday had never travelled outside London, spoke no French or Italian, was shy and uncouth, and probably embarrassed by Jane's high style and evident sexuality.

She in turn may also have found Faraday physically awkward, and even irritating. He was small and stocky – not more than five foot four – with a large head that always seemed slightly too big for his body. His broad, open face was surrounded by an unruly mass of curling hair parted rather punctiliously in the middle (a style he never abandoned). His large, dark, wide-apart eyes gave him a curious air of animal innocence. He spoke all his life with a flat London accent (no match for Jane's elegant Morningside), and had difficulty in pronouncing his 'r's, so that as he himself said, he was always destined to introduce himself as 'Michael *Fawaday*'. In fact none of this prevented him from eventually becoming one of the greatest public lecturers of his generation. But it evidently did not appeal to Jane.[53]

There was no meeting of minds, and Jane simply started to treat Faraday as a valet. She insisted that he travel on the outside of the coach with the luggage and her husband's chemical equipment. A difficult trip followed, despite Davy's pleasing lionisation by French and Italian scientists. Perhaps as a technique of personal survival, Faraday immediately started to keep a daily journal of his adventures (which has survived), and began an extensive and surprisingly humorous correspondence with his friend Benjamin Abbott of the City Philosophical Society back in London.[54]

On 2 November Davy received the Prix Napoléon (worth 6,000 livres) from the Institut de France in Paris. He knew that accepting the award might be unpopular in wartime England, but followed Banks's line at the Royal Society that science should be above national conflicts. He told

Tom Poole: 'Some people say I ought not to accept this prize; and there have been foolish paragraphs in the papers to that effect; but if the two countries or governments are at war, the men of science are not. That would, indeed, be a civil war of the worst description: we should rather, through the instrumentality of men of science, soften the asperities of national hostility.'[55]

This attitude was unpopular at home, and *The Times* attacked Davy's journey as unpatriotic in a time of war. Even the liberal-minded Leigh Hunt wrote a long editorial in the *Examiner* for 24 October 1813, defending the international dignity of science, but also criticising Davy for indulging in 'paltry vanity' among French admirers in Paris. Hunt wittily imagined his triumphal progress down the Parisian boulevards: '*Ah, there is the grande philosophe, Davie!*' – '*See here the interesting Chevalier Humphrey!*'[56]

In fact Davy carefully avoided an audience with Napoleon himself, and referred to him contemptuously as 'the Corsican robber'.[57] Jane refused to adopt Parisian fashions, and was once jeered at by a crowd in the Tuileries for her small English hat. They were both appalled by all the looted works of art in the Louvre (then renamed the Musée Napoléon), and pretended to admire only 'the splendid picture-frames'. But Davy was deeply impressed by the Jardin des Plantes and the Bibliothèque Nationale, fully aware that there was still no equivalent in London.

He was warmly received by Cuvier, Ampère and Berthollet, but got into an awkward priority dispute with the gifted young chemist Joseph Gay-Lussac. Gay-Lussac, Davy's exact contemporary, had made a popular name in France with his intrepid ballooning exploits, and had been hard on Davy's heels with potassium and sodium experiments. Both were now given by the Académie des Sciences a newly isolated substance to analyse: a strange violet crystal recently found as a byproduct of gunpowder manufacture. The competitive nature of this gesture was unmistakeable. Davy had only his small trunk of portable chemical apparatus to work with, but accepted the challenge with alacrity. He delayed his departure from Paris for a month, closeting himself with Faraday, filling his hotel rooms with acrid fumes and 'very bright greenest yellow' gas, much to Faraday's delight, Jane's irritation and the management's alarm.[58]

The rival analytic papers were submitted to the Académie almost simultaneously in December. Both identified the crystals as a new element, which could also be extracted from seaweed. Gay-Lussac's short paper was actually presented and published first, on 12 December. Davy, taken by surprise, presented his to the Académie on 13 December, but unblushingly antedated it to 11 December, and had it published as such in the *Journal de Physique.* He claimed, perhaps justly, that he had previously shared his key ideas with Gay-Lussac.[59] He also wrote to Banks claiming the entire discovery for himself, and had a full account swiftly published by the Royal Society. In consequence his analysis and naming of the unknown substance as 'iodine' was generally accepted, although the priority remains disputed by the French to this day.

Davy's surprising sharpness in such a 'priority' controversy was noticed by Faraday, but writing to Abbott he loyally dismissed it as patriotism rather than ambition: 'Sir H has not been idle in experimental chemistry... his example did great things in urging the Parisian chemists to exertion... He first showed that [iodine] was a simple body. He combined it with chlorine and hydrogen, and latterly with oxygen, and thus has added three acids of a new species to the science... It confirms all Sir H's former opinions and statements, and shows the inaccuracy of the labours of the French chemists on the same subject.'[60]

In early spring 1814 Davy's party travelled south to the Pyrenees, examining extinct volcanoes on the way, and making a leisurely cultural detour to visit Avignon, the Pont du Gard, Nîmes and Montpellier, with its strong tradition of Vitalist thought. Here Davy was an honoured guest of the ancient university for a month, continued his experiments on iodine, and wrote a number of descriptive poems about the South: the foaming rivers of Vaucluse, the shifting lights on Mont Canigou in the Pyrenees, the classic 'Mediterranean Pine' at Montpellier, and the ghosts of the Roman engineers at the massive stone aqueduct of the Pont du Gard.

> Work of a mighty people, of a race
> Whose monuments, with those of Nature, last.
> The Roman mind in all its projects grasped
> Eternal Empire, looked to no decay,
> And worked for generations yet unborn,
> Hence was its power so lasting.[61] ♣

♣ Not the least fact that may have impressed Davy was the amazing scientific accuracy of the Roman engineering. In carrying the water by canal and six main aqueducts from Uzès to Nîmes, a distance of over fifty kilometres, they exploited the very small fall in

The party then doubled back along the coast, through what would soon become the English 'Riviera', and slipped over the Alps into Italy for the summer and autumn. Taking an open boat from Genoa, they were nearly swamped in a violent squall off Lerici: just eight years later Shelley would drown there. Faraday noted mischievously that Lady Davy fell blessedly silent as she gazed at the waves. In Milan they interviewed the ageing Volta, and discussed the expanding mystery of electricity: 'his view rather limited', thought Davy, 'but marking great ingenuity'.

In Florence, while the guest of the Grand Duke, Davy performed an impressive carbon-based experiment which proved that the most apparently precious of objects – the diamond – could also be the product of nature's simplest processes. With the Duke's permission, he commandeered the huge solar magnifying lens at the Florentine Cabinet of Natural History, and subjected an uncut diamond to intense and continuous heat. The diamond eventually burst into flame, leaving a fine crust of black carbon, thereby proving, against all the commonsense evidence, that the clear, hard, glittering crystal was really little more than a lump of coal. Both were varieties of carbon, laid down by nature over millennia.

In was perhaps reflections on this kind of mystery that led Davy to write a long, self-questioning passage in his journal about the limits of contemporary scientific research. Significantly, he picked out three leading disciplines – astronomy, chemistry, and geology. This entry of March 1814 is surrounded by sketches of birds diving into water.

> Our artificial Science has relation to the *forms of Nature*; but yet that which is most important in Nature – *Life* – is above our Science. The Astronomer vainly asserts the perfection of his Science because he is capable of determining the motions of 7 planets and 22 satellites; but comets & meteors which even move in our system are above his reach, & even this solar system is a speck in the immensity of space & suns and worlds are beyond our reach.

---

land levels – required to make the water flow smoothly southwards – by consistently achieving gradients of between ten and twenty centimetres over one kilometre: a fantastic feat of both measurement and construction. The canal successfully delivered 50,000 cubic metres of fresh water to Nîmes every day for 300 years. Though the canal was built in less than a generation under the Emperor Augustus, and renowned throughout Europe, the names of the individual Roman engineers were by Davy's time unknown. This too must have struck him in his reflective mood.

> Our Science [chemistry] refers to the globe only, & in this there is an endless field for investigation: the interior is unknown; the causes of Volcanoes. We have just learnt some truths with respect to the surface: but there is an immensity beneath us. – Geology in every sense of the word is a superficial science.[62]

This expressed in private, perhaps as a result of conversations with young Faraday, a scepticism about scientific knowledge that was very different from the confident assertions of Davy's *Chemical Philosophy*. But it also prophesied the very developments being pioneered by Herschel in sidereal astronomy, and by Hutton in deep-time geology.

They visited Rome, where Davy was enchanted by the Coliseum by moonlight, and Naples, where they climbed a sinister and smoking Vesuvius. In Venice they gingerly sampled the gondolas. All these places would eventually reappear, strangely transformed, in Davy's last book, *Consolations in Travel*. During the summer heat they pushed back over the Alps into Switzerland, Bavaria and the Austrian mountains, where Davy found remote, fast-flowing rivers for trout- and salmon-fishing, to which he vowed to return. His most favoured region, in the Balkan peninsula, rejoiced in the magical name of Illyria. All the time Davy and Faraday were working on chemical experiments: iodine, chlorine, dyes, gases, and the electricity produced by the 'torpedo' or conger eel. Was the eel nature's voltaic battery, and did it hold a clue to Vitalism? – a question which would come to haunt Davy.

In autumn 1814 they travelled slowly south again into Italy, planning to winter in Rome. Riding through Florence in October, they heard of some strange natural gases escaping from rock formations in the Apennines at Pietra Mala, near Lucca. They rode over to investigate just as the autumn storms (later to be celebrated by Shelley in 'Ode to the West Wind') broke. Faraday wrote an exuberant account of this field expedition, undertaken for several hours in the pouring rain, while Lady Davy sat back patiently in the carriage. Davy forgot about everything when they discovered that the mysterious gas could be stirred up from mud with a stick, and could be ignited even in the heavy rain, burning beautifully with a 'very pale' blue flame, like methylated spirits or 'the flame of spirits of wine'.

They stood in the downpour, gazing at the weird blue pool of fire shimmering at their feet. Faraday watched Davy's quick investigative mind observing, eliminating, racing ahead: an unusually cool flame (not like burning oxygen); almost no smell (unlike hydrogen); not at all like volcanic gas (as everyone assumed it was); possibly not from a local source, but percolating from 'a great distance' beneath the earth; possibly therefore 'originating from a mine of fossil charcoal'. And then the careful inductive *caveat*: 'but everything is conjecture & it still remains a source of investigation'.[63]

They succeeded in bottling it, and took it to Florence, where they were again the guests of the Grand Duke. After dining with the Duke (much to Lady Davy's satisfaction), Davy precipitately left the table and commandeered the Duke's laboratories. Here he succeeded in identifying the gas as fire-damp (or methane), very similar to the gas that endangered British coalmines – a prophetic discovery. The investigation was a model of the three-part inductive method he had expounded in *Elements of Chemical Philosophy*: observation, experiment, analogy.

Davy's notebooks for this period also suggest a new pattern of philosophical speculation, almost approaching German *Naturphilosophie*. Some of his observations would have been recognised by Coleridge: 'The aspirations for immortality are movements of the mind similar to those which a bird makes with its wings before they are furnished with feathers.'[64] Others were more closely tied to his laboratory work. On the subject of scientific 'analogy', for instance, he wondered if there was a wider principle at work. 'Probably there is an analogy in *all existence*: the divided tail of the fish is linked in a long succession of like objects with the biped man. In the *planetary system* it is probable man will be found connected with a higher intellectual nature; and it is possible that the *monad*, or soul, is constantly undergoing a series of progressions.'[65]

Davy would later come back to the question of man's progressing towards extraterrestrial intelligences in the future. He reflected too on the past: on the way it was continuously redrawn by a partial present. 'Our histories of past events are somewhat like the wrecks upon the sea-beach: things are often thrown up because they happen to be light, or because they have been entangled in sea-weed: i.e. facts are preserved which suit the temper or party of a particular historian.'[66]

Throughout, Faraday continued his series of long letters to his friend Benjamin Abbott in London, extolling the 'glorious opportunities' of the trip. 'The constant presence of Sir HD is a mine of inexhaustible knowledge.' Yet after a lively account of his adventures with 'Sir H', including climbing Vesuvius in the dark and gathering coal gas in the rain, he began to admit that the months of travelling with Lady Davy were less than happy. In January 1815 he wrote: 'I should have but little to complain of were I travelling with Sir Humphry alone, or were Lady Davy like him, but her temper makes it often times go wrong with me, with herself & with Sir H. She is haughty & proud to an excessive degree and delights in making her inferiors feel her power.' It is not known what Jane herself made of the young Faraday, a tongue-tied companion, hating the foreign food, uninterested in politics or architecture and still unable to speak French or Italian (though learning). He admitted himself that he was gauche and needed to become 'more acquainted with the manners of the world', and learn to 'laugh at her whims'. But perhaps it did not occur to him that Lady Davy might be jealous of his relationship with her husband.[67]

Another letter in late February threw slightly different light on the situation. Much of the difficulty arose from the fact that Sir H had refused to employ a valet throughout the trip, 'being accustomed from early years to do for himself'. Lady Davy had taken advantage of this (although she had her own maid). She liked to 'show her authority', and was 'extremely earnest in mortifying' Faraday with humiliating requests. Perhaps Jane might have described this more as 'teasing' the extremely earnest Faraday. However, after occasional confrontations and even 'quarrels' with her (which Faraday felt he consistently won) she now behaved 'in a milder manner'.[68] By the end of the trip they seem to have rubbed along reasonably well, and Faraday had given up a secret threat (made to Abbott) to abandon chemistry altogether, 'and return to my old Profession of Bookseller'. Davy seemed largely unaware of these domestic difficulties, trying to '*remain neuter*', as Faraday put it in a curious phrase. Perhaps this was not a good sign for the future.

In March 1815, news of Napoleon's escape from Elba cut short a Grand Tour that had already lasted for seventeen months. Davy had originally planned to take his party as far as Greece and Turkey, but perhaps it was a relief to hurry home. On 23 April 1815 he was back in London, making sure that Faraday (now twenty-three) was promoted to a permanent post as Assistant to the Laboratory at the Royal Institution on £75 per annum, and encouraging him to start giving his own chemical lectures at the City Philosophical Society. He also urged him to publish his first scientific papers. All this marked a distinct advance under Davy's patronage. Davy himself was appointed Vice-President of the Institution, which gave him access to the laboratory, and allowed him to continue acting as Faraday's unofficial tutor.

But as he established one protégé, he lost another. Davy's beloved brother John, now twenty-five, had decided to give up medical research and join the army as a military doctor. He would be stationed abroad, largely in the Mediterranean, for the next twenty years. Davy would spend much time trying to entice him back so they could work together again, but would only achieve this under dismaying circumstances. John had been Davy's closest confidant, and his link with Penzance and his happy past. Now that was increasingly lost.

With Jane, Davy bought a beautiful new house at 23 Lower Grosvenor Street, in the heart of Mayfair, and began dining regularly with Banks and other members of the Royal Society. The couple were much in demand socially. A visiting American scholar from Cambridge, Massachusetts, breakfasted with Davy at Lower Grosvenor Street in June, just after they had moved in. George Ticknor knew of the Bakerian Lectures, and Davy received him in his most expansive mood.

> London 8 June 1815. I breakfasted this morning with Sir Humphry Davy, of whom we have all heard so much in America. He is now about thirty-three, but with all the freshness and bloom of five and twenty, and one of the handsomest men I have seen in England. He has a great deal of vivacity, talks rapidly, though with great precision, and is so much interested in conversation that his excitement amounts to nervous impatience and keeps him in constant motion. He has just returned from Italy and delights to talk of it.

This almost aggressive youthfulness and animation (in fact Davy was thirty-six) was not at all what Ticknor had expected from the distinguished chemist, let alone his fascination with Italian art and culture. 'It seemed singular that his taste in this should be so acute, when his professional eminence is in a province so different and so remote.'

But perhaps George Ticknor was a rather earnest academic, for there are signs that Davy began to tease him over the teacups. 'I was much more surprised when I found that the first Chemist of his time was a professed *angler*; and that he thinks, if he were obliged to renounce fishing or philosophy, that he would find the struggle of his choice pretty severe.'

Jane avoided this interview altogether, tactfully sending a message down that she was 'unwell'. When Ticknor did eventually catch up with her, he was impressed by her dark good looks, and what he called 'the choice and variety of her phraseology'.[69]

In a thoughtful mood Davy wrote a new kind of metaphysical poem, 'The Massy Pillars of the Earth'. It reflects on the human condition, and suggests that since nothing is ever destroyed in the physical universe, only transformed (the First Law of Thermodynamics), then man himself must be immortal in some spiritual sense. It also returns in a new way to Davy's early Cornish beliefs about starlight as the source of all energy in the universe:

> Nothing is lost; the ethereal fire,
> Which from the farthest star descends,
> Through the immensity of space
> Its course by worlds attracted bends,
>
> To reach the earth; the eternal laws
> Preserve one glorious wise design;
> Order amidst confusion flows
> And all the system is divine.
>
> If matter cannot be destroyed,
> Then living mind can never die;
> If e'en creative when alloy'd,
> How sure is immortality![70]

Intriguingly, the first stanza appears to anticipate Einstein's General Theory of Relativity (1915), in which light is 'bent' by gravity; and then Eddington's observations of a solar eclipse in 1919, when he recorded starlight actually being bent by the sun. But apparent anticipations of this kind can be deceptive in science, often hiding a more significant contemporary meaning. Here Davy was really expressing a more traditional belief: the sudden confidence that 'eternal laws' govern the universe in a benign and ordered way. In fact this view was largely at odds with the scepticism of his private journals. Instead, it proposes that nothing in the world is lost, or wasted or destroyed. There is 'one glorious wise design' throughout the universe, and ultimately 'all the system is divine'; a belief somewhere between Romantic pantheism and the old Enlightenment deism.

In truth Davy was never 'sure' of individual immortality, which he constantly questions in his laboratory notebooks. Nor was the idea of man's being 'creative' normally any kind of guarantee of it, especially when 'alloy'd' in flesh. What is striking about this poem is its sudden tone of Evangelical self-confidence and its unusually hymn-like form. It could have been written by John Wesley or Isaac Watts, though Davy carefully avoids the words 'God' or 'soul'. It is quite unlike his more private speculative poems, and seems like a deliberate performance. Perhaps he wanted to settle down theologically, as well as socially. But science would never quite allow him to do either.

4

In July 1815 Davy took Jane on another fishing holiday in the Highlands, perhaps in an attempt to revive the happy memories of their honeymoon. But in early August, while at Melrose in the Yarrow valley, they were interrupted by a series of increasingly urgent letters from Dr Robert Gray of the Coal Mines Safety Committee, begging for his assistance. The situation in the mines was becoming critical (another fifty-seven men had died at Success colliery, Newbottles, in June), and 'of all men of science' in England, Sir Humphry was the one who could best bring 'his extensive stores of chemical knowledge to a practical bearing'.

Replying on 18 August, Davy immediately proposed to visit Walls End colliery outside Newcastle, so he could observe the problem of lethal fire-damp on the spot. He determined to apply his pure scientific method: observation, experiment, analogy. He cancelled a visit to Banks at his country house in Lincolnshire, and sent Jane back to London. 'Travelling as a bachelor', he rode down to Walls End, and on 24 August had a long discussion with John Buddle, the Chief Mining Engineer.[71]

Buddle (1773–1843) was a hard and experienced Yorkshireman, a Unitarian and a teetotaller. Neither a miner himself, nor a proprietor, he stood between capital and labour, proud of his professional independence as 'viewer' or engineer. He said that he had once been chaired in triumph by the miners, and another time burnt in effigy. In fact he was dedicated to them. He never married, never drank, lived with his sister, and played the cello in the evenings. In his way, he was a man not unlike

Tom Poole. But initially he had grave suspicions of 'Sir Humphry', the man of science from the South.

Davy was immediately put on his mettle. He knew that the Royal Institution was committed to helping science serve British industry, and this was an important part of Banks's conception of humane progress: the appliance of science. But at Walls End he saw a peculiarly personal challenge, requiring all his experience and skill. In his youth he had explored Cornish tin mines with his lost friend Gregory Watt, and he had a feeling for mining communities and their intense local loyalties. He had never lacked physical courage in his experiments, and had been confidently handling dangerous gases since the Bristol days. He had already had a first encounter with fire-damp in the Apennines with Faraday. Above all, here was a chance for Davy to fulfil his greatest ambition: to show that a man of science could serve humanity – *and be a genius.*

What an outsider like Davy had to encounter in the Northumberland mines was described by a local journalist: 'It would require all the fortitude of nature to refrain from fear, and to examine everything with calmness and precision. The immense depth [sometimes 600 feet], the innumerable windings and the dark solitary wastes of a coalmine are truly astonishing, and create a sensation of horror in the imagination.'[72] ♣

Buddle later recalled: 'After a great deal of conversation with Sir Humphry Davy, and he making himself perfectly acquainted with the nature of our mines, and what was wanted, just as we were parting he looked at me and said, "I think I can do something for you." Thinking it was too much ever to be achieved, I gave him a look of incredulity; at that moment it was beyond my comprehension. However, smiling,

♣ Some impression of what early-nineteenth-century mines were like can still be gained from a visit to the National Coal Mining Museum, near Wakefield in Yorkshire, which offers access to 400 metres of restored underground mineshaft (not to be undertaken by those with claustrophobia). The harshness of the conditions, the crude simplicity of the available mining equipment, and the lethal effect on the general health and life expectation of miners – who would often begin work as children – are sobering. More than this, Sir Humphry Davy's visit to such a mining community as Walls End (now a peaceful suburb of Newcastle) would have produced an extraordinary clash of social cultures, behaviour and even language (all potentially hostile), so that the trust he established there – and particularly the friendship he achieved with John Buddle – must count as one of the most remarkable achievements of his career.

he said, "Do not despair, I think I can do something for you in a very short time." '[73]

From the start, Davy approached each stage of his solution with great originality, and also hectic speed. The Accidents Committee had considered that the prevention of explosions was essentially a problem of designing better ventilation for the mineshafts, rather as Davy had already done in Newgate Prison. Buddle wondered if a different kind of gas could be pumped down to neutralise the fire-damp. But Davy quickly grasped that something far more fundamental was required: safe light.

All miners needed to carry lights (candles or oil lamps) to every part of a mine. How could this be done without exploding the lethal fire-damp gas, and moreover without living in permanent fear of such an explosion? The solution must be simple, inexpensive, robust and absolutely reliable: a miner's 'safe lamp'. Here Davy took his first original step. Instead of starting with the lamp, as every other inventor had done, he started with the gas. The first step was not the technology of the lamp, but a complete scientific analysis of the gas and all its properties. Buddle undertook to send samples of the fire-damp to London as soon as it could be safely gathered and bottled.

Davy went to ground in Durham for over three weeks, and neither Jane nor any friend in London (except Faraday) knew where he was. He visited numerous mines, talked to miners and overseers, silently observing, analysing and reflecting. He borrowed Dr Clanny's bellows lamp for a day, but was not impressed. Then he suddenly seems to have made up his mind. He hurried back to London, and precipitately took over the Royal Institution laboratory on 9 October 1815, which he was not really authorised to do. He ordered glass and metal apparatus, capable of 'withstanding an explosion', from the Institution's instrument-maker, John Newman, and summoned Michael Faraday to his assistance.[74]

They remained closeted in the basement laboratory almost without interruption for three months, pursuing a feverish series of experiments and issuing ongoing reports to the Royal Society. Faraday said he was only let out to attend the weekly meetings of the City Philosophical Society. He later modestly recalled: 'I was a witness in our laboratory to the gradual and beautiful development of the train of thought and experiments which produced the Safety Lamp.'[75]

Davy first began a minute analysis of the properties of fire-damp, quickly confirming that it was 'light carburetted hydrogen' (methane), with unusual combustion characteristics. He discovered that explosions would only occur when methane reached a critical ratio of gas to air (approximately one to eight parts). It then became true explosive fire-damp. Once ignited – a mere lick of a candle flame would do this – it produced an accelerated reaction, spreading with an intense flame that rapidly reached a critical temperature and then exploded with extreme violence. He noted that this critical temperature at which the explosion occurred was surprisingly high: much higher, for example, than for that of the hydrogen used in Charlier balloons.

Paradoxically, fire-damp was also capable, under certain conditions, of burning with a cool flame which did not explode. Accordingly, Davy next tried igniting it in various closed containers. If a glass tube was used, it instantly exploded. But when confined to a narrow metal tube, it would only burn with the cool, slow blue flame he had observed in the Apennines. He established that the reason for this was that the surface of the metal tube, if sufficiently narrow ('less than an eighth of an inch'), had the peculiar property of conducting away the heat and continuously cooling the methane flame, thus keeping it below the critical explosion temperature.

With his analysis of the gas complete, Davy turned his attention to the lamp. He began designing the first model of an enclosed, airtight safety lamp, using a sealed glass chimney round the wick, with a system of narrow metal tubes to let in the air at its base. Methane mixed with air would not explode in these tubes. Davy's hasty sketches were transformed into neat technical drawings by Faraday. The prototypes were then constructed overnight by the Institution's temperamental engineer, John Newman, at nearby Lisle Street, so that Davy could immediately try them out the next morning in large glass containers filled with fire-damp. 'After many disappointments from the instrument-maker', and some spectacular rows, several possible models began to emerge. This trial-and-error process was a new type of teamwork for Davy, which caused some friction. But crucially it allowed him and Faraday to work very rapidly, and on several concepts at once.[76]

Despite some fearful explosions, Davy already had at least three working prototypes of a 'Safe Lantern' ready by the end of October. All of these were sealed lamps, using various forms of metal tubes or 'fire sieves' as air inlets. He summarised his researches in a letter to Banks on 27 October, and a week later sent the lamps to the Royal Society, with a detailed scientific paper which was officially read on 9 November. He also copied his summary in a 'private communication', not to be released, to Dr Gray at the Safety Committee.[77] Not surprisingly, news of at least one prototype was soon leaked to the Newcastle newspapers, which would later lead to confusion about the exact mechanism Davy had discovered, and a bitter priority dispute.

Banks was triumphant. On 30 October he wrote one of his most flamboyant missives to Davy, bubbling with emphatic capital letters, from Revesby Abbey in Lincolnshire. Davy's 'brilliant' discoveries had given him 'unspeakable Pleasure', and would exalt the reputation of the Royal Society throughout the 'Scientific world'. His personal achievement was nothing less than heroic: 'To have come forward when called upon, because no one else could discover the means of defending Society from a Tremendous Scourge of Humanity; and to have by the application of Enlightened Philosophy found a means of providing a Certain Precautionary Measure [the lamp] effectual to guard Mankind for the future against this alarming & increasing Evil, cannot fail to recommend the Discoverer to much Public Gratitude, & to place the Royal Society in a more Popular Point of View than all the abstruse discoveries beyond the understanding of unlearned People could do. I shall most certainly direct your paper to be read at the very first day of our meeting.'[78]

But Banks's congratulations were premature. The lamps with tubes were only *relatively* safe, as Davy discovered after further trials. Here his true genius as a man of science – his impetuosity, his imagination, his ambition and his seething energy – were demonstrated. Davy would not rest, nor would he let Faraday rest. Obsessively pursuing his researches into December, and largely ignoring Christmas, to Jane's evident dismay, he remained closeted with his assistant. In late December or early January he made a further technical breakthrough, which he reported to the Royal Society in a hurried but triumphant paper of 11 January 1816.[79]

What he had discovered was this. *Fine-gauge iron mesh* would work even better than thin metal tubes in preventing an explosion. Indeed, it replaced the need for an airtight glass chimney (easily broken) entirely. The fine apertures in the mesh or 'gauze' provided the equivalent of hundreds of tiny metal cooling tubes ('784 apertures to the inch'). The function of tubes and gauze was 'analogous'. This application of metal gauze or 'tissue' was the key discovery that no other researcher had hit upon.

The methane passed freely through the iron gauze to the naked flame inside the lamp, ignited there and burnt vividly. 'The lower part of the flame is green, the middle purple, and the upper part blue.' But it could not pass back at sufficient temperature to ignite and explode the fire-damp outside in the mine. Even when the gauze became red hot, the flame would not pass through it. Moreover, provided the lamp was entirely enclosed in the iron gauze, it did not have to have an airtight glass chimney. It was much less vulnerable, and could be redesigned as a much cheaper and more robust instrument. Davy wrote dramatically of confining 'this destructive element flame like a bird in a cage'.[80] Holding an iron gauze over a Bunsen burner, and observing that, against all expectation, the flame does not pass through it, is now one of the elementary experiments performed in school chemistry classrooms. It is easy to forget how startling this effect is on seeing it for the first time.

The final version of the lamp was wonderfully simple and surprisingly small. It was a standard uninsulated oil lamp, approximately sixteen inches high, with an adjustable cotton wick, enclosed in a tall column or 'chimney' of fine iron mesh. Astonishingly, the lamp required no other protection. In later models Davy added various improvements, largely designed to withstand rough use in the mine.

Yet the fundamental notion that flame would not pass through gauze appeared so unlikely, so completely counter-intuitive, that Davy had to lay out the stages of his discovery with absolute clarity, step by step. The result was a new kind of scientific narrative. The uncertainty and false starts of the experimental laboratory disappeared. Faraday's sketches showed that trial models had originally included a piston-bellows lamp, a spring-valve lamp and a hinged lamp, none of which was subsequently mentioned.[81] Instead the account was transformed into a gripping, single-track narrative of progressive, seemingly inevitable, discovery.

> In trying my first tube-lamp in an explosive mixture I found that it was safe; but unless the tubes were very short and numerous, the flame could not well be supported . . . I arrived at the conclusion that *a metallic tissue*, however thin and fine, of which the apertures filled more space than the cooling surface, so as to be permeable to air and light, offered a perfect barrier to explosion . . . In plunging a light surrounded by a cylinder of fine wire gauze into an explosive mixture I saw the whole cylinder become quietly and gradually filled with flame, the upper part of it soon

> appeared red hot; yet *no* explosion was produced ... I immediately made a number of experiments to perfect this invention, which was evidently the one to be adopted ... I placed my lighted lamps in a large glass receiver, and by means of a gasometer filled with coal gas, I made the current of air which passed into the lamp more or less explosive, and caused it to change rapidly or slowly at pleasure, so as to produce all possible varieties of inflammable and explosive mixtures: and I found that iron wire-gauze ... was safe under all circumstances ... and I consequently adopted this material in guarding lamps for the coal mines, where in January 1816, they were immediately adopted, and have long been in general use.[82]

When he republished the papers, Davy remarked: 'Every step was furnished by experiment and induction, in which nothing can be said to be owing to accident, and in which the most simple and useful combination arose out of the most complicated circumstances.'[83] In this way he insisted on the Baconian method of stage-by-stage, logical scientific induction, while tacitly admitting the existence of 'complicated' versions of the lamp which he had tried and rejected.

This refusal to allow anything to chance, 'accident' or good fortune was exactly the same as Herschel's insistence that chance played no part in his discovery of Uranus. Coleridge had taken this up as one of the key philosophical problems associated with science, in an essay provokingly entitled 'Does Fortune Favour Fools?', which he republished in *The Friend* in 1818. Here he described Davy, perhaps mischievously, as 'the illustrious Father and Founder of Philosophic Alchemy'. But he praised his great discoveries without reservation, and denied that his scientific research could ever have depended on 'accident' or 'luck'. Yet this left him in a philosophical quandary: did that imply that 'genius' and 'inspiration' had no place in Davy's science?[84]

The essay was originally written in 1809, in response to Davy's work with the voltaic battery. Coleridge argued that Davy's discoveries always depended on 'preconcerted mediation ... evolved out of his own intellect', never on external accident. Davy's scientific method was always conscious, skilful and deliberate, the fruit of deep knowledge and experience. But the essay raised other issues about scientific research. Coleridge's way of describing the experimental process betrayed a certain uneasiness. Chemical experiments – using fire or electricity – contained a kind of violence. Davy's aim was 'to bind down material Nature under the Inquisition of Reason, *and force from her, as by torture*, unequivocal answers to prepared and preconceived questions'. Coleridge also wondered if scientific laws could ever truly 'bind down' all the phenomena of nature. Newtonian laws could define the phases of the moon, for instance, but could they ever define the movements and appearance of *clouds*? 'The number and variety of their effects baffle our powers of calculation: and that the sky is clear or obscured at any particular time, we speak of, in common language, as a matter of *accident*.'[85]

5

The Davy Safety Lamp, the greatest public achievement of his career, would soon be in use all over Britain and Europe. The prototype gauze-enclosed lamp ('the Davy') was presented to the Royal Society on 25 January 1816, after being successfully tested at Walls End, Hebburn and several other collieries that month.[86]

John Buddle, who had witnessed the full horror of several earlier explosions at Walls End, and understood the deep, suppressed fears of miners working in shafts 600 or 1,000 feet beneath the surface, never forgot his first trial with the new Davy Lamp. He later gave a verbatim account to a Parliamentary Committee: 'On the strength of [Davy's] authority I took this lamp, without hesitation, into an explosive mixture. I first tried it in an explosive mixture on the surface, and then took it into a mine; and to my astonishment and delight, it is impossible for me to express my feelings at the time when I first suspended the lamp in the mine, and saw it red hot; if it had been a monster destroyed, I could not have felt more exultation than I did. I said to those around me, "We have at last subdued this monster!" '[87]

Davy went up to Northumberland in March to observe the lamps in action in the mines, and to work on refinements. These would include a platinum coil which relit the wick when it was extinguished by pure methane ('one of the most beautiful and magical experiments in the science of chemistry experiment', remarked Faraday), tin draught shields, double gauze at the top of the chimney, and a reinforced open iron frame to protect the gauze if the lamp was struck or dropped.

He went down 'G' pit at Walls End, spent some two hours beneath the surface, and according to Buddle delivered an impromptu fifteen-minute lecture on using the lamp safely, stressing the need to avoid strong air currents or clouds of coal dust, which could still risk freak explosions. He also pointed out that the state of the flame indicated the presence, and even the strength, of fire-damp in a shaft. His lamp not only caged the flame, it transformed it into a canary.[88]

During this visit Davy received a deputation from the mine-owners, with a public letter of thanks describing his lamp as 'a discovery unparalleled in the history of mining'. It was hoped that 'this great and unrivalled discovery for preserving the lives of our fellow creatures, will be rewarded by some mark of national distinction'.[89] Many individual miners also signed tributes or letters of thanks. In September 1816 'we, the undersigned miners at the Whitehaven Collieries' thanked Davy for his 'invaluable discovery of the safe lamps, which are to us life preservers'. They humbly wished it was in their power to offer more than this 'tribute of gratitude'. The wording of the letter was obviously drawn up by an overseer, but the signatures were genuine, and must have moved Davy. The crumpled paper was laboriously signed by eighty-two miners, forty-seven of whom were illiterate, and put 'x' against their names.[90]

John Buddle, now entirely won over by Davy, was also concerned about a reward. By August there were 144 safety lamps 'in daily use' at Walls End, and they were rapidly spreading to all the other collieries in the North-East.[91] Buddle urged Davy to take out a patent, pointing out that he could not only make his fortune but control the quality of the lamps issued to miners. Davy consistently refused, although he knew his colleague William Wollaston had made a fortune with a patent on processing platinum. Yet Davy was hugely proud of his achievement, and was never modest about it. On Banks's recommendation he received the Rumford Medal from the Royal Society in 1817, and the following year was made a baronet by the Prince Regent. Davy designed his own coat of arms, showing the safety lamp encircled with a Latin motto which announced: 'I Built the Light which brings Safety'.[92]

His reputation was now international. He received acknowledgements from miners in Alsace, Flanders, Austria and Poland. Some years later the Tsar of Russia, Alexander I, sent him a large silver goblet. At home the *Edinburgh Review* ran an enormous article in praise of his work, written by none other than the brilliant geologist who had once paid court to Jane Apreece, Professor John Playfair. 'It may fairly be said that there is hardly in the whole compass of art or science a single invention of which one would rather wish to be the author.' Playfair described the discovery as the result of pure inductive science, 'in no degree the effect of accident', and 'as wonderful as it is important'. Its historic significance was unmistakeable. 'This is exactly such a case as we should choose to place before Bacon, were he to revisit the earth, in order to give him, in a small compass, an idea of the advancement which philosophy has made, since the time when he pointed out to her the route which she ought to pursue.' Here the word 'philosophy' was used exclusively to mean 'science' in the modern sense: what Playfair defined as 'the immediate and constant appeal to experiment'.[93]

Davy published in 1818 a beautiful account of his discovery, *On the Safety Lamp for Coal Miners, with Some Researches into Flame.* Edited from the series of papers he had sent so hurriedly to the Royal Society, this has some claims to be one of the prose masterpieces of English Romanticism. Davy transformed his feverish, often chaotic work at the Royal Institution laboratory in the winter of 1815–16 into a classic piece of scientific storytelling. The prose is clear, pointed, and sometimes of poetical intensity.

The treatise begins with a dispassionate account of the terrible series of explosion accidents in the mines, the human suffering they had caused over decades, and the way they had terrorised the mining communities in the north of England: 'The phenomena are always of the same kind. The miners are either immediately destroyed by the explosion, and thrown with the horses and machinery through the shaft into the air, the mine becoming as it were an enormous piece of artillery, from which they are projected; or they are gradually suffocated, and undergo a more painful death, from the carbonic acid and azote [nitrogen] remaining in the mine, after the inflammation of the firedamp; or what, though it appears the mildest, is perhaps the most severe fate, they are burnt or maimed, and often rendered incapable of labour, and of healthy enjoyment of life.'[94]

Davy then moves to the sequence of experiments he performed in the laboratory in London, producing a narrative as logical and thrilling as a detective story. He describes the previous work of Clanny and Humboldt, his experiences with Faraday in the Apennines,[95] his assembling of his laboratory equipment, the meticulous process of chemical analysis (often highly dangerous), the varied appearances of slow flame and violent explosions, and the final triumphant sight of the prototype gauze lamp burning brightly and safely within the huge glass flask of lethal methane.

By relating the human predicament to the scientific solution, Davy produced one of the great demonstrations of scientific 'Hope'. He showed that applied science could be a force for good previously unparalleled in human society, and might gradually liberate mankind from untold misery and suffering. Deliberately echoing Bacon – as Lavoisier had once done – he claimed that scientific knowledge was a disinterested power for good: 'The results of these labours will, I trust, be useful to the cause of science, by proving that even the most apparently abstract philosophical truths may be connected with applications to the common wants and purposes of life. The gratification of the love of knowledge is delightful to every refined mind; but a much higher motive is offered in indulging it, when that knowledge is felt to be practical power, and when that power may be applied to lessen the miseries or increase the comforts of our fellow-creatures.'[96] This would become the central credo of the next generation of young Victorian scientists, and notably of Michael Faraday.[97]

But the story of the invention was exemplary of future science in another way. Davy's high-minded claims produced a bitter priority dispute. In the spring of 1816 the engineer at the Killingworth mine, just north of Newcastle, George Stephenson, challenged Davy's precedence, and accused him of plagiarising his own 'Geordie Lamp'. This was a solid glass and metal lamp, conical in shape, using tubes and perforations, of which he had made many practical trials. He had finally tested a working version in the Killingworth colliery on 21 October 1815. When he saw the first premature model of Davy's 'tube lamp', published in the Newcastle papers in November, he naturally suspected plagiarism.

They did indeed look very similar, since Davy's gauze lamp had not yet been published – or indeed invented. At a meeting of the Newcastle Literary and Philosophical Society on 5 December 1815, at which both Clanny's bellows lamp and Stephenson's conical lamp were examined ('it resembles a wine decanter', remarked the *Newcastle Chronicle* jovially), the questions of priority and pirating were first raised.

The Newcastle Society showed its admirable objectivity by presenting examples of the truc gauze lamp, as used by Buddle at Walls End, at its meeting of 6 February 1816. It was immediately clear to unbiased observers that the Stephenson and the Davy were very different instruments. But nothing could prevent the major public row now brewing, with letters to the newspapers, polemical pamphlets, and wide controversial comment in the journals. Not all of this was favourable to Davy, and there was a clear evidence of a North-South split as sides were taken.

Furious pamphlets were written against him by a Sunderland lawyer and journalist, J.H. Holmes, who had been writing about mining accidents to the *Morning Chronicle* since July 1815. The Director of Mines at Seaton colliery, James Heaton, gave a demonstration to the Society of Arts at which he made a 'Davy' explode by repeatedly throwing handfuls of coal dust at its gauze.[98] There was also a good deal of general mockery of rival 'inventors', and anonymous letters to the papers signed with provocative pseudonyms like 'Aladdin Lamp' and 'Simple Wire Gauze'.[99]

In 1817 Stephenson published two pamphlets calmly setting out his claims, and showing detailed illustrations of both lamps. He said that his was the result of 'mechanical principles', while Davy's depended on 'chemical' ones – a fair distinction. He also pointed out that he did not have the expensive facilities or 'beautiful instruments' of Sir Humphry's London laboratory, perhaps a further sign of North-South rivalry and class bitterness. Stephenson signed himself defiantly 'Inventor of the Capillary Tube Lamp'.[100]

George Stephenson (1781–1848) was a gifted, self-educated engineer, and later the designer of the early railway steam engine, the famous 'Stephenson Rocket' which brought him international fame. He was an inventor of genius, an honest man and no fraud. He was to be the hero of one of Samuel Smiles's outstanding industrial biographies in 1859. It is clear that he was genuinely misled by the premature November announcements of Davy's prototype lamp. He admitted later that he never understood the scientific analysis of methane, or the principles behind the final iron gauze Davy Safety Lamp. He merely maintained that his tube lamps were the result of practical ('mechanical') trial and error, had been introduced before Davy's, worked safely, were cheap and robust, and had been loyally adopted by many Newcastle miners who fondly referred to them as home-grown 'Geordies'.

In private, Davy reacted very bitterly to these claims. In February 1817 he wrote to Buddle complaining of Stephenson's 'miserable pilfering lying & equivocating pamphlet'.[101] After he had seen one of Stephenson's lamps, he dismissed it contemptuously: 'there is no analogy between his *glass exploding machine*, and my metallic tissue, permeable to light and air, and impermeable to flame'.[102] He had several rival lamps sent to the Royal Institution in 1816, and stored as evidence in the basement (where they still remain).[103] But he made no attempt to get in touch with Stephenson himself, and he never acknowledged that the over-hasty publication of his early prototypes had caused much of the problem.

Davy's intense anxiety to establish scientific priority, already witnessed in France over the iodine débâcle, fuelled much of this debate. He showed no professional generosity towards Stephenson. Above all he demonstrated his driving desire to be seen as the miners' sole saviour. As he proclaimed publicly in Newcastle in September 1817: 'the highest ambition of my life has been to deserve the name of friend to humanity'.[104]

The dispute also became politicised. Lord Lambton, the Whig mine-owner who had been a pupil of Dr Beddoes and had known Davy in Bristol, enthusiastically supported him. But local Tory mine-owners decided to back Stephenson, and presented him with £1,000 in cash and an engraved silver tankard. Many local Newcastle miners also supported Stephenson as 'one of them'.♣

Faraday, who knew more about the actual sequence of Davy's discoveries than anyone else, always remained staunchly loyal to his patron's priority and originality. If he disguised anything, it may have been his own role in writing up the experiments, and creating Davy's dedicated team at the Royal Institution.[105] But this can never be known, as, almost uniquely for a major piece of research, Davy left no original laboratory notes.[106] However, he did acknowledge Faraday's contribution in the Introduction to his published account of 1818. 'I am myself indebted to Mr. Michael Faraday for much able assistance in the prosecution of my experiments.' This was the first, historic, mention of Faraday in print, and it effectively launched his scientific career.[107]

♣ 'Thus was set up, from the beginning,' observes Frank James in his detailed study of the controversy, 'the dynamic for a priority dispute between knight and worker, chemist and engineer, savant and artisan, theory and practice, metropolis and province.' See Frank A.J.L. James, 'How Big is a Hole?', *Transactions of the Newcomen Society* (2005). Something similar had arisen during the controversy over the John Harrison chronometers. The whole question of 'scientific priority' has become a major preoccupation in modern science. See for example the race over the structure of DNA between Crick and Watson at Cambridge, and Rosalind Franklin at King's, London, as described in James Watson's classic *The Double Helix* (1968) and Brenda Maddox's biography *Rosalind Franklin: The Dark Lady of DNA* (2002). Carl Djerassi's play *Oxygen* (2001) beautifully dramatises an earlier eighteenth-century priority dispute between Priestley, Scheele and Lavoisier.

In October 1817, at the height of this drama, Davy received a triumphant reception at the Queen's Head Hotel, Newcastle. He was given a banquet, a presentation of silver plate worth over £2,500, and a commemorative portrait. In an effusive speech Lambton praised his 'brilliant genius' and the 'immortal fame' of his discovery. Carefully balancing his words, he said that science had secured both 'the property of the coal owner' and the 'safety of the intrepid miner'. For two years Davy's lamps had protected 'hundreds of miners in the most dangerous recesses of the earth' without a single fatality. (Except, it appeared, one 'foolhardy' miner who had tried to light his pipe through the gauze.[108])

Davy gave a heartfelt speech in reply, trying – not altogether successfully – to appear the soul of modesty. 'I am overwhelmed by these reiterated proofs of your approbation. You have overrated my merits. My success in your cause must be attributed to my having followed the path of experiment and induction discovered by philosophers who have preceded me ... It was in pursuing those methods of analogy and experiment, by which mystery had become science, that I was – fortunately – led to the invention of the Safety Lamp.'[109]

Yet he could not help also referring to his bitterness about Stephenson and his supporters, and here his feelings rang completely true. 'It was a new circumstance to me that attempts to preserve human life, and to prevent human misery, should create hostile feelings in persons who professed to have similar objects in view.'[110] It was also perhaps an ill omen of a different kind that Jane did not travel with him to Newcastle at this controversial time, but retreated with her own friends to Bath, and read about events in the newspapers.

Banks now intervened on Davy's behalf, and wrote a thunderous public letter to *The Times* and other papers, dating it from Soho Square, 20 November 1817. The letter was countersigned by the three leading chemists of the Royal Society, William Wollaston, Charles Hatchett and Thomas Brande. It stated that they had examined all Stephenson's published claims, and his lamps, followed the entire course of Davy's experimental work, and *his* lamps, and concluded that there could be no doubt whatever that Davy was the sole inventor, 'independent of all others', of the safety lamp. This authoritative judgement of his peers, clearly intended to silence all further debate *ex cathedra*, gave huge satisfaction to Davy. He described it as 'heavy artillery used to destroy bats and owls'.[111] But it would have been no surprise to the ageing Banks, now a past master of scientific diplomacy, that after a brief respectful pause the controversy grumbled on, and has never really ended.[112]

The Newcastle Literary and Philosophical Society, true to its impressive standards of objectivity, refused simply to back their local man, but in a noble attempt to smooth the waters, unanimously elected both Davy and Stephenson as Honorary Members, simultaneously, on 2 December 1817.[113]

The tough, sceptical Yorkshireman John Buddle remained Davy's fiercest supporter, and became a lifelong friend. Whenever he came to London from Newcastle he stayed at Grosvenor Street. Twenty years after his first meeting with Davy he was the star witness at the historic Parliamentary Select Committee on Mining Accidents of 1835.[114] George Stephenson was also there, and gave strong and moving evidence, though no longer accusing Davy of plagiarism. In fact the Committee refused to rule on absolute priority, but suggested various new and positive perspectives. In their view, it was undoubtedly the gauze Davy Lamp, unrestricted by patent, which had become the model for all later improvements, such as the big Upton Roberts Lamp (which combined a glass chamber with a gauze chimney).[115] It was also the genius of Davy which had first championed a wholly new way of applying pure science to industry. But the safety lamp itself they regarded as, in some senses, a joint discovery. They framed this opinion with great diplomacy, if not entirely scientifically: 'The principle of its construction appears to have been *practically* known to the witnesses, Clanny and Stevenson [sic], previously to the period when Davy brought his powerful mind to bear upon the subject, and produced an instrument which will hand down his name to the latest ages.'[116]

There were other surprises in this Parliamentary Report. Not the least was the revelation that small 'Davy boys' were now put in charge of the lamps, to save them from harsher physical labour in the coal seams. The Committee were appalled to learn that these labouring children were often under eight years old. So Davy's lamps were now saving children in a different way. They began to cast a new and wholly unexpected kind of light, and the Victorian movement to ban child labour in the mines began with this Committee.[117]

6

Exhausted by this mixture of triumph and controversy, Davy decided to embark on a two-year European tour with Jane. In part this was a last-ditch attempt to save their marriage. Success and celebrity had put a new kind of strain on their increasingly tempestuous relationship, which had degenerated into a series of well-publicised dinner-party feuds and jealous scenes. Sydney Smith wrote cattily of the chemical 'decomposition' of their marriage despite plentiful 'crucible money' provided by Lady Davy, and her evident 'disappointment and fury' at Davy's lack of personal 'powers of chemistry'. Even Walter Scott shook his head over their behaviour. 'She has a temper and Davy has a temper, and these two tempers are not one temper, and they quarrel like cat and dog, which may be good for stirring up the stagnation of domestic life, but they let the world see it, and that is not well.' Then he added: 'But then, pour soul, she is not happy.'[118]

Davy's loyal brother John was no longer on hand to help them, now being based in Corfu. Yet even he came to feel that their marriage had been based on a delusion: 'it might have been better for both if they had never met'. Neither had domestic virtues or easy temperaments. At home Jane was irritable, highly strung and demanding: 'her ample fortune made her perhaps too independent and self-willed'. In society she was vivacious, generous with friends, and savagely witty. But she would never be, thought John, the '*placens uxor*' – the soothing wife.

John said less about his brother's evident shortcomings: that Davy was difficult, short-tempered, obsessed with his scientific work, and overfond of aristocratic parties and endless field sports. He had also become dangerously hungry for praise and recognition. John did however remark on their childlessness, as something sad for them both. He thought wistfully (but perhaps wrongly) that the marriage would have been happier with children, 'For he was of a loving disposition, and fond of children, and required the return of love – required (who does not?) to be beloved, to be happy.' This wistful remark may also have relevance to certain events towards the end of Davy's life.[119]

It is clear that Davy hoped that being away from London, and distracted by a mixture of travel, sight-seeing and social engagements, he and Jane might regain their marital equilibrium. He also had a number of scientific projects up his sleeve, including the commission to find a chemical method of unrolling and deciphering a number of calcinated papyri from Herculaneum. He intended to give this period of reconciliation a proper trial. The Sun Fire Office (Guildhall) holds a special insurance policy taken out to cover their house at 23 Grosvenor Street during a prolonged absence, dated 4 June 1818.[120]

Davy and Jane left on 26 May 1818, again in their own carriage but now without Michael Faraday in attendance. Instead, Jane firmly took her own maid, who had no scientific ambitions. This time they took the easterly route into Italy, travelling by easy stages along the Rhine and down through the Austrian Alps. They were the honoured guests at several mines in Flanders and Germany, where Davy's lamp and fame had preceded him. Davy was also testing a theory about water temperature, and why mists form over river waters, which allowed him to spend much time alone on every available riverbank.

Jane persuaded Davy to remain for several weeks in Vienna. But eventually they pushed further south into the Austrian Tyrol, and Davy was able to continue his exploration of the Austro-Italian border country called Illyria and Styria. The magical names, half-remembered from Shakespearian romances, were strangely enchanting to him. He found a remote and beautiful land of alpine meadows, deep wooded valleys and fine wild rivers like the Traun, where he could ride and shoot and fish to his heart's content. Yet his fame had reached even here, for on passing through Aussee (in Styria) he was called to a local salt works where several miners had recently been killed by an underground explosion. Davy summoned the local engineer, and personally supervised the construction of several gauze safety lamps for immediate use. They were received 'with gratitude and surprise', and no further explosions occurred.[121]

This remote region of the Balkans, lost between Austria, Italy and Slovenia, was to become Davy's favourite retreat. Its little provincial capital, Laibach (the modern Ljubljana, capital of Slovenia) on the river Sava, was surrounded by deep forests and mountains. It also had an excellent sportsman's inn run by the Dettela family, and few English visitors to bother Davy. For Jane there was the society centred on a small baroque opera house, and an elegant concert hall built in 1701. They remained here for several weeks, happily enough it would seem, until Davy was gradually overcome by a mysterious and curiously haunting fixation.

He found he was strangely struck by Herr Dettela's fifteen-year-old daughter Josephine, a cheerful and sweet-natured girl with bright blue eyes, a high rosy complexion and 'long nut-brown hair' who waited at table and helped with the housekeeping.[122] Davy felt that she constantly reminded him of some woman he had once met long ago, though he could not say whom or where. This erotic echo was strangely upsetting to him, but he finally explained it to himself as relating to the sort of hallucination or feverish 'vision' he had had when very ill during his second series of Bakerian Lectures in 1808. Initially he dismissed it as insignificant, and probably did not mention it to Jane. 'Ten years after I had recovered from the fever, and when I had almost lost the recollection of the vision, it was recalled to my memory by a very blooming and graceful maiden fourteen or fifteen years old, that I accidentally met during my travels in Illyria; but I cannot say that the impression made upon my mind by this female was very strong.'[123] Jane may have noticed it all the same, and perhaps she was used to such things. At all events Lady Davy was relieved when autumn came and they moved south to visit Lord Byron in Ravenna, and then to settle in Naples for the winter. They arrived there at approximately the same time as Percy Bysshe Shelley and his family.

Unrolling the papyri from Herculaneum was not a success. But they made expeditions to Vesuvius and Paestum, and Davy theorised about volcanoes and eruptions. He would later write about these wild landscapes, and other odd encounters, in lightly disguised fictional form in *Consolations in Travel.* In spring 1819 they rode restlessly north again into the Apennines, where Davy wrote a striking series of poems, under the general heading 'Fireflies', at the Bagni di Lucca. Officially he was testing the mineral waters for their peroxide and iron-oxide content, but the setting of most of these pieces is night-time and moonlight, suggesting perhaps the long and probably solitary after-dinner walks he was taking along the banks of the river Serchio.

Not all is melancholy in these meditations. Indeed the fireflies dancing over the dark water, though ephemeral, filled him with delight and even perhaps reminded him of his own safety lamps.

> Ye moving stars that flit along the glade!
> Ye animated lamps that 'midst the shade
> Of ancient chestnut, and the lofty hills
> Of Lusignana, by the foaming rills
> That clothe the Serchio in the evening play!
> So bright your light, that in the unbroken ray
> Of the meridian moon it lovely shines!
> How gaily do you pass beneath the vines
> Which clothe the nearest slopes! How through the groves
> Of Lucca do ye dance! . . .

This thrilling 'animation' of the fireflies he describes, like Erasmus Darwin, as commanded by 'the voice of Love', to which he can still respond. He presents his own heart as lonely, 'by sickness weakened and by sorrow chilled', but not yet 'broken or subdued'. Most of all he confides in the moon herself, and longs for her to prolong his sense of youth and hope, and hasten the birth of 'new creative faculties and powers'.

Davy was now forty, and like every man of science and every poet, he hoped against hope that original work and 'powers of inspiration' still lay ahead in his maturity. His description of these longings was nakedly Romantic, and surely recalled his moonlit walks along the banks of the Avon some twenty years before.

> Though many chequered years have passed away
> Since first the sense of Beauty thrilled my nerves,
> Yet still my heart is sensible to Thee,
> As when it first received the flood of life
> In youth's full spring-tide; and to me it seems
> As if thou wert a sister to my soul,
> An animated Being, carrying on
> An intercourse of sweet and lofty thoughts,
> Wakening the slumbering powers of inspiration
> In their most sacred founts of feeling high.[124]

It is intriguing to compare these clumsy but curiously expressive poems with those written by Shelley at almost exactly the same period at Naples, Pisa and the Bagni di Lucca. Shelley was a permanent exile, without anything like the public recognition that Davy had achieved, and his moods were much more extreme, yet he responded to the same Italian landscapes and the same inner tides of hope and despair. These writings include some of his most beautiful short lyrics, such as the 'Stanzas Written in Dejection in the Bay of Naples', 'To the Moon' and 'The Aziola'. There are also striking similarities of phrase between Davy's poems and Shelley's confessional outpourings about love, beauty and sexual longing in 'Epipsychidion':

> There was a Being whom my spirit oft
> Met on its visioned wanderings, far aloft,
> In the clear golden prime of my youth's dawn . . .

Then, from the caverns of my dreamy youth
I sprang, as one sandalled with plumes of fire,
And towards the lodestone of my one desire,
I flitted, like a dizzy moth, whose flight
Is as a dead leaf's in the owlet light . . . [125]

But as far as is known, Davy and Shelley never met. It was a pity, perhaps.

Back in Venice, the Davys again called on Lord Byron, this time in his rented palazzo on the Grand Canal. They were introduced to his new Venetian mistress, the beautiful, bosomy Teresa Guiccioli. Byron later gave an amusing account of trying to explain to Teresa the exact nature of Davy's experimental genius. 'I explained as well as an oracle his skill in gases, safety lamps, and in ungluing the Pompeian MSS. "But what do you call him?" said she. "A great *chemist*" quoth I. "What can he do?" repeated the lady. "Almost anything" said I. "Oh, then, *mio caro*, do pray beg him to give me something to dye my eyebrows black." ' Byron added that this was at least better than the reaction of the average 'English blue-stocking'.[126]

Byron was fascinated by Davy's enthusiastic conversation, and his unrestrained boasting about the safety lamp. He put him – with Mungo Park and the polar explorer William Parry – into the first canto of his satirical new poem, *Don Juan*, as one of the signs of the times:

This is the patent Age of new inventions
For killing bodies, and for saving souls,
All propagated with the best intentions:
Sir Humphry Davy's lantern, by which coals
Are safely mined for (in the mode he mentions);
Timbuctoo travels; voyages to the Poles;
Are always to benefit mankind: – as true,
Perhaps, as shooting them at Waterloo.[127]

Davy, in turn, began reading all Byron's poetry, and found its elegance and worldly irony now rather more to his taste than the Coleridge and Wordsworth of his youth. But his poetical reflections were cut short when Jane announced that she was ill and exhausted after so much travelling. She insisted that she would have to convalesce in Paris. Davy accompanied her there, but then heard of another illness, that of Sir Joseph Banks.

9

Sorcerer and Apprentice

1

Sir Joseph Banks had been getting older and more infirm, and he hated it. After one particularly bad episode of gout in summer 1816, when he was seventy-three, he grumbled from his retreat at Spring Grove: 'I fear its probable that I shall be obliged to spend the greater part of my Future Life in a Prostrate Posture... For these 12 or 14 years past my legs have Swelled towards evening... I am so effectually confined to my bed that I am not even allowed to be carried downstairs & placed on a Coach.' Later he added with grim humour: 'The name of Nestor seems likely to adhere to me.'[1] Nevertheless, Lady Banks could rarely keep him away from his scientific breakfasts at Soho Square for more than a week at a time.[2]

His friends too were scattered, ailing or dead. John Jeffries the balloonist had settled back to earth in America. Mungo Park now existed only as a two-volume *Memoir*, published in 1815, although the Africa Association continued to send military explorers along the Niger on his trail. Sir Humphry Davy seemed to be more and more frequently abroad. In January 1820 Banks had received a long, rambling missive from him in Naples. Banks summarised its contents to Charles Blagden: 'Vesuvius has been in Eruption ever since he arrived & has given him opportunity of trying many chemical experiments on the Liquid Lava.' This could have been a sly reference to Lady Davy, though Banks added with all due gravity that Davy's theories on the causes of volcanic fire 'appear to favour the Plutonists'.[3]

Banks hardly ever saw William Herschel now, finding that the old astronomer preferred to stay close to his great forty-foot telescope at Slough, and lived there, thought Banks regretfully, 'so much like an Hermit'.[4] But there was good news of young John Herschel, winning all the prizes at Cambridge, and becoming Senior Wrangler (that is, the top mathematician in his year) in 1813. John had published a first paper 'on analytical formulae' in *Nicholson's Journal* for October 1812 – like young Davy, just before his twenty-first birthday. Banks accordingly exercised his patronage, and had young Herschel elected to the Royal Society the following year.[5] He promised great things for the future.

To other protégés, like the thirty-seven-year-old zoologist Charles Waterton, about to depart on yet another expedition to South America, Banks was more solicitous than of yore. 'I cannot say that I felt the Satisfaction I used to do in hearing that you intend embarking for the *Ninth* time to encounter the dangers and privations of the Trackless forests of Guiana. You are not so young as you used to be ... the old Saw tells us that the pitcher that goes often to the well is cracked at last. May heaven defend you from all Evil results is the sincere Prayer of your Old Friend!'[6] This was not the way he used to cajole Mungo Park.

Banks urged Waterton to come home safely, settle down and write a book about his 'numerous journeys'. Such a book would 'extend materially' the bounds of natural science, and 'put the Public in possession of your discoveries'. More and more Banks saw this as one of the prime duties of the man of science: to collect and explain his findings, to publish them, and put them in the public domain. It is exactly what he himself had failed to do with his own *Endeavour Journal*, nearly fifty years before.[7] In Waterton's case Banks's kindly advice would result in a popular masterpiece, *Wanderings in South America* (1825).

With the arrival of peace in Europe in 1815, international communications had improved, and scientific reports were now pouring in on Soho Square. There was a new emphasis on technology and applied science. Coal-gas pipes now snaked (above ground) through the London streets, so that Westminster Bridge and the Houses of Parliament were illuminated with the new gaslights, 'most Brilliant', Banks noted approvingly.[8] There were paddle ships powered by steam engines, which could ply the Thames against the tide, and make all-weather crossings to France.

These began to appear in Turner's pictures, and even in one of Coleridge's late poems, plangently entitled 'Youth and Age', expressing sentiments that Banks would certainly have recognised:

> This breathing house not built with hands,
> This body that does me grievous wrong,

O'er aery cliffs and glittering sands
How lightly then it flashed along:–
Like those trim skiffs, unknown of yore,
On winding lakes and rivers wide,
That ask no aid of sail or oar
That fear no spite of wind or tide!
Nought cared this body for wind or weather
When Youth and I lived in it together.[9]

From further afield, there came reports of climate change: huge sheets of thawing pack ice were sighted off Greenland, melting snowcaps seen in Alpine mountains, and unprecedented river spates and flooding were recorded throughout Europe. Banks was not disposed to panic at these strange phenomena. 'Some of us flatter ourselves that our Climate will be improved & may be restored to its ancient state, when grapes ripened in Vineyards here.'[10]

In fact much of this was the spreading global consequence of the eruption of the Tambora volcano in Indonesia in April 1815. By releasing a mass of ash into the circulation of the upper atmosphere, it brought the 'sunless summer' of 1816 throughout Europe, with a sinister red haze in the sky at midday, and blood-red apocalyptic sunsets. This delighted the Plutonists, but also brought a renewed awareness of nature's power and mystery, just as had happened after the Lisbon earthquake of 1755, when Caroline Herschel remembered being so frightened in Germany.

Pink snow fell in Italy, and the harvest failed in France, Germany and England. Byron, exiled from Britain and passing this summer on Lac Leman with Shelley, wrote his poem 'Darkness', reflecting on the possibilities of a future cosmological catastrophe, as hinted at by Herschel's late papers.

I had a dream which was not all a dream.
The bright sun was extinguished, and the stars
Did wander darkling in the eternal space,
Rayless, and pathless, and the icy earth
Swung blind and blackening in the moonless air ...[11]

Banks continued to send out his regular, encyclopaedic scientific correspondence throughout the globe. His letters might concern planting crops in South Australia, collecting antiquities in Egypt, surveying the ice pack towards the North Pole, breeding dogs in Newfoundland, or even capturing giant sea snakes off Scandinavia (later to appear in Tennyson's poem 'The Kraken'). But he also found time for some delicate gestures, such as sending packets of strawberry seedlings by the night coach to Paris for the former Madame Lavoisier, in her new incarnation as the ex-Countess Rumford. 'They are Roseberry Strawberries, the kind I most approve of for Quantity of produce & for Flavour.' On another occasion Madame Lavoisier – a great favourite – got a beautifully scented 'climbing Ayrshire rose' which Banks had 'well-rooted in a basket'.[12]

Banks had always admired clever women. He had been instrumental in obtaining the royal salary for Caroline Herschel. He adored his own unconventional sister Sophia (with her collections of coins, cards and balloon memorabilia), and was devastated when she died after a coach accident, aged seventy-four, in 1818. Even now the old Tahitian libertine occasionally resurfaced, as when he upbraided the elegant socialite the Duchess of Somerset for mocking a woman friend who was carrying on an affair. Banks – then a respectable seventy-year-old – exploded in a private letter. 'Tremendous is the punishment inflicted by the Class of *Virtuous Women* on those who err & stray from the paths of Propriety... It is surely a more severe destiny than that of immediate death.'[13] Perhaps he was remembering the free-spirited Sarah Wells, who had given such lively dinner parties in the old days.

Yet, perhaps inevitably, his own views had become increasingly conservative. He would never consider having women elected to the Royal Society. He now tended to growl at all bluestockings (including Lady Davy). His attitude to young Lord Byron and his amorous adventures was indicative. Banks naturally admired Byron, as an aristocrat and an independent spirit. He had been much struck when Byron once attended an open meeting of the Royal Society and listened to a physiological paper based on extensive use of vivisection.

It was a horrific paper – 'the suffering of the animals on which [Dr Wilson] operated produced a most marked disgust' – and many listeners walked out. But Byron stayed to the end, listening calmly to the evidence, and saying nothing. Only then did he make his objections known, confronting Banks directly in person. 'Some people left the room. Lord Byron who was admitted that evening, came to me to say: "Surely, Sir Joseph, this is *too much.*" '[14] Banks liked this style, gentlemanly but undaunted. When Byron went into exile in 1816, Banks was careful to have the Parisian publisher Galignani post over all his latest poetry.[15]

However, on receiving an early copy of the first canto of Byron's *Don Juan* in 1819, Banks was outraged. 'I never read so Lascivious a performance. No woman here will Confess that she has read it. We hitherto considered his Lordship only as an Atheist without morals. We now must add to his respectable Qualifications that of being a Profligate.'[16] Yet had Banks lived to read the tenth canto (1821), he might well have been amused by His Lordship's nimble mockery of Newton and the story of the falling apple, which of course Byron associates with Adam and Eve in the Garden of Eden.

> When Newton saw an apple fall, he found
> In that slight startle from his contemplation –
> 'T is said (for I'll not answer above ground
> For any sage's creed or calculation) –
> A mode of proving that the earth turn'd round
> In a most natural whirl, called 'gravitation;'
> And this is the sole mortal who could grapple,
> Since Adam, with a fall or with an apple.

Byron went on to praise the achievements of post-Newtonian science in his own elegant and bantering way.

> Man fell with apples, and with apples rose,
> If this be true; for we must deem the mode
> In which Sir Isaac Newton could disclose
> Through the then unpaved stars the turnpike road,
> A thing to counterbalance human woes:
> For ever since immortal man hath glow'd
> With all kinds of mechanics, and full soon
> Steam-engines will conduct him to the moon.

Most remarkable of all, in the next stanza Byron light-heartedly connected the discovery and daring of contemporary science with that of contemporary poetry. Both should be dauntless, and 'sail in the wind's eye'.

> And wherefore this exordium? – Why, just now,
> In taking up this paltry sheet of paper,

My bosom underwent a glorious glow,
And my internal spirit cut a caper:
And though so much inferior, as I know,
To those who, by the dint of glass and vapour,
Discover stars and sail in the wind's eye,
I wish to do as much by poesy.[17]

Banks's conservatism showed in other ways. The end of the Napoleonic Wars had raised the question of the future role of science in the growing British Empire and colonies. What loyalty did science owe to the state? Officially in wartime Banks had taken a patriotic line when required, maintaining that national and commercial interests must lead, though producing scientific advantages. His enthusiasm for the Australian penal settlements around Sydney Cove was based on his belief that the tough colonial life would redeem their inhabitants, and ultimately benefit the Empire. Yet they were also more than justified, in his view, by the mass of scientific data and botanical specimens that were constantly sent back to London by their early governors and explorers, like Macquarie, Flinders and Bligh.

Despite his personal experiences in the Pacific, and the reports of Mungo Park from West Africa, Banks would not commit the Royal Society to support the abolition of slavery in the black colonies. Indeed he was inclined to be satirical about the Abolitionists, once remarking to his confidant Sir Charles Blagden how 'Saint Wilberforce is just returned [from the Antipodes]; he carries with him 4 Persons Tried and Proved in all religious Points up to the standard of Beatification'.[18]

Yet, paradoxically, he would support abolition in his own way. The slave trade, he believed, should be dismantled for purely commercial reasons. It was simply scientifically inefficient. Rivalry with the French in the West Indies, where there was a huge sugar industry based on black slaves, proved that the labour of 'freemen' was more productive than that of slaves. But this, he maintained, was not a moral position. 'A struggle almost equal to an Earthquake must take place & Slavery must be abolished not on moral principles, which are in my opinion incapable of being maintained in argument, but on Commercial ones which weigh equally in moral & immoral minds.'[19]

Certainly by 1815, when the black revolutionary movement had established itself on Haiti, Banks could write excitedly to 'Saint' Wilberforce with all his old, boyish and Romantic enthusiasm. 'Was I Five and Twenty, as I was when I embarked with Capt. Cook, I am very sure I should not lose a day in Embarking for Hayti. To see a sort of Human Beings emerging from Slavery & making the most rapid Strides towards the perfection of Civilization, must I think be the most delightful of all Food for Contemplation.'[20] The new King of Haiti – perhaps a more superior 'sort of Human Being' – never stopped sending Sir Joseph Banks specimens for Kew, and inviting him to make a ceremonial visit to the island.

2

If it was obvious to Banks what a young man should do at five and twenty, it was not so clear to John Herschel. In November 1813, at the age of twenty-one, John had the most serious disagreement of his life with his old father, Sir William. It was over his choice of profession.

John's career at Cambridge had been a meteoric success. As his aunt Caroline put it dotingly in her journal: 'from the time he entered the University till his leaving he had gained all the first Prizes without exception'.[21] Caroline had remained in his confidence. She was eagerly introduced to his glittering Cambridge friends, among them the mathematician Charles Babbage, future Lucasian Professor, and the Lancashire geologist William Whewell, future Master of Trinity. She had been a guest of honour at his twenty-first birthday dinner, when John presented her with a 'very handsome' silver necklace. Typically, she almost immediately gave it away to a niece, 'I being too old for wearing such ornaments,' as she remarked coquettishly.[22]

She was particularly impressed when John and Babbage formed the Analytical Society in 1812. It was dedicated to replacing Newton's fluxions with the Continental calculus. This was the very subject that she remembered William and his brother Jacob arguing over all those years ago in the little house in Hanover. Now her nephew was a Fellow of the Royal Society, and in her view he had the world at his feet.

But being the only child, the prodigy, the apple of his father's eye, John found it very difficult to speak his mind to either parent. He wrote confidentially to Babbage: 'God knows how ardently I wish I had ten lives, or that capacity, that enviable capacity, of husbanding every *atom* of time, which some possess, and which enables them to do ten times as much in one life.'[23]

Finally, shortly before the Christmas vacation of 1813, John plucked up courage and wrote a long letter from Cambridge to his father in Slough. In it he made it clear that he wished either to remain permanently as a Fellow at Cambridge, doing research in pure mathematics, or else to support himself as a lawyer in London. Here he thought he would have ample time to pursue more practical science – notably chemistry and geology – in the evenings and during law vacations. He knew that he was probably heir to a considerable fortune, from both his mother's and his father's side. But he believed that it was his duty to acquire an independent livelihood, and that 'a man should have some ostensible means of getting his bread, by the labour of his head or hand'.

William Herschel was dismayed by this outburst, and wrote back reproachfully. His son was ungrateful, and did not have 'a just idea' of his privileged situation. William could approve of neither of John's proposed careers. He was scathing about the law: 'It is crooked, tortuous and precarious ... Your studies have been of a superior kind.' The idea that the mere routine of the law would allow 'unbounded scope' for science was 'a most egregious error'. Herschel was also subtly undermining about Cambridge. 'You say that Cambridge affords you the society of persons of your own age, and your own way of thinking; but know my dear son, that the company and conversation of older, experienced men, of sound judgement, whose way of thinking will often be different from your own, would be much more instructive, and ought to be carefully frequented.'

Finally he urged his son to become – of all things – a clergyman. John must have been astonished to receive a long, passionate letter in support of a career in the Church. But perhaps he also saw the unconscious humour of this recommendation. The real advantages, as his father solemnly enumerated them, emerged in a list of ever ascending and increasingly secular importance: 'A clergyman ... has time for the attainment of the more elegant branches of literature, for poetry, for music, for drawing, for natural history, for short and pleasant excursions of travelling, for being acquainted with the spirit of the law of his country, for history, for political economy, for mathematics, for astronomy, for metaphysics, and for being an author upon any one subject in which ... [he is] qualified to excell.'[24]

John quickly replied, explaining frankly that he could not believe in Anglican doctrine. But Herschel was quite equal to this apparently insuperable objection. 'You say the church requires the necessity of keeping up a perpetual system of self-deception, or something worse for the purpose of supporting theological tenets of any set of men. The most conscientious clergyman may preach a sermon full of sound morality, and no one will enquire into theological subtleties.'

This attitude infuriated John, and he remonstrated with his father, coming close to accusing him of hypocrisy. Herschel's dismay now turned to anger. 'You say [you] cannot help regarding the source of church emolument with an evil eye. The miserable tendency of such a sentiment, the injustice and the arrogance it expresses, are beyond my conception.' There was now a grave risk of a serious breach between father and son, as there would be so frequently among a whole later generation of Victorian families over exactly such matters.

After four days' reflection, the old astronomer – William Herschel was now aged seventy-five, and increasingly fragile – suddenly recognised the fatal gleam of filial disaffection, the risk of a real rupture. Perhaps he remembered the arguments with his own family years ago in Hanover; or perhaps Caroline reminded him of them. At all events, he wrote again to John in a chastened and forgiving mood. Nothing he had said was intended 'to breathe the spirit of bitterness' against his son. He simply wished to hear 'everything you have to say on the subject'. He loved him unconditionally. 'I can as little doubt your sincere attachment to your old philosophical father, as he does of your perfect returning affection.' His mother added soothing postscripts to Herschel's letters, assuring John that his father was not 'really angry', and adding pathetically: 'Cannot you dine on Xmas day? You would make us all happy.'[25]

In the end John consented to come down from Cambridge and have a long, frank discussion with his father about the future. Herschel reassured him that there need be no profound disagreement, carefully pointing out in one of his letters that he had deliberately never previously discussed religion with his son: 'I wished to leave you at liberty to follow your own sentiments.' He did not believe they could possibly disagree on that subject, being 'two unprejudiced persons with natural good sense'. In effect, he had no more belief in Anglican doctrines than his son had.

Wisely, Herschel decided to give John his head, and bide his time. His son would not have to pursue the Church. Instead, he could go to London with Babbage, attend regular meetings of the Royal Society, and try out a legal career at Lincoln's Inn. As Herschel suspected, the trial of

London legal life did not last long, and was succeeded by a mathematical tutorship and then a full Fellowship back at St John's, Cambridge. 'I am determined,' John wrote rather desperately to Babbage in March 1815, 'as the profession is of my own choosing, much against the wish of my parents, that I will pursue it in good earnest.'[26] But he was unhappy and drifting.

In the summer of 1816 John took a holiday with his father, now aged seventy-seven, to Dawlish on the idyllic Devon coast. For several evenings they sat out under the stars, with rebellious Queen Cassiopia very bright overhead, and gently talked things over. Eventually John submitted. With that extraordinary lifelong determination of his, William Herschel had once again quietly achieved his real objective: to have John come home, and pursue a full-time career in science. On this basis, Herschel agreed to provide his son with an immediate and generous private income, so that he would be free to pursue pure research in whatever field he chose. In return John found himself volunteering to assist his ageing father with his astronomical work, and take over the running of the forty-foot and the observatory at Slough.

Caroline had faithfully supported John throughout these discussions and waverings. Very often John's afternoons would end in her lodgings at Slough, where he could take tea, let off steam and discuss the leading scientific questions of the day. Then at dusk they would all three – father, son and comet-hunting aunt – meet at the foot of the great telescope, and work would begin. In December 1819 John presented his first paper to the Royal Society, correcting Newton on the subject of polarised light. Joseph Banks observed that it caused a stir among the mathematicians, and 'much interest among the Polarizers'. A new man of science was launched.[27]

3

William Herschel's reputation was now spreading rapidly among other young men. Percy Bysshe Shelley, fascinated by science since his earliest days at Eton and Oxford, was driven to revolutionary thoughts. At the age of eighteen, Shelley had been expelled from university for publishing a pamphlet, 'The Necessity of Atheism'. At twenty-one, he drew on Herschel's work (as well as Godwin and several French *philosophes*) to write a series of free-thinking prose 'Notes' which he appended to his epic poem *Queen Mab*, published in 1813. This technique of adding long, discursive notes to the poetry was imitated from Erasmus Darwin's *Botanic Garden*, though in Shelley's case they were often angry and polemical.

Shelley used Herschel's vision of an open-ended solar system, and an unimaginably expanded universe, to attack religious belief. His arguments went as follows. The cosmos as revealed by science must contain many thousands of different nebular systems, and therefore millions of habitable planets, so it was impossible to sustain a narrow, religious concept of one Almighty Christian Redeemer. Since there would be so many other 'fallen' worlds to redeem, the idea of God being born and crucified on each planet became absurd. As Shelley put it provokingly, 'His Works have borne witness against Him.' He wrote a particularly fierce note 'On the Plurality of Worlds' in *Queen Mab*:

> The indefinite immensity of the universe, is the most aweful subject of contemplation ... It is impossible to believe that the Spirit that pervades this infinite machine begat a son upon the body of a Jewish woman ... The works of His fingers have borne witness against him ... Sirius is supposed to be 54 trillion miles from the Earth ... Millions and millions of suns are ranged around us, all attended by innumerable worlds, yet calm, regular, and harmonious, all keeping the paths of immutable Necessity.[28]

Shelley's later prose writings, still little-known, continue these materialist ideas, and explore the implications of contemporary scientific research with candour and ferocity. In his 'Essay on a Future State' (1819), he argued that the scientific and anecdotal evidence for the total cessation of all mental and bodily functions after death was definitive. There *was* no Future State.[29]

In his teasing 'Essay on the Devil and Devils', Shelley used the ideas of Herschel and Laplace to satirise beliefs in a geocentric cosmology. 'Are Earthlings or Jupetrians more worthy of visitations by the Devil ... ?' He was also amused to see that Herschel always believed that the sun was inhabited, and asks if this was, after all, the most sensible location for Hell.[30] Many other scientific ideas also appeared in his poems of 1819–21, especially in *Prometheus Unbound*, which revels in Herschel's new cosmology and Davy's chemistry.

In Act I, the earth speaks of her own birth struggles, witnessed by the rest of the galaxy:

> Then see those million worlds which burn and roll
> Around us; their inhabitants beheld
> My sphere'ed light wane in wide heaven ...[31]

In Act II, Asia describes the earliest, painful emergence of human tribes upon the planet, in terms that recall Davy's account of man before the advent of science and hope:

> ... and the unseasonable seasons drove
> With alternating shafts of frost and fire,
> Their shelterless, pale tribes to mountain caves:
> And in their desert hearts fierce wants he sent ...
> Prometheus saw, and waked the legioned Hopes.[32]

In Act IV, Panthea describes electrical energy in terms that almost seem to anticipate the notion of an atomic nucleus surrounded by electrons:

> A sphere, which is as many thousand spheres,
> Solid as crystal, yet through all its mass
> Flow, as through empty space, music and light:
> Ten thousand orbs involving and involved,
> Purple and azure, white, and green, and golden,
> Sphere within sphere; and every space between
> Peopled with unimaginable shapes ...[33]

But perhaps most striking of all is the love song that Shelley gives to the moon to sing to planet earth. Though a pure, traditional love lyric, this elegantly includes scientific notions of gravitational orbit, tidal attractions and magnetic fields. Moreover, the moon's lyric is given an extraordinary kind of hypnotic *humming* sound – the sound of spinning through space.

> Thou art speeding round the sun
> Brightest world of many a one;
> Green and azure sphere which shinest
> With a light which is divinest
> Among all the lamps of Heaven

To whom light and life is given;
I, thy crystal paramour
Borne beside thee by a power
Like the polar Paradise,
Magnet-like of lovers' eyes;
I, a most enamoured maiden
Whose weak brain is overladen
With the pleasure of her love,
Maniac-like around thee move
Gazing, an insensiate bride,
On thy form from every side . . . [34] ♣

4

As the impact of the new Romantic science spread through Regency England, Banks was much concerned with securing the reputation of the Royal Society. He had struggled to maintain its pre-eminence in British science, and had fought to prevent the splitting away of new, separatist bodies like the Geological Society (1807) and the Astronomical Society (1820). 'I see plainly that all these new-fangled Associations will finally dismantle the Royal Society, and not leave the Old Lady a rag to cover her,' he wrote in 1818.[35] He accepted an honorary membership in the Geological, but pointedly resigned it two years later, making his displeasure at its independent policies known.

Banks felt that the new Astronomical Society would certainly steal his thunder with new discoveries. When the Duke of Somerset accepted the first presidency, Banks called him to breakfast, and convinced him to resign even before he had taken up the presidential chair. Other Royal Society members were sufficiently intimidated to send Banks notification of their invitations to join the Astronomical Society, with copies of their refusals annexed.[36]

♣ It might be too much to consider this as Shelley's tribute to Herschel and his faithful, orbiting assistant Caroline. But it can be said that the view Shelley imagines of the 'green and azure sphere' seen from the moon is exactly that enshrined in the famous 'Earthrise' photograph of December 1968 (see page 161).

But Banks was trying to hold back a tide of history. It was no coincidence that it was the young men from Cambridge, John Herschel and Charles Babbage, who were leading the astronomers away from the Royal Society. The increasing separation and professionalisation of the individual scientific disciplines had begun at the universities. It would become the general hallmark of Victorian science. Nor could Banks have imagined that it would be a woman who would first identify this development, and grasp its opportunities, in a short, incisive book, *On the Connexion of the Physical Sciences* (1834). It was written by Mary Somerville (1780–1872), whose husband was a Fellow of the Royal Society. Though she lived to be ninety-one, and would have an Oxford college named after her posthumously, Somerville herself was never elected.

Banks retained a noble Enlightenment vision of a unified science, but his Romantic instincts had steadily given way to conservative policies. Under him, the election of the Royal Society's members had steadily ossified. Over 10 per cent were now clergymen (including a large number of bishops), and nearly 20 per cent were members of the landed aristocracy. The Council itself consisted of 40 per cent such members.[37] Neither of these groups necessarily excluded true men of science, but among younger members there was a growing feeling of stifling consensus, cautious propriety and snobbish exclusion, which did not reflect the spirit of the age.

This was particularly felt in the increasingly flourishing philosophical societies of the provinces, and especially the great manufacturing cities of the Midlands and the North. To John Herschel and Babbage it seemed astonishing that a chemist like John Dalton, from Manchester, should not have been elected, or that Michael Faraday should have received no medal for his work. Many younger members now referred scathingly to Banks as a 'courtier'. This did not prevent Babbage from asking him for a personal recommendation when applying for the Chair of Mathematics at Edinburgh in 1819 (difficult because he was not a Scot). Banks ended his letter of reference with genuine warmth: 'Adieu my dear Sir, believe me Anxious for your success & with real Esteem and Regard'.[38]

Banks felt the pressures of age and unpopularity, and, increasingly weak and immobilised, wondered if he should continue. He secretly admitted that his eyes were no longer good enough even to look through a microscope. Gout inflamed his arm joints, and uric acid formed kidney stones which regularly passed through his urethra with agonising spasms.

In November 1819 he wrote uncertainly to his confidant Blagden: 'Our Election approaches. I almost feel uneasy at again offering myself a Candidate. If I am again elected it will be the 42nd time. Enough I think to satisfy the ambition of any man.' Blagden noted that the President's arithmetic was also weakening: it would be his forty-first election.[39]

His election was, in the event, a triumph. Confirmed by acclaim, he was 'unanimously replaced in the Chair'. But it was not a good winter. 'The cold Weather disagreed with me & I think paralysed all the activity of Science. Now the death of the late King [George III] & the dangerous indisposition of Geo IV has brought all things to a Stand Still.'[40]

Yet still Banks schemed and dreamed with his protégés. He had arranged for young Lieutenant William Edward Parry to mount a polar expedition through Baffin Bay, to make one more attempt on the elusive North-West Passage, from the Arctic to the Pacific oceans. The twenty-eight-year-old Parry, manfully suppressing his unseamanlike nerves, had been summoned to one of the by-now legendary breakfasts at 32 Soho Square, and left a vivid record of the event and Banks's bluff and hospitable style.

> At ten precisely Lady and Mrs [Sophia] Banks made their appearance, to whom I was introduced in form, and without waiting for Sir J (who was wheeled in, five minutes after) we sat down to breakfast. Sir J shook hands with me very cordially, said he was glad to become acquainted with a Son of Dr Parry's, for whom he entertained the highest respect, and was glad to find I was nominated to serve on the Expedition to the North West. Having breakfasted, I wheeled Sir J into an anteroom which adjoins the library, and, without any previous remark, he opened the map which he had just constructed, and in which the situation is shown, of that enormous mass of ice which has lately disappeared from the Eastern coast of Greenland ... He desired that I would come to him as often as I pleased ('the oftener the better') and read or take away any books I could find in his library that might be of service to me. He made me take his map with me ... Having obtained *carte blanche* from Sir J, I shall of course go to his library without any ceremony, whenever I have occasions ...[41]

Throughout his last spring Banks waited anxiously for reports of 'our Polar adventurers', and news of their progress. Parry's specially constructed ship HMS *Hecla*, 'fitted as strong as wood & Iron can make her', would take two years to pass through the ice, and Banks was dead before this young protégé returned. Parry was the first to sail right through the perilous Lancaster Sound, and had named a remote and icy promontory at the far end, adjacent to the Beaufort Sea, Banks Island, after his patron.[42]

One of Banks's last pet projects was to find some brilliant young astronomer to set up a major observatory in South Africa, at the Cape, so the southern sky could be explored as William Herschel had explored the northern. He never gave up looking for this man, although in fact he was close by all the time.[43]

Banks became very ill with jaundice in the spring of 1820. His last letters were written from Soho Square to Blagden in Paris. In one of them, very brief and signed 'in haste', he showed that he had lost none of his ranging interests. He commented on a new thermometer used to calculate the strength of alcoholic spirits; on the notorious 'Lancashire Black Drop' opium, 'said to resemble Morphium very much and produce the same effects of Depression'; and on two delightful Newfoundland puppies he was sending to Blagden on the Paris mail coach, very eager to meet him, but waiting for a suitable passenger to take them over.

They may never have met their new master. To Banks's dismay and grief, Charles Blagden died, while drinking coffee with Berthollet and Laplace, a fortnight later. It was perhaps the greatest professional blow Banks had sustained since the death of his old shipmate and scientific comrade Daniel Solander.[44]

In late May 1820 Sir Joseph Banks wrote in a firm hand from Soho Square to offer his resignation to the Royal Society, being 'so far impaired in sight and hearing' as to be unable to carry out his presidential duties. The Society unanimously rejected his resignation. Possibly the last letter he read was from the Director of the Botanic Gardens at Glasgow. It enclosed a list of their most sought-after rare plants, including no fewer than ten in the family of Banksia. If he was childless, yet he had a numerous offspring.

Sir Joseph Banks died on 19 June 1820, nursed by his faithful and long-suffering wife.[45] With his death, after over forty years as President of the Royal Society, there was the sense that a distinctive era in British science had come to an end. Within a decade this had sharpened into a growing feeling of uncertainty and crisis.

5

At first it seemed that Sir Humphry Davy, called back from his European wanderings, was the likeliest successor. Davy arrived in London on 16 June 1820, three days before the death of Banks. The presidency of the Royal Society was now vacant, and Davy saw this as the natural summit of his professional ambitions, as he told his mother in a confidential letter. He sent her a beautiful Italian shawl, posted down to Cornwall, and coral necklaces to his sisters. At this critical juncture he was alone in London, for Jane had remained in Paris. They were both aware that the second Continental tour had not healed the rifts in their marriage.

Yet it was now more than ever important to establish a workable *modus vivendi* with Jane. Davy urged her to return, and never considered divorce, largely because of the Royal Society. In a curious way they were both trapped by the requirements of their public lives. They agreed to accompany each other to official events, but to travel and entertain separately as far as possible. With this in mind, they sold the house in Grosvenor Street early that summer, and bought an even larger one in Park Street, on the more fashionable side of Grosvenor Square, nearer Hyde Park. Here, with large suites of rooms and separate staircases, Jane and Davy could conduct more independent lives, but still present themselves as the first scientific couple of the nation.

Davy threw all his energy into lobbying Fellows to support his candidature for the presidency, with private letters and discreet dinner invitations. His old friend Davies Giddy acted as his unofficial party manager. His high public profile, his baronetcy and his reputation at home and abroad as the inventor of the safety lamp attracted what appeared to be an unassailable majority. Yet there were rumours of dissent. Aristocratic members were uneasy at Davy's Cornish background (so different from Banks's Eton and Oxford), while younger members, on the contrary, wondered if his social ambitions had overtaken his scientific ones.

An alternative candidate emerged. The shy, mild, supremely dedicated and meticulous chemist Dr William Hyde Wollaston (who had been appointed caretaker President) found himself being championed by the young Turks, and especially the group of Cambridge men including Babbage, Whewell and John Herschel. It was felt that Wollaston represented British science at its purest, while Davy, for all his fame, was a contentious figure. John Herschel expressed this view vigorously in a private letter to Babbage in June 1820: 'The reasons for wishing that Davy should be opposed are grounded solely on his personal character, which is said to be arrogant in the extreme, and impatient of opposition in his scientific views, and likely, if power were placed in his hand to oppose rising merit in his own line, and not patronise it in others, and in particular to involve the Royal Society in controversies of much personal acrimony with other learned European bodies.'

These *caveats* made clear reference to Davy's treatment of Faraday, and the awkward priority dispute with Gay-Lussac and the French Académie des Sciences.[46] As Herschel did not know Davy personally at this stage, much of this was hearsay and gossip. Yet it was precisely the sort of thing that Wollaston dreaded, and, appalled at the notion of open wrangling between scientific men, he abruptly withdrew his candidature in favour of Davy. The vote was set for November 1820.

Davy and Jane now accepted an invitation to spend the later part of that summer (the grouse-shooting season) in Scotland with Walter Scott, recently made a baronet by the newly crowned George IV. They travelled to the manse at Abbotsford separately, but both enjoyed mingling with the Scottish aristocracy and literary men like Scott's son-in-law John Lockhart and Henry Mackenzie (author of *The Man of Feeling*), who took a fancy to Jane and travelled in her carriage during the endless hunting expeditions. Davy managed to spend most of his days shooting on the moors, and his evenings in Scott's smoking room. With considerable diplomacy Scott had also invited Wollaston, who proved himself a keen fisherman, so that he and Davy were soon on good terms, teasing each other with piscatorial arcanae.

Lockhart later wrote an amusing account of Davy striding out at dawn in his full fishing gear, his white wide-brimmed hat stuck with innumerable fly-hooks and his enormous green waders far in excess of what any tinkling Scottish burn could possibly require. Yet Davy would also recite from memory passages of Scott's *Lay of the Last Minstrel*, while sipping whisky during a moorland picnic. Lockhart recalled one of the Scottish ghillies whispering to him when Davy and Scott had kept the party up with their 'rapt talk' round the log fire, long after midnight: ' "*Gude* preserve us! This is a very superior occasion! Eh Sirs!" – then cocking his eye like a bird – "I wonder if Shakespeare and Bacon ever met to *screw ilk other up*?" '[47]

Back in London, Davy was elected President of the Royal Society unopposed in November 1820. In his acceptance speeches he tried very hard to smooth over old differences, and presented an uplifting vision of 'The Progress and Prospects of Science' to the assembled Society in December. He recalled the great tradition of 'experiment, discovery and speculative science', from the time of Hooke and Newton to that of William Herschel and Cavendish. If he was now in some sense their 'general' and leader, he announced, 'I shall always be happy to act as a private soldier in the ranks of science.' Perhaps there were some ironic smiles at that.

Prophetically, Davy picked out the fields of research in which the most exciting new work would be done: astronomy, polar exploration, the physics of heat and light, electricity and magnetism, geology, and the physiology of plants and animals. He was careful to mention by name the work of Wollaston, Dalton, John Herschel, the young Scottish physicist David Brewster and a number of French chemists. He urged the Fellows to be guided by the spirit of Bacon and Newton, to work soberly by 'the cautious method of inductive reasoning', but with passion, and 'to exalt the powers of the human mind'. Finally he issued an exhortation, with a kind of challenge and warning attached. 'Let us then labour together, and steadily endeavour to gain what are perhaps the noblest objects of ambition – acquisitions which may be useful to our fellow creatures. *Let it not be said, that, at a period when our empire was at its highest pitch of greatness, the sciences began to decline . . .* '[48] That last sentence would come back to haunt the Society.

Davy's initial determination to recover the support of the younger men was shown in several ways. He made attempts to befriend John Herschel (now twenty-eight) over Park Street dinners, and voted money for Charles Babbage's first prototype of his famous 'difference engine', or calculating machine. In 1821 he made sure that the annual Copley Medal was awarded to young Herschel for his work on polarised light (as Banks had once assured it to his father for discovering Uranus). The award was accompanied by a handsome speech. 'You are in the prime of life, in the beginning of your career, and you have powers and acquirements capable of illustrating and extending every branch of physical enquiry . . . May you continue to exalt your reputation, already so high.' He concluded with a reference to the work of Sir William, now an almost legendary figure. John was urged to follow 'the example of your illustrious father, who full of years and of honours, must view your exertions with infinite pleasure; and who, in the hopes that his own imperishable name will be permanently connected with yours in the annals of philosophy, must look forward to a double immortality'.[49] Though obviously well intentioned, this must surely have sounded more than a little heavy-handed to John, and more than a little ludicrous to Babbage.

Davy had also been curiously undiplomatic about his old patron and predecessor in the Chair, Sir Joseph Banks. It was surely a moment for generosity, especially considering the support Banks had given him over the Bakerian Lectures and the safety lamp. Yet he circulated a sketch of Sir Joseph which seemed unnecessarily grudging and critical: 'He was a good humoured and liberal man, free and various in conversational power, a tolerable botanist, and generally acquainted with natural history. He had not much reading, and no profound information. He was always ready to promote the objects of men of science; but he required to be regarded as a patron, and readily swallowed gross flattery. When he gave anecdotes of his voyages, he was very entertaining and unaffected. A courtier in character, he was a warm friend to a good King. In his relations to the Royal Society he was too personal, and made his house a circle like a court.'[50]

So Davy dismissed Banks as a dilettante and a patriarch. There was little recognition of the huge scientific network that he had established, nor of the way in which he had kept British science alive and international in a time of war. Above all there was no recognition of Davy's personal debt to him, let alone of the heroic way he had battled against personal illness and disability. Perhaps this can be partly explained by Davy's immense anxiety to assert his own authority at this juncture. Perhaps, too, he wanted to be seen as speaking for the younger generation of professional men of science. But it was a hurtful and puzzling document all the same.

That summer Davy went down to Penzance, to glory a little in his own immortality. He was given a public dinner by the Mayor, interviewed and toasted, presented at a provincial ball, all to the immense satisfaction of his ageing mother. He was the guest of honour of the newly founded Geological Society (to which he made a handsome contribution), and was congratulated by its jocular new President, John Ayrton Paris. Paris may have mentioned that he had ambitions to become Sir Humphry's eventual biographer. At all events he shrewdly noted that Lady Davy

could not be persuaded to accompany the great man on this filial visit.*
Davy, aged forty-two, innocently revelled in the role of local boy made good, and wrote dreamily to his old friend Tom Poole in Somerset: 'I am enjoying the majestic in Nature, and living over again the days of my infancy and early youth ... I am now reviving old associations, and endeavouring to attach old feelings to a few simple objects.'[51]

For a time Davy felt that he had achieved his greatest ambition in becoming President of the Royal Society. Yet his peremptory genius in the laboratory made him something of a tyrant in the Committee Room. Over the next three years, to his dismay and astonishment, he found that his popularity, while still immensely high with the general public, was increasingly resented at Somerset House. His social ambitions and snobbery were easily mocked, and he did not have the gift of drawing out hidden talents. As John Herschel had feared, he was irascible, and easily drawn into feuds.

This was particularly marked in the case of Michael Faraday, who was after all Davy's star pupil, now aged twenty-nine. In 1821 Faraday had married very happily, after a tender two-year courtship sparking numerous love letters, including some rather neat light verse.[52] His bride was Sarah Barnard, a pretty, quietly-spoken girl and fellow Sandemanian who was happy to move into his modest set of rooms above the Institution, thus enabling him to continue with his formidable daily workload in the laboratory below. Faraday lectured, published papers, and developed a strong connection with the French physicists, notably Pierre Hachette and André Ampère.[53]

However, there was no increase in his salary, and in 1823 Davy took the extraordinary step of blackballing Faraday's election to a Fellowship of the Royal Society. This was all the more surprising since Faraday had just been elected to the scientific Accademia in Florence, and to the Académie des Sciences in Paris. Davy's explanation was that Faraday had offended Wollaston, by plagiarising some experiments concerning electro-magnetic rotation and falsely claiming priority in results. But Faraday had

---

* Curiously, vague feelings against Lady Davy have always remained in the collective folk memory of Penzance, probably because she never deigned to visit this remote Cornish seaside town during her lifetime. I was told on several occasions that the large stone statue erected to Davy, dominating Market Jew Street, showed his frock-coat with a missing button 'because Lady Davy was a bad wife and would never sew it back on'.

already established his own authority in this field (which he would soon revolutionise), and anyway he was always meticulous in such matters. Evidently no deliberate plagiarism was intended, and Wollaston himself (as mild as ever) was much inclined to dismiss the whole affair. Davy seemed incapable of recognising Faraday's rising star in the world of international science.

Faraday's supporters, of whom there were an increasing number, thought that Davy seemed motivated, at least unconsciously, by jealousy of his old pupil. Some may also have believed that his painful experience of the safety lamp controversy had made him oversensitive about priority disputes. Yet others suggested darkly that Sir Humphry was influenced by a different kind of magnetism, the negative polarity of Lady Davy towards her erstwhile 'valet'. Again, all this was much as John Herschel had foretold. Faraday's election became an embarrassing *cause célèbre*, with notices pinned up and torn down from the Royal Society's noticeboard. Faraday's seconders in the ballot eventually included the names of John Herschel, Babbage, Charles Hatchett, Peter Roget (from the old Bristol Pneumatic Institute days), Dr Babington (one of Davy's fishing cronies), Davies Gilbert (his campaign manager), and even Wollaston himself.

In the end Faraday's election had to be proposed no fewer than eleven times, a proceeding without precedent, and was not ratified until January 1824. The Royal Society's minutes noted that finally there was only one vote cast against him, but according to the club rules it remained anonymous. It was obviously Davy's own. It seems that the President had been isolated and humiliated.[54]

There was even a story put about that Davy had deliberately encouraged Faraday to undertake a potentially lethal chemical experiment, which had nearly blinded him. It was said that one Saturday evening in March 1823, looking in at the Royal Institution's basement laboratory, Davy had casually suggested to Faraday that he try a further analysis of chlorine crystals by heating potassium chlorate with sulphuric acid in a sealed glass tube. (The properties of potassium chlorate – used to produce medical chlorine – were one of Davy's great chemical discoveries.) After Davy left, Faraday did so, and the subsequent explosion, in Faraday's own words, 'blew pieces of glass like a pistol-shot through the window', lacerating his face and filling his eyes with tiny fragments of glass. Sarah Faraday spent the rest of the evening tenderly sponging them out with cold water.

This sinister story gained credibility as it circulated. So much so, that thirteen years later Faraday was still being asked about it, and was not entirely inclined to exculpate his old professor. Perhaps Davy – knowing very well the properties of potassium chlorate – had set him a kind of pedagogical object lesson; or, frankly, a trap. Perhaps he wanted to underline just how much Faraday still had to learn in chemical matters. 'I did not at that time know what to anticipate, for Sir H. Davy *had not told me his expectations*, and I had not reasoned so deeply as he appears to have done. Perhaps he left me unacquainted with them to try my ability.'[55]

Though apparently disingenuous, this is a surprisingly damaging suggestion, and does imply some ill-will on Davy's part. Yet it also reveals that Faraday still thought of himself (married and aged thirty-one) as Davy's naïve apprentice, whose 'ability' might very reasonably be put to the test. Faraday also fails to mention the fact that the chlorine experiments took place over several days, and produced not one but several explosions. The first was a minor one that surprised, but did not harm, him. He then deliberately pursued the course of the experiments, presumably now forewarned of what might occur. The explosion that nearly blinded him was actually the third to shake his laboratory. Neither Faraday nor Davy wore the 'safety spectacles' that are now *de rigueur* in laboratory work. It all throws light on a new and highly significant human relationship that was emerging in professional science: that between the director and his research assistant, between master and pupil, between sorcerer and apprentice.[56] ♣

Davy was more successful in forming new friendships outside the Royal Society, notably with the rising politician Robert Peel, then Home Secretary. Like Banks before him he tried to raise the government's awareness of science and technology. With Peel he became a Trustee of

---

♣ The lively ambiguity of this relationship continues in modern research laboratories, where the line between assistant and collaborator remains easily blurred. A protocol has emerged in the joint signing of scientific papers for journals such as *Nature*; and in many British universities it is obligatory for a Director of Studies to allow his postgraduate assistants to co-sign research studies. But there are still many anomalies. It is currently the view that Edwin Hubble owed a great deal more to his assistant Milton Humason, a genius with stellar photography, than was originally recognised in his historic papers on red-shift. The examples of William Lawrence with Abernethy, Gay-Lussac with Berthollet, and most of all perhaps, Caroline Herschel with her brother, are even more subtle and complicated.

the British Museum, and helped develop the Great Russell Street site, which included organising George III's great bequest of books, which became the famous King's Library. The collection also included the Greek and Egyptian statuary which had inspired Shelley, Keats and Leigh Hunt to write their fine, thoughtful sonnets about Nature, Time and Empire. Shelley's 'Ozymandias', a meditation on the enormous stone head of Ramses II, was one of the last poems he wrote before departing for Italy. It might be described as a meditation on imperial *hubris*.[57]

Davy wanted to establish a stronger scientific presence at the Museum, and make it more open to the general public. He suggested reorganising it as three main departments under separate management: 'a good Public Library – a Gallery of Art – a Gallery of Science'. After four years of frustrating committee meetings – not his forte – he had made little headway, writing in 1826: 'I have been to the British Museum, but I despair of anything being done for Natural History. The Trustees think of nothing but the Arts, and money is only obtained for these objects.'[58] ♣

He took over another of Banks's pet projects, the foundation of the Zoological Society with Sir Stamford Raffles, and drafted its Prospectus, proposing a zoological garden in Regent's Park. He agreed with Peel that it should aim to rival the Jardin des Plantes in Paris, and have at its heart the collection of wild animals from all over the world, and finding ways of adjusting their natural habitats to a northern climate. He could not forbear to add that these might perhaps include 'eight or ten races of partridges'. He also pursued Banks's enthusiasm for polar research, and took over his protégé Parry. In July 1826, Maria Edgeworth, staying as his guest

♣ The foundation of the Natural History Museum, in South Kensington, was achieved in 1881, and the Science Museum in 1885. The New British Library on Euston Road, opened in 1996, took over the King's Library, which now forms the central architectural feature of the building, as a huge and dramatic glass bookcase, rising six storeys high through the central core of the building. Curiously, the New British Library fulfils much of Davy's original vision, containing both science and humanities reading rooms, as well as rare books, maps and manuscripts, and two art galleries with changing displays. Near the main staircase is a bronze bust of Faraday; but none of Davy. In the courtyard is Eduardo Paolozzi's gigantic statue of Newton (1995), an iron man seated on a plinth, leaning forward to take the measure of the world with his dividers. The image wonderfully combines several contradictory versions of science: a noble Enlightenment Newton, reminiscent of Rodin's *The Thinker*; a satanic, calculating, anti-Romantic Newton, based on William Blake's engraving of 1797; and finally, more than a hint of Dr Frankenstein's outcast Creature of 1818.

in Park Street, came down to breakfast to find 'Sir Humphry with a countenance radiant with pleasure and eager to tell me that Captain Parry is to be sent on a new Polar expedition'.[59]

Davy also became a founder member of the all-male Athenaeum, where he could gracefully withdraw from Lady Davy's company in the evenings. As the club was in another part of Somerset House, this was very convenient, and it became virtually an extension of his own presidential study room. He insisted to his fellow founder, the Tory MP and Secretary to the Admiralty John Wilson Croker, that scientific members should be put up with literary and artistic ones, and candidates should be drawn equally from the Royal Academy and the Royal Society. His own list of personal recommendations included John Herschel.

But there was another contretemps when on Davy's advice Michael Faraday was appointed the first Club Secretary. Faraday assumed that at last this was a mark of social acceptance. However, he soon found that the position was really a low-grade, time-wasting clerical appointment, an affair of lists and envelopes, and carried the £100 salary of a clerk. He quietly withdrew. It was hard to know if his old patron Davy had intended a professional kindness or a social slight. Perhaps, concluded Faraday, he barely knew himself. But from now on Faraday, who was very happy in his own drawing room, would make his own way.

Davy's relations with his protégé would remain enigmatic and uneasy, until in 1825 Faraday was eventually proposed as Director of the Royal Institution. Here Sir Humphry Davy was forced to give his approval, and gravely sealed the appointment. So at long last Michael Faraday – modest, unworldly and utterly unlike his patron – was finally appointed to the position that would soon make him world-famous.

6

For John Herschel it was the shadow of his father, and the great forty-foot telescope, which seemed longest. By 1820 it was clear to John that his father was failing. William, now aged eighty-one, could no longer handle the larger telescopes, and was fretful and forgetful over his scientific papers. He grew petulant and anxious if his son was away from Slough too long, and uneasy when John and Babbage made their first extended Continental tour together, visiting France and Italy for four long months between July and October 1821.

In Paris they met the great Alexander von Humboldt, who inspired them with his tales of the South American forests and mountains, which he had visited during his legendary five-year expedition between 1799 and 1804. His *Personal Narrative* had been published in 1805, and translated throughout Europe. It included his visions of the great Amazon river, and his famous account of how he nearly died trying to climb the 20,700-foot Mount Chimborazo (he reached 19,309 feet). They were much struck by his dynamic philosophy of science: 'To track the great and constant laws of Nature manifested in the rapid flux of phenomena, and to trace the reciprocal interaction – the struggle, as it were – of the divided physical forces.'[60]

Humboldt had become a central figure at the great Berlin Academy of Sciences, which Herschel and Babbage particularly wished to emulate. He knew and greatly admired William Herschel's work, but was inclined to underestimate his son's potential. 'John Herschel appears to me inferior to the originality of his father, who was astronomer, physicist and poetical cosmologist all at the same time ... The science of the Cosmos must begin with a description of the heavenly bodies and with a geographical sketch of the universe: or perhaps I should say with a true *mappa mundi*, such as was traced by the bold hand of William Herschel.'[61] But Humboldt, now fifty-two, treated the young men in kindly, avuncular fashion, told them how much he admired English science, and how he had heard Joseph Banks lecture in London long after his return from the round-the-world voyage of 1768–71. So the torch of Romantic inspiration was passed on.[62]

In Switzerland Herschel and Babbage made geology an excuse for adventurous scrambling in the Alps, and wanderings over the glaciers of Chamonix in the footsteps of Dr Frankenstein's Creature. They also made meteorological studies of the mountain storms and cloud formations, and climbed everywhere with telescopes, thermometers, geological hammers and a 'mountain barometer', supposed to warn them of impending storms.[63]

On his return to Slough, John found his unconquerable aunt Caroline had become the sole person who could manage William's daily regime, and understand his increasingly rambling scientific requests. She also helped John develop new sweeping techniques with the cumbersome forty-foot, and once again began acting as astronomical assistant, still able – to John's admiration and amazement – to sustain long nights in the shed beneath the telescope scaffolding. When he formed the Royal Astronomical Society with Charles Babbage in 1820, their first Honorary Member was his aunt Caroline, and this gesture sealed the bond between them. John had strong views about science being open to women – the Society's second Honorary Member was to be Mary Somerville.

At Slough, the old observations workshops were falling into disuse, and masses of equipment and unfiled papers accumulated. William retreated to his study or his day-bed, but occasionally sent Caroline on quixotic missions to recover sheets of calculations or copies of papers once sent to the *Philosophical Transactions.* She alone could do this, but it caused her endless frustration and heartache.

Ill health now came to plague both Herschel and Caroline. The long nights of observation had gradually stricken him with crippling arthritis, while she began to suffer from an eye infection that a local doctor (not James Lind) casually diagnosed as leading to inevitable blindness. After several terrifying weeks, spent largely alone convalescing in her darkened lodgings, fearing she would never be able to see the stars again, Caroline recovered and slowly began using her telescope once more. The experience shook her profoundly, and reminded her of her isolation. The disease was almost certainly ophthalmia, which was rife among poorer households in the Thames Valley at this period. Several years before, Percy Shelley, living nearby across the river at Great Marlow, had also caught it while taking food and blankets to destitute families, as part of one of his many philanthropic projects.[64]

In the last months of his life Herschel had become increasingly weak and immobile. Yet he was loath to give up his stars. During the summer he wrote in a trembling hand on a tiny slip of paper, one of his last surviving notes. '*Lina* – There is a great comet. I want you to assist me. Come to dine and spend the day here. If you can come soon after one o'clock we shall have time to prepare maps and telescopes. I saw its situation last night – it has a long tail.' Caroline meticulously filed away this note, and years later annotated it in her neat, precise script: 'I keep this as a relic! Every line now traced by the hand of my dear brother becomes a treasure to me.'[65]

Towards the end, William asked Caroline to unearth a copy of his late 'Sidereal' paper, together with a print of his forty-foot telescope, to present to a friend who had asked for a special memorial gift. Close to tears, Caroline hurried to his chaotic library of papers to find it. After a long, miserable, dusty search, she finally discovered it, but was then too upset to read it through: '*For the universe* I could not have looked twice at what I had snatched from the shelf,' she recalled. She returned and put it into her brother's hands. 'When he faintly asked if the *breaking up of the Milky Way* was in it, I said "Yes!", and he looked content. I cannot help remembering this circumstance, it being the last time I was sent to the Library on such an occasion.'[66]

On 25 August 1822, Sir William Herschel, knighted and recognised by learned societies around the world, died in his room overlooking the great forty-foot telescope. He was quietly buried in the little church of St Lawrence, Upton, where he had been married. Just as he had feared, his son John had been abroad and had not been at his deathbed. But on his return, at Caroline's urging John wrote a long epitaph to be carved in marble above his father's tomb, and had it translated into elegant Latin by the Provost of Eton. It contained a wonderful phrase: '*Coelorum perrupit claustra*' – 'He broke through the barriers of heaven'; or as a later friend translated, '*He o'er-leapt the parapet of the stars.*'[67]

Herschel's long and distinguished obituary appeared in *The Times*, and across four columns in the September issue of the *Gentleman's Magazine*: 'As an Astronomer he was surpassed by no one of the present age, and the depth of his research, and extent of his observations, rendered him perhaps second only to the immortal Newton.' The magazine added punctiliously: 'In these observations, and the laborious calculations into which they led, he was assisted by his excellent sister, Miss Caroline Herschel, whose indefatigable and unhesitating devotion in the performance of a task usually deemed incompatible with female habits, surpassed all eulogium.' No doubt Caroline was pleased with that mention, though she doubtless objected to astronomy being referred to as a task 'usually incompatible with female habits'.[68]

This obituary was immediately followed in the same issue of the *Gentleman's Magazine* by a short notice of the death of one Percy Bysshe Shelley, son of the Whig MP for Horsham. 'Supposed to have perished at sea, in a storm, somewhere off Via Reggio, on the coast of Italy... Mr Shelley is unfortunately too well-known for his infamous *novels and poems*. He openly professed himself an atheist. His works bear the following titles: *Prometheus Chained* [sic] ... etc.'[69] For good measure a London daily newspaper, the *Courier*, added: 'Shelley, the writer of some infidel poetry, has been drowned: now he knows whether there is a God or no.'[70]

The poet Thomas Campbell, who had interviewed Herschel at Brighton a decade before, wrote a long appreciation of his life in the October issue of the *New Monthly Magazine.* It included a summary of the way Herschel had changed the layman's view of the cosmos: how the solar system was larger and more mysterious than Newton ever supposed; how the creation of the stars had taken place in inconceivable gulfs of time and space, and was still developing and unfolding; how our Milky Way was probably just one galaxy (or island universe) among millions; and how this galaxy – our beautiful home in space – would inevitably wither and die like some fantastic but ephemeral flower. Campbell carefully avoided raising any theological implications, and instead played wittily on the late, mad King George's (perhaps apocryphal) remark: 'Herschel should not sacrifice his valuable time to crotchets and quavers.'[71]

Among many other honours, a new constellation was proposed, *Telescopium Herschelii*, The Telescope; and thus it appears in James Middleton's beautiful *Celestial Atlas* of 1843, located 10 degrees above Castor and Pollux, close to where Uranus first swam into his ken.

Caroline now seemed strangely detached. Despite everything that John could urge, she took the surprising decision to return at once to Hanover, although she had not been there for nearly fifty years. She was now seventy-two, and set about briskly winding up her affairs, making John the executor of her Will.[72] William had left her an annual £100 pension for life, but she immediately made it over to John in quarterly instalments. It was as if she wanted to bring the circle of her life in England to an abrupt close.

The one thing that would have kept her in England, she told John, was if she had been able to 'offer of my service for some time longer to you, my dear nephew' as astronomical assistant. But she felt too old and infirm to do this.[73] She took with her to Germany a large, comfortable English bed, some astronomy books, and the beautiful seven-foot Newtonian 'sweeper' telescope which William had made for her all those years ago in 1786. 'It shall stand in my room and be my monument – as the Forty-Foot is yours.'[74]

On 16 October there was a final reception for her at Bedford Square, London, hosted by Lady Herschel and John. Charles Babbage rode down from Cambridge, arriving at the very last minute. Caroline's parting message to him, an unspoken one, was about John. 'I could find no time for any conversation with [Babbage]; but just by a pressure of the hand recommended my Nephew (in incoherent whispers) *again* to the continuance of his regards and Friendship.'[75]

Caroline was destined to live on for another twenty-six years, her mind sharp and her memories vivid and sometimes bitter. 'I did nothing for my brother,' she once confided, 'but what a well-trained puppy dog would have done, that is to say, I did what he commanded me.'[76] She began to send the first version of her *Memoirs* back from Hanover in the year following William's death, 1823. It was written up slowly, carefully withheld from all her German relatives and friends, and posted in secret instalments to John in England, with many hesitations and *caveats*. She wrote poignantly: 'As my thoughts are continually fixed on the past, I was as it were *conversing* with you on paper, not choosing to trust them to anyone about me [in Hanover]. For I know none who would understand me, or whom it can concern what my own private opinion and remarks have been about the transactions that continually passed before my eyes. But there can be no harm in telling them to my own dear Nephew.'[77]

She and John corresponded regularly for the next twenty years. Very rarely she wrote about her personal feelings, but sometimes there were sudden glimpses, like clouds clearing on a good observation night. 'I am grown much thinner than I was six months ago; when I look at my hands they put me so in mind of what your dear father's were, when I saw them tremble under my eyes, as we latterly played at backgammon together.'[78]

She read all John's Royal Society papers as they appeared, and took huge pleasure in his successes, as if her beloved brother were still alive. She kept him supplied with all the new technical books and papers published in German, and recommended he read the philosopher Schelling. 'You must give me leave to send you any publication you can think of, without mentioning anything about paying for them.' Like many old people, she was fierce if he did not reply to her letters immediately.

Caroline's explanation for her generosity was characteristic. 'It is necessary that every now and then I should lay out a little of my spare cash ... for the sake of supporting the reputation of being a *learned lady* ... for I am not only looked at for such a one, but even *stared* at here in Hanover.'[79] She assembled and recalculated for John a huge new *Star Catalogue* of the 2,500 nebulae. Should he ever escape on the expedition he was starting to dream about, he could add to it while observing the stars of the southern hemisphere. Herschel and Babbage made sure she was awarded the Astronomical Society's Gold Medal for this in 1828.

It sports her name on a beautiful medallion showing William Herschel's forty-foot telescope, and the Society's motto *Quicquid Nitet Notandum* – 'Let Whatever Shines be Noted'.[80] For herself there was now little chance of star-gazing: 'Two or three evenings a week are spoiled by company. And *at the heavens is no getting*, for the high roofs of the opposite houses.'[81]

When John visited her in Hanover he found Caroline to be more energetic than ever. After all those years of stellar observation, she was still essentially a night bird. 'She runs about town with me, and skips up her two flights of stairs. In the morning until eleven or twelve she is dull and weary, but as the day advances she gains life, and is quite *fresh and funny* at ten o'clock pm, and sings old rhymes, nay even dances! to the great delight of all who see her.'[82]

It was Caroline who worried about *his* health, and urged him not to let science drive him too hard. He must not become obsessive about his work, or allow himself to become remote or unfeeling. Here she was clearly looking back on her brother William's career: 'I wish often that I could see what you were doing, that I might give you a caution (if necessary) not to overwork yourself like your dear father did. I long to hear that the Forty Foot instrument is safely got down ... I know how wretched and feverish one feels after two or three nights waking, and I fear you have been too eager at your Twenty-Foot ... I should be very sorry on your account, for if I should not live long enough to know you comfortably married ... if you can meet with a good-natured, handsome and sensible young lady, pray think of it, and do not wait till you are old and cross.'[83]

7

In 1823, pursuing his idea of raising the national profile of science, Davy had accepted a commission from the Royal Navy to solve a major problem with their new steam-powered warships. This was the rapid corrosion of their copper hulls in sea water, which also encouraged their fouling with weed and barnacles. After a relatively short period at sea, the combined effect could drastically reduce the ships' speed through the water and manoeuvrability in action. The commission was widely publicised in the press, and Davy threw himself into the task, hoping to achieve a public success comparable to his invention of the safety lamp in the winter of 1815.

For this work he no longer asked for Faraday's help. He solved the corrosion problem quickly and brilliantly, by analysing the corroding (oxidising) effect of salts on copper, and through a series of experiments finding that it could be neutralised with the use of small cast-iron plates placed along the length of the ship's hull. The more rapid oxidising of the iron produced a charge of 'negative electricity' along the hull, which prevented the oxidising of the copper. He wrote excitedly to his brother John of his 'most beautiful and unequivocal' results.[84] He read a paper on this discovery to the Royal Society in January 1824, and went on naval trials aboard HMS *Comet*, one of the Navy's latest steam paddle-ships, to Scandinavia to demonstrate the results. The work was greeted by a fanfare of approval in the newspapers when he got home.

To crown his achievement, Davy announced with a flourish that, as with the safety lamp, he would refuse to take out a patent. 'I might have made an immense fortune by a patent for this discovery; but I have given it to my country, for in everything connected with interest, I am resolved to live and die at least *sans tache*.'[85] If not the Napoleon of science, he would be the Nelson.

But Davy's claims for the new process were premature. Within months it was found that the unoxidised copper hulls attracted weeds and barnacles far more quickly and heavily than before. By October accusing paragraphs were appearing in the Portsmouth papers, and sarcastic letters in *The Times*. The navy was disgruntled, the Royal Society was embarrassed, and the press was derisive. Davy's reputation was tarnished, not to say barnacled, by this episode, and his unpopularity at the Royal Society increased.[86] It was also noted that while he was touring Scandinavia, his wife was altogether elsewhere on the Continent, travelling through Germany and charming the aged Goethe at Weimar, in a party organised by one of her aristocratic friends, the gossiping Lord Dudley.

Ironically, Davy's science was perfectly correct, only the practical application was faulty. After several years of further sea-trials an adaptation of his iron-plate techniques did keep the Royal Navy's copper hulls perfectly clean. It was largely his impetuosity, his premature publication of results and his increasing hunger for glory that had betrayed him. Moreover, pure science was not the same as applied science. Successful laboratory experiments did not always transfer smoothly to actual conditions in the field. He wrote touching letters to his mother trying to explain all this, and insisting he was right. 'Do not mind any of the lies you may see in the newspapers ... about the failure of one of my experiments. All the experiments *are successful*, more even than I could have hoped.'[87]

But Davy's reputation was now increasingly vulnerable. Robert Harrington had again mocked him in a widely circulated pamphlet as 'a self-styled Hercules ... seated on the shoulders of Sir Joseph Banks'.[88] In 1824 he was attacked by the new magazine *John Bull* in its satirical series 'Humbugs of the Age'. He was pilloried not as a scientist, but as a snob and a socialite (No. 1 was De Quincey, No. 2 was a worldly prelate, and No. 3 was Davy). 'The poor fellow fancies himself irresistible among the girls, and is evidently pluming himself while conversing with them ... about the last new novel, or the set of china, or the pattern of a lace, or the cut of a gown – not at all about chemistry. *O! he is a universal genius. You never, my dear, would take him for a great philosopher.*'[89]

Davy was still attempting to secure his position with the younger Fellows of the Royal Society. He had John Herschel appointed as one of the two Society Secretaries in 1824, but then undermined the reformist implications of this by refusing to have Charles Babbage elected as the other. The irascible Babbage accused Davy of temporising and trimming, while Davy let it be known that the combination of two Cambridge University mathematicians in two such key appointments would, in his opinion, unbalance the Royal Society's traditional composition. Unbalancing the traditional composition, with its predominance of 'slumbering' gentlemen amateurs, was of course exactly what Herschel and Babbage had intended.

Babbage began to reflect angrily on the minatory phrase from Davy's inaugural address, the potential 'decline of British science'. Here was a possible line of attack. But how could 'decline' be *inductively* demonstrated? For example, how many scientific papers or lectures, he wondered, had each Fellow *actually published*? No one had ever considered something so ungentlemanly as gathering such data from the *Philosophical Transactions*. But it might be a good empirical question to ask. He and Herschel had, after all, already published well over fifty papers between them.[90]

Over the next three years Davy spent most of his summers travelling outside London – usually to go shooting or fishing – in Wales, the Lake District, Ireland and Scotland. He joined the house parties of aristocratic acquaintances, but was rarely accompanied by Jane. Older friends like

Wordsworth and Scott noted that his health was weakened. He walked less (though he still climbed Helvellyn), and he drank and talked more.

In September 1826 his mother Grace died in Penzance after a short illness. This had a profoundly upsetting and undermining effect on Davy, from which he never entirely recovered. It was Grace who had sustained him from the earliest days in Borlase's pharmacy, and followed all his triumphs so faithfully. It was now that the hollowness of his marriage left him emotionally unsupported. He attended his mother's funeral in Penzance with his sisters and his brother John, who had returned swiftly from Corfu. But he was not accompanied by Jane, who remained in London. Friends and family thought she was unbelievably callous; but Davy had almost certainly asked her not to come. It had long been agreed between them that his Penzance life was his own.

From this time Davy began to suffer from feelings of exhaustion, pains in his shoulder and right arm which he attributed to rheumatism, and palpitations in his throat. In fact he was suffering from progressive heart disease, which had prematurely killed many on the male side of his family. In October, during his final Bakerian Lecture, he had to admit that his work on the copper sheathing of ships had not been immediately successful.[91] At the annual general meeting of the Royal Society he barely got through his official address, sweating profusely, and returned home to Park Street without attending the official dinner.

In December 1826, while on a shooting party in Sussex with Lord Gale, Davy suffered a series of strokes, and to his horror found himself partially paralysed down his right side. He was taken back to Park Street, where Jane (who had as usual spent Christmas in London) proved herself efficient and kindly in organising nurses and doctors. Davy was only forty-eight, and could not forget his father's premature death at the same age. His friend and physician Dr Babington recommended exercise and diet, and gradually he began to recover the use of his arm, and some rather stiff movement in his leg. By January 1827 he was able to write again, and he found to his immense relief that he could still cast a fishing line and shoot tolerably well. But he tired easily, and became deeply depressed and irritable. Babington suggested a long holiday on the Continent.

Fitting out his carriage with books and hunting gear, Davy set out with his dogs and his brother John in January. For Jane this must have been a decisive moment, but the old intimacies of the Highlands could not be recovered on either side, even in this extremity. She decided she could not travel happily with her husband, and so would remain in London, looking after his affairs at Park Street, entertaining the more sympathetic Royal Society Fellows, and keeping up her wide circle of aristocratic correspondence. It was John alone who travelled with him over the snow-bound Alps, and remained with him at Ravenna until recalled to his post as military doctor on Corfu in late spring. It was a painful farewell for both of them.

From now on the whole tenor of Davy's life would profoundly and permanently change. He became much more like the solitary boy who had roamed the wilds of Cornwall in his youth. He was aware of his fatal illness, and knew that he could drop dead at any moment, and that no medication existed that could help him: a terrifying prospect. He was also aware of insidious psychological enemies: chronic depression, alcoholism, morphine addiction, or simply spiritual despair. He had little to cling to but his belief in science.

John later wrote movingly of his brother's predicament: 'The natural strength of his mind was very clearly manifested under these circumstances. Dependent entirely on his own resources; no friend to converse with; no one with him to rely on for aid, and in a foreign country, without even a medical advisor; destitute of all the amusements of society; without any of the comforts of home – month after month, he kept his course, wandering from river to river, from one mountain lake and valley to another, in search of favourable climate; amusing himself with gun and rod, when sufficiently strong to use them, with "*speranza*" [hope] for his rallying word.'[92]

In July Davy wrote stoically from the shores of Lake Constance: 'My only chance of recovery is in entire repose, and I have even given up angling, and amuse myself by dreaming and writing a very little, and studying the natural history of fishes. Though alone, I am not melancholy ... I now use green spectacles, and have given up my glass of wine per day.'[93]

At Ravenna he wrote a series of short meditative poems, simply entitled 'Thoughts'. He was anxious not to fall into easy, consoling delusions; and in this the man of science came to the aid of the poet. Often the literary effect is severe, sceptical and coldly metaphysical. There is none of the showy confidence, or the assertive music, of his earlier hymn 'The Massy Pillars of the Earth'. Yet Davy's own voice remains clear.

We trace analogies; as if it were
A joy to blend all contrarieties,
And to discover
In things the most unlike some qualities
Having relationship and family ties.

Thus life we term a spark, a fire, a flame;
And then we call that fire, that flame, immortal,
Although the nature of all fiery things
Belonging to the earth is perishable.

But sometimes he allowed himself a great outburst of feeling, an uprush of longing for survival and consolation and love.

Oh couldst thou be with me, daughter of heaven,
Urania! I have no other love;
For time has withered all the beauteous flowers
That once adorned my youthful coronet.

With thee I still may live a little space,
And hope for better, intellectual light;
With thee I may e'en still in vernal times
Look upon nature with a poet's eye,

Nursing those lofty thoughts that in the mind
Spontaneous rise, blending their sacred powers
With images from mountain and from flood,
From chestnut groves amid the broken rocks

Where the blue Lima pours to meet the wave
Of foaming Serchio ... [94]

Many of these poems led him back to one of his consoling rivers. As a distraction in the evenings, Davy decided to begin writing a book about fishing. It would recount a series of piscatorial adventures and conversations, in the spirit of Izaak Walton, but adding a good deal of natural history and fishing folklore. He entitled it *Salmonia, or Days of Fly-Fishing.*

Davy's scientific writing had always been admirably plain, factual and direct, though in his lectures he prided himself on being able to produce the clever analogy or the uplifting overview. He now tried something quite different, the play of dialogue and contrasting viewpoints. For this he invented four fictional fishermen, amalgamating elements of himself and several of his friends, including his faithful doctor Babington, Professor Wollaston from the Royal Society, and a composite literary figure who might have been part Coleridge and part Walter Scott. These he compounded into four allegorical figures: 'Ornither', an expert on birds and field sports; 'Poietes', the literary man who is also 'an enthusiastic lover of nature'; 'Physicus', who is 'uninitiated' as an angler, but who has a shrewd scientific approach to natural history, and a taste for metaphysics; and 'Halieus', a fully accomplished fly-fisherman.

Davy's first attempt at fiction was not entirely successful. The first three of his fishermen are not easily distinguished from each other, and their role seems largely to give the fourth, Halieus, a chance to show off his knowledge of natural history – at stunning length. Halieus is a convincing, if unintentional, portrait of a scientific pedant. Yet on close examination, the book is full of intriguing and unexpected digressions, especially when Halieus is unexpectedly contradicted. In an early section ('Day One') Davy investigates the mysterious memories of fish, which he regarded as quite as interesting a phenomenon as those of human beings. For example, once a trout was caught and thrown back into the river, could it remember being hooked? Could it remember the *pain* of being hooked? Could it feel – or remember – pain at all? And if so, was trout fishing inherently cruel? This is an astonishingly modern question, and one which hauntingly recalls Davy and Coleridge's forgotten speculations about pain and anaesthetics.

Halieus, the self-confident and assertive fisherman, tries to dismiss the question as essentially absurd. 'If all men were Pythagoreans, and professed the Brahmin's creed, it would undoubtedly be cruel to destroy *any* form of animated life; but if fish are to be eaten . . . ' This would appear to be Davy's own dismissal of the issue, until the metaphysical Physicus intervenes. 'But do you think nothing of the torture of the hook, and the fear of capture, and the misery of struggling against the powerful rod?' Halieus tries to dismiss this on anatomical grounds. Fish do not have feeling in the gristle of the mouth. But again Physicus returns to the charge, from another angle: 'Fishes are mute, and cannot plead, even in the way that birds and quadrupeds do, their own cause ...'[95] Here Davy gives a surprising picture of two different kinds of sensibility in debate. Many other philosophical questions are raised in this indirect manner, and Davy slowly began to expand the work.

The idea of being useful, and leaving a scientific inheritance, came increasingly to preoccupy him. He describes in *Salmonia* ('Day 4') an incident that had occurred years before during a day's fishing at Loch Maree in the Highlands. Two adult eagles were teaching their young to fly above the loch, climbing in ever widening circles 'into the eye of the sun'.[96] He expanded this into one of his most striking and symbolic poems, 'The Eagles'. Coleridge had often talked to Poole of the natural symbolism of eagles (images of pride, power and independence), and described himself as an eagle who could not soar. Davy's poem moves in a different direction, towards the idea of eagles representing initiation and apprenticeship.

He depicts himself watching in rapture the two adult grey-tailed eagles in the bright sunlight, followed by their young offspring. This moment is transformed into an image of Davy the man of science, hoping to inspire his young scientific protégés to ever greater discoveries.

> The mighty birds still upward rose
> In slow but constant and most steady flight.
> The young ones following; and they would pause,
> As if to teach them how to bear the light
> And keep the solar glory full in sight.
> So went they on till, from excess of pain,
> I could no longer bear the scorching rays;
> And when I looked again they were not seen,
> Lost in the brightness of the solar blaze.
>
> Their memory left a type and a desire:
> So should I wish towards the light to rise
> Instructing younger spirits to aspire
> Where I could never reach amidst the skies,
> And joy below to see them lifted higher,
> Seeking the light of purest glory's prize.[97]

Of course, the poem has a certain irony. Davy's greatest protégé, his young eagle Michael Faraday, had not flourished under his patronage and now flew increasingly on his own. Yet perhaps Davy acknowledged the necessity of this, for in *Salmonia* the all-knowing Halieus comments: 'Of these species [of eagle] I have seen but these two, and I believe the young ones migrate as soon as they can provide for themselves; for this solitary bird requires a large space to move and feed in, and does not allow its offspring to partake its reign, or to live near it.'[98]

Writing the book was not easy going. 'This paper is stained by a leach which has fallen from my temples whilst I am writing,' he noted.[99] The work went better when he took rooms at Herr Dettela's inn at Laibach in Illyria. He hardly recognised Josephine, now a young woman of twenty-five, but with the same bright blue eyes and nut-brown hair. 'I hope it is a good omen that my paper by accident is *couleur de rose*,' he joked.[100]

7

In November 1827 Davy returned briefly to London to resign his presidency of the Royal Society. He later gave a moving glimpse of his disillusion with his own scientific career on this sad return: 'In my youth, and through the prime of manhood, I never entered London without feelings of pleasure and hope. It was to me as the grand theatre of intellectual activity, the field of every species of enterprise and exertion, the metropolis of the world of business, thought, and action . . . .I now entered the great city in a very different tone of mind, one of settled melancholy . . . My health was gone, my ambition was satisfied, I was no longer excited by the desire of distinction; what I regarded most tenderly [my mother], was in the grave . . . My cup of life was no longer sparkling, sweet, and effervescent . . . it had become bitter.'

In a wonderfully sardonic aside, Davy added that his metaphor of the 'cup of life' was scientifically derived from the chemical fermentation of 'the juice of the grape', and then after a certain lapse of time, its oxidisation and acidification.[101]

Rather than remaining with Jane, he spent Christmas with his old friend Tom Poole at Nether Stowey. As Davy clambered painfully out of his carriage in Lime Street he greeted Poole with a weary smile: 'Here I am, the ruin of what I was.'[102] But soon memories of the happy Bristol days – with Beddoes, Southey, Gregory Watt and Coleridge – were revived, and Davy considered taking a large country house in the Quantocks for his retirement. With this in mind they rode over to visit Andrew Crosse at

Fyne Court, near Broomstreet, on the eastern escarpment of the hills. Crosse was a wealthy and eccentric bachelor who had spent most of his fortune on installing 'an extensive philosophical apparatus' with which he later claimed to have generated spontaneous life forms. It was later suggested that he was another 'original' of Mary Shelley's Dr Frankenstein.

Crosse's huge, chaotic laboratory was installed on the ground floor, in what had originally been the ballroom of Fyne Court. It contained large, gleaming electrical condensers, which were linked to a network of copper wires strung through the trees round the whole estate. These were designed to pick up massive charges of static or 'atmospheric' electricity. The largest condenser was marked with a blasphemous warning notice: '*Noli Me Tangere*' – that is, 'Do not Touch Me' – because of the possible electric shock. The phrase is famous from the Gospels: the risen Christ's first words to Mary Magdalene.

Poole noticed that Davy, for the first time, became animated and cheerful. 'As we were walking round the house very languidly, a door opened and we were in the laboratory. He threw a glance around the room, his eyes brightened in the action, a glow came over his countenance, and he looked like himself, as he was accustomed to appear twenty years ago.'[103] Davy did not take the house, but put the finishing touches to *Salmonia*, and told Poole: 'I do not wish to live as far as I am personally concerned, but I have views which I could develop, if it pleased God to save my life, which would be useful to science and mankind.'[104]

In spring 1828 Davy departed once more for the Alps and lower Austria, again leaving Lady Davy behind, according to their agreement. He was writing, fishing and taking morphine. Throughout this summer and autumn he wrote a series of enigmatic letters to his wife, discussing his health and his scientific researches, but always making vague references to Josephine Dettela, the innkeeper's daughter.[105] In June he wrote from Laibach: 'The first time since my illness, I have found a month pass too quickly here. The weather has been delightful, and I have had enough shooting ... and my pursuits in natural history respecting the migration of birds, have given me some new and curious results. I must not forget the constant attention and kindness of my "Illyrian maid", I mean poetically and really. The art of living happy is, I believe, the art of being agreeably deluded; and faith in all things is superior to Reason, which, after all, is but a dead weight in advanced life, though as the pendulum to the clock in youth.'[106]

In July he went down to the coast to collect some specimens of the electrical ray or torpedo fish at Trieste. He had renewed his interest in Vitalism and the mysteries of animal electricity. But he hurried back to Laibach, again writing to Jane almost teasingly: 'I am just returned to my old quarters & my pretty Illyrian nurse, after an excursion of a fortnight to Trieste ... I succeeded in my projected experiment on the Torpedo and I have I think been able to establish a new principle with respect to the species of Electricity which will be a [gain] in Nat Science.'[107]

He worked on throughout the summer, trying to believe he was convalescing, and remained at Laibach as long as possible, until the autumn weather broke and the snowclouds began to gather in the mountains. In November, forced to go to Rome for the winter, he was already looking back wistfully. He confessed to Jane: 'I remained at Laybach till October 30. I left that place with regret, kindness makes the sunshine of life in a sick man & that kindness is not less agreeable because it is given by a blooming and amiable maiden – I shall ever be grateful to my charming Illyrian nurse.'[108] He now admitted that he was suffering from low spirits, 'too feeble to bear general society', and greatly missing Josephine. 'I fear I shall find no Illyrian nurse here, such as the spirit that dispelled my melancholy at Laybach.'[109] In December he wrote more hopefully, and a little more explicitly, to his brother John in Corfu. 'Perhaps in the spring you could come to me at Trieste & see me in Illyria. I would then show you my dear little nurse, to whom I owe most of the little happiness I have enjoyed since my illness.'[110]

It is strange that Jane did not react to all these hints, but in the New Year she eventually responded with a light-hearted question about the identity – real or imagined – of the mountain 'nymph' who danced attendance on him so charmingly in Illyria. She noted, without irony, that she had seen that the ageing Goethe had his youthful female followers too. Davy seemed pleased to answer. 'If you mean my little nurse and friend of Laybach, I shall be very glad to make you acquainted with her. She has made some days of my life more agreeable than I had any right to hope. Her name is Josephine or Pappina.'[111] John later tactfully recalled: 'Laybach, which had peculiar attractions for him ... might be considered his headquarters in this region. The attractions were, its situation near a fine river ... and, not least ... a kind little nurse, the daughter of the innkeeper.'[112]

Was this all the fantasy of a dying man? At Laibach sometime in that summer of 1828 Davy wrote rough drafts of two short love poems to

Josephine. They occupy three pages of his scientific notebook, and are much crossed out and difficult to read. They reveal a little more about their relationship. He nicknamed her 'Pappina', a tender diminutive he used in the first poem, which is headed 'Laybach August 16 1828. To Josephine Dettela'. It begins:

> Kiss me Pappina, kiss me again!
> Thy kisses will become a gage:
> They waken in my heart a hope
> Which was not of my early age...

No other poem that Davy had ever written has this simple, erotic directness. It is all the more touching, perhaps, because of its clumsy, childlike syntax. It is an open appeal for tenderness and love. Yet Davy does not use the word 'love'. Instead he repeats the word 'hope'. In the next stanza he feels the hope 'a blessed father feels', the hope 'a much-loved brother knows', a hope that 'heaven reveals'. The little lyric continues:

> But when thy angel form I see
> And gaze upon that bright blue eye
> And watch thy calm [unclouded] smile
> And know thy virgin purity
>
> Thy lips' warm pressure wakes no Thought
> Unworthy of the virgin's name
> But gives me hopes allied to heaven
> Which will preserve thy earthly frame...

The remaining five lines are largely deleted or corrected, though the meaning is clear. 'And in thy kiss... And in thy kiss I seem to share... And oft I shed a tear... It is a tear of happiness... Thy innocence I seem to share... And sure I share thy happiness.'

The second poem consists of only five lines, mostly crossed out and rewritten, and crossed out again. It is titled simply 'To the Same'. It insists again on the purity of his feelings, and the theme is the 'Vestal fire' in Josephine's eye, a calm, innocent 'sacred light' which 'never glitters through a tear'. It is the 'source and hope of heavenly bliss'. The last line has only one word: the rhyme for 'bliss' – 'kiss'. Here the draft of the second lyric breaks off. Davy wrote nothing further about Pappina in his notebooks of 1828. Though, judging by his letters, he seems never to have stopped thinking about her, or hoping to get back to Illyria.[113]

John Davy, who must have seen both poems subsequently, chose never to publish them. They are after all very slight, unpolished and unfinished. They certainly do not suggest any grand passion. But they do suggest Davy's longing for tenderness, and a sweet reciprocation on Josephine's part; and that can take many forms. In Regency slang the name 'Pappina' would refer admiringly to her breasts. Perhaps the haunting phrase about Davy's feeling for her, 'which was not of my early age', also suggests something – a desire, a fulfilment – which had been denied him ever since his Cornish childhood.

In fact it is possible that Davy briefly set up house with Josephine in that last summer. Although to Jane he always speaks of staying at the Laibach inn, he did in reality rent a private lodge at a little village just outside the town, as local enquiries more than a century later revealed: 'In Podkoren, on the Wurzen Pass, and just into Slovenia, a house that he rented for fishing bears a blue plaque in English and Slovenian. The village is very small; the house is one of the best in it, but it is by no means large – rather like what a country doctor near Penzance might have lived in. In front are the fields, and behind the beech woods and streams where he must have walked. Nearby is the Sava, or Save River, where he loved to fish. The view of the Julian Alps must be still much as he saw it.'[114]

Davy never forgot the uplifting view of those life-giving Alps, which brought back so many memories. 'They surround the village on all sides, and rise with their breast of snow and crests of pointed rock into the middle of the sky. The source of the Save is a clear blue lake surrounded by woods, and the meadows are as green as those of Italy in April, or of England in May.'[115]

8

Meanwhile, *Salmonia* was published in England in 1828, and favourably reviewed by Walter Scott in the *Quarterly*, thanks to Lady Davy pulling strings with her cousin. With considerable insight, she understood how much a little literary glory would soothe Davy at this juncture. Indeed, greatly encouraged, he began working on a second expanded edition. He now also embarked on a last work, which was intended as a summation of a lifetime's thoughts and beliefs, *Consolations in Travel, or The Last Days of a Philosopher*. He planned to dedicate it not to any of his grand aristocratic friends or patrons, but to his old confidant from the West Country days, Tom Poole. To Poole he confided his hopes for the book: 'I write and philosophize a good deal, and have nearly finished a work with a higher aim than [Salmonia] ... which I shall dedicate to you. It contains the essence of my philosophical opinions, and some of my poetical reveries. It is like the Salmonia, an amusement of my sickness; but *paulo majora canamus*.* I sometimes think of the lines of Waller, and seem to feel their truth –

> ' "The soul's dark cottage, batter'd and decay'd,
> Lets in new light through chinks that Time has made" '[116]

The *Consolations* is one of the most extraordinary prose books of the late Romantic period. Its title links it to the tradition of Boethius' medieval *Consolation*, a form of renouncing the world before death. But Davy mixes philosophy and autobiography with highly original sections of science fiction, some visionary travel writing, various theories of history, race and society, and an important apologia for science. It also contains unexpected speculations about the nature of evolution, and the future of the human species.

The *Consolations* is divided into six Dialogues, with the fragment of a seventh, never completed; but the rather stilted exchanges of *Salmonia* are greatly improved upon. Though still using various semi-fictionalised figures, the whole book is intensely confessional. The early chapters are inspired by memories of Davy's encounters and reflections during his various visits to Rome, including the Coliseum by moonlight, to Naples and the top of Vesuvius, and to the ruins of Paestum, especially during the two-year tour of 1818–20. The later chapters, which become steadily more intimate, draw on the two long summers spent in Austria and Illyria in 1827–28. The book goes far beyond a travelogue. There is a

---

*'Eventually we must sing of greater things.' The book had run to nine editions by 1883. The French edition, edited by the great Parisian science writer Camille Flammarion, supplied a long and dramatically expressive title: *Les derniers Jours d'un Philosophe. Entretiens sur la Nature, les Sciences, les Métamorphoses de la Terre et du Ciel, l'Humanité, l'Ame, et la Vie eternelle.* That certainly covered it.

Young Humphry Davy
Davy, recently arrived in London and beginning to make a stir as the new Professor of Chemistry at the Royal Institution, and its star lecturer. Portrait by Henry Howard, oil on canvas, 1803.

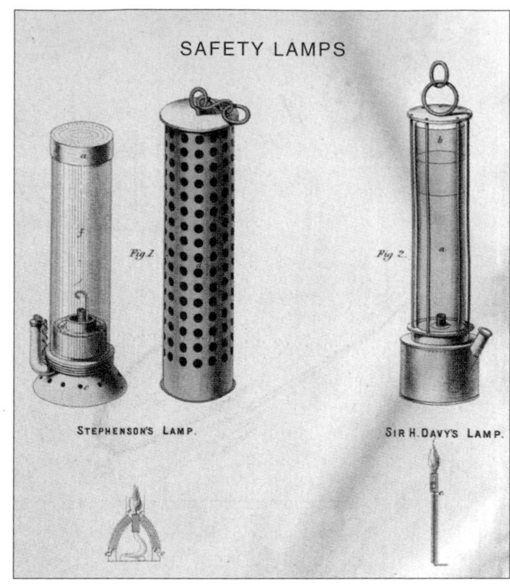

Rival safety lamps
Coloured diagrams of rival safety lamps designed by George Stephenson and Sir Humphry Davy, modified versions c.1816–18. Though the shapes appear very similar, the glass chimney and ventilation perforations of the Stephenson (left) are clearly different from the simple gauze cowl of the Davy (right).

Sir Humphry Davy, President of the Royal Society
Davy as the new young PRS, still the working man of science, immersed in his papers, and proudly displaying his safety lamp on the right. Portrait by Thomas Phillips, oil on canvas, 1821.

Sir Humphry Davy, PRS
The confident, dashing figure of the established President now outshines his safety lamp, which is set back in shadow on the left-hand side of the picture. Portrait by Sir Thomas Lawrence, c.1821–22 or later.

Scientific Researches! New Discoveries in PNEUMATICKS! or An Experimental Lecture on the Powers of Air

Gillray cartoon of a lecture at the Royal Institution, demonstrating the supposed effects of laughing gas (nitrous oxide). On the podium are Dr Garnett administering gas, an impish Davy holding the bellows, and to the right Count Rumford eagerly observing proceedings. Unidentified figures in the audience may include Banks, Cavendish, Coleridge and members of Davy's female fan club, several of whom are taking notes. Published by Hannah Humphrey, 1801.

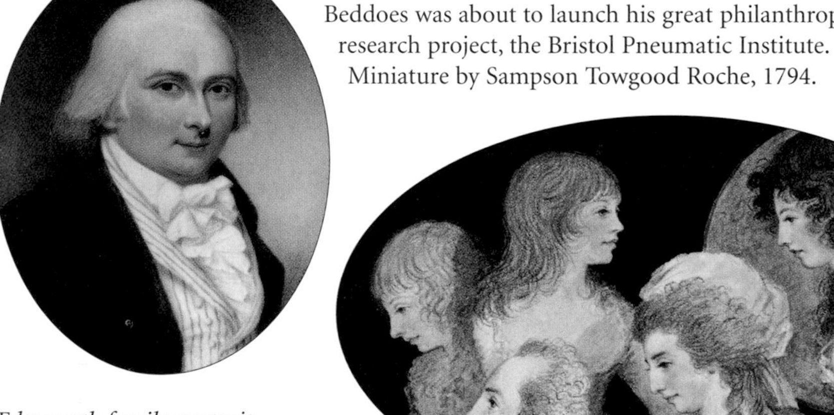

Dr Thomas Beddoes

Beddoes was about to launch his great philanthropic research project, the Bristol Pneumatic Institute. Miniature by Sampson Towgood Roche, 1794.

Edgeworth family portrait

Detail from a group portrait of the extensive Edgeworth family by Adam Buck, 1787. Anna Beddoes aged sixteen, the only figure in right profile, is characteristically isolated from the rest of her family. Immediately below her are her father Richard Lovell Edgeworth and her stepmother; on the right her pretty brunette half-sister Honoraria.

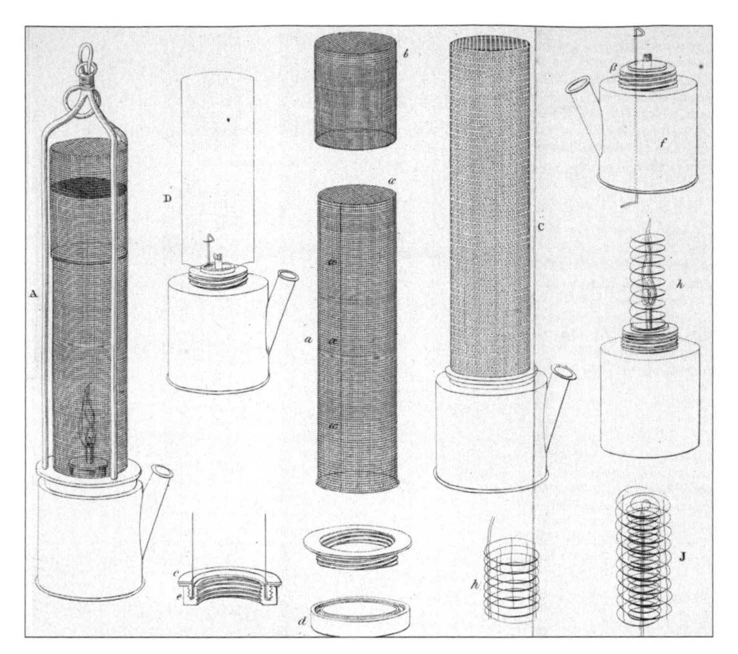

The Davy safety lamps
Analytic drawings, based on those made by Michael Faraday, to illustrate Davy's historic presentation of his safety lamp to the Royal Society in January 1816. They also show his later design of the platinum 'self-lighting' wick, and his protective refinements to the gauze cowl, in 1817. Published in *Collected Works of Humphry Davy*, Volume 6 (1840).

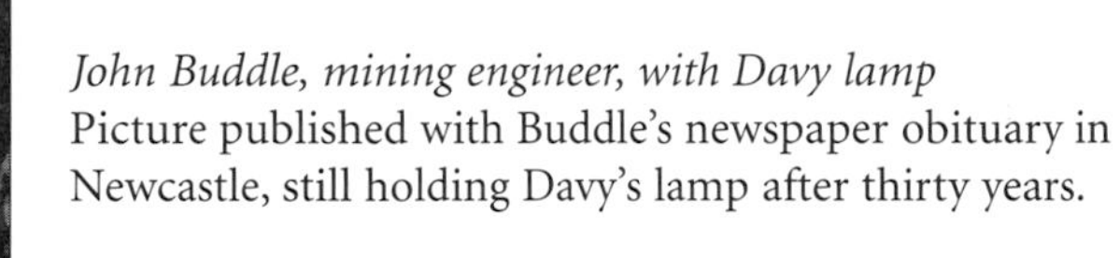

John Buddle, mining engineer, with Davy lamp
Picture published with Buddle's newspaper obituary in Newcastle, still holding Davy's lamp after thirty years.

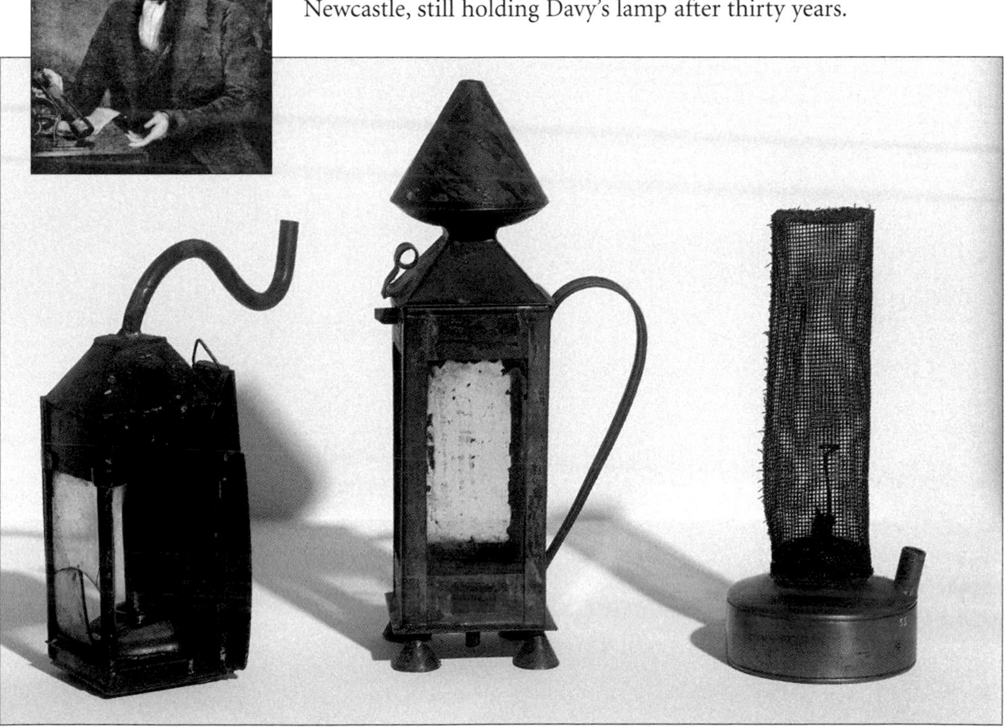

Three prototype safety lamps
These rough prototype safety lamps, constructed in the laboratory of the Royal Institution, 1815–16, show the beautiful simplicity of Davy's invention (on right). Photograph, The Royal Society.

An anonymous author
A mysterious and glamorous unidentified female author, painted in the year of the reissue of the novel *Frankenstein*, when Mary Shelley was thirty-three and had published three other novels and several collections of essays. The locket indicates the precious memory of a loved one: Percy Shelley had been drowned nine years previously. Portrait by Samuel John Stump, oil on canvas, 1831.

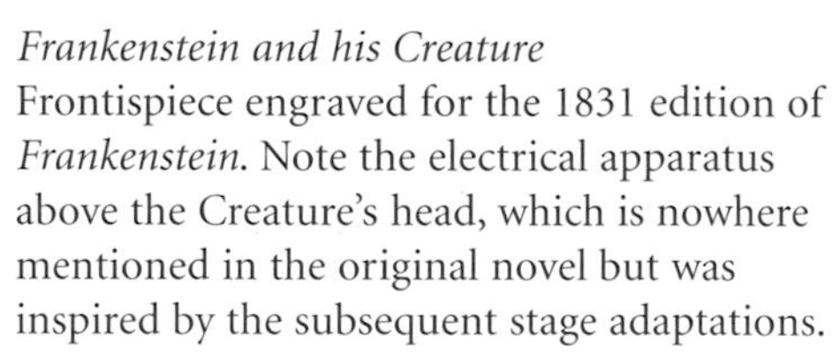

Frankenstein and his Creature
Frontispiece engraved for the 1831 edition of *Frankenstein*. Note the electrical apparatus above the Creature's head, which is nowhere mentioned in the original novel but was inspired by the subsequent stage adaptations.

Mary Shelley
Portrait by Richard Rothwell, 1840, at the time when she was editing Shelley's *Collected Poems*.

Young John Herschel
Portrait by Muller of John Herschel aged about seven in 1799, shortly before being sent to Eton. With the kind permission of John Herschel-Shorland.

Caroline Herschel's gold medal
The Royal Astronomical Society's Gold Medal presented to Caroline Herschel in 1828. The medallion shows William Herschel's forty-foot telescope, and the Society's motto – 'Let Whatever Shines be Noted'.

Michael Faraday
Faraday's wide-eyed look of wonder particularly irritated Lady Davy, and delighted Coleridge. Drawing by William Brockedon, 1831.

Clockwise from top left

John Herschel
Drawing by Henry William Pickersgill, c.1835.

David Brewster
Lithograph after Daniel Maclise, c.1830.

Charles Babbage
Detail from a daguerreotype by Antoine Claudet, c.1847–51.

Charles Darwin
Albumen print by Maull & Polyblank, c.1855.

Mary Somerville
Drawing by Sir Francis Leggatt Chantre, 1832.

Louis de Bougainville
A commemorative postage stamp to remind us that the French got there before Captain Cook.

Charles Waterton
The Yorkshire explorer Waterton brought back the stuffed bird (a Guiana red cotinga) and the distinctly resentful wild cat from his wanderings in South America. Portrait by Charles Willson Peale, 1824.

Nature Unveiling Herself Before Science
Nature – beautiful but increasingly vulnerable? Two bronzes by Louis Ernest Barrias, 1890.

Isaac Newton
Bronze statue by Eduardo Paolozzi, 1995, installed in the main courtyard of the new British Library, Euston Road, London. Based on an image by William Blake, and perhaps a memory of Mary Shelley's *Frankenstein*.

Andromeda
Image from the Hubble Telescope showing Andromeda, the nearest spiral galaxy or 'island universe', which is steadily approaching our Milky Way.

sense of open, passionate debate of ideas, and the necessity of unveiling the truth, whether in science or religion.

Davy never forgot that he was dying, and that his time had almost run out. He wrote unguardedly to Jane that he would therefore make it his 'best work', and would fearlessly reveal truths vital to the 'moral and intellectual world' which could never be recovered if lost by his death: 'I may be mistaken on this point, yet it is the conviction of a man perfectly sane in all the intellectual faculties, and looking into futurity with prophetic aspirations, belonging to the last moments of existence.'[117] The deliberate testamentary phrasing was no coincidence. But behind it lay the poignant anxiety that heavy opium-dosing might have distorted his 'intellectual faculties' and his reason.[118]

The book includes much strange, visionary material and imaginary voyaging. One chapter starts on planet earth and ends on Saturn. An early draft suggests that Davy wished to establish a spiritual Guide, a beautiful woman who would lead him (as Beatrice led Dante, perhaps) through all his scientific writings and reflections. In the final text he reduced her to a disembodied voice, though she still seems feminine: 'A low but extremely distinct and sweet voice, which at first makes musical sounds, like those of a harp.'[119]

This trope of the beautiful female guide had appeared in several of Davy's poems of the 1820s, as it had in Shelley's 'Epipsychidion' (1821). It is again curious that Shelley, quartered in Rome in the spring of 1819, had also written a visionary story set in the Coliseum by moonlight. But then, Davy and Shelley were both part of a whole generation of exiled Romantic wanderers in Italy, looking for health, love and imaginative inspiration – including Goethe, Humboldt, Lord Byron, Trelawny, Mary Shelley, Walter Savage Landor and John Keats.

Dialogue I, subtitled 'The Vision', begins with Davy's moonlit dream in the Coliseum at Rome. Left alone by two companions amidst the ruins, he finds himself addressed by an invisible presence, 'which I shall call that of Genius'. Davy is told a sort of scientific creation myth, a Promethean version of man's growing material dominion over the earth. From his primitive tribal beginnings, art and technology lifts man above the wild animals, until such global developments as chemistry, engineering, medicine and the 'Faustian' invention of the printing press bring an advanced Western civilisation.[120]

This account by Genius includes some racial theory about the 'superiority of the Caucasian stock' of a type familiar to students of Blumenbach.

But Genius also makes an uneasy prophecy of colonial persecutions, of the kind that Banks would have recognised. 'The negro race has always been driven before the conquerors of the world; and the red men, the aborigines of America, are constantly diminishing in number, and it is probable that in a few centuries their pure blood will be entirely extinct.'[121]

These increasingly unsettling visions culminate in a visit to a society of extraterrestrial beings on Saturn and Jupiter, by means of a shuttle network of comets. This device of comet-hopping appears in much earlier speculative literature such as Voltaire's story 'Micromégas' ('The Tiny Huge', 1752), and it is also reminiscent of Caroline Herschel's humorous fantasy of comet-travelling. Davy sees the moon and the stars go by 'as if it were in my power to touch them with my hand ... it seemed as if I were on the verge of the solar system'. The double ring of Saturn appears to him as 'I have heard Herschel often express a wish he could see it'.[122]

On Saturn the super-intelligent inhabitants are said to include 'the monad or spirit of Newton ... now in a higher and better state of planetary existence, drinking intellectual light from a purer source and approaching nearer to the infinite and divine Mind'. These intelligences appear to float around Davy rather like enormous angelic seahorses, with wings made of 'extremely thin membranes ... varied and beautiful ... azure and blue'. They gently explain that the entire solar system, including Mars and Venus, is full of life forms, and these do not stop at Uranus. After physical death, human beings are transported into 'higher or lower' planetary forms, depending on their 'love of knowledge or intellectual power'.[123] Davy, like the reader, is left stunned and disturbed by these revelations.[124]

The second Dialogue takes place one evening on the summit of Vesuvius. It moves into a lively discussion of the variety of religious experience on earth, comparing Christian, Jewish and Islamic beliefs. Older theologies are considered as incomparably cruel. 'To the Supreme Intelligence, the death of a million of human beings, is the mere circumstance of so many spiritual essences changing their habitations, and is analogous to the myriad millions of larvae that leave their coats and shells behind them, and rise into the atmosphere, as flies on a summer day.'[125]

But Davy's own beliefs are more optimistic, and like *Naturphilosophie* tend to suggest some form of spiritual evolution taking place on earth. They compare human destiny to that of 'a migratory bird' looking by instinct for a higher existence. The outlook is hopeful, and mankind is young. 'We are sure from geological facts, as well as from sacred history, that man is a recent animal on the globe.'[126]

Davy's 'visionary maiden' recurs in the second Dialogue, where one character (Philalethes), who is evidently Davy, claims to have first seen her in his feverish dreams two decades before, but denies that she was originally based on any actual woman. Nonetheless, her Illyrian reappearance is clearly based on Josephine Dettela. The others gently tease Philalethes about this angelic apparition, but he insists on her physical reality and importance to him.

– 'All my feelings and all my conversations with this visionary maiden were of an intellectual and refined nature.'
– 'Yes, I suppose, as long as you were ill.'
– 'I will not allow you to treat me with ridicule ... to her kindness and care I believe I owe what remains to me of existence ... Though my health continued weak, life began to possess charms for me which I thought were for ever gone; and I could not help identifying the living angel with the vision which appeared as my guardian genius during the illness of my youth.'
– 'I dare say any other handsome young female, who had been your nurse in your last illness, would have coincided with your remembrance of the vision, even though her eyes had been hazel and her hair flaxen ...'[127]

But, sadly perhaps, no further intimacies appear in this Dialogue.

The third Dialogue takes place during a dawn visit to the temples of Paestum, and introduces a new figure called The Unknown, who is discovered wandering through the ruins, clad in rough travelling clothes, with a pilgrim's hat and staff, and a vial of medical chlorine (Davy's discovery) round his neck to guard against marsh fever. Part man of science and part mystic, The Unknown is yet another projection of Davy's secret myth of himself, now the pilgrim scientist on his last journey.

Much of this Dialogue is based on geology and ideas of the earth's evolution. It is notable that while Davy revels in the idea of Herschel's 'deep space', he finds it difficult to accept the concept of 'deep time' that had been argued by Hutton, and would soon be developed by Charles Lyell. Even the sceptical figure (Onuphrio) finds it hard to accept the 'absurd, vague, atheistical doctrine' of evolution, though he describes it surprisingly succinctly. 'That the fish has in millions of generations ripened into the quadruped, and the quadruped into man; and that the system of life *by its own inherent powers* has fitted itself to the physical changes in the system of the universe.'[128]

The talk then drifts intriguingly into a discussion of ghosts, visions and nightmares. Some of these seem to reflect the horrors of Davy's own illness, such as the dream of a group of murderous robbers breaking silently into his bedroom, and one of them 'actually putting his hand before my mouth to ascertain if I was sleeping naturally'.[129] John Davy would later say that one of his brother's most painful and irrational obsessions in these last months was the fear of being buried alive.

The Unknown refers to dreams which bring back the events of a whole life, such as that of Brutus in his tent before battle. 'I cited the similar vision, recorded of Dion before his death by Plutarch, of a gigantic female, one of the fates or furies, who was supposed to have been seen by him when reposing in the portico of his palace. I referred likewise to my own vision of the beautiful female, the guardian angel of my recovery, who always seemed to be present at my bedside.'[130] These are the reflections of a haunted man, whose belief in his own powers of reason is increasingly under siege.

The fourth Dialogue contains more thinly veiled autobiography, referring to Davy's own illness and his mother's recent death in Cornwall. It draws on his more recent travels in the Austrian Alps and Illyria.[131] There is a dramatic account of being swept away in a fishing coracle down the river Traun, being carried through boiling passages of white water, and finally being hurled over the great Traun waterfall itself and losing consciousness. 'I was immediately stunned by the thunder of the fall and my eyes were closed in darkness.' He wakes to find himself being mysteriously pulled to safety. 'I was desirous of reasoning ... upon the state of annihilation of power and *transient death* which I had suffered when in the water.'[132] Whether this terrifying accident actually occurred to Davy is never made clear, but the entire episode seems symbolic of his whole life being swept away towards death.

Appropriately, the scientific theme of this Dialogue is the nature of the Life Principle, its analogies with electricity, and the whole Vitalism debate. Davy also puts forward the sustaining idea that men of science like Archimedes, Bacon and Galileo had actually advanced human civilisation far more than statesmen, religious leaders or artists. This is a position that he had frequently argued in his later lectures, deliberately contradicting Coleridge, who had said that the 'souls of 500 Newtons' had gone into the making of a single Shakespeare. Davy said emphatically that as benefactors of mankind, he held Bacon far above Shakespeare, and Newton far above Milton. 'At that time, when Bacon created a new world of intellect, and Shakespeare a new world of imagination, it is not a question to me which has produced the greatest effect upon the progress of society – Shakespeare or Bacon; Milton or Newton.'[133]♣

In the fifth Dialogue, which he entitled 'The Chemical Philosopher', Davy set out his hopes for the future of chemistry. It embodied all his passionate belief in science as a progressive force for good, both in its practical results and its impact on the mind. This would be widely accepted as a credo by the next generation of young scientists: 'Whilst chemical pursuits exalt the understanding, they do not depress the imagination or weaken genuine feeling; whilst they give the mind habits of accuracy, by obliging it to attend to facts, they like wise extend its analogies; and, though conversant with the minute forms of things, they have for their ultimate end the great and magnificent objects of Nature… And hence they are wonderfully suited to the progressive nature of the human intellect… It may be said of modern chemistry, that its beginning is pleasure, its progress knowledge, and its objects truth and utility.'[134]

♣ What Coleridge actually wrote was this. 'My opinion is this – that deep Thinking is only attainable by a man of deep Feeling, and that all Truth is a species of Revelation. The more I understand of Sir Isaac Newton's works, the more boldly I dare utter to my own mind… that I believe the Souls of 500 Sir Isaac Newtons would go to the making up of a Shakespeare or a Milton… Mind in his system is always passive – a lazy Looker-on on an external World. If the mind be not passive, if indeed it be made in God's Image, and that too in the sublimest sense – the image of the Creator – there is ground for suspicion, that any system built on the passiveness of the mind must be false, as a system' (23 March 1801, *Letters*, Vol. 2, p.709). This saying of Coleridge's has a peculiar power to outrage men of science, even modern ones. In November 2000 there was a special day-long seminar organised at the Royal Society by the then President Sir Aaron Klug, on the subject of 'The Idea of Creativity in the Sciences and the Humanities'. Among its twenty distinguished participants were Richard Dawkins, Matt Ridley, Carl Djerassi, George Steiner, Lisa Jardine and Ian McEwan. This citation from Coleridge proved more contentious than any other single proposition, and eventually goaded an eminent scientist (none of the above) to cry out in exasperation: 'That is complete and utter *balls*… We don't have to put up with such rubbish.' Equilibrium was restored when it was pointed out that the idea of computing the contents of '500 souls' was possibly Coleridge's idea of a mathematical joke.

Davy claimed chemistry as the crown of a 'liberal education', and assumed that a serious chemist would begin with an elementary knowledge of mathematics, general physics, languages (it is interesting that he included Latin, Greek and French), natural history and literature. He should write up his experiments in 'the simplest style and manner'. But his imagination 'must be active and brilliant in seeking analogies ... The memory must be extensive and profound.'[135] He was not above adding a little perilous glamour to the pursuit: 'The business of the laboratory is often a service of danger, and the elements, like the refractory spirits of romance, though the obedient slave of the Magician, yet sometimes escape the influence of his talisman, and endanger his person.'[136]

The sixth and last Dialogue ('Pola, or Time') ends on a mystical note, with an almost Blakean speculation about angelic intelligences. This bursts out on the final pages with a salute to Herschel's views of a dynamic and ever-evolving universe: 'There is much reason to infer, from astronomical observations, that great changes take place in the system of the fixed stars; Sir William Herschel, indeed, seems to have believed that he saw nebulous or luminous matter in the process of forming suns ... It is, perhaps, rather a poetical than a philosophical idea, yet I cannot help forming the opinion, that genii or seraphic intelligences may inhabit these systems, and may be the ministers of the eternal mind.'[137] With characteristic precision, Davy refused to add capital letters to those last two words.

This strange book, part philosophy and part science fiction, was to have a surprising hold on the younger generation of Victorian scientists. What it suggested was that chemistry was the most awe-inspiring and visionary of the sciences, and that 'to study it was to catch the ultimate forces of nature itself' at work.[138] It was frequently referred to by Charles Babbage, John Herschel and Charles Darwin. Though clearly fitting into a recognisable pattern, in which a highly rational man develops intense mystical longings towards the end of his life, it carried a true sense of humanity and hope. In a later *éloge*, Georges Cuvier called it, with pardonable exaggeration, 'in some respects the last words of a dying Plato'.[139]

Consolations in Travel was timely in emphasising the progressive nature of science as an expression of man's 'immortal' spirit, and the particular qualities required by a scientist, both by training and by temperament. It did not reveal much about Davy's personal relations – there is nothing specifically about his childhood, his family, his wife, or the problematic subject of Michael Faraday. But it carried a haunting sense of his career, so marked by both exceptional achievement and bitter disappointment. It could perhaps claim to be the first ever scientific autobiography in English. It certainly belongs to the new Romantic genre of memoir, that includes in various ways Wordsworth's *Prelude* (1805–50), Coleridge's *Biographia Literaria* (1816) and Thomas De Quincey's *Confessions of an English Opium-Eater* (1821).

Though he was solitary during this whole period (apart from his dogs), Davy had the support of a servant, whom he refers to with mournful humour as his Caliban. There was also his godson, a young medical student, John Tobin, son of his Bristol friend James Tobin, who had once tried laughing gas. Young Tobin's main employment seems to have been reading to the great man in the evenings. These could be demanding sessions, covering contemporary novels, much poetry (especially Byron), the *Arabian Nights*, and on one occasion a Shakespeare reading that Tobin claimed lasted for nine hours on end.

Though he was useful for valeting and taking dictation, young Tobin was no Faraday; though like Faraday he found it difficult to maintain equable relations with his moody, reclusive employer: 'Sir Humphry... frequently preferred being left alone at his meals; and in his rides, or fishing and shooting excursions, to be attended only by his servant. Sometimes he would pass hours together, when travelling, without exchanging a word, and often appeared exhausted by his mental exertions.'[140]

It is noticeable that in Lady Davy's absence, Davy gradually seemed much softened towards her, and began writing a stream of increasingly tender letters, of which she kept at least forty-eight carefully done up in ribbons.[141] In one he writes: 'I think you will find me altered in many things – with a heart still alive to value and reply to kindness, and a disposition to recur to the brighter moments of my existence of fifteen years ago, and with a feeling that though the burnt-out flame can never be rekindled, a smothered one may be. God bless you! From your affectionate, H Davy.'[142]

Davy was now less inclined to boast of his achievements, but sadly lamented how little they had been recognised: 'I have been used so ill by the public when I have laboured most to serve them, and injured my body and mind in exertions for their good (witness safety lamp, copper bottoms, Royal Society...).'[143]

Vitalism still held his intense interest. He stubbornly sent off another scientific paper to the Royal Society, on the 'animal battery' contained in the body of the torpedo or electric eel. It was published on 20 November

1828. He had now submitted forty-six papers to the Society, his first on the voltaic battery long ago in June 1801, and his most famous one on the safety lamp in 1816. He did not want the torpedo to be his last, and he continued to investigate the mystery of 'animal electricity' and its possible connection with the universal principle of life. John Herschel would be particularly struck by this paper, which compared the electric eel to a voltaic battery, asked whether the eel could exert this 'most wonderful power' at will, and speculated whether the human brain itself might be 'an electric pile, constantly in action'.[144]

Nature held other analogies, too. A late autumn 1828 entry in Davy's private journals reads: 'Bees, wasps and various winged insects, which appeared to me to be of the *Vesper* or *Apes* families were feeding in almost every flower, their tongues searching the honey. They were all languid, it was a cold evening though the sun was bright, and some of them appeared to me actually to die whilst in the act of feeding on their last meal of ambrosia! Happy beings . . . '[145] Perhaps he had hoped that something like that might have happened to him at Laibach.

At last Davy reluctantly left the enchantments of Pappina and Illyria, and went to winter in Rome. He felt increasingly weak and ill, but continued to work spasmodically on the final sections of *Consolations*. In February 1829 he suffered another devastating stroke, and summoned his brother John. John was now working as a military surgeon in Malta, but instantly talked himself aboard a Royal Navy frigate, and rapidly made his way to Naples, and then by horse to Rome.

Convinced that he was dying, Davy had also begged Jane to join him from London. She finally agreed to do so, hoping 'to arrive not quite useless', and having been detained for several days by her own doctors. She sent ahead a curiously formal letter, pledging to Davy 'all the faith and love I have ever borne to you', but searching in vain for that touch of intimacy or tenderness that had long eluded them both. Its last sentence read: 'I cannot add more than that your fame is a deposit, and your memory a glory, your life still a hope.'[146]

But when Jane finally arrived in Rome in early April, she did something which gave Davy immense pleasure. From her chaotic suite of trunks, bags and hatboxes she produced with a flourish the second, expanded and corrected, edition of *Salmonia*, hot off the press, and with beautiful new steel engravings added throughout. Nothing could have pleased him more, a sort of proof of his literary immortality. He immediately began rereading it.[147]

Davy continued gallantly to dictate sections of the *Consolations* to John. Sometimes he was feverish, his pulse rate rising to 150. As in the old days, John took over the dissection of the torpedo fish, and they gently debated whether 'animal electricity' was the intrinsic source of its life, or a mere physiological mechanism for paralysing prey or for self-protection. 'The greater part of the day I sat by his bedside, reading the "Dialogues", stopping occasionally to discuss particular parts. His mind was wonderfully cheerful and tranquil, and clear, and in a very affectionate and most amiable disposition . . . He had lost all the irritable feeling to which he was very liable before . . . It was difficult to conceive such power of mind, when the body was near dissolution: medically it seemed incompatible.'[148]

At the end of April Davy smelt the spring blowing in over the *campagna*. He announced that he wished to travel again before he died. John arranged for a slow coach journey northwards towards Switzerland, with many stops to admire the spring countryside and gaze at the rivers and waterfalls. Jane tactfully went ahead to arrange for accommodation in Geneva. On 28 May 1829 Davy arrived at the Hôtel de la Couronne, overlooking the tranquil lake where Byron and Shelley and young Dr Frankenstein had once sailed. He took tea, and gazed down from his window at the sunset. He carefully questioned the Swiss waiter about the varied species of fish that the lake contained. To John, with a wistful smile, 'he expressed a longing wish to throw a fly'. He took his evening dose of morphine, and John read him to sleep. That night at 3 a.m., Sir Humphry Davy had another stroke and died.

Davy had no children, and left considerable wealth to a nephew, his sister's boy, Humphry Millett, whom he barely knew. All his scientific papers went to his faithful brother John, though Lady Davy retained family letters and journals. John did not see eye to eye with his sister-in-law, and quickly disappeared back to his adventurous medical career with the army, which took him to the Ionian isles, Ceylon and the West Indies. He was made a Fellow of the Royal Society, married, and eventually settled at Ambleside in the Lake District, where he became the Wordsworths' family doctor.

Jane made no attempt to publish anything of Davy's, or about him, though she dined out for the next twenty years on her amusing tales of 'dear, great Sir Humphry'. But John, partly encouraged by Wordsworth, worked doggedly on his brother's papers for more than twenty-five years.

He first published a two-volume *Life* in 1836, hurried out in reply to a hostile anecdotal biography assembled by the voluble J.A. Paris of the Penzance Geological Society (2 vols, 1831). Later John produced a nine-volume *Collected Works* in 1839–40, with a carefully rewritten *Memoir* of his brother's life, attached as a Preface to Volume I. Finally, when settled in the Lakes, he issued a slim but revealing volume of *Fragmentary Remains* in 1858, which contains much of Davy's poetry. No other major edition of his papers, letters or journals has so far been produced. Perhaps John's most intimate tribute was his own book about fishing in the Lake District, *The Angler and His Friend* (1855).

Sir Humphry Davy's Will included endowments for a Davy Medal to be administered by the Royal Society; and for the maintenance of Penzance Grammar School, which celebrates a Davy Holiday to this day. The remainder of his estate was left to Jane, except for a bequest of '£100 or 1,000 florins' for Josephine Dettela, daughter of the innkeeper of Laibach, Illyria, Austria. In March 1829, a few weeks before he died, Davy added a codicil leaving Pappina a further £50. Lady Davy was made the sole executor of this Will, a duty she carried out faithfully. Despite the urgings of Walter Scott, she never published her own memoirs, which might have described what it was really like to live with a man of science – who knew he was a genius.

10

Young Scientists

1

By the end of the 1820s British science had lost its three international stars, the three scientific knights whose names had been renowned throughout Europe. The deaths of Joseph Banks in 1820, William Herschel in 1822, and finally of Humphry Davy in 1829, marked the passing of an age. The idea that they had between them created a distinctively British science was itself part of Banks's great bequest to the nation. But with these departures its future seemed uncertain, and its reputation undefended. Who among the younger generation would take British science forward? And who would fund it? It was a time of great uncertainty. *The Times* helpfully announced that an age of scientific giants had passed away.[1]

The questions became more insistent. Was the Royal Society fulfilling its role? Was British science itself in decline, compared to France and Germany? Did science have a recognised social and moral role in society? Ever since the Vitalism debate, such questions were no longer limited to a small circle of experts and academics. Public concern about the role of science in society was now widespread. The thirty-four-year-old Thomas Carlyle, newly arrived in Edinburgh and freshly bearded for the fight, was just beginning to make his name as a polemical essayist and an aggressive social commentator. His first influential tract, *Signs of the Times*, dominated almost an entire issue of the *Edinburgh Review* in spring 1829. Here Carlyle announced the demise of Romanticism and the relentless arrival of 'the Age of Machinery'.

Carlyle made the problematic role of the man of modern science a central issue. He attacked the dehumanising effects of utilitarianism, statistics and the 'science of mechanics', and opposed the world of the laboratory to those of art, poetry and religion. Though he did not name the Royal Society or the Royal Institution, he came very close to it. 'No

Newton, by silent meditation, now discovers the system of the world from the falling of an apple: but some quite other than Newton stands in his Museum, his Scientific Institution, and behind whole batteries of retorts, digesters and galvanic piles imperatively "interrogates Nature" – who, however, shows no haste to answer.'[2] Four years later, warming to his theme, Carlyle would announce definitively: 'The Progress of Science ... is to destroy Wonder, and in its stead substitute Mensuration and Numeration.'[3] ♣

In the Royal Society's presidential election of 1829, John Herschel became the natural candidate of the young scientists, despite his own deep personal misgivings. At thirty-seven he was recognised as a polymath at the height of his powers. He had been Secretary of the Society for five years, and had published over a hundred papers on subjects ranging from astronomy to zoology. He was known to be developing a philosophy of 'pure inductive science', heralded as the true heir to Baconian thought. Moreover, he was wealthy and settled. In March 1829 he had heeded his aunt Caroline's advice and married a very beautiful and gifted Scottish girl, Margaret Brodie Stewart. Above all, he was the son of his father, Sir William.

But Herschel soon found himself drawn into a public debate about the personalities and administration of science, quite unlike anything his father had experienced. The unworldly Michael Faraday could not be persuaded to stand. The mercurial Charles Babbage was regarded as unreliable and unsuitable. Both Wollaston and Thomas Young were dead, while the aristocratic candidate was the charming but ineffective Duke of Sussex, brother to King George IV, who knew nothing about science at all – although this was considered by some more traditional Fellows to be an overwhelming advantage.

♣ The troubling image of a shy, reluctant, persecuted female Nature who is crudely questioned and even physically assaulted by an exclusively male Science now begins to appear. It slowly replaces the older Romantic image of a mysterious and seductive Nature, at least a goddess, who is infinitely more powerful than her merely human petitioners and questioners. The rhetoric of assault, molestation, penetration and even rape of Nature by 'Science' develops, though partly unconsciously, throughout the nineteenth century, and was keenly identified by twentieth-century feminist criticism. See for example Anne K. Mellor, 'A Feminist Critique of Science' (1988). It was also popularised, as well as vulgarised, in various other art forms, as for example in the sculpture of the *fin-de-siècle* French artist Louis Ernest Barrias. His pair of metre-high bronze statues, *Nature Unveiling Herself Before Science* (1890), one partly shrouded and the other completely nude, won the Grand Prix at the Exposition Universelle for 1905.

After a good deal of gentlemanly infighting, during which Herschel threatened to withdraw his candidature, the Duke of Sussex was elected in 1830 by a very narrow majority: 119 votes to 111. Babbage, checking his statistics, noted with disgust that less than 33 per cent of the membership had voted. This unsatisfactory result led to a breakaway movement by a handful of young scientists around Herschel. They began to think of circumventing the Royal Society entirely, and appealing to a wholly different constituency: the 'amateur' men (and women) of science who belonged to the provincial scientific or 'philosophical' societies and institutions outside London. As if to soothe Herschel, he was promptly knighted, on a recommendation that many thought came from the Duke of Sussex, anxious to placate his rival. If so, it did not have the desired effect.

Between 1829 and 1831 a series of publications by John Herschel, his friend Charles Babbage and the Scottish science writer David Brewster (who had done fine research work on polarised light) pursued the emotive theme of the supposed 'decline of Science in Britain'. The debate was taken up by the leading journals, rapidly moved beyond the Royal Society, and became one about national culture and the role of the man of science in society. It was no coincidence that all this took place at the same time as the national self-questioning reflected in the violent political debates surrounding the Great Reform Bill.

2

The first salvo was fired by Charles Babbage, when he released a slim but carefully targeted volume, provocatively entitled *Reflections on the Decline of Science in England*, in the spring of 1830. Two years previously Babbage had been appointed Lucasian Professor of Mathematics at Cambridge, Newton's old chair, and he had considerable influence. It was known that his lectures on astronomy at the Royal Institution in 1817 had won the approval of Sir William Herschel, and that his research work had been supported by Sir Humphry Davy. He was wealthy, and had a large house in London at Dorset Square. Here he was working on his famous 'Difference Engine No. 1', a prototype computer which would require 25,000 brass cogs to function. After expending over £17,000 (a colossal sum) of his own money on it, he was understandably keen on the notion of government funding for such projects. This gave added energy, or bias, to his attack.

Babbage's prototype computer later became one of the legends of Victorian science, and a parable about the failure of government research funding. At the point when he ran out of money in 1832, Babbage had succeeded in constructing one self-contained section of his Difference Engine No. 1, employing 2,000 brass components, which still exists and works impeccably as an automatic calculator. A more sophisticated 'Analytical Engine' using a punched-card input and mechanical 'store' on 50,000 brass cogs, the genuine equivalent of a modern computer's RAM 'memory', was designed but never constructed. No one knows if this would have worked. However, Babbage's Difference Engine No. 2, designed in the 1840s to use 4,000 brass cogs, was actually constructed by the Science Museum in 1991, and with some minor alterations works to this day, capable of calculating to thirty-one places of decimals – an impressive power. It weighs three tons and cost £300,000 – considerably cheaper, in relative terms, than the original.[4]*

Babbage's outspoken book was a polemical exposé of weak British scientific institutions and casual attitudes to research. He compared these with the culture of scientific research fostered by the great Continental Academies of Science, in Paris and Berlin. Though 'eminently distinguished for mechanical and manufacturing ingenuity', Britain was shamefully 'below other nations' in pure sciences. While he referred respectfully to the achievements of both Sir Humphry Davy and Sir William Herschel, Babbage implied that times had changed radically.

He instanced the lack of government funding for research, the fact that there had so far been no honours for distinguished scientists such as Faraday and the meteorologist Beaufort, and the lack of recognition for the chemical work of John Dalton and Wollaston. He criticised the weakness of science teaching in the universities (apart, evidently, from his own field, mathematics) and the failure of the Royal Society to fund large

---

* Unlike Harrison's chronometer, Herschel's telescope or Davy's voltaic battery, Babbage's 'computer' had no immediate application that officialdom could see or even imagine, though Babbage claimed correctly that it would transform the calculations for logarithms, astronomical tables, engineering construction models, map-making and marine data. Coleridge once said that radically new poetry 'must create the taste whereby it is appreciated'. Perhaps Babbage believed the equivalent of radically new science. See Jenny Uglow and Francis Spufford, *Cultural Babbage: Technology, Time and Invention* (1996).

scientific projects, or promote the public understanding of science in Britain. Despite its ringing motto, *Nullius in Verba*, the Society fostered no generally agreed philosophy of science.

Babbage's attack on the Royal Society became increasingly contemptuous. Where, he asked, were the British equivalents of Berzelius (Sweden), Humboldt (Germany), Oersted (Denmark) or Cuvier (France)?[5] He claimed that the Society's members were lazy, elitist, ignorant and largely dedicated to club dinners. In a devastating early use of statistical analysis, he showed that only 10 per cent of the 700 members had published two or more scientific papers.[6] He also jeered that British scientific societies were so easy for an amateur to join, that for the expenditure of precisely 'ten pounds, nine shillings and nine pence ha'penny' he had calculated that anyone could obtain 'a comet's trail of upwards 40 letters' as initials after his name – like FRS, for example.[7]

Babbage described the present Royal Society with a simile drawn pointedly from Herschel's work. It was utterly devoid of 'bright stars', and 'only visible to distant nations, as a faint Nebula in the obscure horizon of English science'.[8] He also urged a critical attitude to 'publication of experimental data', and the necessity for peer-reviewing – not, up till then, considered quite fair play. As a further provocation, he gleefully introduced such ungentlemanly terms as '*hoaxing, forging, trimming and cooking results*', which he claimed should be applied very strenuously.[9]

Babbage concluded the book with a suggestive comparison between the contrasted scientific styles of Wollaston and Davy. The first had been a meticulous, patient scientist, utterly without worldly ambition, and modest and private in his profession. He was primarily interested in getting precise results that avoided all possibility of bias or error. The second was a restless scientific enquirer, rapid and ambitious in all his work, superb at popularising and explaining his projects, driven by the desire to pursue and establish the truth, *and to be the first to do so at whatever cost.* Wollaston, he concluded, was a pure, saint-like man of science, while Davy was also a publicist and visionary: 'Wollaston could never have been a poet; Davy might have been a great one.' In the future, Babbage seemed to imply, British science would need both types.[10]

He added, for good measure, a section describing John Herschel at work in his laboratory at Slough, analysing 'the dark lines seen in the solar spectrum by Fraunhofer'.* Babbage perhaps intended a sort of parable of science for the new generation. His story went as follows. When Babbage first peered carefully at the shimmering solar image projected through Herschel's prism, he could not see these dark Fraunhofer lines, though he knew they were there. Herschel then commented to him: 'An object is frequently not seen, from *not knowing how to see* it, rather than from any deficit in the organ of vision... I will instruct you how to see them.'[11] After some time spent re-examining and refocusing the image, Babbage could see them perfectly. The point was that science must always be more than the simple observation of phenomena or data. It was simultaneously a subjective training in observational skills, self-criticism and interpretation: a complete education. This was of course precisely what William Herschel had said forty years before, about learning to see with a telescope.

To add a final sting in the tail, Babbage slipped in an Appendix enthusiastically praising the 1828 conference of the Berlin Academy of Sciences, which he had attended. It had great scientists – like its President Humboldt, who had delivered an address praising Goethe – and great visions of the future. It would next meet in Vienna in 1831. He now proposed a new 'Union of Scientific Societies' in Britain, to follow this admirable German model, with annual meetings in cities outside London. By all means the Royal Society could send participants, if it should so bestir itself. But who else would rally to the cause? Babbage's subversive tract was the first manifesto for what in 1831 would become the British Association for the Advancement of Science.[12]

Michael Faraday would not be drawn into this whirlpool of controversy. Instead he encouraged a Dutch chemist, Gerard Möll, to write a gentle reply and reproof to Babbage, 'By a Foreigner'. Möll observed that 'the English have quite enough of their natural and foreign political

* The identification of Joseph Fraunhofer's lines – similar to a supermarket barcode – was the first stage towards spectography, the method by which astrophysicists would eventually analyse the chemical composition of the stars. Particular elements – e.g. hydrogen – occupy particular places in the spectrum of starlight, and can thereby be identified across enormous distances in space; in fact across the entire visible universe. The implications of spectography are beautifully explored in the 'Barcodes in the Stars' chapter of Richard Dawkins' *Unweaving the Rainbow* (1998), which ends with a long quotation from James Thomson's poem 'To the Memory of Sir Isaac Newton' (1727).

enemies, without waging a civil-scientific war between themselves ... The Barons of the French Institute will be highly amused ... A neutral foreigner cannot help seeing with regret Englishmen scoff and rail at things which ought to have been looked upon as the pride of their country.'[13]

John Herschel was not to be deterred by this appeal to his patriotic and gentlemanly instincts. He followed his friend with a quite different and much subtler line of attack. He decided to put forward a progressive view of British science, and hold out the possibilities of a golden future.

3

Herschel's quietly phrased but immensely authoritative book *A Preliminary Discourse on the Study of Natural Philosophy* was published as the first volume in a popular series, *Lardner's Cabinet Cyclopaedia*. Despite its anodyne title, deliberately chosen to offset Babbage's style of provocation, this became a hugely popular work which would run into many new editions throughout the early Victorian period.

John Stuart Mill would recall in his *Autobiography* how, after his nervous breakdown and therapeutic immersion in the poetry of Wordsworth and Coleridge, it was Herschel's book that showed him how far he had recovered his intellectual grasp by 1837. 'Under the impulse given me by the thoughts excited by Dr Whewell, I read again Sir J. Herschel's *Study of Natural Philosophy*, and I was able to measure the progress my mind had made, by the great help I now found in this work.'[14]

Herschel first looked back at the great triumphs of Romantic science, very properly including work done in France and Germany, and appealed for the public understanding of 'professional science' in Britain. It was a profession first proposed by Bacon, based on the fundamental value of free enquiry.[15] Herschel defined its field as a rapidly expanding arc of scientific disciplines: the classical ones – mathematics, astronomy and optics – now joined by the study of electricity, chemistry, magnetism, geology, botany and gases.[16] He argued that common to all of them was the three-part 'inductive' method. First, the precise gathering of quantitative data by observation and experiment; second, the emergence of a general 'hypothesis' from this data; and third, the testing of this hypothesis once more by experiment and observation, to see if it could be disproved.[17] This inductive discipline was central to all sciences, and led on to the first aim of free scientific enquiry: the investigation of the unknown. 'The immediate object we propose to ourselves in physical theories is the analysis of phenomena, and the knowledge of the hidden processes of Nature in their production, so far as they can be traced by us.'[18] Nature was still hidden and mysterious, alive with 'processes' and powers, though Herschel was careful to avoid any hint of *Naturphilosophie*, or any speculation about the 'Power and Intelligence' that might ultimately maintain it. Nevertheless, nature revealed continuously 'wonder upon wonder'.[19]

This was greeted as the first attempt since Francis Bacon's *Novum Organum, or New Instrument* (1620) to write a popular treatise on the inductive philosophy of science. It had an engraving of Bacon (with both microscope and telescope – *micromegas*) on the title page, and began with a Latin epigraph from Cicero: *In primis, hominis est propria VERI inquisitio atque investigatio*'. This was translated for the reader as 'Above all other things, Man is distinguished by his pursuit and investigation of TRUTH' – an interesting assertion. Of course the whole text was written in English, though Herschel chose the shrewd device of organising it in numbered paragraphs, as well as conventional literary chapters. Indeed it emerged that Herschel, unlike his father, could write fluently, and sometimes with great imaginative force. (One other effect of his Cambridge education was that throughout his life he wrote admirable light verse, and later completed a translation of Virgil's *Aeneid.*) In one passage he argued the necessity for clarity and precision in the use of scientific terms with almost poetic originality.

> For example, the words – square, circle, a hundred etc convey to the mind notions so complete in themselves, and so distinct from everything else, that we are sure when we use them we know the whole of our own meaning. It is widely different with words expressing natural objects and mixed relations.
>
> Take, for instance, IRON. Different persons attach very different ideas to this word. One who has never heard of magnetism has a widely different notion of IRON from one in the contrary predicament. The vulgar, who regard this metal as incombustible, and the chemist, who sees it burn with the utmost fury, and who has other reasons for regarding it as one of the most combustible bodies in nature; – the poet, who uses it as an emblem of rigidity; and the smith and the engineer, in whose hands it is plastic, and moulded like wax into every form; – the jailer, who prizes it as an obstruction, and the electrician who sees in it

> only a channel of open communication by which – that most impassable of objects – air may be traversed by his imprisoned fluid, have all different, and all imperfect, notions of the same word.
>
> The meaning of such a term is like a rainbow – everybody sees a different one, and all maintain it to be the same.[20]

That final embracing reference to 'everybody's' rainbow was a deliberate act of inclusion: Newton's rainbow, but also Wordsworth's and Keats's and Goethe's are all implied.*

Herschel went on to praise the intellectual and even spiritual value of the true scientific outlook. Everything in nature became interesting and significant, nothing was beneath notice. The most 'trifling natural objects', such as a soap bubble, an apple or a pebble, could reveal a scientific law (respectively, the laws of aerostatics, gravitation or geology).

> To the natural philosopher there is no natural object unimportant or trifling... A mind that has once imbibed a taste for scientific enquiry has within itself an inexhaustible source of pure and exciting contemplations. One would think that Shakespeare had such a mind in view when he describes a contemplative man finding
>
> Tongues in trees – books in the running brooks
> Sermons in stones – and good in everything
>
> Where the uninformed and unenquiring eye perceives neither novelty nor beauty, *he* walks in the midst of wonders.[21]

---

* Goethe's *Treatise on Colour* (1810), which criticised Newton's 'mechanical' analysis of the rainbow spectrum, remained a totem of German *Naturphilosophie*, though it caused increasing irritation in empirical British scientific circles. Yet Goethe explored such suggestive ideas as 'the sensory-moral effects of colour', the 'spiral tendency in vegetation', and the effect of weather (clouds, sunlight, changing barometric pressure) on mental states and moods. Goethe was wonderfully perceptive about what he insisted was the unity of the scientific and artistic sensibility. He wrote an outstanding short essay on the delicate balance between 'objective' and 'subjective' observation of data: 'Empirical Observation and Science' (1798). 'The observer never sees the pure phenomenon with his own eyes; rather, much depends on his mood, the state of his senses, the light, the air, the weather, the physical object, how it is handled, and a thousand other circumstances.' See Goethe, *Collected Works*, vol 12: *Scientific Studies* (1988). Humboldt also praised him: 'Goethe, whom the great creations of the poetic Fancy have not prevented from penetrating the arcana of Nature' (Berlin Academy conference, 1828).

It is intriguing that Herschel was quoting from Shakespeare's *As You Like It* (Act II, scene i), a scene which takes place in the idealised and magical Forest of Arden. Herschel evidently saw the 'contemplative' man of science naturally inhabiting such a sylvan world, a place of visions and transformations, where all turns out for the good. So among the triumphs of contemporary science he listed a series of simple discoveries and technological inventions that had hugely improved human safety: among them the lightning conductor, the lighthouse lens, the safety lamp, iodine and chlorine disinfectant (the last three being Davy's).[22]

Like Davy, Herschel chose chemistry as the exemplary discipline of the Romantic period. Developing from the errors of alchemy and phlogiston theory, chemistry had been 'placed in the ranks of the exact sciences – a science of number, weight and measure'. It had produced practical applications in every sphere: medicine, agriculture, manufacturing, aerostation and meteorology, for example. But it had also advanced pure science: the doctrines of oxygen, latent heat, atomic weight, polar electricity and the prime elements (of which more than fifty were now known). Moreover, this was the achievement of an international group: Lavoisier, Black, Dalton, Berzelius, Gay-Lussac and Davy.[23]

In ten brilliantly clear and even thrilling pages (paragraphs numbered 368–77), Herschel gave an international history of fifty years' researches into electricity, from Franklin and Galvani to Davy and Oersted. From early vague ideas of some mysterious natural fluid – a 'wonderful agent' – seen in lightning strikes, the Aurora Borealis or 'the crackling sparks which fly from a cat's back when stroked', he traced the experimental path which led to increasingly precise and sophisticated concepts of electrical current, conductors, positive and negative poles, batteries, charge and discharge, animal electricity ('an unfortunate epithet'), nervous circuitry, chemical affinity (Davy's 'total revolution') and 'the wonderful phenomenon of electro-magnetism', which awaited further exploration.[24]

Herschel prophetically implied that electricity and electro-magnetism still hid many secrets, and that their investigation would become the leading science of the new age. This would indeed be Faraday's coming field of triumph. He summarised (paragraph no. 376) this pursuit in the image of a great and noble sea voyage of exploration. 'There is something in this which reminds us of the obstinate adherence of Columbus to his notion of the necessary existence of the New World; and the whole history of this beautiful discovery may serve to teach us reliance on those general analogies and parallels between great branches of science by which one strongly reminds us of another, though no direct connection appears.'[25]

This notion of a great network or connection of sciences, beginning to form a single philosophy and culture, was crucial to his book. In the same positive vein Herschel argued that science, while often going against common sense or intuition, expanded the human imagination with previously inconceivable ideas of movement or magnitude. The examples he gave were the speed of starlight, the movement of a gnat's wings, or the vibrations of colour frequency. Finally he promoted the moral value of science. It was a source of clarity and intellectual excitement, and (perhaps more controversially) of philosophical calm in troubled times. In all these ways John Herschel sought to give 'the man of science' a new and central place in English society – and not just the Royal Society.

Faraday himself wrote appreciatively to Herschel from the Royal Institution, in one of his breathless, enthusiastic screeds. 'When your work on the study of Nat. Phil. came out, I read it as all others did with delight. I took it as a school book for philosophers and I feel it has made me a better reasoner and even experimenter and has altogether heightened my character and made me if I may be permitted to say so a better philosopher.'[26]

Many others felt the same. For one undergraduate at Cambridge the book was like a summons to arms. 'Humboldt's *Personal Narrative* and Herschel's *On Natural Philosophy* stirred up in me a burning zeal to add even the most humble contribution to the noble structure of Natural Science. No one of a dozen other books influenced me nearly so much as these two.' The undergraduate was twenty-two-year-old Charles Darwin, and his humble contribution was to be *On the Origin of Species* (1859).[27]

4

It was now the turn of David Brewster (1781–1868). Educated in Edinburgh, Brewster was a physicist who had contributed widely to scientific journals and encyclopaedias. His field was applied optics, such as lighthouse lenses, and he invented the kaleidoscope; but he was also inventing the new career of science journalism. A Calvinist who had abandoned the Church, he was a natural evangelist for science. He decided that a campaign rather than a book was needed, and now published specific proposals for a new national scientific association in a number of magazines, including the *Quarterly Review*. He wrote urgently to Babbage in February 1830: 'I wish you could spare ten minutes to my equation... and would it not be useful to organize an Association for the purpose of protecting and promoting the *secular* interests of Science? A few influential noblemen and MP's would give great help in forwarding such an object.'[28]

Such an Association was to meet annually, as Babbage had suggested, on the German model, at different provincial cities – but not London, being the territory of the Royal Society. It was to draw its membership primarily from the universities, the House of Commons and the local 'Literary and Philosophical' societies in the great northern cities. There had been fewer than ten of these when Banks had begun at the Royal Society in the 1780s, among the earliest being Manchester, Derby and Newcastle upon Tyne. There were some thirty in existence by the time Davy was elected in 1820, and nearly seventy by the time Charles Darwin came back from the Galapagos islands in 1836. This was the beginning of the historic expansion of Victorian science.[29]

There was much campaigning, recruiting and arguing throughout 1830–31. Babbage in London, Brewster in Edinburgh and Whewell in Cambridge led the drive. A typical missive from Whewell read: 'I can see abundance of good things that such a Society may do: one matter which requires multiplied and extensive fagging is meteorology, which I hope Dalton may do ... Sedgwick is still hammering in Wales. Darwin ... is just on the point of setting out as a naturalist with Captain Fitzroy who is to complete the survey of the south end of America. I expect he will bring you home the tip of Cape Horn ... '[30]

Faraday still remained elusive, and Herschel – mindful of his position as Secretary to the Royal Society – tactfully explained in an immensely long letter that he could only send 'sincere good wishes for its utility and consequent success'. He did however recognise 'the want in this country and in the actual state of science, of a great, central and presiding power to give an impulse and direction to enquiry'.[31] By autumn 1831 it was still hoped that a few other 'scientific *lions* may be allowed to perambulate the country'.

Finally, a somewhat depleted first meeting of the British Association for the Advancement of Science took place at York in October 1831. Undaunted, the members vigorously discussed comets, railways, geological strata, the Aurora Borealis, marsupial mating habits, and subversively drank Joseph Priestley's health (a reproach to the Royal Society and a greeting to America).[32] A combative keynote speech about the development of science in Britain was delivered by the first President, William Rowan Hamilton, but this was not felt to have quite the reach or impact of Humboldt in Vienna. There was some lively disagreement (which was to continue for many years) over the correct balance between dinners and lectures, or 'feasting versus philosophy'. However, in the absence of figures like Faraday and Herschel, the whole thing slipped away almost entirely unreported in the press.

A second meeting of the fledgling Association took place at Oxford in 1832. A fine theatrical performance from Professor William Buckland on the subject of geology and the courtship of primitive reptiles received some praise. This time *The Times* deigned to notice the occasion, but loftily dismissed it as 'a mere unexplained display of philosophical toys', and pointed out that Buckland sometimes seemed to forget that he lectured 'in the presence of ladies'.[33]

But with the third meeting in June 1833 the British Association really began to make a national impact. It was held at Cambridge, itself considered a major coup, and the capture of the heartland of progressive rational thought in Britain. Cambridge was also Newton's shrine, and the base of the powerful 'Trinity and John's' group of scientific academics. This time the list of those attending included almost all those who would soon become the rising stars in the firmament of early Victorian science: Michael Faraday, Sir John Herschel, John Dalton, Charles Babbage, Sir David Brewster, Adam Sedgwick, William Whewell, Thomas Chalmers, Thomas Malthus and William Somerville. The only notable absentee was Charles Darwin, just then botanising in Uruguay during the *Beagle*'s voyage.[34]

Some of 'the ladies' were also pressing for admittance, including several powerful scientific wives, like Margaret Herschel and Mary Somerville. They pretended to be fully engaged in hosting receptions and choosing the menus, while unofficially they listened at the back of the lecture halls, took notes, and critically judged the quality (and appearance) of the speakers. The major debate was on the nature of the Aurora Borealis, which symbolically called upon a wide range of scientific interests including meteorology, optics, electricity, magnetism, polar exploration and solar astronomy. It was held at the heart of the university, in the Cambridge Senate House, on King's Parade. The main luncheon, a cold collation for 600 members, was staged at Trinity, with guests drifting across Great Court to toast the statue of Newton. Then came fireworks, and a 'botanical barge' energetically punted up the Cam. One other noticeable participant, now ill and frail, but still intellectually formidable, was Samuel Taylor Coleridge, aged sixty.

Coleridge was put up in a friend's rooms at Trinity itself, and remarked appreciatively that his bed was 'as near as I can describe it a couple of sacks full of potatoes tied together ... Truly I lay down at night a man, and arose in the morning a bruise.' This, rather than opium, might explain why he was never able to rise till the afternoon, though he always had 'a crowded levee' at his bedside. Nevertheless he stayed for three days, attended many of the meetings, and always found undergraduates and professors crowding round to talk to him. He certainly was one of the lions, though from a disappearing age.

All his old enthusiasm for scientific matters came sweeping back, and he was soon in the thick of it, boldly announcing that 'Lyell's system of geology is half truth – but not more'; while Descartes' vortices 'were not a *hypothesis*: they rested on no facts at all ... Your subtle fluid etc is pure gratuitous assumption.' Then he delighted everyone by suddenly saying: 'That fine old Quaker philosopher Dalton's face was like – like All Soul's College.' This was a very *Oxford* joke in Cambridge.[35]

He was up to the minute with Herschel's *Natural Philosophy*, and gave an impressively Coleridgean account of the role of 'hypothesis or theory' in the inductive philosophy. 'The use of a Theory in the *real sciences* is to help the investigator to a complete view of all the hitherto discovered parts relating to it; it is a Collected View, θεωρια [*Theoria*], of all he yet knows *in one*. Of course whilst any facts remain unknown, no theory can be exactly true, because every new part must necessarily displace the relation of all the others. A theory therefore only helps investigation: *it cannot invent or discover*.'[36]

Memories of Humphry Davy must have come flooding back, in all the glow of his Bristol youth, for Coleridge got on particularly well with the young Michael Faraday. Unlike Lady Davy, he was favourably impressed by Faraday's fine open face, with its mop of curling hair and gazing wide-apart eyes, and his modest manner, with its peculiar directness and intensity. 'I was exceedingly pleased with Faraday, he seemed to me to have the true temperament of Genius – that of carrying on the spring and freshness of youthful, nay boyish, feelings into the mature strength of manhood.'

This was a signal recognition by Coleridge, who had defined such ageless energy as a characteristic of literary genius some seventeen years before, in Chapter 4 of his *Biographia Literaria* (1816). In a passage describing the poetry of Wordsworth, he wrote: 'To carry on the feelings of childhood into the powers of manhood; to combine the child's sense of wonder and novelty with the appearances which every day for perhaps forty years had rendered familiar – *with sun and moon and stars throughout the year, And man and woman* – this is the character and privilege of genius, and one of the marks which distinguish genius from talent.'[37] He was now applying these literary criteria to a man of science. In his last published work, *On Church and State* (1830), he had included men of science as an essential part of what he christened 'the clerisy': that is, the diffuse body of thinkers, writers, teachers and opinion-formers who made up the intelligentsia or informing culture of a nation.[38]

At one meeting, chaired by William Whewell, Coleridge was drawn into a passionate discussion of semantics. It revolved around the question of what exactly someone who works 'in the *real sciences*' (as he had phrased it) should be *called*. This is how Whewell reported the British Association debate in the *Quarterly Review* of 1834:

> Formerly the 'learned' embraced in their wide grasp all the branches of the tree of knowledge, mathematicians as well as philologers, physical as well as antiquarian speculators. But these days are past ... This difficulty was felt very oppressively by the members of the BAAS at Cambridge last summer. There was no general term by which these gentlemen could describe themselves with reference to their pursuits.
>
> 'Philosophers' was felt to be too wide and lofty a term, and was very properly forbidden them by Mr. Coleridge, both in his capacity as philologer and metaphysician. 'Savans' was rather assuming and besides too French; but some ingenious gentleman [in fact Whewell himself] proposed that, by analogy with 'artist', they might form 'scientist' – and added that there could be no scruple to this term since we already have such words as 'economist' and 'atheist' – but this was not generally palatable.[39]

The analogy with 'atheist' was of course fatal. Adam Sedgwick exploded: 'Better die of this want [of a term] than bestialize our tongue by such a barbarism.' But in fact 'scientist' came rapidly into general use from this date, and was recognised in the OED by 1840. Sedgwick later reflected more calmly, and made up for his outburst by producing a memorable image. 'Such a coinage has always taken place at the great epochs of discovery: like the medals that are struck at the beginning of a new reign.'[40]

This argument over a single word – 'scientists' – gave a clue to the much larger debate that was steadily surfacing in Britain at this crucial period of transition 1830–34. Lurking beneath the semantics lay the whole question of whether the new generation of professional 'scientists' would promote safe religious belief or a dangerous secular materialism. Hitherto, either austere intellectual Deism, held for example by William Herschel, or else the rather more picturesque Natural Theology conveniently accepted by Davy (at least in his public lectures) had disguised this problem, whatever the revelations of astronomy or geology, or the inspired ragings of Shelley.

For many Romantic scientists, with a robust intellectual belief in the 'argument by Design', there was no immediate contradiction between religion and science: rather the opposite. Science was a gift of God or Providence to mankind, and its purpose was to reveal the *wonders* of His design. This indeed was the essence of 'natural' religion, as promoted for example by William Paley in his *Natural Theology* (1802), with its famous analogy with the divine watchmaker. It was the faith that brought Mungo Park back alive from his first Niger expedition. It was the faith that inspired Michael Faraday to become a Deacon in the Sandemanian Church in July 1832.

But public faith often differed from private beliefs. Whatever he said in his famous lectures, Davy's poetry and his posthumous writings, such as *Consolations in Travel*, suggested a kind of science mysticism that certainly precluded a Christian God, and possibly even any kind of Creator at all. Others, like William Herschel, had been content to rely on an instinctive, perhaps deliberately unexamined, belief in a benign Creator somewhere distantly behind the great unfolding scheme of nature. Though in Herschel's case, his own observations had shown how extremely – *appallingly* – distant, both in time and space, that Creator must be. Moreover, his sister Caroline never once mentioned God anywhere in her journals.[41] As for Joseph Banks, his sister Sophia had had no high opinion of his natural piety.

Yet with the growing public knowledge of geology and astronomy, and the recognition of 'deep space' and 'deep time', fewer and fewer men or women of education can have believed in a literal, Biblical six days of creation. However, science itself had yet to produce its own theory (or myth) of creation, and there was no alternative Newtonian Book of Genesis – as yet. That is why Darwin's *On the Origin of Species* appeared so devastating when it was finally published in 1859. It was not that it reduced the six days of Biblical creation to myth: this had already been largely done by Lyell and the geologists. What it demonstrated was that there was no need for a divine creation at all. There was no divine creation of species, no miraculous invention of butterflies' wings or cats' eyes or birds' song. The process of evolution by 'natural selection' replaced any need for 'intelligent design' in nature. Darwin had indeed written a new Book of Genesis.*

Over the following five years, the well-meaning 8th Earl of Bridgewater would commission a whole series of booklets by the leading men of science, intended to show how British scientific research and discovery unfailingly underpinned Christian – and specifically Anglican – belief. They were to illustrate what might have been called an unproven hypothesis: 'The Goodness of God as Manifested in the Creation'. The thankless task of composing these *Bridgewater Treatises* (1830–36) was

* There was a premonition in an anonymous 'evolutionary' book, *Vestiges of the Natural History of Creation*, which caused a sensation in 1844. But Darwin had worked by John Herschel's rules of pure induction: assembling a mass of precise data (e.g. the evolution of finches' beaks) until the simplest and most convincing hypothesis emerged. Consequently the great mainstay of so many scientists – Natural Theology and the Argument by Design – was worse than untrue: it was unnecessary. The spiritual upheavals this caused devout Victorian scientists were famously described by Edmund Gosse in *Father and Son* (1908). But it was the earlier, preliminary impact of geology, on ordinary thinking men and women, which was recorded by Tennyson in several sections (56 and 102) of *In Memoriam* (1833–50). The subject and inspiration of this poem was his Cambridge friend Arthur Hallam, who died in exactly this year of the third BAAS meeting.

> 'So careful of the type?' but no.
> From scarped cliff and quarried stone
> She cries, 'A thousand types are gone:
> I care for nothing, all shall go . . .'
> (*In Memoriam*, Section 56)

piously or sportively undertaken by Chalmers (on astronomy), the humorous Professor Buckland (on geology), Whewell (on physics), Charles Bell (on anatomy) and several others of lesser note. Thanks to the Duke of Bridgewater's bequest, they were all outstandingly remunerated at £1,000 each, plus all profits.[42]

Reading Buckland on geology, Mary Somerville mournfully observed: 'facts are such stubborn things'. Faraday, a lifelong Sandemanian, refused to make any comment. Charles Babbage threatened to write a ninth and scathing last treatise, but he never finished it.[43]

On a more whimsical note, William Sotheby, Coleridge's old friend and the translator of Dante, celebrated this third conference with a long, prismatic piece of light verse, 'Lines on the 3rd Meeting of the BAAS at Cambridge, 1833'. He set out a new tradition, the roll-call of the great 'scientists'. Among others he saluted Bacon, Newton, William Herschel, Wollaston, Davy, Faraday, Dalton, John Herschel, Babbage, Roget, Hutton, Playfair and Lyell. But he only mentioned one woman: not Caroline Herschel, but Mary Somerville; and she was noticed, ironically, for her official absence.

> Why wert thou absent? Thou whose cultured mind,
> Smoothing the path of knowledge to mankind
> Adorn'st thy page deep stored with thought profound...
> While Cambridge – glorying in her Newton's fame –
> Records with his, thy woman's honoured name,
> High-gifted Somerville!...[44]

Later meetings of the British Association took place, as planned, rotating round the great provincial capitals, but studiously avoiding London. There was now increasing competition to be the host metropolis, as it was realised that the Association was beginning to attract both international recognition and a considerable local boost to city finances. Edinburgh was chosen in 1834, followed by Dublin in 1835, Bristol in 1836, Liverpool in 1837, Newcastle in 1838, Birmingham in 1839 and Glasgow in 1840. By this time over 2,000 people were attending each year, the press coverage was huge, and the official membership had risen to over 1,000.

But the early press reception – now increasingly important in British science – was surprisingly rough, and revealed all sorts of class and cultural anxieties. *The Times* leaders thundered out disapproval annually from 1832 to 1835: 'It is the necessary consequence of the Spirit of the Age ... The principle of humbug, the principle of Penny Magazines, and Mechanics Institutes, the principle of spreading the waters of knowledge over a large surface without caring how shallow they may be – The Association, we prophesy, will soon see its end.'[45] To emphasise its unimpeachable accuracy, *The Times* consistently spelt Michael Faraday as 'Farraday'.

The magazine *John Bull* added to the chorus in 1835: 'Amongst the extensive Humbugs which so eminently distinguish this very extraordinarily enlightened Age, none perhaps is more glaring than the Meeting of what is called the British Association for the Advancement of Science ... With the aid of concerts and dancing, fireworks and fine women, sound claret and strong whisky, the Sages make out remarkably well.'[46]

Scenting a good story, Charles Dickens launched a satirical series in *Bentley's Miscellany* in 1838, entitled 'The Full report of the First Meeting of the Mudfog Association for the Advancement of Everything'. It was supplied with mocking cartoons by his gifted illustrator George Cruikshank, who had achieved such a success with *Sketches by Boz.* In it Dickens invented some early fictional scientists: Professor Snore, Professor Doze and Dr Wheezy, though all of them were more benign and ineffectual than Dr Victor Frankenstein.[47]

5

While these public battles raged, Michael Faraday quietly continued his experiments at the Royal Institution. He was now released from Davy's oppressive shadow, yet still clearly inspired by his memory. He worked immensely hard, giving his first Bakerian Lecture to the Royal Society in 1829, and also accepting a simultaneous post as Professor of Chemistry at the Royal Military Academy, Woolwich. He expanded his work on electromagnetism, and began the construction of the first electrical generators, by producing an 'alternating' electrical current. This would lead to electrical dynamos that would ultimately revolutionise industry as much as James Watt's steam engine. His experiment with magnetic coils and a galvanometer (which was made to move without physical contact), carried out at the Institution's laboratory on 29 August 1831, was said to have ended 'the Age of Steam' at a stroke, and begun the new 'Age of Electricity'.[48]

Faraday also took on from Davy the great task of educating the public in scientific matters. In 1826 he began his series of Friday Evening Discourses, in which a whole range of scientific topics were carefully presented and vividly explained to a general audience. From this grew perhaps his greatest innovation, his Christmas Lectures for Children, which are still given annually (and now televised). The classic example became his brilliantly clear and inventive series 'The Chemical History of a Candle'. This started with the simple notion of flame and combustion, the very process that had so entranced the young Davy. It was beautifully followed out, step by step, into an entire panorama of natural processes: human and animal respiration, plant growth, and the entire global carbon cycle. Faraday would talk and explain with quiet, gentle authority, occasionally bursting out with some delighted exclamation. '*Wonderful* is it to find that the change produced by respiration, which seems so injurious to us – for we cannot breathe air twice over! – is the very life and support of plants and vegetables that grow upon the surface of the earth.'[49]

These lectures were perhaps Faraday's best tribute to his great and difficult patron, and one of the last great documents of Romantic science. 'The Chemical History of a Candle' was eventually adapted by Dickens, without any satirical intent, for his family magazine *Household Words* in 1850.

6

Other important things had been stirring in the world of science writing. David Brewster had begun to work on the first ever biography of Isaac Newton, designed not only to explain the work, but to draw an analytical portrait (within certain limits of propriety) of the great man's mind and temperament. Mary Somerville, wife of a Royal Society Fellow, had also set herself to become a science-populariser, starting with an English translation and adaptation (1831) of Laplace's *Mécanique Céleste*, and with general essays and reviews of the different scientific disciplines.

The geologist Charles Lyell began in 1830 to bring out his classic work *Principles of Geology*, which would finally use scientific evidence to reject the Biblical account of short-scale creation of the earth, as maintained by every authority from Cuvier and Paley to Buffon and Buckland. Lyell's proposal of a 'deep time' corresponded to the 'deep space' cosmology of

William Herschel. It would ultimately provide the supportive authority for Charles Darwin, his great friend, to accept the deep time necessary for evolution by natural selection to take place.

But during these five years of intense controversy from 1829 to 1834, it was the publication of four literary works that contributed most powerfully to the debate about what 'a scientist' really was, or should be. They were all published in popular series aimed at the general public, such as Murray's Family Library. Such collections were intended to put contemporary ideas into general circulation and to reach the public at large. They reflected democratic stirrings, and the sense that ordinary people should be aware of what was being done in their name. These works helped to form the first public image of science, and the ambiguous feelings about scientists themselves.

Humphry Davy's influential *Consolations in Travel, or The Last Days of a Philosopher* had already sharpened these discussions. The expanded posthumous edition issued in Murray's Family Library in 1831 brought it to a much wider general readership, and made it one of the first popular works of scientific autobiography and speculation. It was regarded as a stimulating and eccentric book, which revealed the unexpected inner workings of a scientist's imagination. Writing from aboard the *Beagle*, off the Río de la Plata in May 1833, Charles Darwin begged his sister to send it, alongside Hutton on geology, Scoresby on Arctic regions, and Paul Scrope on volcanoes.[50]

Davy's strange and unforeseen speculations about the nature of social evolution, and the 'planetary' future of the human species, deeply impressed some, while they shocked others. When an American edition was issued, it was carefully edited with pious footnotes pointing out where Davy's views were theologically unorthodox, and suggesting proper corrections. The work was referred to extensively by Charles Babbage and John Herschel in their own books. In his Preface to his *Principles of Geology*, Lyell mentioned Davy's scientific speculations, but argued that the geology of 'the great chemist' was already fatally out of date, so swiftly was science now developing.[51] Later, a copy appeared in Chapter 15 of Anne Brontë's *The Tenant of Wildfell Hall* (1848), where it lies on the drawing-room table like a guarantee of serious intent in the household.

David Brewster's *Life of Sir Isaac Newton*, the first ever major scientific biography in Britain, was also issued in Murray's Family Library in 1831.

It deliberately set out to hold up a triumphant and inspiring image of British science to the nation at large, presenting Newton as a secular saint, 'the high priest of science' and a man of universal genius. It emphasised the creative importance of Newton's boyhood, and the intense originality of his mind, although it carefully eschewed the wonderful story of the falling apple and universal gravity, as told originally by William Stukeley in 1727. Brewster had in fact visited the orchard at Woolthorpe in 1814, to him a sacred site, and inspected the legendary apple tree, and even attempted to take a graft from it. But he carefully restricted himself to mentioning this piece of unscientific idolatry in a footnote. Years later, however, in an expanded 1860 edition of his biography, he flamboyantly told the whole tale, which by now had become the most glorious and perhaps misleading Eureka story in British science.[52]

Throughout, Brewster emphasised the cultural importance of science in society. In Chapter I he presented a survey of British scientific discoveries, ending with a summary of William Herschel's work, showing how a brilliant mind – even if originating abroad – could flourish in England when properly recognised and properly funded. He also emphasised the importance of biography for understanding 'the scientific process by which a mind of acknowledged power actually proceeds in the path of successful enquiry'. Brewer added significantly: 'The history of science does not furnish us with much information on this head, and if it is to be found at all, it must be gleaned from the biographies of eminent men.'[53]

Perhaps his greatest achievement was to popularise Newton's most famous remark about the process of scientific discovery: 'I do not know what I may appear to the world; but to myself I seem to have been only like a boy playing on the seashore, and diverting myself in now and then finding a smoother pebble or a prettier shell than ordinary, whilst the great Ocean of truth lay all before me.' It was a modest and yet thrilling image, which would be carried by thousands of Victorian schoolchildren – and their parents – to the holiday beaches and sea-bathing that were just becoming popular.[54]

The delayed – but increasingly formidable – impact of Mary Shelley's novel *Frankenstein, or The Modern Prometheus* had exactly the opposite effect. It clearly demonised science. The second edition was issued as a single pocket volume, in Bentley's Standard Novels in 1831. It now had as its frontispiece an engraving of the ghastly Creature rearing up in the shadows of Dr Frankenstein's laboratory, the gross limbs ill-fitted and distorted, the head bent forward and half-twisted from the shoulders, and the face carrying an expression of horror and disgust at its own monstrous existence.

The epigraph, from Adam's lament in Milton's *Paradise Lost*, read:

> Did I request thee, Maker, from my clay
> To mould Me man? Did I solicit thee
> From darkness to promote me . . . ?[55]

This edition first contained Mary's memorable and haunting Introduction, which describes her conversations with Byron and Shelley about science at the Villa Diodati in 1816, the work of Erasmus Darwin, and the waking nightmare in which she first conceived the novel. Now, in this new Preface, she added her own retrospective commentary on the notorious passage in which the 'hideous' Creature comes to life. She presented this as a moment of terrible, blasphemous and irreversible scientific *hubris*.

> I saw the pale student of unhallowed arts kneeling beside the Thing he had put together. I saw the hideous phantasm of a man stretched out, and then, on the working of some powerful Engine, show signs of life and stir with an uneasy, half-vital motion. Frightful must it be, for supremely frightful would be the effect of any human endeavour to mock the stupendous mechanism of the Creator of the world. His success would terrify the artist; he would rush away from his odious handiwork, horror-stricken. He would hope that, left to itself, the slight spark of Life which he had communicated would fade; that this Thing, which had received such imperfect animation, would subside into dead matter; that he might sleep in the belief that the silence of the grave would quench forever the transient existence of the hideous corpse which he had looked upon as the cradle of life. He sleeps; but he is awakened; he opens his eyes; behold! the horrid Thing stands at his bedside, opening his curtains and looking on him with yellow, watery, but speculative eyes.[56] ♣

♣ Entire books have been dedicated to following through the minatory influence of Frankenstein's Creature over the last 190 years, especially through films and popular journalism. We may expect a minor earthquake on the bicentenary of publication in 2018. Suffice it to note here that the current discussion of GM crops – undoubtedly vital to sustain global harvests and reduce dependency on crop-spraying – often refers to them as 'Frankenstein foods' (for example, the leading article from *Country Life*, April 2008); and that the *Guardian*'s excellent column 'Bad Science' has an image of Frankenstein's Monster as its logo.

Three years later a very different woman writer entered the field and took up the defence of science. Mary Somerville's *On the Connexion of the Physical Sciences* appeared in 1834, and was published in Murray's Family Library. Its frontispiece showed drawings of Herschel's nebulae. Though more didactic and pious in tone than the other books, it was a significant attempt to bring together new developments in the fields of astronomy, physics, chemistry, botany and geology as a single, ongoing scientific project of discovery. 'The progress of modern science,' Somerville wrote, 'especially within the last five years, has been remarkable for a tendency to simplify the laws of nature, and to unite detached branches by general principles.'[57]

This search for unifying laws, as emphasised by John Herschel, is a central theme of Somerville's study. So, for example, 'Light, heat, sound and the waves of fluids, are all subject to the same laws of reflection, and indeed their undulatory theories are perfectly similar.'[58] This allows her to discuss the action of sunshine, rain, frost, steam, clouds, steam engines, musical instruments and even 'squeezing water out of a sponge' in the same chapter, headed simply 'Heat'.[59]

Newton remains the presiding genius of the book, though there is extensive discussion of the works of the Herschels, Faraday and Davy. A few European scientists are also included in the 'whole circle of the sciences',[60] notably Alexander von Humboldt and Laplace (as has been noted, Somerville had herself popularised his difficult *Mécanique Céleste* as *The Mechanism of the Heavens* in 1831). But great Continental names like Lavoisier, Lamarck, Berzelius, Linnaeus, Buffon and Cuvier do not appear at all, not even in her Index, a truly astonishing omission. There is a sense of a more exclusively British science emerging.

In general Somerville is conventional in her piety, with many reverent references to the 'conspicuous goodness of the First Cause' and the 'magnificence' of divine creation which science reveals. Yet she has a number of passages which might alert the reader to a more sceptical and enquiring view. Her reflections on stellar astronomy are one such, clearly echoing William Herschel. She quietly suggests that 'not only man, but the globe he inhabits – nay the whole system of which it forms so small a part – might be annihilated, and its extinction be unperceived in the immensity of creation'.[61] The reader is left to ask – Unperceived by God? Or without any God to perceive it?

Again, the question of the traditional Biblical age of the earth is gently passed over with the observation that geologists (notably Lyell) were now regularly producing 'traces of extreme antiquity', which contradicted the idea of any special creation, and simply made the formation of the earth 'contemporaneous with that of the rest of the planets'. Presumably this was because the Creator made no difference between 'one day and a thousand years'.[62]

In one remarkable passage, entitled 'Errors of the Senses', Somerville confronts the counter-intuitive nature of science. She even seems to suggest that science underwrites philosophical scepticism, by suggesting that none of man's physical perceptions is ultimately capable of yielding any objective account of the surrounding universe at all: 'A consciousness of the fallacy of our senses is one of the most important consequences of the study of nature. This study teaches us that no object is seen by us in its true place, owing to aberration; that the colours of substances are solely the effects of the action of matter upon light; and that light itself, as well as heat and sound, are not real beings, but modes of action communicated to our perceptions by the nerves. The human frame may therefore be regarded as an elastic system, the different parts of which are capable of... vibrating in unison with any number of superposed undulations, all of which have their perfect and independent effect. Here our knowledge ends; the mysterious influence of matter on mind will in all probability be for ever hid from man.'[63]

Again, the coming crisis in Victorian religious beliefs, a new kind of wonder born out of radical doubt, seems obscurely glimpsed in such passages. Nonetheless, the book was respectfully reviewed by the highly orthodox William Whewell, and went into numerous editions. It was notable because it was written by a woman, but not particularly *addressed* to women readers – let alone children. This pointed up the paradox that women were not yet accepted as equals by the male scientific community, although in the crucial field of interpretation and explanation to a general public, they were already the pioneers.

The first official woman member of the BAAS was not accepted until 1853, though this was not entirely through want of trying. Charles Babbage wrote archly, before the Oxford meeting of 1832: 'I think that *ladies* ought to be admitted at some kind of assembly: remember the dark eyes and fair faces you saw at York and pray remember we absent philosophers sigh over the eloquent descriptions we have heard of their enchanting smiles... If you will only get up an evening *converzazione* for them at Oxford, I will try and start a ball for them at Cambridge.'[64]♣

In October 1834 it was a sign of the times that Coleridge's obituary appeared in the same edition of the *Gentleman's Magazine* as its first full report of the highly successful fourth BAAS meeting in Edinburgh. As many as 1,200 members attended, including 400 women, though these were still only permitted at suitably selected sessions. The geologist Professor Adam Sedgwick gave the plenary address on the future role of science, which was fully quoted in the *Gentleman's Magazine*'s summary. The open seminars, embracing the main scientific disciplines (astronomy, geology, chemistry, physics, botany and statistics) lasted for a week. It was not dull, but it was becoming professional Victorian science. There were concerts, balls, steam-train rides and fireworks. David Brewster talked about his latest scientific toy, the kaleidoscope. Professor Buckland, the geologist, gave another admirable lecture on Fossil Reptiles, and called attention to God's sense of humour in his grotesque creations: 'He convulsed his audience with laughter... with his numerous comical hits.'[65]

7

Erasmus Darwin's grandson, Charles Darwin, had gone up as an undergraduate to Christ's College, Cambridge, in autumn 1827. Initially he seemed bumbling and directionless, struggling to escape from the oppressive shadow of his grandfather. But he was soon inspired by his tutor, the kindly Professor of Botany, John Henslow, and began a microscopic study of pollen grains. He steadily came under the influence of the young science group based at Trinity and St John's, was befriended by the Lancashire polymath William Whewell, and taken on a vigorous geological expedition to North Wales by the muscular Christian Adam Sedgwick (a disciple of Wordsworth's).

'No opinion can be heretical but that which is not true,' declared Sedgwick stoutly at the Geological Society. 'Conflicting falsehood we can

♣ The romantic tale of Paulina Jermyn, the beautiful seventeen-year-old botanist who fell in love at the 1832 British Association meeting at Oxford, perhaps deserves wider currency. See David Wooster, *Paula Trevelyan* (1879).

comprehend; but truths can never war against each other. I affirm, therefore that we have nothing to fear from the results of our enquiries, provided they be followed in the laborious but secure road of honest induction.'[66] Darwin would never forget that declaration as, for thirty years, he struggled with the implication of evolution by natural selection.

With Henslow he read and discussed the papers of Charles Babbage and John Herschel, becoming aware of the subtle implications of the inductive philosophy, and also of the rumbling dissatisfactions with the Royal Society. Inspired by Herschel's *Natural Philosophy*, he heavily underlined a passage beginning: 'To what, then, may we not look forward ... what may we not expect from the exertions of powerful minds ... building on the acquired knowledge of past generations?'[67]

But above all Darwin had begun to dream of a great tropical sea expedition. He studied the voyages of Bougainville, Cook and Banks, along with the *Personal Narrative* of Alexander von Humboldt. By April 1831, the end of his third and final year at Cambridge, he was dreaming of escape, as he confided to his sister Caroline. 'All the while I am writing now my head is running about the Tropics; in the morning I go and gaze at Palm trees in the hothouse and come home and read von Humboldt: my enthusiasm is so great that I can hardly sit still on my chair ... I never will be easy till I see the peak of Tenerife and the great Dragon tree; sandy dazzling plains, and gloomy silent forest are alternately uppermost in my mind.'[68] At the age of twenty-two, and in the shining wake of Joseph Banks, Charles Darwin had departed aboard HMS *Beagle* in December 1831.

8

John Herschel's marriage of 1829, according to his aunt Caroline's prescription, had given him both emotional stability and independence, but did not cramp his scientific ambitions. While Margaret produced a large family, Herschel continued to plan the astronomical expedition to the southern hemisphere, now including his wife and children as an essential part of the scheme. In 1832 he turned down repeated offers of government sponsorship, determined to avoid any imperial implications of the kind that had been so fatal to Mungo Park's second expedition.

He also briskly rejected a proposal from the Royal Society to underwrite part of his expenses. He wished to make himself 'responsible to no one for the results of my expedition', and to retain 'the unconditional power of prosecuting it or abandoning it at any moment that it may suit my *caprice*'. He would not even consider sailing in a Royal Navy ship, except in the unlikely event of a declaration of war with another maritime power. 'But on the other hand, in that event, the King's ships would have other fish to fry than landing stargazers at the world's end.'[69]

Like Banks before him, John Herschel had the freedom of action that belonged to a wealthy man. He had inherited £25,000 under his father's Will ten years before, and further lands and property were now left him by his mother, Lady Herschel, on her death in 1832.[70] So he confidently committed all his own resources to the project. After considering the peripatetic possibilities of South America, with thoughts of Banks and Humboldt in mind, he finally decided to set up a full-scale observatory and scientific station in South Africa.

On 13 November 1833 John and his family left Portsmouth for passage to Cape Town. The dismantled twenty-foot telescope was put aboard in a series of padded packing cases, and his declared intention was a major astronomical expedition to observe and map all the stars of the southern hemisphere, just as his father Sir William had done for the northern. Perhaps it was no coincidence that this was the very scheme that Sir Joseph Banks had been dreaming about in the last months before his death.

The Herschels remained at Cape Town for four years, mapping and cataloguing the stars and nebulae, and botanising in the hills above Cape Town. Their packed notebooks show a ceaseless family activity: daily meteorological observations, zoological and botanical notes, and hundreds of beautiful plant drawings made, with infinite care, using a *camera lucida*.[71] Throughout this time their correspondence with Caroline was never discontinued, and John confided to her that these were the happiest years of his whole life. The young and vivacious Lady Herschel also wrote frequently to her 'aunt'. She acted as hostess to numerous scientific visitors, and often proudly recalled her father-in-law Sir William Herschel, and 'his tough little German sister'.[72]

One of their most notable visitors was young Charles Darwin, on his way back from the Galapagos islands in June 1836. He wrote to his sister as the *Beagle* docked at the Cape of Good Hope: 'I have heard so much about [Herschel's] eccentric but very amiable manners, that I have a high curiosity to see the Great Man.'[73] He was not disappointed. Always on the lookout for fine specimens, Darwin tracked Sir John to his 'most retired charming situation' six miles up country from the main settlement, in a remote clearing surrounded by fir and oak trees, with the twenty-foot installed like some heathen totem pole at the centre.

Herschel himself was never still, an intense, animated figure obsessively bustling about with innumerable projects and observations – in fact just like his father. He appeared 'to find time for everything', even collecting rare Cape bulbs and carpentering bits of furniture. Darwin, who always valued a tranquil and ruminative lifestyle, initially found Herschel's ceaseless activity intimidating and 'rather awful'. But gradually he saw that the Great Man was 'exceedingly good natured', that his wife Lady Herschel was kindness itself, and that the whole Cape project was truly astonishing. He counted this meeting with Sir J. at this early moment in his career 'a memorable piece of good fortune'.[74]

Herschel's expedition to the Cape came to represent for Darwin the important ideal of the independent working scientist, which inspired the rest of his life. On his return to London, his friend Charles Lyell wrote to Darwin: 'Don't accept any official scientific place, if you can avoid it, and tell no one that I gave you this advice ... My question is, whether the time annihilated by learned bodies is balanced by the good they do? Fancy exchanging Herschel at the Cape for Herschel as President of the Royal Society – which he so narrowly escaped being! ... Work exclusively for yourself and for science ... Do not prematurely incur the honour or penalty of official dignities.'[75]

9

Several times Caroline Herschel – by then in her eighties – imagined sailing out to join John's family with her seven-foot telescope, hoping she might 'shake off some 30 years from my shoulders that I might accompany you on your voyage'. It would be like reviving the old days with her brother at Bath. Her sense of frustration expressed itself in a deliberate, comic return to the broken English of her first years in England. 'Ja! If I was 30 or 40 years *junger* and could go too? *In Gottes namen!*'[76]

Caroline did discover, however, a startling new skill in the art of public relations. She learned to feed the local Hanover newspapers with scientific tales from the Cape, in such a way that they were soon being picked up by the international press. Thus Herschel's work had a following right across Europe. Perhaps she had learned the importance of good publicity from her old friend Sir Joseph Banks in Soho Square. One of her earliest *coups* appeared in *The Times* for 27 June 1834.

> The *Hamburg Correspondent*... has the following from Hanover. The friends of astronomy will be pleased to learn that Sir John Herschel has written from the Cape of Good Hope to his aunt, Miss Caroline Herschel, resident here. He has already fixed his Astronomical instruments, especially his 20 foot telescope, and ere now has begun his observations... He resides in the country, about five miles from Cape-Town, near the Table Mountain, in an enchanting valley; lofty trees, rare and beautiful shrubs and flowering plants surround his dwelling; his eye gazes upon clear and cloudless skies, studded with those innumerable stars that are the objects of his elevated pursuits. He is sanguine in his hopes of making important discoveries.[77]

Sometimes these news stories moved slightly beyond Caroline's control. The following year, on 25 August 1835, the *New York Sun* ran a huge splash scoop that Sir John Herschel had finally proved one of his father's most daring astronomical speculations to be true. *Herschel had discovered life on the moon!* The highly dramatic story held the front page of the newspaper for four days, doubled its circulation, and set off a frenzy of excitement from the east coast to the west. Each day the *New York Sun* gave more and more details of Herschel's observations: mighty forests growing in the lunar craters, strange plants, fishes, beaver-like animals (all enormous because of the low lunar gravity), and finally, small ape-like creatures with highly intelligent faces and convenient bat-like wings, flitting through the tenuous lunar atmosphere.[78]

Before the Great Moon Discovery story was blown, a mid-West preacher was collecting subscriptions to send a crate of Bibles to the poor benighted lunar men, and Edgar Allan Poe in Baltimore was considering the possibilities of a whole new genre of fiction: the science fiction hoax (he would launch it with a vivid – but entirely fictitious – account of the first balloon crossing of the Atlantic the following year).[79] Herschel privately dismissed the whole affair as 'incoherent ravings', and calmly refuted it in an Olympian open letter to the Parisian astronomer François Arago, published in the *Athenaeum*.[80]

But Margaret Herschel was more amused. She called the story 'a very clever piece of imagination', and wrote appreciatively to Caroline. 'The whole description is so well *clenched* with minute details of workmanship . . . that the New Yorkists were not to be blamed for actually believing it as they did for 48 hours. – It is only a pity that it *is not true*: but if grandsons stride on as grandfathers have done, *as wonderful things may yet be accomplished*.'[81]

John Herschel's time in South Africa, as significant in its own way as Charles Darwin's *Beagle* voyage, confirmed him as the greatest astronomer and general scientist of his generation. On his return to England in May 1838 he was made a baronet in time to attend Queen Victoria's coronation in Westminster Abbey. Sir John Herschel was elected President of the Royal Society, awarded a second Copley Medal, and by the 1850s was recognised as the leading public scientist of mid-Victorian England. His kindly face, encircled by a sunlike corona of white hair, was famously photographed by Julia Margaret Cameron, using a process that he himself had partly invented.♣

10

The great forty-foot was eventually dismantled at Slough on New Year's Eve, 1840. It had become the relic of a past age, and besides, it shook dangerously and moaned as the winter wind blew through its ancient timbers and rigging, like a ship heading out into a stormy sea.

Sir John Herschel did not forget all the hopes it had symbolised, the great names it had attracted, and the celebrations it had inspired. Having had the scaffolding safely removed, he laid the huge, battered old tube out on the frosty grass, and held a last party inside it, with drinks and toasts and candlelight.[82]

♣ This benign and eccentric image defined the Victorian ideal of the scientist, just as the later faintly surreal images of Albert Einstein – riding a bicycle or putting his tongue out – defined the twentieth-century one. The current images of Stephen Hawking, brilliant but paralysed and gargoyle-like in his wheelchair, perhaps better express the uncertainty of contemporary attitudes to science. The wheelchair itself takes us back to Dr Strangelove, but also eventually returns us to Sir Joseph Banks, rolling briskly into one of his scientific breakfasts in Soho Square, keen to meet his next young protégé and launch a new project 'for the Benefit of all mankind'.

He marked its departure not with an elegant mathematical calculation, but with a boisterous chant, 'Elegy for the Old Forty-Foot':

> In the old Telescope's Tube we sit
> And the shades of the Past around us flit!
> His Requiem sing we with shout and din
> While the Old Year goes out and the New comes in.
> Merrily, merrily, let us all sing
> And make the Old Telescope rattle and ring!

Epilogue

I was fifty-four when I gave my first lecture at the Royal Institution, Albemarle Street. It was a formal Friday Night Discourse, with an invited audience in evening dress, and I was asked to put on an unaccustomed dinner jacket and bow tie. My announced subject was 'The Coleridge Experiment'. The aim was to explore that particularly controversial meeting between science and poetry when Humphry Davy, shortly after starting the Bakerian Lectures in 1808, had gallantly risked his reputation by bringing Coleridge – then in the depth of opium addiction and a fierce marital crisis – to give an extended series of fourteen lectures on the Imagination, before a distinguished invited scientific audience at the Royal Institution. My own lecture was intended to describe the utter chaos that had ensued, but also the few wonderful visionary moments that had been sparked by Coleridge, and which had subsequently shaped much of the modern concept of creativity, and the notion of the imaginative leap.*

Just before starting, I stood behind the closed double doors to the historic lecture theatre, trembling slightly as I heard the solemn growl of the audience on the other side. I was very conscious that I was about to step out onto the very dais where Davy, Faraday and Coleridge himself had once lectured. The Director, standing quietly by my elbow, whispered encouragingly to me. He also wondered, in passing, if I had been told about *the atomic clock*? No, I had not been told about the atomic clock.

The Director explained that there was an atomic clock which buzzed loudly in the lecture theatre after exactly fifty minutes. Lecturers were expected to end their talks on this signal. With the first stirrings of real panic, I murmured that this could presumably be treated as a sort of early-warning system for prolix speakers. Well, yes, indeed it could; but it was rather more a question of *desirable scientific precision.* Indeed, the tradition was that the speaker should fit his lecture to *exactly fifty minutes*, no longer and no shorter, and should immediately wind up his talk when the buzzer sounded.

The Director now looked rather quizzically at my loose bundle of lecture notes. I wondered if the memory of Coleridge's notorious prolixity had never

* My account may be found in *The Proceedings of the Royal Institution of Great Britain*, Vol. 69, 1998.

been quite erased from the Institute's collective consciousness. He added reassuringly that in his experience most of his distinguished scientific lecturers had contrived to be saying their *very last sentence* at exactly the moment when the atomic clock went off. It was all rather elegant: *Talk – Buzz – Stop – Applause.* And of course there would be applause. With that, the Director stepped briskly forward and threw open the large double doors, to reveal the steep tiers of bench seats, crowded with expectant faces, and the growing silence of an atomic clock, noiselessly ticking away...

Indeed, there is a particular problem with finding endings in science. Where do these science stories really finish? Science is truly a relay race, with each discovery handed on to the next generation. Even as one door is closing, another door is already being thrown open. So it is with this book. The great period of Victorian science is about to begin. The new stories are passed into the hands of Michael Faraday, John Herschel, Charles Darwin...and the world of modern science begins to rush towards us.

But science is now also continually reshaping its history retrospectively. It is starting to look back and rediscover its beginnings, its earlier traditions and triumphs; but also its debates, its uncertainties and its errors. No general science history would now be considered complete without a sense of the science achieved centuries ago by the Greeks, the Arabs, the Chinese, the Babylonians. It is no coincidence that the last few years have seen the foundation, in numerous universities across Europe, Australia and America, of newly conceived 'Departments of the History and Philosophy of Science'. The earliest pioneering ones began at Cambridge (UK), and Berkeley (California), with others quickly following at Paris X (Nanterre), Melbourne, Sydney, Toronto, Indiana, Caltech and Budapest (1994). Similarly, it seems to me impossible to understand fully the contemporary debates about the environment, or climate change, or genetic engineering, or alternative medicine, or extraterrestrial life, or the nature of consciousness, or even the existence of God, without knowing how these arose from the hopes and anxieties of the Romantic generation.

But perhaps most important, right now, is a changing appreciation of how scientists themselves fit into society as a whole, and the nature of the particular creativity they bring to it. We need to consider how they are increasingly vital to any culture of progressive knowledge, to the education of young people (and the not so young), and to our understanding of the planet and its future. For this, I believe science needs to be presented and explored in a new way. We need not only a new history of science, but a more enlarged and imaginative biographical writing about individual scientists. (I make some suggestions in the Bibliography that follows, under the heading 'The Bigger Picture'.) Here the perennially cited difficulties with the 'two cultures', and specifically with mathematics, can no longer be accepted as a valid limitation.♣ We need to understand how science is actually made; how scientists themselves think and feel and speculate. We need to explore what makes scientists creative, as well as poets or painters, or musicians. That is how this book began.

The old, rigid debates and boundaries – science versus religion, science versus the arts, science versus traditional ethics – are no longer enough. We should be impatient with them. We need a wider, more generous, more imaginative perspective. Above all, perhaps, we need the three things that a scientific culture can sustain: the sense of individual wonder, the power of hope, and the vivid but *questing* belief in a future for the globe. And that is how this book might possibly end.

♣ I am encouraged to see that my old teacher and early mentor Professor George Steiner, starting from an entirely different premise, has recently come to a similar conclusion: 'Hence my conviction that even advanced mathematical concepts can be made imaginatively compelling and demonstrable when they are presented *historically*...It is via these great voyages and adventures of the human mind, so often charged with personal rivalries, passions and frustrations – the Argosy founders, or gets trapped in the ice of the insoluble – that we non-mathematicians can look into a sovereign and decisive realm...Locate this quest...and you will have flung open doors on "seas of thought" deeper, more richly stocked than any on the globe.' See 'School Terms', in *My Unwritten Books* (2008). The imagery behind this splendid passage comes, of course, from Romanticism: Wordsworth on Newton from *The Prelude*, and Caspar David Friedrich's painting of 1825, *The Sea of Ice*, in which the explorer's tiny, gallant ship is foundering amidst enormous polar ice-floes – but, hope against hope, may yet survive.

Cast List

(Shorter entries imply that more material can be found in the chapter indicated)

JOHN ABERNETHY, 1764–1831. Physician and surgeon at Bart's Hospital, London, he became President of the Royal College of Surgeons. Coleridge was one of his many patients. (See Chapter 7)

MARK AKENSIDE, 1721–70. Poet, whose major work *The Pleasures of the Imagination* (1744) set out the traditional eighteenth-century view of the cosmos, including the idea that the universe was approximately 6,000 years old, and that the stars were spread overhead in a 'concave' dome or heavenly temple (see for example Book I, lines 196–206).

ALEXANDER AUBERT, 1730–1805. FRS. Wealthy and independent-minded British astronomer living at Deptford, London, who set up a fine private observatory at Highbury House, Highbury Fields. Friend and supporter of William and Caroline Herschel, especially in the 1780s, when their early findings were criticised by members of the Royal Society. In 1788 he presented them with a beautiful Shelton long case astronomical clock, with brass compensated pendulum (private archive, John Herschel-Shorland, Norfolk).

CHARLES BABBAGE, 1791–1871. FRS 1816. Brilliant young mathematician, Lucasian Professor of Mathematics at Cambridge, close friend of Herschel's son and Caroline's nephew John Herschel. Irascible and outspoken critic of the Royal Society under the ageing Banks and the ailing Davy, supporter of the fledgling BAAS, and inventor of various Difference Engines (mechanical computers). (See Chapter 10)

SIR JOSEPH BANKS, 1743–1820. President of the Royal Society. (See Chapter 1 and *passim*)

ANNA BARBAULD, 1743–1825. Poet, educationalist and bluestocking, she was greatly interested by scientific ideas. She was a close friend of Joseph Priestley, witnessed many of his early experiments, and wrote a poem in the voice of one of his laboratory mice. Her epic poem 'Eighteen Hundred and Eleven' (1812) predicted a crisis of Empire and intellectual life in Britain enveloped in 'Gothic night', and the rise of American power. A formidable editor, she produced a fifty-volume edition of contemporary British novelists. (See Chapter 6)

FRANCIS BEAUFORT, 1774–1857. Sailor, hydrographer and inventor of the Beaufort wind scale, one to twelve (hurricane). He wrote some interesting accounts of the 'after-death' experiences of drowning sailors.

ANNA BEDDOES, 1773–1824. Volatile younger half-sister of the novelist Maria Edgeworth, wife of the physician Thomas Beddoes, and possibly Humphry Davy's lover at the Pneumatic Institute, Bristol, 1799–1801. Shortly afterwards she had an affair with Beddoes's friend Davies Giddy in London, though she returned to nurse her husband when he was dying of heart failure. Anna had four children: Anna (1801), Thomas (1803), Henry (1805) and Mary (1808). Neither of these first two may have been legitimate. Davies Giddy acted as their legal guardian after Beddoes's premature death. Anna's son Thomas Lovell Beddoes (1803–49) became a poet and political activist, the author of several macabre poetic dramas including *The Last Man* (1823) and *Death's Jest Book* (1850), lived exiled in Germany, and committed suicide in Switzerland. Anna herself went to live abroad, moving restlessly to Belgium, then France, then Italy, and finally dying in Florence, aged fifty. (See Chapter 6)

THOMAS BEDDOES, 1760–1808. Physician, chemist, philanthropist and political radical. Davy's mentor at Bristol, and close friend of the leading members of the Lunar Society in the Midlands. His experimental use of drugs and gases, and the antics of his wife Anna, undermined his public reputation. With the collapse of the Bristol Pneumatic Institute as an experimental centre, he transformed it into the philanthropic Preventative Medical Institute for the Sick and Drooping Poor. He had an early concept of a free national health service, providing particular help for women with children. A heroic but marginalised figure, he was never supported by Banks at the Royal Society. (See Chapter 6)

CLAUDE BERTHOLLET, 1748–1822. FRS 1789. Leading French chemist, friend of Lavoisier's, head of the scientific expedition – including balloon section – that accompanied Napoleon to Egypt in 1789. Later an admirer of Davy's, and friend of Banks's confidant Blagden. His glamorous pupil and protégé was Joseph Gay-Lussac.

JACOB BERZELIUS, 1779–1848. Outstanding Swedish chemist, Professor of Chemistry and Medicine at Stockholm 1807. His pioneering work in electrochemistry included the first accurate table of atomic weights, establishing twenty-eight elements (1828), and giving them their internationally accepted 'initial letter' symbols, as H_20. Warmly congratulated Davy on the Bakerian Lectures and the safety lamp, but from 1815 was increasingly challenging his dominance. Married late, aged fifty-six, when his best scientific work was done, to a woman thirty-two years his junior.

XAVIER BICHAT, 1771–1802. French physician and anatomist, worked at the celebrated Hôtel Dieu hospital in Paris, and led a brief life of great intensity and self-sacrifice, inspired by French Revolutionary ideology. Developed analysis of human tissue types, histology, and materialist theory of life. His influential textbook *On Life and Death* was posthumously translated into English in 1816, and fed the Vitalism controversy in Britain.

CHARLES BLAGDEN, 1748–1820. FRS 1772. Physician, bureaucrat, francophile and outstanding scientific gossip. He trained as a naval surgeon, and for some years worked as scientific assistant to Henry Cavendish. Under Banks he became the influential Secretary to the Royal Society from 1784 to 1797. Despite occasional rows, he remained the great supporter and personal confidant of Banks until his death in Paris, a few weeks before Banks's own. To some degree his friendship replaced Solander's for Banks.

JEAN-PIERRE BLANCHARD, 1753–1809. French inventor and aeronaut who first crossed the Channel in a balloon, and founded a ballooning school in Vauxhall, London. (See Chapter 3)

JOHANN FRIEDRICH BLUMENBACH, 1752–1840. FRS 1793. Influential German anatomist based at the University of Göttingen, who founded the science of anthropology and the pseudo-science of craniology, and developed an early classification of racial types. His famous collection of skulls was known as 'Dr B's Golgotha'. Friend of Banks, and famous lecturer at Göttingen heard by students from all over Europe, including Coleridge, William Lawrence and Thomas Lovell Beddoes.

JOHANN ELERT BODE, 1747–1826. German astronomer and Director of the Berlin Observatory. Designed the most authoritative *Celestial Atlas* (1804), which finally superseded John Flamsteed's of 1729. (See Chapter 2)

LOUIS-ANTOINE DE BOUGAINVILLE, 1729–1811. French naval commander and explorer, circumnavigated the globe and landed on Tahiti a year before James Cook. (See Chapter 1)

DAVID BREWSTER, 1781–1868. Scottish physicist and campaigning science journalist. An early promoter of the BAAS with John Herschel. His researches included work on polarised light and lighthouse lenses, and he invented the kaleidoscope. He wrote an influential first biography of Sir Isaac Newton (1831), eventually expanded through several editions (1860). (See Chapter 10)

COMTE DE BUFFON (GEORGES-LOUIS LECLERC), 1707–88. French geologist and naturalist who developed early theories of the earth's rapid, catastrophic changes through flood (supported by Neptunists) and volcanic action (Plutonists). He was director of the Jardin du Roi, the modern Jardin des Plantes, Paris, and wrote a forty-four-volume *Natural History* (1804). His studies of mountains and glaciers were referred to by Shelley in his poem 'Mont Blanc' (1816).

FANNY BURNEY, MADAME D'ARBLAY, 1752–1840. Novelist, journal writer, friend of the Herschels through her father Charles Burney (FRS 1802), the musicologist. Fascinated but sceptical about scientific advances, she praised Caroline Herschel's comet-finding work, and wrote to Banks wondering why women could not be Fellows of the Royal Society. Survived radical breast surgery without anaesthetic (Paris, September 1811), and wrote a long, courageous account of the experience. (See Chapter 7)

GEORGE GORDON, LORD BYRON, 1788–1824. Poet with a lively but sceptical interest in science and voyages. Looked through Herschel's telescope, and met Davy both in London and in Italy. His poem 'Darkness' (1816) reflects current cosmological speculation, and several passages of *Don Juan* (1818–21) comment on scientific research and the vanity of 'progress'. (See Chapter 9)

SAMUEL TAYLOR COLERIDGE, 1772–1834. Poet, critic, essayist and philosopher. Closely involved with Davy's early scientific work in Bristol and London, 1799–1804. Later wrote about the history and philosophy of Romantic science in his newspaper *The Friend* (1809–19) and his *Philosophic Lectures* (1819), and became involved in the Vitalism debate, writing his *Theory of Life* (1816–19) to discuss the issues. Attended and spoke at the historic third meeting of the BAAS at Cambridge in 1833, at which the term 'natural philosopher' was first replaced by the word 'scientist'. (See Chapters 6 and 10)

WILLIAM COWPER, 1731–1800. Poet who suffered from disabling depression all his life, but whose lively letters and long, rambling poem *The Task* give a vivid response to the scientific advances of the day, especially Banks's voyage and the balloonists. (See Chapter 1)

BARON GEORGES CUVIER, 1769–1832. FRS 1806. The leading French zoologist and comparative anatomist of his day, he taught at the Museum of Natural History, Paris. He disagreed with Lamarck, rejecting the concept of evolution, and proposing a theory of biological development through global catastrophe. He published twenty-two volumes on ichthyology (fishes).

JOHN DALTON, 1766–1844. Chemist, meteorologist and early theorist of atomic weights, producing a pioneering Table of 20 Elements in 1808, and laws on the thermal expansion of gases. A shy, retiring personality, born in Cumberland and working in Manchester, he was reluctant to join the Royal Society in London. Herschel and Babbage thought he was shamefully neglected, but 40,000 people attended his funeral in Manchester.

ERASMUS DARWIN, 1731–1802. FRS 1761. Physician, poet, polymath and inventor. Moving spirit of the Lunar Society at Birmingham, which met each month on the night of the full moon (in theory so they could walk home safely). A close friend of James Watt and Matthew Boulton, he described much. of the new science of the day in his long and remarkable poem *The Botanic Garden* (1791). Its extensive and highly informative prose notes on cosmology, geology, meteorology, chemistry and physics – a didactic method later used by Southey and Shelley – provide an encyclopaedic account of the state of science at the turn of the eighteenth century.

HUMPHRY DAVY, 1778–1829. President of the Royal Society 1820–27. (See Chapters 6, 8 and 9)

MICHAEL FARADAY, 1791–1867. Chemist and physicist of genius, inventor of the electric motor, dynamo and transformer. Director of the Royal Institution, London, for over thirty years. He was Davy's great protégé, and unlike his patron one of the most popular figures in British science. (See Chapters 8, 9 and 10)

BARTHÉLEMY FAUJAS DE SAINT-FOND, 1741–1819. French geologist and traveller, a specialist in volcanoes. He was a great anglophile, and in his *Travels in England and Scotland in Examining the Arts and Sciences* (1799) he gave a vivid account of interviewing Herschel and Caroline at work. He also wrote with enthusiasm about ballooning.

JOHANN GEORG FORSTER, 1754–94. German botanist and travel writer. With his father he joined Cook's second Pacific expedition (the one that brought back Omai), and he subsequently published a vivid and somewhat scurrilous account of it in English, *A Voyage Round the World* (1777). He was appointed Professor of Natural History at Kassel, and corresponded frequently with Banks. His father Johann Rheinhold Forster, who had published a more sober *Observations during a Voyage Round the World* (1778), outlived him.

BENJAMIN FRANKLIN, 1706–90. FRS 1756. Physicist and statesman, he was American Ambassador to France 1776–85, and proved an invaluable source of information about French science for Banks, notably on mesmerism and ballooning 1783–84. He specialised in work on the properties of electricity: the static charge, the lightning surge and the lightning conductor. (See Chapters 3 and 7)

LUIGI GALVANI, 1737–98. Italian physician, Professor of Anatomy at Bologna University. His dramatic claim to have discovered reanimation or 'animal electricity', when dead specimen frogs were fixed with metal pins, was disproved in a celebrated paper sent to the Royal Society by Volta in 1792. Nevertheless the term 'galvanism' remained loosely applied to a wide range of electrical phenomena, including the 'galvanometer' used to detect electrical currents from 1820.

JOSEPH GAY-LUSSAC, 1778–1850. Outstanding French analytic chemist, the pupil of Berthollet, and Davy's great rival in Paris when working on pneumatics, the expansion of gases and the properties of boron and iodine. A glamorous figure, famed for his high-altitude balloon ascent in 1804 (to 7,000 metres) and his marriage to a beautiful seventeen-year-old shopgirl whom he saw reading a chemistry book between serving customers. (See Chapter 8)

DAVIES GILBERT (*NÉ* GIDDY), 1767–1839. FRS 1791. Wealthy mathematician, MP and science administrator. He was a pupil of the radical doctor Thomas Beddoes at Oxford, and later befriended young Humphry Davy in Penzance. He had an extended affair with Beddoes's volatile wife Anna in London, and later acted as her children's guardian. He married respectably and changed his name to Gilbert in 1817, and became President of the Royal Society in 1827–31, following the crisis caused by Davy's resignation.

JOHANN WOLFGANG VON GOETHE, 1749–1832. German heavyweight boxer, went ten rounds with the ghost of Sir Isaac Newton, referees still out. (See Chapters 7–10 *passim*)

LUKE HOWARD, 1772–1864. The first British meteorologist and student of clouds and weather phenomena. Gave his first important paper at the Quaker Askeian Society, London, in 1802. His pioneering work *On the Modification of Clouds* was published in 1804, and was read throughout Europe. Goethe asked him to compile an autobiography, translated several of his papers, and wrote four long poems on the subject of clouds inspired by Howard's classification system. (See Chapter 3)

ALEXANDER VON HUMBOLDT, 1769–1859. Major figure in European science, famous for his expedition to South America with the mathematician Aimé Bonpland. Traveller, botanist, zoologist, geologist, meteorologist, cosmologist – the universal man of Romantic science. His *Personal Narrative* was published in 1806. His great unfolding work was *Cosmos* (1845). He knew both William and John Herschel, but never had the chance to meet Caroline. His brother, the scholar and philologist Wilhelm von Humboldt, helped found Berlin University. (See Chapters 9 and 10)

JAMES HUTTON, 1726–97. Scottish physician, trained in Holland, who effectively founded the modern discipline of geology. In studying the stratification of rock, and particularly its erosion by rivers, he came to reject the Biblical creation myth and the catastrophe theories of Buffon and Cuvier, and to argue for an infinitely slow evolution of the earth, 'with no vestige of a beginning, no prospect of an end'. His highly technical and ill-written work was made accessible by his disciple Professor John Playfair of Edinburgh University, and prepared for the evolutionary geology of Charles Lyell. His notion of 'deep time' supported Herschel's notion of 'deep space', and ultimately Charles Darwin's theory of evolution.

CHRISTIAAN HUYGENS, 1629–95. Great Dutch physicist and astronomer, cosmologist and manufacturer of astronomical instruments, including the 'compound pendulum' clock. His wave theory of light (in contrast to Newton's particle theory), his telescope studies of Saturn's rings and its moon Titan, and his notion of a large, densely populated universe, all inspired William Herschel to think afresh about interstellar space. A wildly imaginative mind, Huygens also believed that the inhabitants of Jupiter built spaceships.

EDWARD JENNER, 1749–1823. Naturalist and physician. Remembered for his invention of smallpox 'vaccination' in 1796, by using the counter-intuitive method of infecting a healthy patient with *vaccinia* or cowpox matter, and thus provoking the production of antibodies effective against the far more deadly smallpox. Jenner experimented quietly in Gloucestershire, scratching the skin of his patients with a thorn. Attacked and ridiculed by cartoonists like Gillray, his technique was eventually championed by the Royal Society, taken up across Europe, and by 1853 was compulsory in Britain. Though often described as a 'mere country doctor', Jenner had trained in London under the great surgeon John Hunter, worked in Soho Square as an assistant to Banks and Solander, and wrote many expert papers on birdsong and migration, notably concerning the cuckoo.

IMMANUEL KANT, 1724–1804. First-magnitude German philosopher, who set out a number of brilliant and influential ideas about the possible structure of the cosmos in *A Natural History of the Heavens* (1755), and later on man's subjective perception of physical reality. He speculated on the notion of galactic 'island universes' and extraterrestrial life. Kant's analysis of the human concepts of 'space, time and causation' are particularly relevant to the problematic idea of scientific objectivity, and the way we all observe – but also subjectively imagine – the world around us. Powerfully influenced a whole generation of Romantic thinkers, from Herschel to Goethe and Coleridge.

JOHN KEATS, 1795–1821. Poet and medical student at Guy's Hospital, London. His scientific training shaped much more of his poetry than is usually realised. (See Chapters 7 and 9)

ANTOINE LAVOISIER, 1743–94. The greatest chemist of the French Enlightenment, his *Traité Élémentaire* (1789) inspired young scientists across Europe, but especially Davy. He was executed by Robespierre for tax fraud: 'The Revolution does not need chemists.' His brilliant young wife Anne-Marie Paulze, scientific illustrator and translator, survived to dazzle many European men of science, but chose to marry Count Rumford, which turned out to be a mistaken experiment.

SIR WILLIAM LAWRENCE, 1783–1867. Physician, surgeon, anatomist and early anthropologist. He was a leading surgeon at Bart's Hospital, and Professor of Anatomy at the Royal College of Surgeons. For some time he was Shelley's doctor. His theoretical and personal rivalry with his mentor John Abernethy during the Vitalist controversy made him a national figure. (See Chapter 7)

JAMES LIND, 1736–1812. FRS 1777. Physician, traveller and astronomer. Sailed to China in 1776, and accompanied Banks to Iceland in 1772. Settled at Windsor, where he became one of the King's consultant physicians, and treated Caroline Herschel's leg wound in 1783. He later befriended young Shelley at Eton, and gave him the run of his Oriental collection and radical library. Tall, thin and eccentric, he had, according to Fanny Burney, 'a taste for tricks, conundrums, and queer things'. (See Chapter 4)

CARL LINNAEUS, 1707–78. Great Swedish natural historian, Professor of Botany at Uppsala, where he established a world-famous botanical garden, widely imitated – for example by Banks at Kew. His system of botanical taxonomy (usually binomial: a generic Latinesque name followed by a species adjective) became standard in the eighteenth century and is retained to this day, for example in plant encyclopaedias in most European languages. The Linnaean Society of London was founded in 1788. Many other systems – such as Howard's classification of clouds – imitated Linnaean taxonomy. (See Chapter 1)

VINCENZO LUNARDI, 1759–1806. Glamorous Italian aeronaut, he popularised ballooning in Britain, but was much criticised for risking the life of his cat during his first ascent. (See Chapter 3)

JANE MARCET, 1769–1858. One of the great early science popularisers for young readers. Her *Conversations on Chemistry* (1806, 1811) ran to sixteen editions, and inspired the teenage Michael Faraday. She used a dialogue format to explore ideas through simple question and answer, a method originally derived from Plato. Married to a wealthy Swiss émigré, Alexander Marcet FRS, she got to know many of the leading scientists of the day, including Humphry Davy and Jacob Berzelius. Her *Conversations* became a winning publishing formula, which she successfully expanded to cover several other topics, notably in her *Conversations on Natural Philosophy* (1819).

NEVIL MASKELYNE, 1732–1811. FRS. Mathematician and Astronomer Royal, who produced a valuable Astronomical Almanac for mariners. He supported Herschel at the Royal Society, and later became a loyal and kindly friend of Caroline's, having her stay on her own at the Royal Observatory, Greenwich. He sat on the Board of Longitude, and was subsequently vilified – perhaps unjustly – for his treatment of the chronometer-maker John Harrison. He had a complex relationship with Sir Joseph Banks, whom he did not think knew enough mathematics.

WILLIAM NICHOLSON, 1753–1815. British chemist and early experimenter with electrolysis, who famously repeated Lavoisier's experiment decomposing water into hydrogen and oxygen, thereby demonstrating that it was not a primary 'element'. He was the founder and editor of *Nicholson's Scientific Journal*, an influential monthly publication, comparable to today's *New Scientist*, which published many of Davy's early papers.

MUNGO PARK, 1771–1806. Scottish physician, explorer and travel writer, famous for his two expeditions to West Africa, following the river Niger. Some of his travel kit and medical instruments are still kept in the museum at Selkirk. (See Chapter 5)

THOMAS LOVE PEACOCK, 1785–1866. Satirical novelist and poet. He explored the aberrations of many writers and intellectuals, among them Byron, Shelley, Coleridge and Lord Monboddo. His long essay *The Four Ages of Poetry* (1820) compared imaginative writing with non-fiction and scientific prose, and provoked Shelley's *Defence of Poetry* (1821). Unbelievably, he failed to poke fun at balloons. See *Nightmare Abbey* (1818) and *Crotchet Castle* (1831).

JEAN-FRANÇOIS PILÂTRE DE ROZIER, 1754–85. The first aeronaut in the world to make a successful balloon flight, travelling in a huge hot-air Montgolfier with his companion the Marquis d'Arlandes in a twenty-five-minute flight across Paris on 21 November 1783. After several other ascents, which made him famous across Europe, he was killed trying to fly across the Channel on 15 June 1785. (See Chapter 3)

JOHANN WILHELM RITTER, 1776–1810. German physicist and lecturer of great brilliance and eccentricity. He trained and taught at the University of Jena, until elected to the Bavarian Academy of Sciences in Munich in 1804. Following Herschel, he discovered the existence of ultraviolet rays at the lower end of the visible spectrum. In Munich he fell under the influence of the occultist Franz von Baader, and developed a theory of 'universal geophysical electricity'. He began various experiments in water divining and 'metal witching', and with reanimating dead bodies. Shelley was probably referring to him among the 'various German physiologists' in the 1818 Preface to *Frankenstein*. Ritter died poverty-stricken and possibly insane in Munich. His *Fragments from a Young Physicist* (1810) was published posthumously. (See Chapter 7)

JAMES SADLER, 1753–1828. The first British aeronaut to make a successful scientific flight in a hydrogen balloon, from Oxford on 4 October 1784. His son Windham Sadler successfully flew the Irish Sea from Dublin to Holyhead, but was later killed in a balloon accident. (See Chapter 3)

FRIEDRICH SCHELLING, 1775–1854. Poet and idealist German philosopher, successively Professor of Philosophy at the universities of Jena, Munich and Berlin. He created the Romantic system of beliefs known broadly as *Naturphilosophie* (Nature Science, or Science Mysticism), in which nature is a single organism instinctively evolving or 'waking' towards the goal of higher self-consciousness. His *System of Transcendental Idealism* (1800) interpreted the natural world as a dynamic system of invisible energies, polar forces (like electricity) and mystical correspondences. Everything in nature, from a lump of coal to a human being, aspires upwards to a higher, more spiritualised form, and will ultimately rise to the *Zeitgeist* (World Spirit). His influence can be traced in Davy, Coleridge and Vitalism, and indirectly he is the father of all forms of 'alternative science'. His ideas can also be traced emerging in such later writers as Teilhard de Chardin or even James Lovelock with his Gaia theory.

MARY SHELLEY, 1797–1851. Novelist, short-story writer and essayist. The godmother of British science fiction with *Frankenstein, or The Modern Prometheus* (1818, with an important Introduction about creativity and scientific ideas added to the 1831 edition). Her work really became known through stage adaptations 1820–30, and much later through film. See also her apocalyptic account of a global plague in *The Last Man* (1826). (See Chapters 7 and 10)

PERCY BYSSHE SHELLEY, 1792–1822. Poet and essayist, fascinated by science, especially in his two long poems *Queen Mab* (1812) and *Prometheus Unbound* (1819), and his prose writings on atheism. He was particularly interested in theories of cosmology, geology, meteorology, mesmerism and electricity. Significant scientific ideas appear in 'Notes to Queen Mab' (1812), 'Mont Blanc' (1816), 'Ode to the West Wind' (1819), 'The Cloud' (1820) and 'The Magnetic Lady to her Patient' (1821). His long poem *Alastor, or the Spirit of Solitude* (1815) reflects his interest in exotic exploration, and particularly in Mungo Park's river journeys.

DANIEL SOLANDER, 1733–82. Swedish botanist, trained under Linnaeus at Uppsala, and scientific assistant at the British Museum. FRS June 1764. Fat, lazy and lovable, he accompanied Joseph Banks on the great *Endeavour* voyage, and remained his great friend and confidant until his premature death at Banks's house in Soho Square, aged forty-eight, in May 1782.

MARY SOMERVILLE, 1780–1872. Mathematician and brilliant interpreter and populariser of science for adults, especially with her broad survey of current scientific trends, *On the Connexion of the Physical Sciences* (1834). She translated (and clarified) Laplace's *Mécanique Céleste* as *The Mechanism of the Heavens* (1831), and with Caroline Herschel was elected one of the first two women Fellows of the Royal Astronomical Society, 1835. She also tutored Byron's daughter Ada Lovelace (1815–52) in mathematics. A powerful hostess in Victorian scientific circles, she was awarded the Victoria Medal of the Royal Geographical Society in 1869. The first women's college in Oxford, Somerville – now co-educational – was named after her.

ROBERT SOUTHEY, 1774–1830. Poet, critic and notable biographer. A good friend to young Davy at Bristol, he eagerly discussed the early relations between Romantic science and poetry, but was soon overtaken by the work and influence of Coleridge. He later wrote a fine all-action biography of Nelson (1813) – one of Davy's heroes – and composed the famous children's story 'Goldilocks and the Three Bears'. Curiously enough, 'Goldilocks' has become a familiar term used by cosmologists to describe the median placing of any planet within a solar system which has the potentiality of life – being 'not too hot, not too cold', and 'not too big, not too small'.

SIR BENJAMIN THOMPSON, COUNT RUMFORD, 1753–1814. Physicist, philanthropist and adventurer. Touring through Europe, he became the lifelong friend of Prince Maximilian of Bavaria, and between 1789 and 1795 was Minister for War and for the Army under Carl Theodor, for whom he created the *Englische Garten* as a public works project to help the unemployed of Munich. He established soup kitchens and a public health programme. His technical inventions included the Rumford stove, the Rumford lamp and the Rumford fireplace. He made a disastrous second marriage to Madame Lavoisier (1805), living beyond his means on the Champs Élysées; then lived obscurely in a Paris suburb with the last of his many mistresses, Victoria, who bore him a son, Charles. He has a statue and a monument in Munich, and a tombstone in Auteuil (Paris). In his Will he left his estate to his daughter Sarah, a gold watch to Davy, and his capital savings to the future Harvard University. Many of his technical models and experiments (including his famous friction cannon) are preserved in the Deutsche Museum, Munich. (See Chapter 6)

ALESSANDRO VOLTA, 1745–1827. FRS and Professor of Experimental Physics, Como, Italy, 1775. He disproved Luigi Galvani's theory of animal electricity in 1792, and went on to produce a historic paper on the first chemical pile or battery, which Banks was quick to publish in the Royal Society journal *Philosophical Transactions*, 1800. This was the basis for future pioneering work by Davy (London), Berzelius (Stockholm) and Gay-Lussac (Paris). He gave his name to the volt, a measure of the force of an electrical current. He was visited by Davy in 1814.

ADAM WALKER, 1731–1821. Inspirational science teacher at Eton College, who taught the use of the telescope and microscope, and believed in a plurality of worlds ('30 thousand suns!'). His science primer, *Familiar Philosophy* (1779), was an early best-seller in the popular science field. During a long and eccentric career he invented the patent empyreal air-stove, the Celestine harpsichord and the eidouranion or transparent orrery, a portable device for projecting an illuminated model of the solar system and the main constellations. His *Course of Lectures on Natural and Experimental Philosophy* (1805) was eagerly read by the young Shelley, and covered the basics of Romantic science including astronomy, chemistry, electricity, geology and meteorology.

JAMES WATT, 1736–1819. Engineer and member of the Lunar Society. In partnership with Matthew Boulton he developed new forms of steam engine, for use in mines and textile manufacture. The international unit of electricity, the watt (a measure of the overall power of an electrical current), was named after him. Helped Davy construct his gas-breathing devices at Bristol. His ailing son Gregory Watt junior was a gifted geologist, and an early friend of Davy's at Bristol until his premature death in 1804.

THOMAS WEDGWOOD, 1771–1805. Chemist and inventor of early photographic method, using 'paintings upon glass' to form images on paper coated with silver salts, which would last 'for a few minutes' (1802). Fragile youngest son of the pottery king and philanthropist Josiah Wedgwood, he was ill for most of his short life and was supplied with opiates by Banks and Coleridge.

WILLIAM WHEWELL, 1794–1866. Geologist and natural historian. The son of a Lancashire carpenter, he eventually became Master of Trinity College, Cambridge. His *Philosophy of Inductive Science* (1859) became the standard Victorian work on the methodology of inductive science, and included an imaginative notion of the 'trial hypothesis'. At Trinity he was celebrated for his less imaginative strictures: no dogs, no cigars, and no women.

GILBERT WHITE, 1720–93. Naturalist and Hampshire clergyman, author of the famous botanical and natural history *Journal* which he kept for over thirty years, and which was published as *The Natural History and Antiquities of Selborne* (1788). Among a myriad other things – swallows, tortoises, snowflakes, birdsong – he was fascinated by balloons, and compared them with bird flight and migration. Widely read by other writers, such as Coleridge and Charles Darwin, he gently championed the notion of precise, patient and exquisite observation of the natural world for its own sake.

WILLIAM HYDE WOLLASTON, 1766–1828. FRS. A chemist and metallurgist, he quietly made his fortune from patenting various forms of malleable platinum. Famous for his patience and precision in the laboratory, and his good nature in society, he refused to become involved in various controversies at the Royal Society stirred up by Davy. John Herschel wrote a revealing sketch of the two men as contrasted scientific personalities. (See Chapter 10)

JOSEPH WRIGHT OF DERBY, 1734–97. Dramatic painter of experimental and industrial scenes, who reinterpreted late-eighteenth-century Enlightenment science as a mysterious, romantic adventure into the unknown. Close friend of Erasmus Darwin and the Lunar men. His most influential pictures were *The Orrery* (1767, frontispiece of this book), *The Air Pump* (1768, National Gallery, London) and *The Alchemist* (Derby, 1770). He also produced some striking, almost apocalyptic industrial scenes of factories and forges (especially at night), and many fine individual portraits.

EDWARD YOUNG, 1683–1765. Poet and clergyman. His major work, *Night Thoughts on Life, Death and Immortality* (1742), a poem in twelve books, was a traditional Christian meditation on the way the universe demonstrated God's design and divine creativity. He announced, 'An undevout astronomer is mad,' though he had some doubts about the size and complication of the cosmos as revealed by Newton's mathematics: 'Perhaps a *seraph's* computation fails!' (Book IX, lines 1, 226–35). A later edition of the poem was superbly illustrated with William Blake's watercolour engravings, a consolation for those terrified by the new cosmology.

Bibliography

The Bigger Picture

(In chronological order of publication)

Thomas Kuhn, *The Structure of Scientific Revolutions*, Chicago UP, 1962–70

Albert Bettex, *The Discovery of Nature* (with 482 illustrations), Thames & Hudson, 1965

James D. Watson, *The Double Helix: A Personal Account of the Discovery of the Structure of DNA*,1968/2001

Arthur Koestler, *The Act of Creation*, Danube edition, 1969

Jacob Bronowski, *The Ascent of Man*, 1973

Adrian Desmond and James Moore, *Darwin*, Penguin, 1992

Lewis Wolpert, *The Unnatural Nature of Science*, Faber, 1992

James Gleick, *Richard Feynman and Modern Physics*, Pantheon Books, 1992

Michael J. Crowe, *Modern Theories of the Universe from Herschel to Hubble*, Chicago UP, 1994

Gale Christianson, *Edwin Hubble: Mariner of the Nebulae*, Farrar, Straus & Giroux,1995

Peter Whitfield, *The Mapping of the Heavens*, The British Library, 1995

John Carey (editor), *The Faber Book of Science*, Faber, 1995

Janet Browne, *Charles Darwin: Volume I: Voyaging*, and *Volume 2: The Power of Place*, Pimlico, 1995 and 2000

Michael Shortland and Richard Yeo, *Telling Lives in Science: Essays in Scientific Biography*, CUP, 1996

Dava Sobel, *Longitude*, Fourth Estate, 1996

Roy Porter, *The Greatest Benefit to Mankind: A Medical History of Humanity from Antiquity to the Present*, HarperCollins, 1997

John Gascoigne, *Science in the Service of Empire*, CUP, 1998

Richard Dawkins, *Unweaving the Rainbow: Science, Delusion and the Appetite for Wonder*, Allen Lane, Penguin Press, 1998

Lisa Jardine, *Ingenious Pursuits: Building the Scientific Revolution*, Little, Brown, 1999

Jonathan Bate, *The Song of the Earth*, Picador, 2000

Ludmilla Jordanova, *Defining Features: Scientific and Medical Portraits 1660–2000*, National Portrait Gallery, London, 2000

Patricia Fara, *Newton: The Making of Genius*, Macmillan, 2000
Mary Midgley, *Science and Poetry*, Routledge, 2001
Thomas Crump, *A Brief History of Science as Seen Through the Development of Scientific Instruments*, Constable, 2001
Oliver Sacks, *Uncle Tungsten: Memories of a Chemical Boyhood*, Picador, 2001
Carl Djerassi and Roald Hoffmann, *Oxygen* (a play in 2 acts), Wiley, New York, 2001
Anne Thwaite, *Glimpses of the Wonderful: The Life of P.H. Gosse*, Faber, 2002
Brenda Maddox, *Rosalind Franklin: The Dark Lady of DNA*, HarperCollins, 2002 Peter Harman and Simon Mitton (editors), *Cambridge Scientific Minds*, CUP, 2002
Arnold Wesker, *Longitude* (a play in 2 acts), Amber Lane Press, 2006
Natalie Angier, *The Canon: The Beautiful Basics of Science*, Faber, 2007
Walter Isaacson, *Einstein: His Life and Universe*, Simon & Schuster, 2007
George Steiner, *My Unwritten Books*, Weidenfeld & Nicolson, 2008

The Scientific and Intellectual Background 1760–1830

Peter Ackroyd, *Newton*, Chatto & Windus, 2006
Madison Smartt Bell, *Lavoisier in the Year One: The Birth of a New Science in the Age of Revolution*, Atlas Books, Norton, 2005
Michael J. Crowe, *The Extraterrestrial Life Debate, 1750–1900*, CUP, 1986
Andrew Cunningham and Nicholas Jardine, *Romanticism and the Sciences*, CUP, 1990
Erasmus Darwin, *The Botanic Garden, A Philosophical Poem with Notes*, 1791
Hermione de Almeida, *Romantic Medicine and John Keats*, OUP, 1991
Adrian Desmond, *The Politics of Evolution: Morphology, Medicine and Reform in Radical London*, Chicago UP, 1989
Patricia Fara, *Pandora's Breeches: Women, Science and Power in the Age of Enlightenment*, Pimlico, 2004
Penelope Fitzgerald, *The Blue Flower* (a novel), HarperCollins, 1995
Tim Fulford (editor), *Romanticism and Science, 1773–1833*, a 5-vol anthology, Pickering, 2002
Tim Fulford and Peter Kitson (editors), *Romanticism and Colonialism: Writing and Empire, 1780–1830*, CUP, 1998
Tim Fulford, Debbie Lee and Peter Kitson, *Literature, Science and Exploration in the Romantic Era*, CUP, 2004
John Gascoigne, *Joseph Banks and the English Enlightenment*, CUP, 1994
James Gleick, *Isaac Newton*, Pantheon Books, 2003

Johann Wolfgang von Goethe, *Scientific Studies* (edited by Douglas Miller), Suhrkamp edition of Goethe's *Works*, vol 12, New York, 1988
Jan Golinski, *Science as Public Culture: Chemistry and Enlightenment in Britain 1760–1820*, CUP, 1992
Richard Hamblyn, *The Invention of Clouds*, Picador, 2001
Peter Harman and Simon Mitron, *Cambridge Scientific Minds*, CUP, 2002
John Herschel, *On the Study of Natural Philosophy*, 1832
J.E. Hodgson, *History of Aeronautics in Great Britain*, OUP, 1924
Penelope Hughes-Hallett, *The Immortal Dinner*, Penguin, 2001
Desmond King-Hele, *Erasmus Darwin and the Romantic Poets*, Macmillan, 1986
David Knight, *Science in the Romantic Era* (essays), Ashgate, 1998
David Knight, *Science and Spirituality*, Routledge, 2003
Trevor H. Levere, *Poetry Realized in Nature: Coleridge and Early Nineteenth-Century Science*, CUP, 1981
Alan Moorehead, *The Fatal Impact: An Account of the Invasion of the South Pacific, 1767–1840*, Hamish Hamilton, 1966, 1987
Alfred Noyes, *The Torchbearers: An Epic Poem*, 1937
William St Clair, *The Reading Nation in the Romantic Period*, OUP, 2004
James A. Secord, *Victorian Sensation: The Extraordinary Publication of Vestiges of the Natural History of Creation*, Chicago UP, 2000
Jenny Uglow, *The Lunar Men: The Friends Who Made the Future, 1730–1810*, Faber, 2002
Jenny Uglow and Francis Spufford, *Cultural Babbage: Technology, Time and Invention*, Faber, 1996

Joseph Banks

Joseph Banks, *The Endeavour Ms Journal 1768–77*, University of New South Wales, Australia, internet copy
The Endeavour Journal of Sir Joseph Banks, edited by J.C. Beaglehole, Public Library of New South Wales, 1962
The Selected Letters of Sir Joseph Banks 1768–1820, edited by Neil Chambers, Imperial College Press, Natural History Museum and Royal Society, The Banks Project, 2000
The Scientific Correspondence of Sir Joseph Banks 1765–1820, edited by Neil Chambers, 6 vols, Pickering & Chatto Ltd, 2007
Hector Cameron, *Sir Joseph Banks*, 1952
Sir Harold Carter, *Sir Joseph Banks 1743–1820*, British Museum, Natural History, 1988

Vanessa Collingridge, *Captain Cook*, Ebury, 2003
Journals of Captain Cook, edited by J.C. Beaglehole, 3 vols, CUP, 1955–74; Penguin Classics, edited by Philip Edwards, 1999
William Cowper, *The Task*, Book One, 1785
Patricia Fara, *Joseph Banks: Sex, Botany and Empire*, Pimlico, 2004
John Gascoigne, *Joseph Banks and the English Enlightenment*, CUP, 1994
Jocelyn Hackforth-Jones, 'Mai', illustrated essay in *Between Two Worlds*, National Portrait Galley catalogue, 2007
John Hawkesworth, *Voyages Undertaken in the Southern Hemisphere*, 1773
Eva Lack, *Die Abenteuers des Sir Joseph Banks* (with rare illustrations of fish, plants and Harriet Blosset), Vienna and Cologne, 1985
James Lee, *Introduction to Botany*, with a Preface by Robert Thornton MD, 1785, 1810
E.H. McCormick, *Omai*, OUP, 1978
Richard Mabey, *Gilbert White*, Century, 1986
Alan Moorehead, *The Fatal Impact*, 1966, 1987
Patrick O'Brian, *Joseph Banks*, Harvill Press, 1987
Sydney Parkinson, *A Journal of a Voyage in the South Seas*, 1773
Roy Porter, 'The Exotic as Erotic', in *Exoticism in the Enlightenment*, edited by G.S. Rousseau and Roy Porter, Manchester UP, 1989
Edward Smith, *Sir Joseph Banks*, 1911
Daniel Solander, *Collected Correspondence*, edited by Edward Duyker and Per Tingbrand, Scandinavia UP, 1995

William and Caroline Herschel

Angus Armitage, *Sir William Herschel*, Nelson, 1962
Helen Ashton, *I Had a Sister*, L. Dickson, 1937
John Bonnycastle, *Introduction to Astronomy in Letters to his Pupil*, 1786 (expanded editions 1788, 1811, 1822)
Claire Brock, *The Comet Sweeper: Caroline Herschel's Astronomical Ambition*, Icon Books, Cambridge, 2007
Lord Byron, *Selected Poems*, edited by A.S.B. Glover, Penguin, 1974
Samuel Taylor Coleridge, *Collected Letters*, 6 vols, edited by E.L. Griggs, OUP, 1956–71
Michael J. Crowe, *Modern Theories of the Universe*, Dover, 1994
Erasmus Darwin, *The Botanic Garden, A Philosophical Poem with Notes*, 1791
James Ferguson, *Astronomy Explained*, with a Preface by David Brewster, 1811

The Herschel Chronicle, edited by Constance A. Lubbock (his granddaughter), 1933

Caroline Herschel, *Memoir and Correspondence of Caroline Herschel*, edited by Mrs John Herschel, Murray, 1876; Cambridge UP, 1935

Caroline Herschel, *Caroline Herschel's Autobiographies*, edited by Michael Hoskin, Science History Publications Ltd, Cambridge, 2003

William Herschel, *Scientific Papers*, 2 vols, edited by J.E. Dreyer, Royal Society and Royal Astronomical Society, 1912

Michael Hoskin, *William Herschel and the Construction of the Heavens*, Osbourne, 1963

Michael Hoskin, *Stellar Astronomy*, Science History Publications, 1982

Michael Hoskin, *The Herschel Partnership as Viewed by Caroline*, Science History Publications, Cambridge, 2003

Derek Howse, *Nevil Maskelyne*, CUP, 1989

Edwin Hubble, *The Realm of the Nebulae*, Constable, 1933

John Keats, *Complete Poems*, edited by John Barnard, Penguin, 1973

Henry Mayhew, *James Ferguson*, 1817

Percy Bysshe Shelley, *Shelley's Prose*, edited by David Lee Clark, Fourth Estate, 1988

Peter Sime, *William Herschel*, 1890

Adam Smith, *The Principles of Philosophical Enquiries, Illustrated by the History of Astronomy*, Edinburgh, 1795

Frances Wilson, *The Ballad of Dorothy Wordsworth*, Faber, 2008

Edward Young, *Night Thoughts* (poem), 1744–45

Specialist Articles

J.A. Bennett, 'The Telescopes of William Herschel' (with illustrations), *Journal for the History of Astronomy* 7, 1976

Michael Hoskin, 'On Writing the History of Modern Astronomy', *Journal for the History of Astronomy* 11, 1980

Michael Hoskin, 'Caroline Herschel's Comet Sweepers', *Journal for the History of Astronomy* 12, 1981

Simon Schaffer, 'Herschel on Matter Theory and Planetary Life', *Journal for the History of Astronomy* 11, 1980

Simon Schaffer, 'Uranus and Herschel's Astronomy', *Journal for the History of Astronomy* 12, 1981

Simon Schaffer, 'Herschel in Bedlam: Natural History and Stellar Astronomy', *British Journal for the History of Science* 13, 1986

Simon Schaffer, 'On the Nebular Hypothesis', in *History, Humanity and Evolution*, edited by J.R. Moore, CUP, 1988

The Balloonists

Thomas Baldwin, *Airopaidia*, 1786 (the narrative of a solo voyage in Lunardi's balloon, including the first aerial sketches made from a balloon basket)
Henry Beaufoy, *Account of an Ascent with James Sadler, from Hackney*, 1811, British Library catalogue B.507 (I)
Henry Beaufoy, *Two Balloon Scrapbooks of Henry Beaufoy, 1783–1843* (Item 57 in the McCormack Collection), Princeton University, USA
David Bourgeois, *L'Art de Voler*, Paris, 1784
Catalogue of Well-Known Balloon Prints and Drawings, Sotheby's, 1962
Tiberius Cavallo FRS, *The History and Practice of Aerostation*, 1785
William Cowper, *The Task* (poem), in *Letters and Poems*, 1785
Le Départ du Rêve, Grand Palais exhibition catalogue, Paris, 1985
Erasmus Darwin, *The Botanic Garden*, 1791
Audoin Dollfuss, *Pilâtre de Rozier*, Association Francaise pour l'Avancement des Sciences, Paris, 1993
Raymonde Fontaine, *La Manche en Ballon*, Paris, 1982
The Gentleman's Magazine – accounts of Lunardi's ascents in 1784–85, and Sadler's ascents in 1810–17
Charles Gillispie, *The Montgolfier Brothers*, Princeton UP, 1983
James Glaisher, with Camille Flammarion, Wilfred de Fonvielle and Gaston Tissandier, *Travels in the Air*, London, 1871
Charles Green, *The Flight of the Nassau Balloon*, 1836
Richard Hamblyn, *The Invention of Clouds*, Picador, 2001
Georgette Heyer, *Frederica* (a novel containing an excellent account of a balloon ascent), E.P. Dutton, 1965
J.E. Hodgson, *History of Aeronautics in Great Britain*, OUP, 1924
Dr John Jeffries, *Narrative of Two Aerial Voyages with M. Blanchard as Presented to the Royal Society*, 1786
Vincent Lunardi, *My Aerial Voyages in England*, 1785; and *Five Aerial Voyages in Scotland*, 1785
Thomas Monck Mason, *Aeronautica*, 1838
Thomas Mayhew, *An Account of a Balloon Flight*, 1855
Edgar Allan Poe, 'The Great Balloon Hoax' (story), *New York Sun*, 1847
Gavin Pretor-Pinney, *The Cloud Spotter's Guide*, Sceptre, 2006
L.T.C. Rolt, *The Aeronauts*, Longman, 1966

James Sadler, *An Authentic Account of the Aerial Voyage*, 1810

James Sadler, *Across the Irish Channel*, 1812

Windham Sadler, *Aerostation*, 1817

Mrs Sage, *A Letter by Mrs Sage, the First English Female Aerial Traveller, on Her Voyage in Lunardi's Balloon*, 1785, British Library catalogue 1417.g.24

Gaston Tissandier, *Histoire des Ballons et Aeronauts Célèbres 1783–1890*, 2 vols, Paris, 1890

Mungo Park

William Feaver, *The Paintings of John Martin*, OUP, 1975. This includes a dramatic full-page colour reproduction of *Sadak in Search of the Waters of Oblivion* (1812, Southampton Art Gallery)

Tim Fulford, Debbie Lee and Peter J. Kitson, 'Mental Travellers: Banks and African Exploration', in *Literature, Science and Exploration in the Romantic Era*, CUP, 2004,

The Gentleman's Magazine, long review of 'Mr Park's Travels', with illustrations from Rennell, August 1799

Georgiana, Duchess of Devonshire, 'A Negro Song', 1799

Lewis Gibbons, *Niger and Mungo Park*, 1934

Stephen Gwynn, *Mungo Park and the Quest for the Niger*, 1932

BH (anon), *The Life of Mungo Park*, 1835, British Library catalogue 615.a.12

John Keats, 'Nile Sonnets', 1818

Kenneth Lupton, *Mungo Park, African Traveller*, OUP, 1979

Mungo Park, *Journals*, 2 vols, edited anonymously, including 'A Journal of Park's Last Voyage', 'Amadi Fatoumi's Journal' and a Memoir by W. Wishaw, 1815

Mungo Park, *Travels in the Interior of Africa*, 1799, 1860; Nonesuch, 2005

Kira Salak, *The Cruellest Journey: 6,000 Miles by Canoe to the Legendary City of Timbuktu*, Bantam Books, 2005

Anthony Sattin, *The Gates of Africa: Death, Discovery and the Search for Timbuktu*, HarperCollins, 2003

Percy Bysshe Shelley, *Alastor* (poem), 1815; and 'Nile Sonnets', 1818

Robert Southey, 'Note on Mungo Park', in *Thalaba*, 1803

Alfred Tennyson, 'Timbucto' (poem), 1827

Joseph Thomson, *Mungo Park and the Niger*, 1890

Charles Waterton, *Wanderings in South America*, 1825

William Wordsworth, rejected passage on Mungo Park, from *The Prelude*, 1805

Humphry Davy

Thomas Beddoes and James Watt, *Considerations on the Medical Use of Factitious Airs*, J. Johnson, 1794, British Library catalogue B. Tracts. 489
Henry Brougham, 'Sir Humphry Davy', in *The Lives of the Philosophers in the Time of George III*, London, 1855
George I. Brown, *Count Rumford: The Extraordinary Life of a Scientific Genius*, Sutton, 1999
Lord Byron, *Don Juan* (poem in 16 cantos), 1819–24
F.F. Cartwright, *The English Pioneers of Anaesthesia*, Simpkin Marshall, 1952
Samuel Taylor Coleridge, *Collected Letters*, edited by E.L. Griggs, vols 1–2, OUP
Humphry Davy, *Collected Works*, edited by John Davy, 9 vols, 1839–40
Humphry Davy, *Fragmentary Remains*, edited by John Davy, 1858
John Davy, *The Life of Sir Humphry Davy*, 2 vols, 1836
John Davy, *Memoirs of Sir Humphry Davy*, in Humphry Davy, *Collected Works*, Vol I, 1839
Sophie Forgan (editor), *Science and the Sons of Genius: Studies on Humphry Davy* (essays), Science Reviews Ltd, 1980
June Z. Fullmer, *Young Humphry Davy*, American Philosophical Society, 2000
James Hamilton, *Michael Faraday: The Life*, HarperCollins, 2002
Harold Hartley, *Humphry Davy*, Open University, 1966
Richard Holmes, *Coleridge: Early Visions*, Hodder & Stoughton,1989
Richard Holmes, *Coleridge: Darker Reflections*, HarperCollins, 1998
David Knight, *Humphry Davy: Vision and Power*, Blackwell Science Biographies, 1992
Davy Lamont-Brown, *Humphry Davy: Life Beyond the Lamp*, History Press, 2004
John Ayrton Paris, *The Life of Sir Humphry Davy*, 2 vols, 1831
Roy Porter, *The Greatest Benefit to Mankind: A Medical History of Humanity*, HarperCollins, 1997
Nicholas Roe (editor), *Samuel Taylor Coleridge and the Sciences of Life*, OUP, 2001,
W.D.A. Smith, *Under the Influence: A History of Nitrous Oxide and Oxygen Anaesthesia*, Macmillan, 1982
Robert Southey, *The Life and Correspondence of Robert Southey*, edited by C.C. Southey, vols 1–2, 1849
Dorothy A. Stansfield, *Thomas Beddoes MD: Chemist, Physician, Democrat*, Reidel Publishing, Boston, 1984
Thomas Thorpe, *Humphry Davy, Poet and Philosopher*, 1896
Anne Treneer, *The Mercurial Chemist: A Life of Sir Humphry Davy*, Methuen, 1963

The Safety Lamp and Controversy

A Collection of all Letters in Newcastle papers relating to Safety Lamps, London, 1817. See British Library catalogue Tracts 8708.i.2 (1)

Humphry Davy, *On the Safety Lamp for Preventing Explosions*, London, 1825 (contains an Appendix on the use of his safety lamps in Europe)

Humphry Davy, *On the Safety Lamp for Coal Miners, with Some Researches into Flame* (6 papers), London, 1818. See also his revised version in *Collected Works*, Vol 6, 1840

J.H.H. Holmes, *A Treatise on Coalmining of Durham and Northumberland and the Explosions of Firedamp in the last 20 Years*, London, 1816, British Library catalogue 726.e.37

Frank A.J.L. James, 'How Big is a Hole? The Problems of the Practical Application of Science in the Invention of the Miners' Safety Lamp by Humphry Davy and George Stephenson in Late Regency England', in *Transactions of the Newcomen Society* 75, 2005, pp.175–227

John Playfair, 'Sir Humphry Davy's Safety Lamp', in *Edinburgh Review* LI, 1816, pp.230–40

'Report of the Select Committee on Accidents in Mines', in *Parliamentary Papers*, 1835, vol 5, September 1835. British Library (Science) Series Parliamentary Papers 1835

Samuel Smiles, *George Stephenson*, 1855

Stephenson's Lamp now at Killingworth compared to Humphry Davy's Lamp (2 pamphlets), London, 1817, British Library catalogue 8708.i.2 (5)

Dr Frankenstein and the Soul

John Abernethy, *An Enquiry into Mr Hunter's Theory of Life Lectures*, 1815

John Abernethy, *A General View of Mr Hunter's Physiology*, 1817

John Abernethy, *The Hunterian Oration for 1819*, 1819

John Abernethy, 'Letters to George Kerr 1814–1822', in *St Bart's Hospital Journal, 1930–1*, vol 38, edited by A.W. Franklin

Xavier Bichat, *Physiological Researches on Life and Death* (translated by F. Gold), 1816

Fred Botting (editor), *New Casebooks: Frankenstein*, Palgrave, 1995

Druin Burch, *Digging up the Dead: The Life and Times of Astley Cooper*, Chatto & Windus, 2007

Fanny Burney, *The Journals and Letters of Fanny Burney (Madame d'Arblay)*, vol 6, edited by Joyce Hemlow, Oxford, 1975

Richard Carlile, *Address to the Men of Science*, 1821

F.F. Cartwright, *The English Pioneers of Anaesthesia*, Simpkin Marshall, 1952
Samuel Taylor Coleridge (with James Gillman and J.H. Green), *Notes Towards a More Comprehensive Theory of Life*, 1816–19; edited by Seth B. Watson MD, 1848
Nora Crook and Derek Guiton, *Shelley's Venomed Melody*, CUP, 1986
Humphry Davy, *Elements of Agricultural Chemistry*, 1814
Hermione de Almeida, *Romantic Medicine and John Keats*, OUP, 1991
Thomas De Quincey, *Confessions of an English Opium-Eater*, 1821
Thomas De Quincey, 'Animal Magnetism' (essay), 1840
Adrian Desmond, *The Politics of Evolution: Medicine in Radical London*, Chicago, 1989
George D'Oyly, 'An Enquiry into the Probability of Mr Hunter's Theory of Life' (The Vitality Debate), in *Quarterly Review*, 1819, vol 43, pp.1–34. Usefully reprinted in Oxford World's Classics edition of Mary Shelley's *Frankenstein*, Appendix B
Tim Fulford, Debbie Lee and Peter J. Kitson, 'Exploration, Headhunting and Race Theory', in *Literature, Science and Exploration*, CUP, 2004
Jan Golinski, *Science as Public Culture: Chemistry and Enlightenment in Britain 1760–1820*, CUP, 1992
Carl Grabo, *A Newton Among Poets: Shelley's Use of Science in Prometheus Unbound*, University of North Carolina Press, 1931
John Keats, 'Lamia' (poem), 1820
William Lawrence, *A Short System of Comparative Anatomy by JF Blumenbach* (translated with an Introduction by Lawrence), 1807
William Lawrence, *An Introduction to Comparative Anatomy: Two Lectures*, 1816
William Lawrence, *The Natural History of Man* (Lectures on Physiology and Zoology), 1819
William Lawrence, 'On Life', *Rees's Cyclopaedia*, 1819
William Lawrence, 'On Man', *Rees's Cyclopaedia*, 1820
Trevor H. Levere, *Poetry Realized in Nature: Coleridge and Early Nineteenth Century Science*, CUP, 1981
Helen MacDonald, *Human Remains: Dissection and its Histories*, Yale UP, 2006
Anne K. Mellor, 'A Feminist Critique of Science', in *Mary Shelley: Her Life, Her Fictions, Her Monsters*, Routledge, 1988
Peter Mudford, 'William Lawrence', in *Journal of the History of Ideas* 29, 1968
Roy Porter and G. Rousseau (editors), *The Ferment of Knowledge*, CUP, 1980
Nicholas Roe, 'John Thelwall's Essay on Animal Vitality', in *The Politics of Nature*, Palgrave, 2002

Sharon Ruston, *Shelley and Vitality*, Palgrave, 2005

Mary Shelley, *Frankenstein, or The Modern Prometheus,* 1818; the 1818 edition reprinted in Oxford World's Classics, edited by Marilyn Butler, 1993; 2nd edition, 1831, reprinted as composite edition, Penguin Classics, edited by Maurice Hindle, 1992

Percy Bysshe Shelley, essays 'On Life', 'On Love', 'On Dreams', 'On a Future State', 'On the Devil and Devils', 'On Christianity' (1814–18), in *Shelley's Prose, or The Trumpet of a Prophecy*, edited by David Lee Clark, Fourth Estate, 1988

Walter Wetzels, 'Johann Wilhelm Ritter: Romantic Physics in Germany', in *Romanticsm and the Sciences*, edited by Andrew Cunningham and Nicholas Jardine, CUP, 1990

Sorcerer and Apprentice; and Young Scientists

Charles Babbage, *The Decline of Science in England*, 1830

David Brewster, *Life of Isaac Newton*, Murray's Family Library, 1831

The British Association for the Advancement of Science: Early Correspondence, edited by Jack Morrell and Arnold Thackray, The Camden Society, 1984

Janet Browne, *Charles Darwin: Volume I: Voyaging*, and *Volume 2: The Power of Place*, Pimlico, 1995 and 2000

Gunther Buttman, *In the Shadow of the Telescope: A Biography of John Herschel*, Lutterworth Press, 1974

Charles Darwin, *Correspondence: Vol I, 1821–1836*, edited by Frederick Burkhardt and Sydney Smith, CUP, 1985

Charles Darwin, *The Voyage of the Beagle, 1831–1836*, edited by Janet Browne and Michael Neve, Penguin Classics, 1989

Charles Darwin, *Autobiography*, edited by Michael Neve, Penguin Classics, 2002

Humphry Davy, *Consolations in Travel, or The Last Days of a Philosopher*, Murray's Family Library, 1829, 1831

Michael Faraday, *Correspondence 1811–1831*, Vol 1, edited by Frank A.L.J. James, Institute of Electrical Engineers, 1991

Marie Boas Hall, *All Scientists Now*, CUP, 1984

James Hamilton, *Michael Faraday: The Life*, HarperCollins, 2002

John Herschel, *A Preliminary Discourse on the Study of Natural Philosophy*, 1831

John Herschel, *Herschel at the Cape: Letters and Journals of John Herschel*, edited by David S. Evans, Texas, 1969

Richard Holmes, *Shelley: The Pursuit*, Weidenfeld & Nicolson, 1974

Jack Morrell and Arnold Thackray, *Gentlemen of Science: The Early Years of the BAAS*, OUP, 1981

Steven Ruskin, *John Herschel's Cape Voyage*, Ashgate, 2004

James Secord, *Victorian Sensation*, Chicago UP, 2000

Mary Shelley, *Frankenstein*, 1st edition, Lackington, 1818; edited by Marilyn Butler, Oxford World's Classics, 1993

Mary Shelley, *Frankenstein*, 2nd edition, Bentley's Popular Library, 1831; reprinted as composite edition, Penguin Classics, edited by Maurice Hindle, 1992

Percy Bysshe Shelley, *Prometheus Unbound: An Epic Poem in 4 Acts*, 1819

Mary Somerville, *On the Connexion of the Physical Sciences*, 1834

Thomas Sprat, *History of the Royal Society*, Kessinger, 2003

David Wooster, *Paula Trevelyan* (Paulina Jermyn), 1879

References

ABBREVIATIONS

CHA – *Caroline Herschel's Autobiographies*, edited by Michael Hoskin, Scientific Publications Ltd, Cambridge, 2003

CHM – *Memoir and Correspondence of Caroline Hesrchel*, edited by Mrs John Herschel, Murray, 1879

HD Archive – Humphry Davy Manuscripts and scientific instruments held at the Royal Institution, London

HD Mss Bristol – Humphry Davy Mss at Somerset Record Office, Bristol

HD Mss Truro – Humphry Davy Mss at the Cornwall Record Office, Truro

HD Works – Humphry Davy, *Collected Works*, edited by John Davy, 9 vols, 1839–40

JB Correspondence – *The Scientific Correspondence of Sir Joseph Banks 1765–1820*, edited by Neil Chambers, 6 vols, Pickering & Chatto Ltd, 2007

JB Journal – Joseph Banks, *Manuscript of the Endeavour Journal 1768–1770*, University of New South Wales (internet transcript). See also *The Endeavour Journal of Sir Joseph Banks*, edited by J.C. Beaglehole, Public Library of New South Wales, 2 vols, 1962; and Joseph Banks, *Endeavour Journal Ms, 1768–70* (facsimile edition, London Library)

JB Letters – *The Selected Letters of Sir Joseph Banks 1768–1820*, edited by Neil Chambers, Imperial College Press, Natural History Museum and Royal Society, The Banks Project, 2000

JD Fragments – Humphry Davy, *Fragmentary Remains*, edited by John Davy, 1858

JD Life – *The Life of Sir Humphry Davy*, by John Davy, 2 vols, 1836

JD Memoirs – *Memoirs of Sir Humphry Davy*, by John Davy, 1839 (included in vol 1 of HD Works)

Park Mss – 'Letters and Papers relating to Mungo Park's last Journey', British Library Add Mss 37232.k and Add Mss 33230.f

WH Archive – Private archive, John Herschel-Shorland, Norfolk

WH Chronicle – *The Herschel Chronicle*, edited by his granddaughter Constance A. Lubbock, CUP, 1933

WH Mss – William Herschel Manuscripts, Cambridge University Library microfilm, from manuscripts held at the Royal Astronomical Society, London

WH Papers – *The Collected Scientific Papers of Sir William Herschel including Early Papers hitherto Unpublished*, edited by J.L.E. Dreyer, 2 vols, Royal Society and Royal Astronomical Society, 1912

Prologue

1 The notion of 'Romantic science' has been pioneered by Jan Golinski, *Science as Public Culture, 1760–1820*, CUP, 1992; Andrew Cunningham and Nicholas Jardine, *Romanticism and the Sciences*, CUP, 1990; Mary Midgley, *Science and Poetry*, Routledge, 2001; Tim Fulford, Debbie Lee and Peter J. Kitson, *Literature, Science and Exploration in the Romantic Era*, CUP, 2004; and Tim Fulford (editor), *Romanticism and Science, 1773–1833*, a 5-vol anthology, Pickering, 2002
2 Samuel Taylor Coleridge, *Philosophical Lectures 1819*, edited by Kathleen Coburn, London, 1949; and *The Friend* 1819, 'Essays on the Principles of Method', edited by Barbara E. Rooke, Princeton UP, 1969. See Richard Holmes, *Coleridge: Darker Reflections*, 1998, pp480–4, 490–4
3 Wordsworth, *The Prelude*, 1850, Book 3, lines 58–64
4 Coleridge, *Aids to Reflection*, 1825; see Holmes, op. cit., pp548–9
5 Plato's wonder as interpreted by Coleridge in 'Spiritual Aphorism 9', *Aids to Reflection*, 1825, p236

Chapter 1: Joseph Banks in Paradise

1 JB Journal, 18 October 1768
2 Ibid., 11 April 1769
3 JB letter to Pennat, November 1768; from Harold Carter, *Sir Joseph Banks*, British Library, 1988, p76
4 JB Journal, 14 April 1769
5 Hector Cameron, *Sir Joseph Banks*, 1952, p6
6 Vanessa Collingridge, *Captain Cook*, 2003, p158
7 JB Journal, 2 May 1769
8 James Cook, Journal, 2 May 1769
9 JB Journal, 2 May 1769
10 JB Journal, 'On the Customs of the South Sea Islands', pp120–50, essay dated August 1769
11 Patrick O'Brian, *Joseph Banks*, Harvill, 1989, p65
12 Ibid.
13 John Gascoigne, *Joseph Banks and the English Enlightenment*, 1994, p17
14 Ibid., p88
15 Lady Mary Coke, *Journals*, August 1771, p437
16 JB letter to William Perrin, February 1768, from Gascoigne, p16
17 JB Journal, 10 September 1768
18 JB Journal, p23
19 O'Brian, p65
20 White, 8 October 1768; from Richard Mabey, *Gilbert White*, Century, 1986, p115
21 JB Journal, 16 January 1769
22 Ibid., 25 March 1769
23 Ibid., 17 April 1769
24 Sydney Parkinson, *A Journal of a Voyage in the South Seas*, 1773, p15
25 JB Journal, 30 April 1769
26 Ibid., 29 April 1769
27 Ibid., 25 April 1769
28 Ibid., 22 April 1769
29 Ibid., 4 June 1769
30 James Cook, Journal, Tuesday, 6 June 1769
31 Parkinson, Journal, from Collingridge, p166
32 JB Journal, 10 May 1769
33 JB Journal, pp120–50, essay dated August 1769
34 JB Journal, 3 June 1769
35 Ibid., 28 April 1769
36 Ibid., 28 May 1769

37 Ibid., 29 May 1769
38 Ibid., 12 May 1769
39 Ibid., 10 June 1769
40 Ibid., 13 June 1769
41 Ibid., 14 June 1769
42 Ibid., 18 June 1769
43 Ibid., 24 June 1769
44 Ibid., 19 June 1769
45 Ibid., 22 June 1769
46 Parkinson, Journal, 1773, p32; and O'Brian, p101
47 James Cook, Journal, 30 June 1769
48 JB Journal, 28 June 1769
49 Ibid., 30 July 1769
50 Ibid., 29 June 1769
51 JB Letters, 'Thoughts on the Manners of the Otaheite', 1773, p332
52 JB Journal, 3 July 1769
53 Ibid., 12 July 1769
54 Ibid.
55 Ibid.
56 JB Letters, 6 December 1771, p20
57 Parkinson, Journal, 1773, p66
58 JB Journal, 'On the South Seas', August 1769, p124
59 Ibid., p128
60 Ibid., p132
61 Ibid.
62 JB Journal, (end) August 1770. Cook's entry of the same date describes the natives as 'in reality... far more happier than we Europeans'
63 JB Journal, 3 September 1770
64 O'Brian, pp145–6
65 JB Letters, 13 July 1771, p14
66 Gascoigne, p46
67 O'Brian, p66
68 Lady Mary Coke, *Journals*, August 1771, from Edward Smith, *Joseph Banks*, p22n
69 O'Brian, p151
70 Robert Thornton MD, Preface to *An Introduction to Botany*, by James Lee, 1810, ppxvii–iii
71 Gascoigne, p17
72 Thornton, 1810, ppxviii
73 Cameron, p44
74 Ibid., p 45
75 Ibid., p46
76 James Boswell, *Journal*, 22 March 1772
77 John Hawkesworth, 'Tahiti', in *Voyages Undertaken in the Southern Hemisphere*, 1773; the section can also be found in Fulford, *Romanticism and Science*, vol 4, pp158–9
78 JB, 'Thoughts on the Manners of the Otaheite', 1773, JB Letters, p330
79 JB letter, 30 May 1772, from O'Brian, p158
80 Lord Sandwich to Banks, 20 June 1772, in JB Letters, Appendix V, p354
81 JB Letters, Appendix V, p355
82 Rev William Sheffield, letter to Gilbert White, 2 December 1772, from O'Brian, p168
83 Daniel Solander, 16 November 1776, *Collected Correspondence*, edited by Edward Duyker and Per Tingbrand, Scandinavia University Press, 1995, p373
84 Carter, p153
85 Gascoigne, p50
86 Tim Fulford, Debbie Lee and Peter J. Kitson, *Literature, Science and Exploration in the Romantic Era*, CUP, 2004, p49
87 O'Brian, p181
88 Reproduced in the exhibition catalogue *Between Worlds: Voyagers to Britain 1700–1850*, National Portrait Gallery, 2007

89 British Academy Conference, 2006, my correspondence
90 William Cowper, 6 October 1783
91 William Cowper, *The Task*, 1784, Book 4, 'The Winter Evening', lines 107–19
92 Ibid., Book 1, lines 654ff
93 John Byng, quoted in Beaglehole, *Journal of Sir Joseph Banks*, 2 vols, 1962, p114
94 Gascoigne, p52
95 Collingridge, *Cook*, 2002, pp405–15
96 Gascoigne, p46
97 Daniel Solander, 5 June 1779, *Collected Correspondence*, op. cit.
98 Gascoigne, p18
99 O'Brian, p308
100 Derek Howse, *Nevil Maskelyne*, 1989, p161
101 Patricia Fara, *Joseph Banks: Sex, Botany and Empire*, 2003, pp136–7
102 Coleridge to Samuel Purkis, 1 February 1803, *Collected Letters* vol 2, p919
103 JB Correspondence I, p331
104 JB Letters, 16 November 1784, pp77–80
105 Carter, p121
106 Gascoigne, p32
107 Baron Cuvier, 'Éloge on Sir Joseph Banks', 1820, from *Sir Joseph Banks and the Royal Society*, anonymous booklet, Royal Society, 1854, pp66–7

Chapter 2: Herschel on the Moon

1 WH Chronicle, p1
2 Account from Herschel's Journal in CHM, p42
3 WH Chronicle, p73
4 Account from CHA
5 WH Papers 1; Armitage, p24
6 Michael J. Crowe, *The Extraterrestrial Life Debate, 1750–1900*, CUP, 1986, p63
7 WH Mss 6279; also WH Chronicle, p76
8 WH Papers 1, pxc; also WH Chronicle, p77
9 Herschel to Maskelyne, 12 June 1780, WH Papers 1, ppxc–xci
10 CHM, p41
11 CHM, p149
12 CHA, pp14–15
13 CHA, pp19–20
14 CHA, p14
15 WH Papers 1, pxiv
16 CHA, p24
17 CHA, p112
18 CHM, p24
19 CHA, p23
20 CHA, p21
21 CHA, p24
22 CHM, p7
23 CHM, p6
24 CHA, p41
25 CHA, p25
26 CHA, p30
27 CHA, p136
28 CHA, p26; CHM, p10
29 CHM, p12
30 CHM, p11
31 WH Papers 1, pxix
32 Angus Armitage, *Herschel*, 1962, p19
33 CHM, p11; also CHA, p108
34 CHA, p110
35 CHA, p109
36 Armitage, p19
37 CHA, p33
38 Helen Ashton, *I Had a Sister*, 1937, pp153–61
39 CHA, p33
40 CHA, p34; Ashton, p161
41 CHA, p37
42 CHM, p20

43 CHA, p37
44 CHA, pp29, 34
45 CHM, p17
46 WH Papers 1, pxvii
47 WH Archive, William and Jacob Mss Letters 1761–63
48 WH Archive Mss Letters March 1761; also WH Chronicle, p18
49 WH Archive Mss Letters May 1761; also WH Chronicle, p26
50 WH Archive Mss Letter October 1761; also WH Chronicle, p28
51 WH Chronicle, p24
52 WH Archive Mss Letter October 1761; also WH Chronicle, p28
53 WH Papers 1, pxc, letter to Nevil Maskelyne
54 Armitage, p21
55 Ibid., p22
56 Ibid., p20
57 CHA, p7
58 CHA, p113; CHM, p18
59 CHA, p36
60 Ian Woodward, 'The Celebrated Quarrel between Thomas Linley and William Herschel', pamphlet printed Bath (British Library catalogue L.409.c.585.1); also WH Chronicle, pp42–3
61 WH Papers 1, ppxx–xxi
62 Armitage, p22
63 Crowe, 1986, pp124–9
64 James Gleick, *Isaac Newton*, 2003
65 Derek Howse, *Nevil Maskelyne*, 1989, pp70–1
66 Howse, pp66–72
67 Michael Hoskin, *The Herschel Partnership*, p21
68 CHM, pp22–3
69 CHA, p24
70 CHM, p25
71 CHM, p27
72 CHM, p32
73 CHA, p53
74 CHA, p123
75 CHM, p33
76 CHA, p51; CHM, p35
77 WH Mss 6278 1/8/8, dated 1784. But the use of the diminutive 'Lina' first becomes evident in manuscripts dating from 1779
78 WH Mss 6290
79 CHA, p52; CHM, p35
80 CHA, p55
81 CHA, p52; CHM, pp36–7
82 CHM, pp37–8
83 CHA, p55
84 WH Papers 1, Introduction
85 WH Mss 6290
86 JB Correspondence 1; Hoskin, p46
87 I owe these acute observations to Dr Percy Harrison, Head of Science, Eton College
88 WH Mss, H W.2/1. 1f.i
89 WH Mss, 'Herschel's First Observation Journal', Ms 6280
90 Michael Crowe, *Extraterrestrial*, 1994, pp42, 74–5. Herschel eventually increased it to 2,500 by 1820, and Edwin Hubble to 17,000 by the mid-twentieth century.
91 Armitage, p22
92 WH Mss 6290 7/8, dated January 1782; also WH Chronicle, p73
93 WH Chronicle, p72
94 WH Mss 6278 1/8/5
95 CHA, p127
96 CHA, p128
97 CHA, p129
98 CHM, p40
99 WH Mss 6290
100 Michael Crowe, *Theories of the Universe*, 1994
101 James Ferguson, *Astronomy Explained*, 1756, p5; and discussed by Michael Crowe, *Extraterrestrial*, 1986, p60

102 Crowe, *Extraterrestrial*, p170; also Crowe, *Theories of the Universe*, 1994, p73
103 CHM, p42
104 CHA, p61
105 CHA, p61
106 WH Papers vol 1, plxxxvii
107 WH Mss W.3/1.4, drafted 1778–79; discussed Crowe, 1986, pp64–5
108 WH Mss 6280, Observation Journal, 28 May 1776; and Crowe, 1986, p63
109 WH Mss W.3/1.4, drafted 1778–79, from Crowe, 1986, p65
110 CHA, p61
111 WH Mss 6280, First Observation Book
112 CHA, p61
113 WH Mss 6280, First Observation Book
114 Ibid., pp31ff, 170ff
115 CHA, p62
116 Simon Schaffer, *Journal of the History of Astronomy*, vol 12, 1981
117 Howse, p147
118 Schaffer, 'Uranus and Herschel's Astronomy', *Journal for the History of Astronomy*, vol 12, 1981, p12
119 WH Papers 1, p36
120 WH Mss 6279; also WH Chronicle, p79
121 WH Mss 6279; WH Chronicle, p81
122 WH Papers 1; WH Chronicle, pp81–2
123 Howse, pp147–8
124 See WH Chronicle, pp78–80
125 WH Chronicle, p86, from Schaffer, *Journal of the History of Astronomy*, vol 12, 1981, 'Uranus and Herschel's Astronomy', p14
126 Watson, letter to Herschel 25 May 1781, in WH Chronicle, p85
127 Howse, *Maskelyne*, p149
128 WH Chronicle, p95
129 'A Letter to Sir Joseph Banks Bart. PRS', 1783, in WH Papers 1, pp100–1
130 WH Mss 6278 1/7, letter 19 November 1781; also JB Correspondence 1, p292
131 JH Mss 6278 1/1/57
132 JH Mss 6278 1/1/63
133 'Account of My Life to Dr Hutton', 1809, from WH Chronicle, p79
134 WH Chronicle, p95
135 John Bonnycastle, *Introduction to Astronomy in Letters to a Pupil*, 1786 (expanded edition 1811), pp354–7
136 Ibid., p241
137 Immanuel Kant, *Universal Natural History and the Theory of the Heavens*, 1755 (translation 1969, British Library catalogue 9350.d.649), Part I, p67. Kant also wrote: 'There is here no end but an abyss of real immensity, in the presence of which all the capability of human conception sinks exhausted, although it is supported by the aid of the science of mathematics.' Part I, p65
138 Erasmus Darwin, *The Botanic Garden*, 1791, Canto 1, lines 100–14, and Note to line 105; see also Canto 2, lines 14–82, and Canto 4, line 34
139 WH Chronicle, p102
140 JB Correspondence 1, p299
141 WH Chronicle, p101
142 JB Correspondence 1, p307
143 WH Chronicle, pp103–4
144 CHM, p45
145 CHM, p46; Howse, p148
146 WH Chronicle, pp115–16
147 Peter Sime, *William Herschel*, 1890, pp259–61
148 WH Chronicle, p116

149 WH Mss 6278 1/8/6, 20 May 1782
150 CHA, pp66–7
151 CHM, pp48–9
152 Holmes, *Coleridge: Early Visions*, 1994, pp18–19
153 Coleridge, *The Ancient Mariner*, Part IV, lines 263–71
154 Andrew Motion, *Keats*, Faber, 1997, pp27, 39, 121
155 WH Papers 1, pxix
156 Herschel to Johann Bode at Berlin, 20 July 1785, WH Mss 6278/11, p134
157 WH Mss 5278 1/4
158 Lucien Bonaparte, Wikipedia
159 WH Papers 1, pxix
160 CHA, p82
161 Samuel Johnson, *Collected Letters*, edited by Bruce Redford, vol III, 25 March 1784, p144
162 CHM, pp50–5
163 Hoskin, pp74–5
164 WH Mss 6281, Observation Journal No. 5, 1782
165 WH Chronicle, p105
166 WH Mss 6268 3/11
167 Ibid.
168 CHM, p52
169 Ibid.
170 WH Archive
171 CHM, p52
172 WH Papers 1, pp261–2; and WH Chronicle, pp222–3
173 CHM, p52
174 CHA, p77
175 CHA, p76
176 CHA, p77
177 Ibid.
178 Ibid.; and CHM, p55
179 WH Chronicle, pp190–5: a risky claim perhaps
180 WH Papers 1, pp157–66
181 Ibid. Illustrated in Armitage and Crowe, 1996, excerpts
182 Michael J. Crowe, *Modern Theories of the Universe from Herschel to Hubble*, Chicago UP, 1994
183 WH Papers 1, p265
184 WH Papers 1, p223
185 WH Papers 1, p225, a phrase repeated at end of this paper, at p259. Other extraordinary descriptions of galaxies evolving like plants growing or humans ageing occur in 'Catalogue of a Second Thousand of new Nebulae', 1789, WH Papers 1, pp330 and 337–8. Also in 'On Nebulae Stars, properly so called', 1791, WH Papers 1, pp415ff. See discussion in Edwin Hubble, *The Realm of the Nebulae*, 1933; and Michael Crowe, *Theories of the Universe*, 1996
186 'On the Construction of the Heavens', 1785, WH Papers 1, pp247–8
187 Ibid., p27
188 Ibid., p25. See J.A. Bennett, 'The Telescopes of William Herschel', *Journal for the History of Astronomy*, vol 7, 1976
189 Bonnycastle, pp341–2
190 WH Papers 1, p256

Chapter 3: Balloonists in Heaven

1 JB Correspondence 2, p299
2 Exchange of Banks–Franklin letters, 1783, Schiller Institute, 'Life of Joseph Franklin' (internet)
3 WH Letters, p62, to Franklin, 13 September 1783
4 Ibid.
5 L.T.C. Rolt, *The Aeronauts*, 1966, p29
6 'Dossier Montgolfier (1)', Musée de l'Air, Le Bourget, Paris
7 Rolt, p 30

8 Schiller Institute, 'Life of Joseph Franklin' (internet)
9 Auduin Dollfuss, *Pilâtre de Rozier*, Paris, 1993, p26
10 Ibid., pp17–22
11 Marquis d'Arlandes's original account given in ibid., pp27–42; '*la redingote verte*', p41. Discussed in Rolt, pp46–9
12 Rolt, p50
13 Dr Robert Charles's original account appears in Raymonde Fontaine, *La Manche en Ballon*, Paris, 1980
14 Dr Charles's original account in ibid. (photocopy)
15 'Dossier Montgolfier (1)', Musée de l'Air, Le Bourget, Paris
16 David Bourgeois, *Recherches sur l'Art de Voler*, Paris, 1784, pp1–3
17 Ibid., p3
18 J.E. Hodgson, *History of Aeronautics in Great Britain*, OUP, 1924, p103
19 Rolt, p31
20 WH Letters, p67, to Franklin, 9 December 1783
21 Ibid., p62, to Franklin, 13 September 1783
22 Ms Album of balloon accounts, British Library catalogue 1890.e.15. See also WH Correspondence 2, p304, Blagden to Banks, 16 September 1784; and Hodgson, p97, footnote
23 Hodgson, p66
24 Samuel Johnson to Hester Thrale, 22 September 1783, *Collected Letters*, vol 4, pp203–4
25 WH Mss 6280, Watson, letter 9 November 1783
26 Horace Walpole, letter to H. Mann, 2 December 1783; see Rolt, p159 and Hodgson, p190
27 Joseph Franklin, letters to Banks, 21 November 1783 and 16 January 1784; see Rolt, p158
28 Gilbert White, 19 October 1784, in *Life and Letters of Gilbert White*, vol 2, pp134–6. See also Richard Mabey, *Gilbert White*, pp195–6. The solo pilot was in fact the Frenchman Jean-Pierre Blanchard
29 Charles Burney, letter, September 1783. See Roger Lonsdale, *Charles Burney*, p385
30 Rolt, p60
31 Horace Walpole, June 1785, from Hodgson, p203
32 Rolt, p65
33 Sophia Banks Ms album, BL 1890.e.15. See also Hodgson, p97, footnote, and broadsheet poem 'The Ballooniad' (1784)
34 Portrait of Lunardi reproduced in *Catalogue of Well-Known Balloon Prints and Drawings*, Sotheby's, 1962, p42. See also 'Le triomphe de Lunardi', a series of six allegorical paintings by Francesco Verini, c.1787, held at Musée de l'Air, Le Bourget
35 Account assembled from Vincent Lunardi, *My First Aerial Voyage in London*, 1784; see also Lunardi, *Five Aerial Voyages in Scotland*, 1785
36 Lesley Gardiner, *Vincent Lunardi*, 1963, pp53–60
37 Amanda Foreman, *Georgiana Duchess of Devonshire*, HarperCollins, 1998, p173
38 Gardiner, p56
39 Charles Burney, letter 24 September 1784, in Lonsdale, 1965, p365
40 Gardiner, p59

41 Johnson, 13 September 1784, *Collected Letters of Samuel Johnson*, edited by Bruce Redford, vol 4, p404
42 Johnson, 18 September 1784, ibid., p407
43 Ibid., p408
44 Johnson, 29 September 1784, ibid., pp408–9
45 Johnson, 6 October 1784, ibid., p415
46 The glamorous threesome were celebrated in a famous coloured lithograph by John Francis Rigaud, *Captain Vicenzo Lunardi, Assistant Biggin and Mrs Sage in a Balloon*, now held in the Yale Center for British Art. In the event, only two actually took off.
47 Mrs Sage, *A Letter by Mrs Sage, the First English Female Aerial Traveller, on Her Voyage in Lunardi's Balloon*, 1785. British Library catalogue 1417.g.24
48 Gardiner, p60
49 Ibid., p44. On p77 she also describes ascending through a snow cloud
50 Tiberius Cavallo, *History and Practice of Aerostation*, 1785
51 Gardiner
52 Kirkpatrick to William Windham, in Hodgson, pp147–8
53 Hodgson, pp143–4
54 Johnson, 17 November 1784, *Letters*, p438
55 Johnson's gift is confirmed in James Sadler's memoir, *Balloon: Aeriul Voyage of Sadler and Clayfield*, 1810. See also Hodgson, pp150, 403n
56 See Foreman and Hodgson
57 John Jeffries, *Narrative of Two Aerial Voyages with M. Blanchard as Presented to the Royal Society*,1786. 'The First Voyage', pp10–11 (the 'Second Voyage' being the historic Channel Crossing). British Library catalogue 462.e.10 (8)
58 Jeffries, *Two Aerial Voyages*, pp55–65
59 Ibid.; but also drawn from a slightly racier account published exclusively for American readers as 'The Diary of John Jeffries, Aeronaut: The First Aerial Voyage across the English Channel', in *The Magazine of American History*, vol XIII, January 1885, and supplied to me as a pamphlet reprint (1955) by the Wayne County Library, USA
60 Photograph supplied by Musée de l'Air, Le Bourget, Paris
61 Jeffries, Diary, p16
62 Jeffries, *Two Aerial Voyages*, p69
63 Jeffries, Diary, p21
64 Erasmus Darwin, *The Botanic Garden*, 1791, Part I, Canto IV (Air), lines 143–76, footnote on Susan Dyer
65 Rolt, p91
66 Darwin, *The Botanic Garden*, Part I, Canto IV (Air), lines 143–76
67 Rolt, pp 99–104
68 James Sadler, *An Authentic Account of the Aerial Voyage*, 1810; see Hodgson, p150
69 Reproduced in Henry Beaufoy, 'Journal Kept by HBHS during an Aerial Voyage with Sadler from Hackney', British Library catalogue B.507 (1); see also Hodgson, fig 36
70 James Sadler, *Across the Irish Channel*, 1812, p16
71 Ibid., p23

72 See Holmes, *Shelley: The Pursuit*, 1974, p149
73 Windham Sadler, *Aerostation*, 1817. British Library catalogue RB.23.a.23973
74 Windham Sadler, 'Progress of Science, while Ballooning neglected', an Appendix to *Aerostation*,1817, p16
75 Richard Hamblyn, *The Invention of Clouds*, 2000, which includes beautiful illustrations of Howard's cloud paintings. Gavin Pretor-Pinney, *The Cloudspotter's Guide*, 2006, suggests cloud study as both a science and an entire philosophy of life
76 Carl Grabo, *A Newton Among Poets: Shelley's Use of Science in Prometheus Unbound*, North Carolina UP, 1931
77 Erasmus Darwin, 'The Loves of the Plants', 1789, from Part II of *The Botanic Garden*
78 Coleridge *Notebooks I*, entry for 26 November 1799; see Holmes, *Coleridge: Early Visions*, pp253–4
79 Wordsworth, *Peter Bell*, 1819, stanza 1, lines 5–6
80 Shelley at University College, Oxford in 1811, as recalled by T.J. Hogg in 'Shelley at Oxford', *New Monthly Magazine*, 1832; republished in his *Life of P.B. Shelley*, 1858

Chapter 4: Herschel Among the Stars

1 WH Mss W.1/5.1; and see 'Description of a Forty-Foot Reflecting Telescope', 1795, WH Papers 1, pp485–527 (with magnificent engravings of the telescope, the gantry, the moving mechanisms and the zone clocks and bells)
2 Michael Hoskin, *The Herschel Partnership as Viewed by Caroline*, Science History Publications, Cambridge, 2003, p79
3 J.A. Bennett, 'The Telescopes of William Herschel' (with illustrations), *Journal for the History of Astronomy*, 7, 1976
4 Hoskin, p79
5 WH Mss W.1/5.1; further details in 'Astronomical Observations' (1814), WH Papers 2, p536, footnote
6 Hoskin, p81
7 *Journal of Mrs Papendiek*, in WH Chronicle, p174
8 WH Chronicle, p145
9 WH Chronicle, p152
10 *Journal of Mrs Papendiek*, in WH Chronicle, pp145–6
11 Ordinance Survey map, Royal Berkshire, 1830, reproduced in Hoskin, p58
12 CHA, p81
13 WH Chronicle, p172
14 John Adams, April–May 1756, *Diaries and Autobiography*, edited by L.H. Butterfield, 1964
15 CHA, p83
16 Ibid.
17 CHA, p86
18 CHA, p89
19 Sketch of 'small' sweeper in CHA, p70
20 Michael Hoskin, 'Caroline Herschel's Comet Sweepers', *Journal for the History of Astronomy*, 12, 1981; and CHA, p70
21 WH Mss C1/1.1, 34–5; and CHA, p88
22 CHA, pp89–90

23 James Thomson, 'Summer', lines 1,724–8, from *The Seasons*, 1726–30
24 Claire Brock, *The Comet Sweeper*, Icon Books, Cambridge, pp150–1
25 WH Mss 6267 1/1/3, for 2 August 1786
26 WH Mss 6267 1/1.1. Memorandum made 2 August 1786
27 Hoskin, p85
28 CHM, p68
29 WH Papers 1, pp309–10
30 Howse, *Maskelyne*, p155
31 Hoskin, p83
32 Fanny Burney, *Diary*, September 1786, from WH Chronicle, p169
33 Ibid.
34 Ibid., pp169–70
35 Ibid.
36 Sophie von La Roche, *Diary*, 14 September 1786, from Brock, pp154–5
37 WH Chronicle, p252
38 Nevil Maskelyne, 6 December 1793; see CHA, p70
39 Pierre Méchain, 28 August 1789; see WH Chronicle, p219
40 Hoskin, pp103–7
41 WH Chronicle, p171
42 CHA, p91
43 CHM, p209
44 CHM, p309
45 Hoskin, p87
46 WH Mss 6278 1/5; and Hoskin, p88
47 CHM, p274; see Patricia Fara, *Pandora's Breeches*, 2004
48 Hoskin, p88
49 Ibid., p90
50 CHM, p209
51 WH Mss 6280; and Hoskin, p89
52 CHM, p211
53 Hoskin, pp88–90
54 CHA, p94
55 Ibid.
56 CHM, p308
57 WH Chronicle, p172
58 OS map from Hoskin, p58
59 *Journal of Mrs Papendiek*, WH Chronicle, p174
60 WH Archive: miniature on ivory of Mary Herschel by J. Kernan, 1805; also reproduced in Hoskin, p97
61 Hoskin, pp91–4
62 WH to Alexander, 7 February 1788, from WH Chronicle, p178
63 Hoskin, p92
64 *Journal of Mrs Papendiek*, WH Chronicle, p174
65 Ibid.
66 CHM, p178
67 WH Chronicle, p175
68 CHM, p79
69 CHA, p96
70 CHM, p79
71 WH Mss 6268 4/3
72 CHA, p57
73 CHM, pp78, 96
74 WH Chronicle, p177
75 Simon Schaffer, 'Uranus and Herschel's Astronomy', *Journal for the History of Astronomy*, 12, 1981, p22
76 Hoskin, p106
77 CHM, p83
78 CHM, p82
79 'Description of a Forty Foot reflecting Telescope' (June 1795), WH Papers 1, pp486, 512–26
80 Ibid.
81 WH Chronicle, p168
82 Ibid.
83 CHM, p168
84 Hoskin, p111
85 Ibid.
86 WH Papers 2 (1815), pp542–6

87 'Catalogue of a Second Thousand Nebulae', 1789, WH Papers 1, pp329–37
88 Simon Schaffer, 'On the Nebular Hypothesis', in *History, Humanity and Evolution*, edited by J.R. Moore, 1988
89 Hoskin, p167
90 Broadsheet cartoon by R Hawkins, Soho, February 1790; reproduced in Hoskin, p107
91 CHM, p95
92 Ibid.
93 CHM, p96
94 Ibid.
95 CHM, p98
96 CHA, p123
97 Barthélemy Faujas de Saint-Fond, *Travels in England and Scotland for the Purpose of Examining the Arts and the Sciences*, vol 1, 1799, pp65–78; see Brock, p173
98 WH Papers 1, p423
99 Erasmus Darwin, *Botanic Garden*, Part I, Canto IV (Air), lines 371–88
100 Ibid., note to line 398
101 Crowe, 1986, pp79–80
102 Pierre Laplace quoted in Simon Schaffer, 'On the Nebular Hypothesis', op. cit.
103 Quoted in Crowe, 1986, p78
104 'On the Nature and Construction of the Sun', 1795, WH Papers 1, pp470–84; and 'Observations tending to investigate the Nature of the Sun', 1801, WH Papers 2, pp147–80. See also discussion in Crowe, 1986, pp66–7
105 See Vincent Cronin, *The View of the Planet Earth*, 1981, p173
106 'On the Solar and Terrestrial Rays that occasion Heat', 1800, WH Papers 2, pp77–146; see Hoskin, p99
107 Humphry Davy to Davies Giddy, 3 July 1800, in J.A. Paris, *Davy*, vol 1, p87
108 Hoskin, p101
109 *British Public Characters of 1798*, 1801, British Library catalogue 10818.d. I
110 WH Chronicle, pp309–11; Beattie, *Life of Campbell*, 1860, vol 2, pp234–9; Sime, pp206–9
111 Hoskin, p106
112 CHM, pp259–60
113 CHM, p259
114 Gunther Buttman, *Shadow of the Telescope*, 1974, p8
115 This wooden plane can be seen in the Herschel House Museum, Bath
116 Buttman, op. cit., p11
117 WH Chronicle, p281
118 Michael Hoskin, *William Herschel and the Construction of the Heavens*, 1963, p130
119 WH Chronicle, pp278–9
120 WH Papers 2, 'On the Proper Motion of the Solar System'
121 WH Papers 2, pp460–97, with illustrations of different nebulae shapes
122 WH Papers 2, 'Astronomical Observations', 1811, p460; and discussed by Armitage, *Herschel*, pp117–20; and Hoskin, *Stellar Astronomy*, 1982, p152
123 WH Papers 1, 'The Construction of the Heavens', 1785; and WH Chronicle, p183
124 Byron, *Detached Thoughts*, 1821
125 Byron, *Letters*, to Piggot, December 1813; and Crowe, *Extraterrestrial*, p170
126 Bonnycastle, *Astronomy*, 1811, Preface, ppv–vi

127 Charles Cowden Clarke, *Recollections*, 1861; see also Andrew Motion, *Keats*, pp108–12
128 I owe this vivid suggestion to Dr Percy Harrison, Head of Science, Eton
129 The idea of a sacred, piercing moment of vision into the true nature of the cosmos is also traditional in earlier eighteenth-century poetry. See the strange prose poem by the Northumberland rector James Hervey, *Contemplations on the Night*, 1747
130 Simon Schaffer, 'Herschel on Matter Theory', *Journal for the History of Astronomy*, June 1980
131 WH Papers 2, pp520–41; and WH Chronicle, p287
132 WH Papers 2, p541
133 William Whewell, *On the Plurality of Worlds*, 1850, edited by Michael Crowe, 2001
134 Herschel to Banks, 10 June 1802, in JB Correspondence 5, p199, where Herschel offers the term 'asteroid' reluctantly – 'not exactly the thing we want' – from a suggestion by the antiquary Rev Steven Weston, though fully aware that the recently discovered Pallas and Ceres were not 'baby stars'. The usage is nonetheless dated to Herschel 1802 by the *OED*.
135 Thomas Campbell quoted in WH Chronicle, p335
136 David Brewster, *Life of Sir Isaac Newton*, 1831

Chapter 5: Mungo Park in Africa

1 Sir Harold Carter, *Sir Joseph Banks 1743–1820*, British Museum, Natural History, 1988, p425; and Gascoigne, *Banks and the Enlightenment*, p19
2 JB Letters, p609n; and Hector Cameron, *Sir Joseph Banks*, 1952, p144
3 Cameron, p88
4 As described in Anthony Sattin, *The Gates of Africa: Death, Discovery and the Search for Timbuktu*, HarperCollins, 2003
5 *The Life of Mungo Park*, by HB (anon), 1835, p284
6 Sattin, pp134–6
7 Ibid., pp136–7
8 Mungo Park, *Travels in the Interior of Africa*, 1799, 1860. The edition used here is *Travels*, Nonesuch, 2005, p16
9 Sattin, p140
10 *Travels*, p19
11 Ibid., p31
12 Sattin, p143
13 Banks to Park, winter 1795, in ibid., p141
14 *Travels*, p95
15 Ibid., p98
16 Ibid., p138
17 Ibid., p141
18 Ibid.
19 *The Life of Mungo Park*, by HB (anon),1835, pp289–90; also Sattin, p168
20 *Travels*, pp168–9
21 Ibid., p169
22 Samuel Taylor Coleridge, *The Ancient Mariner*, 1798, Part IV
23 Joseph Conrad, *Geography and Some Explorers*, 1924, pp28–9
24 JB Correspondence 4, Banks to Sir William Hamilton, 14 March 1798, p540
25 Ibid., no.1484, Banks to Johann Blumenbach, 19 September 1798, p554

26 Ibid., no.1513, Blumenbach to Banks, 12 June 1799, p590
27 Walter Scott's meeting with Park 1804; described in *The Life of Mungo Park*, by HB (anon), 1835, 'Addenda'; and Sattin, p235
28 JB Letters, no. 78, Banks to Lord Liverpool, 8 June 1799, p209
29 Kenneth Lupton, *Mungo Park African Traveller*, OUP, 1979, p146. Lupton was the one-time District Officer at Boussa, and knew the African locations well
30 Ibid., p158
31 *Travels*, 'Journal of Second Journey', pp264–5
32 Ibid., p271
33 Park Mss, Martyn to Megan, 1 November 1805, BL Add Mss 37232.f63
34 *Travels*, 'Journal of Second Journey', p272
35 Park Mss, Park to Lord Camden, 17 November 1805, BL Add Mss 37232.f65; see also Park's letter to Allison Park's father, 10 November 1805, BL Add Mss 33230.f37; and Lupton, p175
36 *Travels*, p274
37 Park Mss, Park to Joseph Banks, 16 November 1805, BL Add Mss 37232.k.f64
38 Alfred Tennyson, 'Timbucto' (poem), 1827
39 Lupton, 'Appendix of Later Accounts' from Isaaco, Amadi Fatouma, Richard Lander and several subsequent Niger explorers
40 Thomas Park to Allison Park, dated Accra September 1827, from Joseph Thomson, *Mungo Park and the Niger*, 1890, pp241–2
41 Richard Lander's report 1827, reprinted in Stephen Gwynn, *Mungo Park and the Quest for the Niger*, 1932, p233
42 Percy Bysshe Shelley, *Alastor, or The Spirit of Solitude*, 1815, lines 140–9
43 Thomas Love Peacock, *Crotchet Castle*, 1830; see Holmes, *Shelley: The Pursuit*, 1974, p292
44 See William Feaver, *The Art of John Martin*, Oxford, 1975; and discussion in Tim Fulford (editor), *Literature, Science and Exploration in the Romantic Era*, 2004, pp97–107
45 '[Ritchie] is going to Fezan in Africa there to proceed if possible like Mungo Park', John Keats to George Keats, 5 January 1818; 'Haydon showed me a letter he had received from Tripoli... Ritchie was well and in good spirits, among Camels, Turbans, Palm trees and sands...', Keats to George Keats, 16–31 December 1818

Chapter 6: Davy on the Gas

1 Described in Davy's letters to his mother Grace Davy, in June Z. Fullmer, *Young Humphry Davy*, American Philosophical Society, 2000, pp328–32
2 JD Fragments, pp2–5
3 Thomas Thorpe, *Humphry Davy, Poet and Philosopher*, 1896, p10
4 Anne Treneer, *The Mercurial Chemist: A Life of Sir Humphry Davy*, 1963, p6
5 Local sources, author's visit to Penzance, May 2006
6 Ibid.
7 JD Memoirs, p68

8 There are various versions of this early poem in the HD Archive: see Paris, vol 1, p29; Treneer, pp4–5; or Fullmer, p13
9 Treneer, p16
10 John Davy quoted in ibid., p21
11 Ibid.
12 Introduction to *Humphry Davy on Geology: The 1805 Lectures*, pxxix, British Library catalogue X421/22592
13 HD Archive Box 13 (f) pp41–50, Mss notebook dated 1795–97
14 HD Archive Box 13 (f) p61
15 The whole poem, no fewer than thirty-two stanzas, is given in JD Memoirs, pp23–7
16 HD Works 2, p6
17 Jan Golinski, *Science as Public Culture: Chemistry and Enlightenment in Britain 1760–1820*, CUP, 1992, pp133–42
18 Ibid., p109
19 Johann Wolfgang von Goethe, 'Maxims and Reflections', from Goethe, *Scientific Studies*, edited by Douglas Miller, Suhrkamp edition of Goethe's *Works*, vol 12, New York, 1988, p308
20 Reprinted in HD Works 9
21 See Madison Smartt Bell, *Lavoisier in the Year One: The Birth of a New Science in the Age of Revolution*, Atlas Books, Norton, 2005. See also J.-L. David's famous romantic portrait, *Antoine Laurent de Lavoisier et sa Femme* (1788)
22 Preface to *Traité Élémentaire*, translated by Robert Kerr, 1790
23 *Consolations*, Dialogue V, in HD Works 9, pp361–2
24 JD Memoirs, p34
25 For the Watt family, see Jenny Uglow, *The Lunar Men: The Friends who Made the Future, 1730–1810*, Faber, 2002
26 Treneer, p24
27 From Beddoes notes made 1793, quoted in Golinski, p171
28 HD Mss Truro, Beddoes letter in Davies Giddy Mss DG 42/1
29 Ibid.
30 Dorothy A. Stansfield, *Thomas Beddoes MD: Chemist, Physician, Democrat*, Reidel Publishing, Boston, 1984, pp162–4
31 HD Mss Truro, Davies Giddy Mss DG 42/8
32 HD Mss Truro, Davies Giddy Mss DG 42/4
33 See Holmes, *Coleridge: Early Visions*
34 John Ayrton Paris, *The Life of Sir Humphry Davy*, 2 vols, 1831, vol 1, p38
35 See David Knight, *Humphry Davy: Vision and Power*, Blackwell Science Biographies, 1992
36 Richard Lovell Edgeworth 1793, quoted in Fullmer, p106
37 Treneer, pp30–1
38 HD Archive Notebook 20a; and Fullmer, p169
39 HD Works 2, p85
40 HD Works 2, p84
41 HD Works 2, pp85–6; see HD Archive Ms Notebook B (1799)
42 HD Archive Mss Box 13(h) pp15–17 and Box 13(f) pp33–47
43 See Fullmer, pp163–6
44 From author's visit and photographs, May 2006. See also John Allen, 'The Early History of Varfell', in *Ludgvan*, Ludgvan Horticultural Society, no date
45 Golinski, pp157–83

46 Reply from James Watt, Birmingham, 13 November 1799, in JD Fragments, pp24–6
47 HD Works 3, pp278–9
48 HD Works 3, pp278–80; on Davy's impetuosity and courage see Oliver Sacks, *Uncle Tungsten: Memories of a Chemical Boyhood*, Picador, 2001
49 Joseph Cottle, *Reminiscences*, vol 1, 1847, p264
50 HD Works 3, pp246–7; James Watt, Birmingham, 13 November 1799, in JD Fragments, pp24–6; equipment partly illustrated in Fullmer, p216
51 Treneer, p72
52 Fullmer, p213
53 Ibid., p214
54 HD Works 3, p272
55 HD, *Researches Chemical and Philosophical chiefly concerning Nitrous Oxide*, London, 1800, p461. See HD Works 3
56 JD Life 1, pp79–82
57 HD Archive Mss Box 13 (c) pp5–6; and Fullmer, p215
58 Treneer, p47
59 HD Archive Mss Box 20 (b) p118
60 HD Archive Mss Box 20 (b) p120
61 HD, *Researches*, 1800, p491
62 Ibid., p492; discussed in Cartwright, pp237–8
63 HD Works 9, pp74–5; comments by Physicus, Day 4, in *Salmonia*, 1828
64 Fullmer, p218
65 Cartwright on Anaesthetics, 1952, pp100–23; Treneer, pp40–8
66 HD Archive Mss Box 20(b) p208
67 HD Archive Mss Box 20 (b) p209
68 HD *Researches*, 1800, pp100–2
69 A premonition of Frankenstein! HD *Researches*, 1800, p102
70 Southey to Tom Southey, 1799, from Treneer, p44
71 *A Memoir of Maria Edgeworth*, edited by her children, 1867, vol 1, p97
72 Treneer, p45
73 Ibid., p43
74 Ibid., p54
75 Southey to William Wynn, 30 March 1799
76 'Unfinished Poem on Mount's Bay', in Paris, vol 1, pp36–9
77 JD Fragments, pp34–5
78 Ibid., pp37–9
79 JD Life 1, p119
80 Treneer, p44
81 Holmes, 'Kubla Coleridge', in *Coleridge: Early Visions*, 1989
82 'Detail of Mr Coleridge', *Researches*, 1800, and HD Works 3, pp306–7
83 Coleridge to Davy, 1 January 1800, *Coleridge Collected Letters*, edited by E.L. Griggs, vol 1; and see Treneer, p58
84 JD Memoirs, pp58–9
85 JD Fragments, p24; Fullmer, pp269–70
86 HD Works 3, pp289–90; and compare Fullmer, pp269–70
87 HD Archive Mss Box 20 (b) pp129–34, dated 26 December 1799
88 HD Archive Mss Box 20 (b) p95
89 JD Memoirs, pp59–66
90 Ibid., pp66–7
91 HD Works 3; Fullmer, p211
92 HD Works 3, pp1–3
93 JD Memoirs, pp54–5
94 Preface to *Researches*, 1800, HD Works 3, p2
95 Joseph Cottle, *Reminiscences of S.T. Coleridge and Robert Southey*, 1847

96 Treneer, p48
97 *The Sceptic*, anon, 1800, British Library catalogue Cup.407.gg.37
98 Golinski, p173
99 Ibid., p153
100 Treneer, p63
101 Paris, vol 1, p58
102 Trevor H. Levere, *Poetry Realized in Nature: Coleridge and Early Nineteenth Century Science*, CUP, 1981, p32
103 See Coleridge to Davy, six letters, 9 October 1800–20 May 1801, *Coleridge Collected Letters*, edited by E.L. Griggs, vols 1–2; see Treneer, pp67–8
104 Coleridge to Davy, 9 October 1800
105 Holmes, p247
106 Coleridge, letter to Davy, 15 July 1800, *Collected Letters*, vol 1, p339. He also added in a chemical vein: 'I would that I could wrap up the view from my House [Greta Hall] in a pill of opium, & send it to you!'
107 Southey to William Taylor, 20 February 1800; from Fullmer, p148
108 Southey to Coleridge, 3 August 1801; from ibid., pp148–9
109 JD Fragments, pp29–30
110 'On the Death of Lord Byron', 1824, Davy, *Memoirs*, pp285–6
111 HD Works 8, p308
112 Fullmer, pp328–32
113 The most revealing evidence is the unpublished letter Anna Beddoes wrote to Davy on 26 December 1806, HD Archive Mss Box 26 File H 9
114 Fullmer, p82
115 Ibid., p281
116 Verse fragments from HD Archive, Ms Notebook 13 J; Box 26 File H; and Fullmer, pp106–8
117 HD Archive Mss Box 26 File H 7
118 HD Archive Mss Box 26 File H 6, 13 and 14
119 HD Mss Bristol, Davy to John King, 14 November 1801, Ms 32688/33
120 HD Archive Mss Box 13 (g) p116
121 HD Archive Mss Box 13 (g) p158
122 See Stansfield, pp 234–5. Some more light is thrown on Anna's enigmatic and volatile character by A.C. Todd, 'Anna Maria, Mother of Thomas Lovell Beddoes', in *Studia Neophilologica*, 29, 1957
123 'Glenarm, by moonlight, August 1806', HD Archive Mss Box 13 (g) p166; printed in JD Memoirs, pp50–1
124 HD Archive Mss Box 26 File H 9 and 10
125 JD Fragments, p150
126 Coleridge to Southey, 1803; see Treneer, p114
127 Treneer, p78
128 JB Correspondence 4, letters 1290–6, cover an exchange between Banks, James Watt and the Duchess of Devonshire about the viability of Dr Beddoes's scheme in December 1794
129 HD Works 3, p276
130 F.F. Cartwright, *The English Pioneers of Anaesthesia*, 1952, p311
131 HD, *Researches*, 1800, p556; and HD Works 3, p329
132 Holmes, pp222–7
133 Coleridge to Davy, 2 December 1800, *Collected Letters*, vol 1, p648
134 Paris, vol 1, p97
135 Cartwright, p320
136 *Bristol Mirror*, 9 January 1847, from ibid., p317

137 JD Memoirs, pp80–1
138 *Philosophical Magazine*, May–June 1801, from Treneer, p78
139 David Knight, essay in the *Oxford Dictionary of National Biography*. It is curious that no essential improvement has taken place in the design of chemical batteries since the nineteenth century, and this is currently the greatest single obstacle to the efficient global use of solar energy from solar panels. (Conversation with Richard Mabey on the banks of the river Waveney, midsummer's day 2008.)
140 Dorothy A. Stansfield, *Thomas Beddoes MD: Chemist, Physician, Democrat*, Reidel Publishing, Boston, 1984, pp120, 234–42; also J.E. Stock, *Memoirs of Thomas Beddoes*, 1811
141 HD Mss Bristol, Davy to John King, 22 June 1801, Ms 32688/31
142 HD Mss Bristol, Davy to John King, 14 November 1801, Ms 32688/33
143 Ibid.
144 Coleridge, *Letters*, 1802
145 HD Works 2, pp311–26
146 Ibid., p314
147 Ibid. pp318–19
148 Ibid., p321
149 Ibid., p323
150 Ibid.
151 Ibid., p326
152 Preface, *Lyrical Ballads*, 1802. See discussion in Mary Midgley, *Science and Poetry*, Routledge, 2001
153 Maria Edgeworth, letter, 8 October 1802; from Lamont-Brown, p59
154 HD Archive Mss Box 13c p32; and Golinski, pp194–7
155 Coleridge to Southey, 17 February 1803, *Collected Letters*, vol 2, p490
156 Davy to Coleridge, March 1804; see Holmes, p360
157 Paris, vol 2, pp198–9
158 Ibid., p199
159 See Nicholas Roe, *Samuel Taylor Coleridge and the Sciences of Life*, 2001, pp142–4
160 Partly reprinted in HD Works 5 and 8; lucidly discussed in Harold Hartley, *Humphry Davy*, Open University, 1966, pp50–74; and Oliver Sacks, *Uncle Tungsten*
161 JD Memoirs, pp116–17
162 'Introduction to Electro-Chemical Science', originally delivered March 1808, HD Works 8, pp274–305
163 HD Works 8, p281
164 HD Works 8; see Hartley, pp50–4
165 Treneer, p111
166 HD Works 5, pp59–61
167 Hartley, p56
168 Beddoes, 17 November 1808, from Stansfield, p239
169 Henry Brougham, 'Three essays on Humphry Davy', *Edinburgh Review*, 1808, vol 11: first pp390–8; second pp394–401; third pp483–90
170 Coleridge to Tom Poole, 24 November 1807
171 Treneer, p104
172 JD Memoirs, p117; HD Works 8, p355
173 HD Archive, quoted in Holmes, *Coleridge: Darker Reflections*, p119
174 'Written after Recovery from a Dangerous Illness', printed in JD Memoirs, pp114–16
175 *Consolations in Travel*, 1830, Dialogue II, HD Works 9, pp254–5
176 Ibid., p255
177 JD Memoirs, pp394, 397

178 *Consolations*, Dialogue II, HD Works 9, pp254–5. The story of Josephine Dettela, 1827–29, will be continued in my Chapter 9
179 Stansfield, pp194–5
180 Davy to Coleridge, December 2008, *Collected Letters*, vol 3, pp170–1; Treneer, p113
181 Stansfield, p 247
182 HD Archive Mss Box 14 (i), note dated February 1829, Rome. See also Stansfield, p249
183 *British Public Characters, 1804–5* (1809), British Library catalogue 10818.d. 1
184 Anna Barbauld, 'The Year 1811' (1812)
185 Coleridge's note, 1809, in *Notebooks*, vol 2, entry no. 1855
186 HD Works 8, p354

Chapter 7: Dr Frankenstein and the Soul

1 Fanny Burney, 'A Mastectomy', 30 September 1811, in the *The Journals and Letters of Fanny Burney (Madame d'Arblay)*, vol 6, edited by Joyce Hemlow, Oxford, 1975, pp596–616
2 Ibid., p600, footnote
3 Druin Burch, *Digging up the Dead: The Life and Times of Astley Cooper*, Chatto & Windus, 2007, p179. Besides much else, Burch has a chastening section on concepts of pain endurance, anaesthesia and surgery at this period, pp172–82
4 JB Correspondence 5, no. 1616
5 Sharon Ruston, *Shelley and Vitality*, Palgrave, 2005, p39
6 See Holmes, *Coleridge: Darker Reflections*, 1998
7 John Hunter, 1794, from Ruston, p40
8 John Abernethy, *Enquiry into Mr Hunter's Theory of Life: Two Lectures*, 1814 and 1815, p38; and Ruston, p43
9 Abernethy, *Enquiry*, pp48–50
10 Ruston, p45
11 Gascoigne, *Banks and the English Enlightenment*, pp157–9
12 See Tim Fulford, Debbie Lee and Peter J. Kitson, 'Exploration, Headhunting and Race Theory', in *Literature, Science and Exploration in the Romantic Era*, CUP, 2004
13 Holmes, *Shelley: The Pursuit*, p 290
14 See *Shelley's Prose*, edited by David Lee Clark
15 Holmes, *Shelley*, pp286–90; also Ruston, pp91–100
16 Ruston, p193
17 William Lawrence, *Natural History of Man*, 1819, pp6–7
18 William Lawrence, *Introduction to Comparative Anatomy*, 1816, pp169–70; and Ruston, p50
19 William Lawrence: *The Natural History of Man* (Lectures on Physiology and Zoology), 1819, p106
20 Ibid., p8; and Ruston, pp15–16
21 Lawrence, *Introduction to Comparative Anatomy*, p174; and Ruston, p16
22 In his letters of 1797–98, and later Notebooks. See Holmes, 'Kubla Coleridge', in *Coleridge: Early Visions*
23 Hermione de Almeida, *Romantic Medicine and John Keats*, OUP, 1991, pp66–73
24 Holmes, 'The Coleridge Experiment', *Proceedings of the Royal Institution*, vol 69, 1998, p312

25 Nicholas Roe, 'John Thelwall's Essay on Animal Vitality', in *The Politics of Nature*, Palgrave, 2002, p89
26 Burch, *Digging up the Dead*, 2007
27 Thelwall, 'Essay towards a Definition of Animal Vitality', 1793, quoted in Nicholas Roe, *The Politics of Nature*, pp89–91
28 Blagden to Banks, 27 December 1802, JB Correspondence 5, no. 1704
29 G Aldini, *An Account of the Late Improvements in Galvanism... Containing the Author's Experiments on the Body of a Malefactor Executed at Newgate*, London, 1803; see Fred Botting (editor), *New Casebooks: Frankenstein*, Palgrave, 1995, p125
30 *Quarterly Review*, 1819, from *Frankenstein*, Oxford World Classics, pp243–50
31 B.R. Haydon, *Diary*, 1817; Penelope Hughes-Hallett, *The Immortal Dinner*, 2000; Mary Midgley, *Science and Poetry*, pp50–5
32 Quoted by Burch, pp154–5. For a darker view of dissection see Helen MacDonald, *Human Remains: Dissection and its Histories*, Yale UP, 2006
33 Holmes, *Shelley: The Pursuit*, pp360–1
34 'Theory of Life' (1816), in *Coleridge: Shorter Works and Fragments*, edited by H.J. and J.R. Jackson, vol 1, Princeton, 1995, p502
35 Holmes, *Coleridge: Darker Reflections*, 1998, p479
36 Hermione de Almeida, *Romantic Medicine and John Keats*, p102
37 Coleridge to Wordsworth, 30 May 1815, *Coleridge Collected Letters* 4, pp574–5
38 Richard Burton quoted in Andrew Motion, *Keats*, p430
39 John Keats, 'Lamia' (1820), lines 229–38
40 Ibid., lines 47–60
41 Ibid., lines 249–53
42 Ibid., lines 146–60
43 Davy's 'Discourse Introductory to Lectures on Chemistry, 1802, HD Works 2, pp311–26
44 *Frankenstein*, 1818, Chapter 2, Penguin Classics
45 *Mary Shelley's Journal*, 25 August–5 September 1814
46 In September 1815 at Great Marlow; see Holmes, *Shelley*, p296
47 Mary Shelley, 'Introduction' to *Frankenstein* 1831 text
48 *Frankenstein*, 1818, Chapter 1, Penguin Classics
49 JB Correspondence 5, no. 1804
50 J.H. Ritter as featured in www.CorrosionDoctors
51 Walter Wetzels, 'Ritter and Romantic Physics', in *Romanticism and the Sciences*, edited by Cunningham and Jardine, 1990. The best account of the extraordinary writer Novalis appears in Penelope Fitzgerald's inspired novel *The Blue Flower*, 1995
52 JB Correspondence 5, no. 1748, pp316–17
53 Ibid., no. 1790, p368
54 Ibid., no. 1799, p387
55 For a wider perspective see 'Death, Dying and Resurrection', in Peter Hanns Reill, *Vitalizing Nature in the Enlightenment*, California UP, 2005, pp171–6

56 *Frankenstein*, 1818, vol 1, Chapter 5, Penguin Classics, p56
57 These connections are further traced by Ruston, pp86–95
58 Lawrence, *Lectures*, 1817, pp6–7
59 *Frankenstein*, 1818, vol 2, Chapter 3, Penguin Classics, pp99–100
60 Ibid., Chapter 8, p132
61 Ibid., Chapter 9, pp140–1
62 Ibid., Chapter 9, p141
63 Ibid., vol 3, Chapter 2, p160
64 Ibid., Chapter 3, p160
65 Ibid., pp164–5
66 *Frankenstein*, 1831 text, pp178, 180, 186. My italics
67 Ibid., p189
68 Text from 1823 leaflet about *Presumption*; see Fred Botting (editor), *New Casebooks: Frankenstein*, Palgrave, 1995. The evolution and impact of the novel is brilliantly disclosed by William St Clair in *The Reading Nation in the Romantic Period*, OUP, 2004
69 Mary Shelley, *The Letters of Mary Wollstonecraft Shelley*, vol 1, edited by Betty T. Bennett, Johns Hopkins UP, 1988, pp369, 378
70 *Frankenstein*, 1818, vol 2, Chapter 5, Penguin Classics, pp116–17
71 Lawrence, *On the Natural History of Man*, 1819, p150
72 Ruston, p71
73 Adrian Desmond, *The Politics of Evolution: Medicine in Radical London*, Chicago, 1989, p112

Chapter 8: Davy and the Lamp

1 Jane Apreece to Walter Scott, 4 March 1811, in 'Lady Davy's Letters', edited by James Parker, *The Quarterly Review*, January 1962; also Lamont-Brown, p94
2 For example: 'Whene'er you speak, Heaven! how the listening throng/ Dwell on the melting music of your tongue! … ' (Valentine's Day 1805), HD Archive Box 26 File H II
3 Treneer, p119
4 See 'iconography' for Lady Davy (Jane Apreece) in *Oxford Dictionary of National Biography*. At the time of going to press I am still searching for a portrait, having exhausted all leads kindly provided by the National Portrait Gallery, London; the Scottish National Portrait Gallery, Edinburgh; and Christie's, London
5 HD Archive Mss Box 25, containing ninety letters from Lady Davy 1811–22
6 HD Archive Mss Box 25/1
7 HD Archive Mss Box 25/3
8 HD Archive Mss Box 25/2
9 Raymond Lamont-Brown, *Humphry Davy: Life Beyond the Lamp*, Sutton, 2004, p94
10 HD Archive Mss Box 25/3; 13; 18; 20
11 HD Archive Mss Box 25/6
12 Coleridge letter of 28 May 1809; also Treneer, p113
13 HD Archive Mss Box 25/5 (1 November 1811)
14 HD Archive Mss Box 25/11; and Treneer, p124
15 HD Archive Mss Box 25/25 (March 1812)
16 HD Archive Mss Box 25/4; also Lamont-Brown, pp96–7
17 HD Archive Mss Box 25/4
18 'Lady Davy's Letters', edited by James Parker, *The Quarterly Review*, January 1962, p81
19 HD Archive Mss Box 25/26
20 HD Archive Mss Box 25/24; further details Lamont-Brown, pp90–105
21 Thorpe, p162

22 Banks to John Lloyd FRS, 31 March 1812; from June Z. Fullmer, 'The Poetry of Sir Humphry Davy', in *Chymia*, 6, 1960, p114
23 Treneer, p126
24 HD Works 2
25 JD Fragments, p158
26 Holmes, *Shelley*, p153
27 Thomas De Quincey, 'The Poetry of Pope', 1848. He gave Newton's *Principia* as an example of Knowledge, and Milton's *Paradise Lost* as example of Power. De Quincey also published a number of essays on scientific subjects, notably 'Animal Magnetism' (1833), 'Kant and Dr Herschel' (1819) and 'The Planet Mars' (1819)
28 HD Works 4, pp1–40
29 Ibid., p20
30 Ibid., pp1–2
31 Golinski, p262
32 *Consolations*, Dialogue V, 'The Chemical Philosopher', HD Works 9
33 Coleridge in Notebook 23 (1812), quoted by Trevor H. Levere, *Chemists in Society 1770–1878*, 1994, pp363–4
34 Coleridge's Marginalia on Jakob Boehme (c.1810–11), from ibid., p357
35 See Coleridge's letter to Lord Liverpool, 28 July 1817, discussing Davy versus Dalton ('atomist'), *Collected Letters*, vol 4, p760
36 JD Fragments, p174
37 Ibid., p175
38 HD Archive Mss Box 25/31
39 Treneer, p134
40 Ibid., p133
41 Ibid., p137
42 Hamilton, pp119, 207
43 Jane Marcet, *Conversations in Chemistry*, 2 vols, 1813, vol 1, p342
44 Treneer, p138
45 HD Archive Mss Box 25/33
46 HD Archive Mss Box 25/27
47 HD Archive Mss Box 25/28
48 HD Archive Mss Box 25/36
49 Kerrow Hill, *The Brontë Sisters and Sir Humphry Davy*, Penzance, 1994, p16
50 HD Archive Mss Box 25/34
51 Paris, vol 2, pp59–72
52 JD Memoirs, p163
53 Michael Faraday, 'Observations on Mental Education', 1859; quoted in James Hamilton, *Faraday: The Life*, HarperCollins, 2002, p1. See also striking portraits and photographs of Faraday dated 1829, 1831 and c.1850 (National Portrait Gallery)
54 Lamont-Brown, pp110–26
55 Paris, vol 1, p261
56 Leigh Hunt, *Examiner*, 24 October 1813
57 JD Fragments, p190
58 Michael Faraday, *Correspondence 1811–1831*, vol 1, edited Frank A.L.J. James, Institute of Electrical Engineers, 1991, p127
59 Maurice Crosland, 'Davy and Gay Lussac', in Sophie Forgan (editor), *Science and the Sons of Genius* (essays), 1980, pp103–8
60 Faraday, *Correspondence*, p124
61 JD Memoirs, pp172–7; and Hartley, p107
62 Hartley, pp107–8
63 Faraday, *Correspondence*, p101
64 HD Works 1, p218
65 Ibid., p217
66 Ibid., p220
67 Faraday, *Correspondence*, p117

68 Ibid., 23 February 1815, p126
69 Treneer, p175; from Ticknor, *Memoirs*
70 HD Works 1, p235
71 Paris, vol 2, p79
72 J.H. Holmes, *Accidents in Coal Mines*, London, 1816, pp141–2
73 'Report of the Select Committee on Accidents in Mines', in *Parliamentary Papers*, 1835, vol 5, September 1835
74 Faraday, *Correspondence*, p136
75 Bence Jones, *Life and Letters of Faraday*, vol 1, p361
76 Paris, vol 2, p95
77 Ibid., p82
78 JB Letters, p317
79 Paris, vol 2, p97
80 Letter to John Hodgson, 29 December 1815, Northumberland Record Office; from Frank A.J.L. James, 'How Big is a Hole? The Problems of the Practical Application of Science in the Invention of the Miners' Safety Lamp by Humphry Davy and George Stephenson in Late Regency England', in *Transactions of the Newcomen Society*, 75, 2005, p197
81 Frank James, pp185–93
82 HD, *On the Safety Lamp, with Some Researches into Flame*, 1818; and HD Works 6, pp12–14
83 HD Works 6, p4
84 Coleridge, *The Friend* (1818 edition), in *The Friend*, vol 1, edited by Barbara E. Rooke, Routledge, 1969, pp 530–1
85 Coleridge, *The Friend* (1809 edition), no. 19, 1809; in *The Friend*, vol 2, edited by Barbara E. Rooke, Routledge, 1969, pp251–2
86 Frank James, p197
87 John Buddle's evidence (2nd day), Report of the Select Committee, 1835, pp153–4
88 HD Works 6, pp116–17
89 Lamont-Brown, p112
90 Thorpe, p203
91 Paris, vol 2, p111
92 'Igna Constructo Securitas ...' Davy's coat of arms illustrated in *The Gentleman's Magazine*, 1829
93 John Playfair, 'Sir Humphry Davy's Lamp', in *Edinburgh Review*, no. LI, 1816, p233; also Thorpe, p204
94 HD Works 6, pp6–7
95 Ibid., p22, footnote
96 Ibid., p4
97 Hamilton, pp121–5; Lamont-Brown, pp128–33
98 James Heaton demonstration at the Society of Arts, 1817, described in Report of the Select Committee, 1835, p213
99 *A Collection of all Letters in Newcastle papers relating to Safety Lamps*, London, 1817. See British Library catalogue Tracts 8708.i.2
100 Letter from George Stephenson, ibid., Tracts 8708.i.2(5)
101 Treneer, p172
102 Lettter to Lord Lambton, October 1816, in Paris, vol 2, p120
103 Frank James, p203
104 Paris, vol 2, p123
105 See Hamilton, pp122–3
106 Frank James, pp183–95
107 HD Works 6
108 Paris, vol 2, p122
109 Ibid., p124–5; and from David Knight, *Davy*, p113
110 HD Works 1, pp209–10
111 Paris, vol 2, p129
112 Treneer, pp173–4; Thorpe, p208

113 Minute Book of Newcastle Literary and Philosophical Society, December 1817, from Frank James, p211
114 'Report of the Select Committee on Accidents in Mines', in *Parliamentary Papers*, 1835, vol 5, September 1835
115 Ibid., pviii
116 Ibid.
117 Davy boys described in ibid., pp97–108, 165–7. See also Samuel Smiles, *Life of George Stephenson*, 1859; and Newcastle Public Record Office
118 Walter Scott, *Journals* 1, 1826, p109
119 JD Fragments, pp141–3
120 Sun Fire Office insurance document, 4 June 1818, found through internet UK Archives Network
121 HD, *On the Safety Lamp for Preventing Explosions*, London, 1825, p151
122 *Consolations*, Dialogue II, HD Works 9, pp254–5
123 Ibid., p255
124 JD Life 2, pp114–15; and JD Memoirs, pp251–3
125 Shelley, *Epipsychidion*, 1820, lines 190–221 (extract)
126 Byron, letter to John Murray, April 1820; see Treneer, p182
127 Byron, *Don Juan* I (1819), stanza 132

Chapter 9: Sorcerer and Apprentice

1 JB Correspondence 6, p286
2 JB, August 1816, ibid., pp208–9
3 Ibid., p382
4 JB, November 1814, ibid., p152
5 Gunther Buttman, *In the Shadow of the Telescope: A Biography of John Herschel*, Lutterworth Press, 1974, p13
6 JB Correspondence 6, p375
7 Ibid.
8 JB Correspondence 6, 1819
9 Coleridge 'Youth and Age' (1825), in *Selected Poems*, Penguin Classics, p215
10 November 1817, JB Correspondence 6, p252
11 Byron, 'Darkness', written at the Villa Diodati, July 1816. See Fiona MacCarthy, *Byron: Life and Legend*, John Murray, 2002, p69; and discussed in *New Penguin Romantic Poetry*, edited by Jonathan and Jessica Wordsworth, Notes to Poems, p909
12 JB Correspondence 6, September and November 1819, pp355, 367
13 Gascoigne, p52
14 JB Correspondence 6, March 1818, p276
15 Ibid., November 1818, p325
16 Ibid., September 1819, p359
17 Byron, *Don Juan* (1821), Canto 10, lines 1–24. The 'glass and vapour' refer to telescopes and steamships, and also possibly balloons. The ringing phrase 'In the Wind's Eye' was used by modern editors as the title of vol 6 of Byron's *Collected Letters*
18 JB Correspondence 6, August 1816, p209
19 Gascoigne, p41
20 Ibid.
21 Buttman, p13
22 CHM, pp119–21
23 John Herschel to Babbage, October 1813, quoted in Buttman, p14
24 William Herschel to John, 10 November 1813, WH Mss 6278 1/11

25 Lady Herschel to John, 14 November 1813, ibid.
26 John Herschel to Babbage, March 1815, quoted by Buttman, p16
27 JB Correspondence 6, p375
28 Shelley, 'Notes to Queen Mab' (1812)
29 Ruston, p154
30 Further discussion in Ruston p208, and Crowe, *Extraterrestrial*, p171
31 Shelley, *Prometheus Unbound*, Act I, lines 163–6
32 Ibid., Act II, lines 52–9
33 Ibid., Act IV, lines 238–44
34 Ibid., lines 457–72
35 Gascoigne, pp257–9
36 JB Correspondence 6, various letters, 1820
37 Gascoigne, pp249–55
38 JB Correspondence 6, August 1819, p352
39 Ibid., November 1819, p367
40 Ibid., February 1820, p379
41 William Edward Parry to 'My Dearest Parents', December 1817; from O'Brian, p300
42 JB Correspondence 6, asking for news of Parry, 1818, pp251, 326, 377
43 Ibid., 20 December 1819, p374. The man was of course John Herschel
44 Ibid., Berthollet to Banks, 27 March 1820, pp383–4
45 See his Will, described in O'Brian, Chapter 12
46 Marie Boas Hall, *All Scientists Now*, 1984, p18
47 Lockhart, *Life of Sir Walter Scott*, vol 2, 1838, pp40–3
48 HD Works 7, pp5–15
49 Ibid., p21
50 JD Life 2, p126
51 Paris, vol 2, p185
52 Faraday, *Correspondence* 1, p183
53 Ibid., pp244–80 passim
54 Hamilton, p192
55 Faraday to Phillips, May 1836, Bence Jones, *Michael Faraday*, 1870, vol 1, pp335–9
56 Discussed in Bence Jones, pp335–9, and James Hamilton, pp186–9
57 Holmes, *Shelley*, p410
58 Hartley, p129
59 Ibid., p130
60 Humboldt, 'Lecture to the Berlin Academy of Sciences', 1805, quoted in Steven Ruskin, *Herschel's Cape Voyage*, 2004
61 Ibid., pp20–2
62 Ibid., p16
63 Many of these instruments, including the 'mountain barometer', in WH Archive; and see Ruskin, p21
64 'The Garden Days: Marlow 1817' in Holmes, *Shelley*. If I had been a novelist I would have described Shelley and Mary making a night visitation to the great forty-foot, and getting Caroline to show them Andromeda and other distant constellations, and planning a comet-flight into deep space. See 'The Witch of Atlas', 1820
65 CHM, p131. The note is actually dated 4 July 1819
66 CHM, p137
67 WH Chronicle, p363. The second translation is mine
68 *Gentleman's Magazine*, September 1822
69 Ibid.
70 Holmes, *Shelley*, p730
71 Sime, pp259–61

72 WH Chronicle, p359
73 CHM, p163
74 CHM, p171
75 WH Chronicle, p366
76 CHM, p167
77 Caroline Herschel to John, April 1827, British Library Ms Egerton 3761.f45/60; and see J.A. Bennett, 'The Telescopes of William Herschel', in *Journal for the History of Astronomy*, 7, June 1976
78 CHM, p163
79 CHM, p 180
80 CHM, p193
81 CHM, p161
82 David S. Evans (editor), *Herschel at the Cape: Letters and Journals of John Herschel*, Texas, 1969, pxxi
83 CHM, p168
84 Thorpe, p222
85 Treneer, p208
86 Ibid., pp206–12
87 Ibid., p208
88 *The Harringtonian System of Chemistry*, 1819, quoted in Golinski, p217
89 'The Humbugs of the Age', in *John Bull Magazine*, 1, 1824, British Library catalogue PP.5950
90 Evans, pxxx
91 Treneer, p207
92 JD Memoirs, p346
93 JD Fragments, p289
94 JD Memoirs, pp334–6
95 HD Works 9, pp13–14
96 *Salmonia*, Day 4, HD Works 9, pp66–7
97 JD Fragments, p258
98 *Salmonia*, Day 4, HD Works 9, p66
99 HD Archive Mss Box 25/51
100 HD Archive Mss Box 25/61
101 *Consolations*, Dialogue IV, HD Works 9, pp314–15
102 Paris, vol 2, p306
103 Tom Poole to John Davy, c.1835, in Paris, vol 2, p307
104 Paris, vol 2, p309
105 HD Archive Mss Box 25/73, 74, 75. On 25 January 1829: 'I hope I may wear on till the spring & see May in Illyria. I have now constant pain in the region of the heart.' Box 25/84
106 HD Archive Mss Box 25/73; and Lamont-Brown, pp157–63
107 HD Archive Mss Box 25/90
108 HD Archive Mss Box 25/74, letter, 2 November 1828
109 HD Archive Mss Box 25/75, letter, 3 December 1828
110 HD Archive Mss Box 26, File B/17
111 HD Archive Mss Box 25/83
112 JD Fragments, p265
113 Davy's two unpublished poems to Josephine Dettela can be found in HD Archive Mss Box 14 (e) pp128–30
114 Based on local information provided by Professor Dr Janez Batis of the Slovenian Academy of Sciences, for David Knight, *Humphry Davy: Vision and Power*, Blackwell Science Biographies, 1992, pp180, 260
115 JD Fragments, p293
116 Thorpe, p232
117 HD Archive Mss Box 25/87a
118 Fullmer, p350
119 *Consolations*, Dialogue I, HD Works 9, p233
120 Ibid., pp233–6
121 Ibid., pp237–8
122 Ibid., p240
123 Ibid., pp239–47
124 Ibid., pp236–47, 266, 274
125 Ibid., Dialogue II, p266
126 Ibid., pp274, 254–6

127 Ibid., Dialogue III, pp302–3
128 Ibid., Dialogue II, pp304–8
129 Ibid., Dialogue III, p309
130 Ibid., p308
131 Ibid., Dialogue IV, p316
132 Ibid., pp320–1
133 Undated extract from Davy's lecture notebooks, JD Memoirs, p147
134 *Consolations*, Dialogue V, HD Works 9, pp361–5
135 Ibid., pp364–6
136 Ibid., pp365–6
137 Ibid., Dialogue VI, p382
138 Janet Browne, *Charles Darwin*, vol 1, 2003, p30
139 JD Memoirs, 1839
140 John Tobin, *Journal of a Tour whilst accompanying the late Sir Humphry Davy*, 1832, p5
141 JD Fragments, p268
142 JD Fragments, to Jane, September 1827, p296
143 HD Archive Mss Box 25/80, to Jane, 1 September 1828
144 John Herschel, *On the Study of Natural Philosophy*, 1830, pp342–4 and footnote
145 HD Archive Mss Box 14 (M) pp105–6
146 JD Fragments, Jane to Davy, late March 1829, p313
147 John Davy's affectionate account, in JD Memoirs, p412
148 JD Memoirs, p408

Chapter 10: Young Scientists

1 In a series of gloomy articles, e.g. *The Times*, 28 June 1832. See Marie Boas Hall, *All Scientists Now*, CUP, 1984
2 *Edinburgh Review*, 49, 1829, pp439–59; and Hamilton, p270
3 Thomas Carlye, *Sartor Resartus*, 1833
4 Anthony Hyman, 'Charles Babbage: Science and Reform', in *Cambridge Scientific Minds*, edited by Peter Harman and Simon Mitton, CUP, 2002
5 Charles Babbage, *The Decline of Science in England*, 1830, p102
6 Ibid., p152
7 Ibid., p44
8 Ibid., p102
9 Ibid., p174
10 Ibid., pp203–12
11 Ibid., p210
12 Ibid., p200
13 Hamilton, p229
14 J.S. Mill, *Autobiography*, 1870, p124
15 John Herschel, *A Preliminary Discourse on the Study of Natural Philosophy*, 1831, p4
16 Ruskin, pp117–21
17 *Natural Philosophy*, 1830, Part II
18 Ibid., p191
19 Ibid., p4
20 Ibid., p20
21 Ibid., pp14–15
22 Ibid., p55–6
23 Ibid., pp299–303
24 Ibid., pp329–40
25 Ibid., p340
26 Faraday to John Herschel, 10 November 1832, *Correspondence*, vol 1, p623
27 Charles Darwin to W.D. Fox, 15 February 1831, in *Correspondence Volume I, 1821–1836*, CUP, edited by Frederick Burkhardt and Sydney Smith, 1985, p118 footnote 2. See also Charles Darwin, *Autobiography*
28 *Gentlemen of Science: Early Correspondence*, Camden Society, 1984, p26

29 Jack Morrell and Arnold Thackray, *Gentlemen of Science: Early Years*, OUP, 1981, pp12–17
30 *Gentlemen of Science: Early Correspondence*, pp85–6
31 Ibid., pp55–8
32 Morrell and Thackray, pp180–201
33 *The Times*, 23 June 1832, p4, columns 3–4
34 *The Voyage of the Beagle*, June 1833
35 Coleridge, 29 June 1833; *Table Talk*, edited by Carl Woodring, 1990, vol 1, p392 and footnote
36 Ibid., pp394–5
37 Coleridge, *Biographia Literaria*, 1817, Chapter 4
38 Holmes, *Coleridge: Darker Reflections*, p555
39 *Quarterly Review*, 51, 1834, pp54–68. James Secord, *Victorian Sensation*, University of Chicago Press, 2000, pp404–5; see also Richard Yeo, 'William Whewell', in *Cambridge Scientific Minds*, 2000
40 Hamilton, p261
41 Unpublished comment by Mrs Margaret Herschel, in the holograph Introduction to the manuscript of Caroline Herschel's *Memoirs*, in WH Archive, John Herschel-Shorland. It is interesting that this comment was suppressed from the printed Introduction by her publisher John Murray
42 James Secord, *Vestiges of Natural Creation*, Chicago UP, 2000, p47
43 'Fragment of Bridgwater Treatise', Charles Babbage, *Collected Works*, vol 11
44 William Sotheby's poem is reprinted in Tim Fulford (editor), *Romanticism and Science, 1773–1833*
45 *The Times*, 4 September 1835, p3
46 *Gentlemen of Science*, p543
47 *Bentley's Miscellany*, IV, 1838, p209
48 The whole series of experiments is dramatically described in James Hamilton, *Faraday*, 2002, pp245–52, which beautifully explains the construction of early coils and dynamos
49 *On the Chemical History of a Candle*, 1861; *Faber Book of Science*, edited by John Carey, 2003, p90
50 Darwin, *Correspondence* 1, p324
51 Knight, *Humphry Davy*, pp176–7
52 Brewster, *Life of Newton*, 1831, Chapter XI, pp 148–50; and contrast 1860 edition
53 Ibid., Chapter III, pp35–7, and Chapter XI, p336
54 Ibid., Chapter XIX, p388
55 John Milton, *Paradise Lost*, Book 10, lines 743–5
56 'Author's Introduction to the 1831 Standard Edition', *Frankenstein, or The Modern Prometheus*, 1831, px. Introduction dated 15 October 1831
57 Mary Somerville, *The Connexion of the Physical Sciences*, 1834, p4
58 Ibid., p260
59 Ibid., 'Section 24'
60 Ibid., p432
61 Ibid., p2
62 Ibid., p432
63 Ibid., pp260–1
64 *Gentlemen of Science: Early Correspondence*, Camden Society, 1984, p137
65 'Report on the British Association for the Promotion of Science', in

The Gentleman's Magazine, October 1834
66 Janet Browne, *Charles Darwin: Volume 1: Voyaging*, Pimlico, 2003, p137
67 John Herschel, *Natural Philosophy*, pp350–3; and Adrian Desmond and James Moore, *Darwin*, Penguin, 1992, p91
68 Browne, vol 1, p135
69 Letter to John Lubbock FRS, 13 May 1833, quoted in Steven Ruskin, *John Herschel's Cape Voyage*, p51
70 Ibid., p47
71 WH Archive*:* John Herschel's notebooks, drawings and equipment are still preserved by John Herschel-Shorland, Norfolk
72 WH Chronicle, p177
73 Darwin, *Correspondence* 1, p498
74 Ibid., p500
75 Charles Lyell to Darwin, 26 December 1836, ibid., p532
76 Caroline Herschel, letter to John Herschel, British Library Ms Egerton 3761–2; also Claire Brock, *The Comet Sweeper: Caroline Herschel's Astronomical Ambition*, Icon Books, Cambridge, 2007, p205
77 *The Times*, Friday, 27 June 1834, quoted in Evans, *Herschel at the Cape*, p88
78 *New York Sun*, 25–30 August 1835, internet file
79 Edgar Allan Poe, 'The Great Balloon Hoax', 1836
80 Ruskin, *Herschel's Cape Voyage*, p97
81 Evans, pp236–7
82 Ibid., pxix

Acknowledgements

For the use of copyright materials and illustrations, and kind permission to consult and refer to manuscripts, rare editions and archives, my most grateful acknowledgements are due to the British Library, London; the University Library, Cambridge; the Bibliothèque Nationale, Paris; the National Portrait Gallery, London; the Royal Institution, London; the Royal Society, London; the Royal Astronomical Society, London; the Science Museum, London; the London Library; the Whipple Museum, Cambridge; the Herschel Museum, Bath; the National Mining Museum, Wakefield; Somerset County Record Office, Bristol; the Cornwall County Record Office, Truro; la Musée de l'Air et de l'Espace, Le Bourget, Aeroport de Paris; the University of New South Wales, Australia, for permission to quote from their transcript of the manuscript of Joseph Banks's *Endeavour Journal*; to Pickering & Chatto (publishers) Ltd for permission to quote from *The Scientific Correspondence of Sir Joseph Banks, 1765–1820*, edited by Neil Chambers; to the Imperial College Press, Natural History Museum and Royal Society, Banks Archive Project, for *The Selected Letters of Sir Joseph Banks, 1768–1820*, edited by Neil Chambers; to Cambridge Science History Publications Ltd, 16 Rutherford Road, Cambridge CB2 8HH, for permission to quote from *Caroline Herschel's Autobiographies*, edited by Michael Hoskin; to the Royal Astronomical Society for permission to quote from the manuscripts of William and Caroline Herschel; and to John Herschel-Shorland, Harleston, Norfolk, for permission to quote from Herschel manuscripts and for all his kindness in letting me see and refer to Herschel family artefacts in his possession.

In attempting to cross between several scientific disciplines and fields of specialist study, I owe a particular debt to the following scholars and writers whose work has inspired and encouraged me, and whose publications (detailed in my Bibliography) I wholeheartedly recommend to the reader. For Joseph Banks and Pacific exploration: Neil Chambers, Patrick O'Brian and John Gascoigne. For the Herschels and astronomy: Michael Hoskin and Simon Schaffer. For Humphry Davy and chemistry: David Knight, Anne Treneer and Frank A.J.L. James. For Mungo Park and African exploration: Anthony Sattin and Kira Salak. For Victor Frankenstein, Regency medicine and the Vitality debate: Roy Porter and Sharon Ruston. For general overviews of the field of Romantic science and the emerging role of the scientist in society: Tim Fulford, Lisa Jardine and Jenny Uglow. I am also hugely grateful to Professor Amartya Sen, then Master of Trinity College, Cambridge, and the Fellows of Trinity, for giving me two wonderful summers as Visiting Fellow Commoner (2000, 2002), and enabling me (among much else) to spend long evenings talking with mathematicians, chemists, astronomers and astrophysicists – several of them Nobel Prize-winners – which gave me some sense of what science is really about.

My warmest personal thanks are due to my old friend and colleague Professor Jon Cook, to whom this book is dedicated; to Professor Kathryn Hughes and Dr Druin Burch (my medical postgraduate) at the University of East Anglia; William St Clair, Richard Serjeantson and Priya Natarajan (our beautiful astrophysicist) at Trinity College, Cambridge; Professor Christoph Bode at the Ludwig-Maximilians-University, Munich; Roderick Winstrop at the Cambridge Observatory; Jim Saulter (pharmacist) and John Allen at Penzance; Debbie James, Curator at the Herschel Museum, Bath; Lenore Symons, the Archivist at the Royal Institution, London; Celia Joicey and Pallavi Vadhia at the National Portrait Gallery; Pierre Lombarde, Directeur, Centre de Documentation, Musée de l'Air et de l'Espace, Le Bourget; Dr Paul Baronek, then of GlaxoSmithKline, for his advice on drugs and medical procedures; Alan Judd for late-night intelligence at The Reform; Patricia Duncker for discussing the fact and fiction of telescopes; Tim Dee of the BBC for producing our three drama-documentaries, *The Frankenstein Experiment* (Radio 3, 2002), *A Cloud in a Paper Bag* (Radio 3, 2007) and *Anaesthesia* (Radio 4, 2009); my brother Adrian Holmes of Young & Rubicam, and my sister Tessa Holmes of the London College of Printing, for their shrewd help with questions of presentation and design; my late uncle, Squadron Leader David Gordon (RAF Bomber Command), who taught me to build short-wave radios, to understand the principles of flight, and once smuggled me into the cockpit of his Vulcan V bomber (not armed); the West Kent Gliding Club and the Norfolk Hot Air Balloon Co. for some highly instructive airborne moments; Eleanor Tremain for finding Andromeda; Dr Percy Harrison, Head of Science, Eton College, for patiently trying to save me from at least some of my scientific howlers; Mr Glasgow, Department of Orthopaedics, Norfolk and Norwich University Hospital, for discussing anaesthetics in the few seconds before he put me under; Richard Fortey, FRS, for swift, exacting and helpful observations at proof stage; and finally Sir Michael Holroyd, for simply being such an inspiration to an entire generation of biographers (Romantic or otherwise).

I have been very lucky at HarperCollins to have such a truly outstanding team behind this book: Robert Lacey (words), Sophie Goulden (pictures), Louise McLeman (internal design), Julian Humphries (cover design), Helen Ellis (trajectories), Douglas Matthews (the prince of indexers), and above all my dauntless, visionary editor Arabella Pike, who would have done brilliantly aboard the *Endeavour* (although that was a much shorter voyage than this one). Best thanks also to my agent David Godwin, who backed this starry-eyed project from its start. Two other teams have supported me far more than they can ever know: the ever-loving Dominos, and of course those wild Delancey boys. To Rose Tremain, once again: *without you no book.*

R.H.

Index